Ecological Restoration and Management of Longleaf Pine Forests

Ecological Restoration and Management of Longleaf Pine Forests

Edited by
L. Katherine Kirkman
Steven B. Jack

CRC Press
Taylor & Francis Group
Boca Raton London New York

CRC Press is an imprint of the
Taylor & Francis Group, an **informa** business

CRC Press
Taylor & Francis Group
6000 Broken Sound Parkway NW, Suite 300
Boca Raton, FL 33487-2742

First issued in paperback 2020

ISBN-13: 978-1-4987-4818-6 (hbk)
ISBN-13: 978-0-367-65773-4 (pbk)

Library of Congress Cataloging-in-Publication Data

Names: Kirkman, L. Katherine, editor. | Jack, Steven B., 1960- editor.
Title: Ecological restoration and management of longleaf pine forests/ editors, L. Katherine Kirkman and Steven B. Jack.
Description: Boca Raton : CRC Press, 2017. | Includes bibliographical references and index.
Identifiers: LCCN 2017010051 | ISBN 9781498748186 (hardback : alk. paper)
Subjects: LCSH: Longleaf pine--Southern States. | Longleaf pine--Conservation--Southern States. | Pine--Southern States. | Pine--Conservation--Southern States. | Forest ecology--Southern States. | Forest restoration--Southern States.
Classification: LCC SD397.P59 E36 2017 | DDC 634.9/7510975--dc23
LC record available at https://lccn.loc.gov/2017010051

Visit the Taylor & Francis Web site at
http://www.taylorandfrancis.com

and the CRC Press Web site at
http://www.crcpress.com

Dedication

Photograph courtesy of Richard T. Bryant.

This book is dedicated to Dr. Robert J. "Bob" Mitchell (1955–2013). Bob was among the initial cohort of scientists recruited to the Joseph W. Jones Ecological Research Center at Ichauway and was instrumental in helping establish its multifaceted program focused on the ecology, management, and restoration of the longleaf pine ecosystem. As a long-time colleague and collaborator to those of us at Ichauway, Bob's intellectual "thumbprint" and scientific legacy is pervasive throughout this volume. He was a creative senior scientist, dedicating his research program to furthering a scientific understanding of the abiotic and biotic processes that structure longleaf pine ecosystems and applying that knowledge to the development of sustainable management and restoration strategies.

Bob thought broadly and deeply about longleaf pine—studying the ecophysiology, population dynamics, fire ecology, biogeochemistry, roots, mycorrhizal partnerships, and competitive interactions of the ecosystem—and was a prominent figure in the development of ecological forestry approaches for natural resource management. Through his careful reasoning, rigorous investigation, critical evaluation of ideas, and numerous collaborations with other eminent scientists, Bob challenged long-held conventional paradigms about the ecology of longleaf pine forests, and in the process, fundamentally changed the understanding and management of the ecosystem. Bob's numerous contributions to longleaf pine ecology, conservation, and management are widely recognized by scientific colleagues and by many natural resource practitioners throughout the Southeast. Those of us who were fortunate to have worked closely with him recognize the marked influence he had on our investigations, careers, and the research program at the Joseph W. Jones Ecological Research Center.

Kay Kirkman and Steve Jack
–Editors

Contents

PART III Ecosystem Restoration: Linking Ecological Understanding and Management

PART IV Restoration and Practical Issues

PART V Longleaf Pine Ecosystem Restoration Perspectives

Foreword

The longleaf pine ecosystem is one of the most extraordinary of the major forest ecosystems in North America, notable for its incredibly rich biological diversity and the extreme role that fire plays in its maintenance. It is without question the most biologically diverse of any temperate forest ecosystem in North America. The richness of the biological diversity is evident in both its plant and animal life. The ground layer of longleaf pine forests typically includes hundreds of vascular plant species, providing an incredible plant identification challenge to the amateur botanist and even the professional plant ecologist. Vertebrate life in the longleaf ecosystem is rich with reptiles, amphibians, birds, and mammals, including many species of special interest, such as the gopher tortoise and red-cockaded woodpecker.

The longleaf pine ecosystem depends fundamentally upon the frequent occurrence of fire for its sustainability. These frequent fires maintain open, savanna-like forests dominated by longleaf pine and herbaceous ground cover; as a longleaf forester friend of mine once commented, "Jerry, you just need to understand that this ecosystem is a tallgrass prairie with some pine trees stuck in it!" This statement is true to a degree because the longleaf pine ecosystem requires frequent fire, and eliminating fire or reducing its frequency causes the forest to transition to another state with greatly reduced biodiversity, similar to the effects of removing fire from prairie grass systems. In actuality, however, the basic ecological processes and their linkages with structural characteristics are vastly different in longleaf pine forests and tallgrass prairies. Arguably, the longleaf pine ecosystem represents an extreme example of a forest's sensitivity to its disturbance regime; it also exemplifies what happens when frequent fire interacts with a productive forest ecosystem. Hence, my characterization of longleaf pine as the "bookend," or most extreme example of a frequent-fire forest ecosystem.

Some of the characteristics of longleaf pine and of the ecosystem that it inhabits have been known for some time, such as the unique grass stage of the young longleaf pine seedlings and the importance of frequent fire. However, beyond some of the species' silvics and the richness of plants and animals, the longleaf pine ecosystem has received relatively little scientific attention, despite its historic extent in the Southeast. After the virgin stands were logged, longleaf pine was largely rejected as a commercial tree species by production foresters because the trees grew too slowly, making it poorly adapted to short-rotation plantation forestry. Instead, southern forestry focused on the faster-growing and shorter-lived loblolly, slash, and shortleaf pines. With few exceptions, the only advocates for longleaf pine for many decades were a community of hunters and a few foresters and wildlife managers who managed some private estates as hunting grounds, particularly for quail.

Interest in the longleaf pine ecosystem exploded in recent decades, however, particularly among those interested in the maintenance of regional biodiversity, with the red-cockaded woodpecker as a flagship organism. However, there was very little science to inform efforts at restoration and management of the ecosystem as a whole; much of the effort has been focused on individual species (e.g., quail, or more recently, woodpeckers) or managing gamebird habitat with prescribed fire.

Our general perspectives on forests in North America have been dramatically altered by scientists since the 1960s, from thinking of them as a collection of trees to understanding them as rich and complex ecosystems. Much of this research occurred in the hardwood forests of the eastern United States, such as at the Coweeta Experimental Forest in North Carolina, Hubbard Brook Experimental Forest in New Hampshire, and Harvard Forest in Massachusetts. In my home region of the Pacific Northwest, ecosystem-level science largely began with studies in old-growth Douglas-fir and western hemlock forests with support from the International Biological Program. Much of the ecosystem science has been funded by the National Science Foundation. However, during this period, there was no center of research focused on learning about longleaf pine as an ecosystem and using it to develop approaches to restoring and managing such ecosystems.

This changed when the Robert W. Woodruff Foundation established the Joseph W. Jones Ecological Research Center in 1991 on the 29,000-acre Ichauway Plantation. The property had been acquired by Mr. Robert W. Woodruff, long-time CEO of Coca-Cola Corporation, for a quail hunting plantation. After several years of due diligence, the Woodruff Foundation established and has subsequently supported the Jones Center as a facility for research, education, and demonstration of management approaches to longleaf pine ecosystems and embedded wetland habitats.

All of the critical elements necessary for true ecosystem-level longleaf pine research and management were provided by the Jones Center—a dedicated place of fire-maintained, naturally regenerated longleaf pine stands, a critical mass of talented participants, outstanding facilities, and sustained support. The dedicated land base, which must include appropriate areas for both research and management on the subject ecosystem, contributes in multiple critical ways. It provides a common geographical focus for interdisciplinary teams. It also provides for long-term continuity, which, in turn, allows for long-term experiments and demonstrations, creation of essential long-term data sets, and continued knowledge accumulation from diverse scientific studies and management activities. A critical mass of talented interdisciplinary scientists and managers is a second crucial element in developing ecosystem science and applications to management. Such an interactive team has been developed and sustained at the Jones Center with the support of the Woodruff Foundation and the recruiting efforts of Director Lindsay Boring. I have been a scientific advisor to the Jones Center since its inception, and I have watched this program mature over more than two decades. The research program is integrated with a clear mandate for education and outreach, and as a result, the Jones Center has become the "go-to" place for those who want to learn about the longleaf pine ecosystem and how such ecosystems can be restored and maintained.

This volume represents a major synthesis of the last two decades of research and management experience about the longleaf pine ecosystem. It comes at a critical time, with the immensely increased current national interest in restoration of longleaf pine ecosystems. It includes and integrates the science of the trees, the ground cover, the fauna, and the associated streams and wetlands. It informs us about important ecosystem functions and energy and material flows, including carbon and water. Because of the incredible biological diversity of the longleaf ecosystem, it certainly helps us to recognize the multiple and critical habitat functions that result from forest heterogeneity. And, as the "bookend" fire-frequent ecosystem, it provides us with a substantive understanding of how such disturbances structure forest ecosystems.

Perhaps the most important lesson provided by this volume is its demonstration of the potential for collaborations between humankind and nature in restoring and sustaining ecosystems. There is no future for the longleaf pine ecosystem in the 21st century without the active participation of humankind. With sufficient knowledge of and appreciation for the complexity of such ecosystems, it is possible to restore and manage them in ways that will sustain their biological richness and multiple ecological functions while providing the environmental, economic, and cultural goods and services sought by society.

We owe great thanks to the personnel of the Joseph W. Jones Ecological Research Center for providing us with an immensely expanded understanding of the longleaf pine ecosystem, and to the Robert W. Woodruff Foundation for initiating and supporting this critical activity! And we most sincerely hope it is only the first installment on many future decades of research.

Jerry F. Franklin
Professor of Ecosystem Analysis
College of Forest Resources
University of Washington
Seattle, Washington

Preface

The longleaf pine (*Pinus palustris*) ecosystem harbors unparalleled biological diversity in North America, and its sustainability is complexly tied to structure, fuels, and fire. Given the enormous loss of this once-dominant forest type of the southeastern U.S. Coastal Plain, restoration has become a priority among multiple federal agencies in partnership with nongovernmental organizations, private industry, universities, and private landowners. Although planting longleaf pine seedlings is a starting point, ecological restoration of this ecosystem requires much more than merely establishing trees. Understanding the complexities of an intact ecosystem provides insights into recovery and reassembly processes and is therefore fundamental to implementing a process-based restoration approach. The purpose of our book is to integrate basic knowledge about this complex ecosystem with the practical challenges of operational implementation in an effort to examine the opportunities and constraints that challenge conservation management of the longleaf pine ecosystem now and into the future.

We titled the book *Ecological Restoration and Management of Longleaf Pine Forests* to reflect a holistic approach for recovering an ecosystem that has been degraded, damaged, or destroyed and for managing longleaf pine forests for multiple objectives. Ecological restoration approaches focus on reestablishing ecosystem functions by modifying or managing the ecosystem components—including their composition, structure, and spatial arrangement—and by reestablishing the ecological processes that these ecosystems need for sustainability and resilience to disturbance (SER 2004; Stanturf et al. 2014; Franklin et al. 2017). In this sense, restoration requires the integration of basic ecological understanding and practical application. For the longleaf pine ecosystem, we define ecological restoration as reestablishing sustainable longleaf pine-dominated communities, maintaining them with frequent prescribed fire, and establishing conservation of biodiversity as a high priority. We also address management practices that can help to maintain these characteristics even in the face of potentially dramatic changes in climatic conditions.

SETTING THE CONTEXT—WHY THIS BOOK AND WHY NOW?

Because early information on the longleaf pine ecosystem was largely descriptive, much of our understanding of the species was derived from seminal works of botanists and foresters such as Schwarz (1907), Chapman (1932), and Wahlenberg (1946). Longleaf pine was virtually ignored by researchers throughout most of the 20th century because: (1) the remaining acreage of natural longleaf pine stands was quite small, (2) longleaf pine was considered to be a species of special interest and not economically viable for the forest products industry, and (3) longleaf pine regeneration, whether natural or artificial, was often unsuccessful.

Despite these challenges, some studies of longleaf pine and the longleaf pine ecosystem did take place throughout the century. At the Southern Research Station of the U.S. Forest Service, researchers and their cooperators throughout the Southeast conducted trials to examine:

- The silvicultural approaches that could be used to manage longleaf pine for wood products (Boyer and Farrar 1981; Farrar and Boyer 1990; Farrar 1996)
- The best ways to secure natural regeneration (Croker 1975; Croker and Boyer 1975; Boyer 1979, 1993)
- The importance of using prescribed fire to manage longleaf pine forests (Komarek 1974; Wade and Lundsford 1989; Outcalt 1994; Brockway and Lewis 1997; van Lear et al. 2005)

In the 1980s and 1990s, scientists developed methods for producing vigorous seedlings in tree nurseries (Barnett and McGilvray 1997) that greatly improved longleaf pine planting success and allowed the species to be introduced at sites where mature trees were not present. In addition, the late 20th century brought a focus to wildlife species that are endemic to longleaf pine ecosystems (Diemer 1986; Landers et al. 1990; Engstrom 1993; Palis 1997; Ross et al. 1997; Plentovich et al. 1998), as well as the high diversity of the ground cover plant communities that characterize longleaf pine forests (Walker and Peet 1983; Clewell 1989; Hardin and White 1989; Noss 1989; Drew et al. 1998; Kirkman et al. 1998; Hainds et al. 1999). These nontimber considerations were driven by the decreasing acreage of fire-maintained longleaf pine forests in the landscape and the associated losses of suitable habitat.

This interest arose largely through concerns about threatened and endangered wildlife species that are associated with longleaf pine forests and the specific habitat characteristics they provide. The species that epitomizes this group is the endangered red-cockaded woodpecker (*Picoides borealis*), but the group also encompasses many imperiled species including Bachman's sparrow (*Peucaea aestivalis*), gopher tortoise (*Gopherus polyphemus*), Florida pine snake (*Pituophis melanoleucus mugitus*), indigo snake (*Drymarchon couperi*), dusky gopher frog (*Lithobates sevosus*), Florida pocket gopher (*Geomys pinetis*), and Sherman's fox squirrel (*Sciurus niger shermani*). Because conservation efforts for these wildlife species focused on providing habitat that was similar to what was historically found in longleaf pine forests, understanding the structure and function of these forests took on greater importance. Incentive programs for planting longleaf pine, largely justified as providing wildlife habitat, were introduced in the late 1990s and led to a large increase in acreage of plantations across the species' historical range, and to increased interest in the restoration of native ground cover and other key components of these forests.

It was at this point of increased focus on longleaf pine that Jose et al. (2006) published a summary of longleaf pine ecosystem research during the 20th century. The topics covered included a review of the history and demise of longleaf pine ecosystems, research summaries that demonstrated the high diversity in vegetative and animal communities of longleaf pine ecosystems, identification of appropriate silvicultural systems, a discussion of the critical role of fire in maintaining function and structure, and a description of the newly emerging efforts and challenges that are inherent in ground cover restoration.

Considerable research since this publication has added to our understanding of the tightly coupled structural and functional relationships that characterize the longleaf pine ecosystem. In addition, social and political changes since 2006 have had a profound influence on region-wide longleaf pine reestablishment efforts. Greater national awareness of the rarity of longleaf pine forests and their importance for many endemic animal and plant species led to increased conservation interest from a much broader group of "stakeholders" and additional funding opportunities for restoration of longleaf pine across its historical range. One result of the increased consideration for longleaf pine was the launch of America's Longleaf Restoration Initiative in 2009 (see Chapter 1). This program set an ambitious goal to more than double the longleaf pine acreage within the next few decades and institutionalized a broad collaborative arrangement among federal, state, private, and nongovernmental organizations. The financial and technical support available through this collaborative program helped increase the acreage of longleaf pine in its historical range, largely through the addition of longleaf pine plantations; this action has halted, and perhaps reversed, the decades-long decline of longleaf pine acreage. The larger contingent of people working on longleaf pine issues, and especially applied management considerations, also expanded our knowledge on how to restore and manage this threatened ecosystem.

SCOPE AND STRUCTURE OF THIS BOOK

The intensified focus on longleaf pine over the last 15 years generated many new questions that we used to organize this book (see Sidebar). They include questions that demonstrate the close linkages between the basic ecology of longleaf pine ecosystems and applied restoration and management actions. Rather than presenting a strictly scientific treatment of basic ecological principles or a case study documentation of management activities, our objective was to show how the adaptive management approach can blend information gained from applied research, observations of responses to specific management treatments, and scientific studies to provide restoration and management tenets for the longleaf pine ecosystem. Our approach was to synthesize more than two decades of work at the Joseph W. Jones Ecological Research Center, where the organizational objectives are to highlight and strengthen the linkages between basic research and active restoration and management, and then to integrate this knowledge with that of the cumulative efforts from the larger community of scientists and practitioners working to restore longleaf pine ecosystems.

QUESTIONS ILLUSTRATING LINKAGES BETWEEN BASIC AND APPLIED RESEARCH

Plant Community Dynamics and Restoration Approaches

How do season, frequency, and intensity of fire influence plant community dynamics, and how can various prescribed fire regimes be implemented as ground cover restoration tools?

What are controls on species richness across soil types, and how do nuanced influences of competition, seed dispersal, and availability of regeneration niches affect the reassembly of diverse ground cover vegetation in a restoration context?

How does species richness vary across temporal and spatial scales, and what are the implications of this variation for long-term monitoring efforts?

How do local adaptations of ground cover species vary between geographic regions, and how important is seed provenance to successful restoration?

Forest Structure and Functional Relationships—The Critical Pieces of the Restoration Puzzle

How do forest structural and functional properties, such as pyrogenicity, affect longleaf pine regeneration? What canopy configurations best promote the sustainability of a frequently burned multiaged forest?

What is the appropriate sequence of management actions for areas newly planted with longleaf pine to reconstruct the desired structure of a multiaged, heterogeneous forest?

How do site conditions influence successional trajectories and the successful reestablishment of fine fuels that can maintain a frequent fire regime?

What are the important functional roles of ground cover species, and what species should be prioritized for initial reintroduction efforts?

What functional roles do canopy species other than longleaf pine contribute on different types of soil, and how should restoration targets accordingly vary with soils and site conditions?

Ecosystem Processes and Restoration Strategies

What are the influences of fire regimes on biogeochemical cycles, and how do altered soil processes resulting from fire suppression or agricultural land use affect restoration trajectories?

What is the capacity for regional carbon sequestration in longleaf pine forests? Does restoration of longleaf pine promote or increase sequestration?

What are stand-level water budgets of longleaf pine forests across the landscape, and can regional evapotranspiration be reduced with broad-scale restoration? What is the impact of potentially reduced evapotranspiration with expected climate change?

The Landscape Context of Wildlife Habitat

What are the terrestrial/wetland habitat linkages in the longleaf pine ecosystem, and what is the role of prescribed fire in maintaining amphibian diversity?

How are trophic relationships structured, and how does broad-scale habitat or population management impact community dynamics?

What habitat requirements characterize species of concern, and what spatial scale and patch configuration are appropriate for species reintroductions?

What are the ranges of habitat conditions used by the wildlife species that are associated with longleaf pine, and how can these indicator species be used to assess habitat condition and restoration "success?"

During restoration, which temporal scale and sequence of stand development will benefit wildlife populations and communities?

These objectives provide the structure for the book. The first part presents the historical and social context for longleaf pine ecosystem restoration. The second part focuses on a fundamental understanding of longleaf pine ecosystem processes, stand dynamics, community assembly processes, embedded wetlands, and trophic cascades. Part III addresses how our basic understanding of the ecosystem can be applied to restoration and management. Part IV describes the challenges of managing a fire-dependent forest in the Southeast and the role of adaptive management in that effort. The last part examines the potential ecosystem services that are associated with longleaf pine forests and compares the ecology and management of longleaf pine forests with other fire-dependent forests in North America. Finally, the summary chapter examines future challenges and opportunities associated with recovery of this ecosystem. Given the differences in disciplinary standards of measurements, we have deliberately used metric units in some chapters and English units in others, depending on the subject matter and target audience.

In designing the content and presentation in this book, we decided not to include a chapter reviewing the ecology of red-cockaded woodpeckers or other charismatic imperiled fauna and flora because there is considerable information available on these topics elsewhere. Instead, we focus on wildlife management objectives from a landscape context (see Chapters 8, 9, and 12) and discuss the socioeconomic influences of these species of high-profile conservation status on longleaf pine ecosystem management (see Chapters 3 and 17). We also chose not to include a chapter that specifically addresses the application of prescribed fire in longleaf pine forests, although we did include chapters that discuss fuels (see Chapter 6) and smoke management/air quality issues (see Chapter 13). We made this decision because the basic principles to implement prescribed fire in forest settings have been well documented (Wade and Lundsford 1989) and because prescribed fire concepts are interwoven throughout the book to reflect our conviction that the application of fire is central to the restoration and management of longleaf pine ecosystems.

Although most researchers and practitioners agree that there are no adequate and sustainable substitutes for fire, much debate about the most appropriate season for prescribed fire has taken place over the last several decades (Platt, Evans, and Davis 1988; Seamon et al. 1989; Robbins and Meyers 1992; Streng et al. 1993; Brewer and Platt 1994; Hermann et al. 1998; Hiers et al. 2000; Shepherd et al. 2012; Noss 2013; Robertson and Hmielowski 2014). Many of the chapters address specific effects on flora and fauna related to the season in which fire occurs, and this somewhat controversial topic is considered in Chapters 2 and 17.

Too often, basic research produces knowledge of ecological processes that could potentially guide land management practices, but the information is not delivered to land managers, or is not applied and evaluated at operational scales. Our hope is that this volume provides new insights that practitioners, land managers, and researchers can use in implementing innovative restoration and management approaches that are founded on basic ecological principles and observations from a wide spectrum of applied research and adaptive management efforts.

Acknowledgments

We gratefully acknowledge and commend our scientific colleagues at the Joseph W. Jones Ecological Research Center at Ichauway for their collaboration in conceiving and developing this project. We are deeply indebted to the Ichauway resource management staff, as well as many other experienced practitioners on private and public lands, for sharing their practical knowledge about the management and restoration of longleaf pine. Leon Neel, a forestry and wildlife consultant with many decades of experience managing longleaf pine forests, particularly informed our thinking about longleaf pine ecology and management in innumerable ways over many years. Through partnerships and discussions with these professionals, we have gained much insight into the ecology and management of longleaf pine ecosystems.

We also acknowledge the numerous discussions, inspiration, and enthusiastic support provided by the dedicated members of the Joseph W. Jones Ecological Research Center Advisory Committee. We extend our appreciation to the many research collaborators, including numerous graduate students, whose work has contributed to our collective understanding of longleaf pine ecosystems. Several of these researchers have also contributed as chapter authors.

We greatly appreciate the time and effort of numerous external colleagues who graciously provided timely reviews of chapter drafts including: Matt Cohen, Yvette Dickinson, David Dyson, Jack Ewel, William Flatley, Jerry Franklin, Rick Gillam, Jeff Glitzenstein, Jim Guldin, Kyle Harms, Andy Hudak, John Kilgo, Benjamin Knapp, Jennifer Knoepp, John Kush, Keville Larson, Marcus Lashley, Daniel McLaughlin, Bruce Means, Brian Palik, Kyle Palmquist, Kevin Robertson, Betsie Rothermel, Carrie Sekerak, Ge Sun, Jacob Thompson, and Clay Ware. Their critical feedback substantively refined the manuscript. We are appreciative of CRC Press/Taylor & Francis for their support of the book and, in particular, Chuck Crumly, for working with us to navigate the publishing process. We sincerely thank Carol Whitlock for her exceptional and meticulous technical editing efforts. The diligent work of Liz Cox, librarian at the Joseph W. Jones Ecological Research Center, was invaluable in retrieving published materials and in organizing literature cited sections. The assistance of Melanie Kaeser, Lisa Giencke, and Mary Frances Nieminen in assembling a final product and Jessica McCorvey in formatting figures is also gratefully acknowledged. Finally, this book was made possible through the continued support of our employer, Ichauway, Inc. and the Robert W. Woodruff Foundation.

L. Katherine Kirkman
Steven B. Jack

Editors

L. Katherine Kirkman, PhD, is a plant ecologist at the Joseph W. Jones Ecological Research Center at Ichauway, where she has spent the last 25 years conducting basic and applied research in longleaf pine ecosystems and embedded wetlands. She received degrees from the College of William and Mary (BS Biology) and University of Georgia (MS Botany, and PhD Botany). She serves as an adjunct faculty member at the University of Georgia, Auburn University, and the University of Florida, where she has co-advised graduate students conducting research at Ichauway. She has coauthored three books, *Trees of Georgia* (1990), *Native Trees of the Southeast* (2007), and *Natural Communities of Georgia* (2013).

Steven B. Jack, PhD, is an applied forest scientist at the Joseph W. Jones Ecological Research Center at Ichauway. In this position he is involved in the operational resource management of Ichauway, provides an interface between the research and resource management staffs, conducts applied research on conservation-oriented management and restoration of longleaf pine for forest and wildlife objectives, and participates in numerous outreach activities. He received degrees from Erskine College (AB Physics and Math), University of Florida (MS Forest Resources and Conservation), and Utah State University (PhD Silviculture and Forest Ecology). Prior to working at the Jones Center he held a faculty position at Texas A&M University.

Contributors

Carla L. Atkinson
Department of Biological Sciences
University of Alabama
Tuscaloosa, Alabama

Larry Band
Department of Geography
University of North Carolina
Chapel Hill, North Carolina

Michael A. Battaglia
Rocky Mountain Research Station
USDA Forest Service
Fort Collins, Colorado

Seth W. Bigelow
Joseph W. Jones Ecological Research Center
Newton, Georgia

Lindsay R. Boring
Joseph W. Jones Ecological Research Center
Newton, Georgia

Steven T. Brantley
Joseph W. Jones Ecological Research Center
Newton, Georgia

Michael J. Cherry
College of Natural Resources and Environment
Virginia Tech University
Blacksburg, Virginia

L. Mike Conner
Joseph W. Jones Ecological Research Center
Newton, Georgia

James A. Cox
Vertebrate Ecology Program
Tall Timbers Research Station and Land Conservancy
Tallahassee, Florida

Lisa M. Giencke
Joseph W. Jones Ecological Research Center
Newton, Georgia

Joseph J. Hendricks
Department of Biology
University of West Georgia
Carrollton, Georgia

J. Kevin Hiers
Tall Timbers Research Station
Tallahassee, Florida

Steven B. Jack
Joseph W. Jones Ecological Research Center
Newton, Georgia

Stephen T. Jackson
Department of the Interior Southwest Climate Science Center
U.S. Geological Survey and Department of Geosciences and School of Natural Resources and Environment
University of Arizona
Tucson, Arizona

L. Katherine Kirkman
Joseph W. Jones Ecological Research Center
Newton, Georgia

Andrew J. Larson
W.A. Franke College of Forestry and Conservation
University of Montana
Missoula, Montana

E. Louise Loudermilk
Center for Forest Disturbance Science
Southern Research Station
USDA Forest Service
Athens, Georgia

Daniel Markewitz
Warnell School of Forestry and Natural Resources
University of Georgia
Athens, Georgia

Barrett B. McCall
Larson & McGowin, LLC
Mobile, Alabama

Robert A. McCleery
Department of Wildlife Ecology and Conservation
University of Florida
Gainesville, Florida

R. Kevin McIntyre
Joseph W. Jones Ecological Research Center
Newton, Georgia

Mark A. Melvin
Joseph W. Jones Ecological Research Center
Newton, Georgia

Jonathan A. Myers
Department of Biology
Washington University in St. Louis
St. Louis, Missouri

Joseph J. O'Brien
Center for Forest Disturbance Science
Southern Research Station
USDA Forest Service
Athens, Georgia

Stephen D. Pecot
Larson & McGowin, LLC
Mobile, Alabama

Brandon T. Rutledge
Joseph W. Jones Ecological Research Center
Newton, Georgia

Elizabeth M. Schlimm
Maryland Department of Natural Resources
Cantonsville, Maryland

Lora L. Smith
Joseph W. Jones Ecological Research Center
Newton, Georgia

Michael C. Stambaugh
Department of Forestry
School of Natural Resources
University of Missouri
Columbia, Missouri

Amanda L. Subalusky
Cary Institute of Ecosystem Studies
Millbrook, New York

Robert D. Sutter
Enduring Conservation Outcomes, LLC
Savannah, Georgia

R. Scott Taylor
Joseph W. Jones Ecological Research Center
Newton, Georgia

J. Morgan Varner
Pacific Wildland Fire Sciences Laboratory
Pacific Northwest Research Station
USDA Forest Service
Seattle, Washington

James M. Vose
Center for Integrated Forest Science
Southern Research Station
USDA Forest Service
Raleigh, North Carolina

David N. Wear
Center for Integrated Forest Science
Southern Research Station
USDA Forest Service
Raleigh, North Carolina

Reviewers

Mike Battaglia
Forest and Woodland Ecosystems
Rocky Mountain Research Station
USDA Forest Service
Fort Collins, Colorado

Seth Bigelow
Joseph W. Jones Ecological Research Center
Newton, Georgia

Lindsay Boring
Joseph W. Jones Ecological Research Center
Newton, Georgia

Steven Brantley
Joseph W. Jones Ecological Research Center
Newton, Georgia

Matt Cohen
School of Forest Resources and Conservation
University of Florida
Gainesville, Florida

Mike Conner
Joseph W. Jones Ecological Research Center
Newton, Georgia

Yvette Dickinson
School of Forest Resources and Environmental Science
Michigan Technological University
Houghton, Michigan

David Dyson
Larson & McGowin, Inc.
Mobile, Alabama

Jack Ewel
Department of Biology (Emeritus)
University of Florida
Gainesville, Florida

William Flatley
School of Forestry
Northern Arizona University
Flagstaff, Arizona

Jerry F. Franklin
College of Forest Resources
University of Washington
Seattle, Washington

Rick Gillam
Air Resources Division
U.S. Environmental Protection Agency Region 4
Atlanta, Georgia

Jeff Glitzenstein
Tall Timbers Research Station
Tallahassee, Florida

Steve Golladay
Joseph W. Jones Ecological Research Center
Newton, Georgia

James Guldin
Longleaf Pine Restoration and Management
Ecology and Management of Southern Pines
Southern Research Station
USDA Forest Service
Hot Springs, Arkansas

Kyle Harms
Department of Biological Sciences
Louisiana State University
Baton Rouge, Louisiana

Kevin Hiers
Tall Timbers Research Station
Tallahassee, Florida

Andy Hudak
Forest and Woodland Ecosystems
Rocky Mountain Research Station
USDA Forest Service
Moscow, Idaho

John Kilgo
Forest Watershed Science
Southern Research Station
USDA Forest Service
New Ellenton, South Carolina

Benjamin Knapp
School of Natural Resources
University of Missouri
Columbia, Missouri

Jennifer Knoepp
Forest Watershed Science
Southern Research Station
USDA Forest Service
Otto, North Carolina

John Kush
School of Forestry and Wildlife Sciences
Auburn University
Auburn University, Alabama

Keville Larson
Larson & McGowin, LLC (retired)
Mobile, Alabama

Marcus Lashley
College of Forest Resources
Mississippi State University
Starkeville, Mississippi

Louise Loudermilk
Center for Forest Disturbance Science
Southern Research Station
USDA Forest Service
Athens, Georgia

Kevin McIntyre
Joseph W. Jones Ecological Research Center
Newton, Georgia

Daniel McLaughlin
Department of Forest Resources and Environmental Conservation
Virginia Polytechnic Institute and State University
Blacksburg, Virginia

Bruce Means
Coastal Plains Institute
Tallahassee, Florida

Joe O'Brien
Center for Forest Disturbance Science
Southern Research Station
USDA Forest Service
Athens, Georgia

Brian Palik
Center for Research on Ecosystem Change
Northern Research Station
USDA Forest Service
Grand Rapids, Minnesota

Kyle Palmquist
Botany Department
University of Wyoming
Laramie, Wyoming

Stephen Pecot
Larson & McGowin, LLC
Mobile, Alabama

Kevin Robertson
Tall Timbers Research Station
Tallahassee, Florida

Betsie Rothermal
Archbold Biological Station
Lake Placid, Florida

Carrie Sekarak
Ocala National Forest
Seminole Ranger District
USDA Forest Service
Umatilla, Florida

Brandon Rutledge
Joseph W. Jones Ecological Research Center
Newton, Georgia

Lora Smith
Joseph W. Jones Ecological Research Center
Newton, Georgia

Michael Stambaugh
Department of Forestry
University of Missouri
Columbia, Missouri

Robert Sutter
Enduring Conservation Outcomes
Savannah, Georgia

Ge Sun
Eastern Forest Environmental Threat Assessment Center
Southern Research Station
USDA Forest Service
Raleigh, North Carolina

Jacob Thompson
Non-game Program
Wildlife Resources Division
Georgia Department of Natural Resources
Brunswick, Georgia

James Vose
Center for Integrated Forest Science
Southern Research Station
USDA Forest Service
Raleigh, North Carolina

Clay Ware
Southeast Regional Office
U.S. Fish and Wildlife Service
Atlanta, Georgia

Part I

The Background for Ecological Restoration

1 The Fire Forest of the Past and Present

L. Katherine Kirkman, Steven B. Jack, and R. Kevin McIntyre

CONTENTS

INTRODUCTION

Nearly every region of the United States has an "emblematic" forest type that historically has strong links with the local culture. In the Coastal Plain of the Southeast this forest type is the fire-maintained "piney woods." These forests were notable for their relatively open and pine-dominated canopies, a diverse herbaceous ground cover, and a low-density midstory that provided open views for long distances. Depending on soil and moisture conditions, some piney woods also had a network of drainages and isolated wetlands within the forest matrix that further added to the heterogeneous structure and increased biological diversity. This forest structure and composition was maintained by frequent, low-intensity fires that regulated competitive relationships within the plant community. The dominant pine species in most fire-maintained piney woods was longleaf pine (*Pinus palustris*).

The range of longleaf pine once extended from the Piedmont and Coastal Plain areas of the mid-Atlantic states, to the piney woods of Georgia, Florida, and Alabama, through the lowlands of Mississippi and Louisiana, and into Texas (see Chapter 2). Within this range, longleaf pine was originally the dominant species in savannas (<25% canopy cover) and woodlands (25%–60% canopy cover), and a dominant to codominant species in forests (>60% canopy cover). In response to a long history of frequent fire, the longleaf pine ecosystem of the 17th century was an extensive mosaic of vegetation in which embedded wetlands, linear drainages, and riparian floodplains contributed to the rich biodiversity of the region.

In subsequent centuries, however, longleaf pine was reduced to <4% of its original 92 million acres through a history of overharvest for the high-quality timber, conversion to other pine species (often short-rotation plantations), decades of fire exclusion, conversion to nonforestland uses such as agriculture, and, in more recent years, urban encroachment. All of these factors contributed to the current endangered status of this ecosystem (Noss et al. 1995), with only remnant communities remaining that must be maintained with prescribed fire. Numerous publications recount in detail the land use history of post-European settlement to the present and the near demise of this regionally significant vegetation type (Croker 1987; Ware et al. 1993; Earley 2004; Van Lear et al. 2005; Frost 2006).

This chapter reviews the highly variable condition of longleaf pine forests, both remnants and those that have been newly established, and identifies how restoration approaches vary depending on site-specific topoedaphic location and land use legacies. We describe historical and current patterns of land ownership as well as the difficulties in tracking both the region-wide status of forest condition and the social complexities involved in prioritizing regional restoration activities. We summarize a generalized system of classifying minimally disturbed longleaf pine-dominated plant communities that develop through complex interactions of fire frequency, soil characteristics, and topographic position. And finally, we describe the implication of initial starting points for the restoration of remnant and former longleaf pine sites.

LONGLEAF PINE HISTORY AND DECLINE

Longleaf pine-dominated forests were one of the most extensive forest ecosystems in North America. Before European settlement, they covered about 57 million acres, and longleaf pines were a significant component of another 35 million acres of mixed pine-oak forests in the southeastern United States (Frost 2006). As Europeans settled the South Atlantic coastal areas and began to move inland in the 17th and 18th centuries, impacts on the longleaf pine resource were relatively minor and spatially restricted. Harvesting was typically limited to areas where the lumber was used, in addition to commercial harvesting along the waterways that were large enough to float logs to coastal ports. Other uses of longleaf pine forests were production of turpentine and other naval stores, primarily in the Carolinas, and free-range grazing of cattle throughout the range of the species (Croker 1987).

Before the 1850s, the lumber industry in the Southeast was comparatively small-scale in its impact and footprint. After the Civil War, however, cheap land, depletion of red pine and white pine forests in the upper Midwest, and—most importantly—refinement and widespread adoption of rail technology combined to exponentially accelerate the harvesting of virgin longleaf pine forests. The establishment of a viable network of rail lines, combined with the development of small-gauge rail tracks that could be easily moved from one logging site to another, opened huge areas of previously inaccessible forestlands to harvesting (Williams 1989). During the 1870s and 1880s, millions of southeastern acres were acquired by lumber companies and land speculators. Timber harvesting peaked between 1907 and 1909, with annual estimates of harvested pine ranging from 13 to 14 billion board feet, up from just 2 billion board feet in 1875 (Wahlenberg 1946; Williams 1989; Carter et al. 2015). After that point, harvesting operations continued to work their way across the Southeast into the western part of the longleaf pine range, but annual harvested volume began to decline. By the 1930s, most of the once-vast longleaf pine forests had been cut through. In 1932, annual lumber production from pine had decreased to 3 billion board feet (Carter et al. 2015).

The most common source for longleaf pine acreage estimates, both past and present, is the U.S. Forest Service–Forest Inventory and Analysis (FIA) program, which was authorized by the McSweeney-McNary Forest Research Act of 1928, the founding legislation of national inventory and monitoring activities. FIA data collection has evolved from periodic surveys to an ongoing monitoring program in which data are continuously collected from a grid of permanent sample plots that are distributed across the forested U.S. landscapes. The two categories of forests that make up what are commonly considered "longleaf pine acres" both have longleaf pine as the dominant pine species: The longleaf pine forest type is defined as those stands with >50% stocking of pine in which longleaf is the dominant pine species; the longleaf pine/oak forest type is defined as those stands in which pine accounts for 25%–50% of the stocking and longleaf pine is the dominant pine species (Oswalt et al. 2012). The historical estimates described below most commonly reflect acreage numbers for the longleaf pine forest type alone; estimates that also include longleaf pine/oak forests will be noted.

The earliest survey data from 1935 reported approximately 23.4 million acres of longleaf pine (Wahlenberg 1946). Old-growth longleaf pine forests, or those containing an old-growth component, were estimated at about 1.7 million acres. FIA data showed that longleaf pine had declined to 12.2 million acres by 1955, dropping to about 3.77 million acres by 1985 (Kelly and Bechtold 1990). Losses continued through the end of the 20th century as natural longleaf pine forests were converted to loblolly (*P. taeda*) or slash (*P. elliottii*) pine plantations for fiber production (see Chapter 3). Somewhere around the end of the 20th century, the acreage of longleaf pine reached its lowest point, with FIA data from 1995 indicating that about 3 million acres remained (Outcalt and Sheffield 1996). Additionally, there were as many as 600,000 acres of longleaf pine/oak forest type remaining (Frost 2006). By this time, it was broadly recognized that longleaf pine ecosystems were critically endangered (Noss et al. 1995). Even more dramatic was the loss of old-growth longleaf pine stands, with estimates in the early 2000s of only 12,600 acres remaining across the historical longleaf pine range (Varner and Kush 2004).

As longleaf pine acreage reached its lowest point, interest in these ecosystems began to increase. The longleaf pine forest became a focus of federal agencies, state foresters, and conservation nongovernmental organizations (NGOs) mostly because of its increasing rarity, but partially because of the regulatory mandates that accompanied the listing of species under the Endangered Species Act—in particular, the red-cockaded woodpecker (*Picoides borealis*). In the private sector, a growing interest in managing forests for multiple values rather than just timber income attracted many landowners to the management of longleaf pine forests. In 1995, a group of longleaf pine advocates came together to form The Longleaf Alliance at Auburn University with the goal of promoting restoration through education and outreach. With the realization in 2007 that the task of longleaf pine restoration—at a meaningful scale across its historical range—was beyond the scope of any single group, a body of 22 agencies and organizations joined together to develop a range-wide conservation plan called America's Longleaf Restoration Initiative and later to establish the Longleaf Partnership Council. The resulting strategies, aided by the collective efforts of individuals, organizations, agencies, and partnerships, have succeeded in reversing the trend of declining acreage of longleaf pine (Oswalt et al. 2012).

AMERICA'S LONGLEAF RESTORATION INITIATIVE

In 2007, 22 federal agencies, state agencies, and NGOs began a collaborative effort to develop a range-wide conservation plan for longleaf pine. The conservation plan was released as America's Longleaf Restoration Initiative in 2009. The broad goal of the plan is to increase longleaf pine acreage to eight million acres, at the same time improving the condition of existing longleaf pine forests and creating higher-quality habitat for longleaf pine-associated wildlife species.

In the conservation plan, 17 focal areas—called Significant Geographic Areas (SGAs)—across the historical range of longleaf pine were identified to concentrate longleaf pine restoration efforts (Figure 1.1). Criteria for SGAs were as follows:

- A landscape of >100,000 acres with one or more longleaf pine natural community types and related ecosystems (as identified from FIA data, Natural Heritage Program element occurrence data, and other sources)
- A core of permanently protected lands with intact natural longleaf pine
- The potential to maintain and increase connectivity across a large area of public and private ownership
- Minimal pressures from land use change that might constrain the ability to continue to protect and manage longleaf pine into the future and management (restrictions on fire management) into the foreseeable future

In addition to the SGAs, which have the highest potential for developing large ecologically functional landscapes that will persist into the future, the conservation plan also supports longleaf pine restoration outside of these areas on smaller parcels.

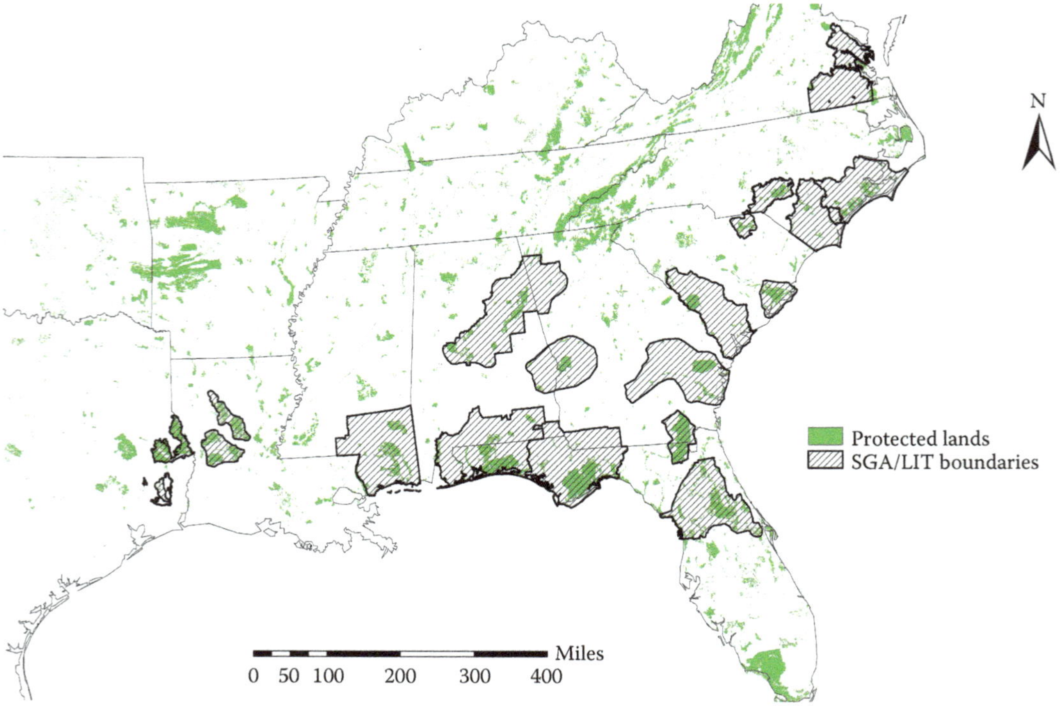

FIGURE 1.1 Distribution of Significant Geographic Areas (SGAs) and local implementation teams (LIT) from America's Longleaf Restoration Initiative, a range-wide conservation plan for the species, and their relationship with protected lands within the historical range of longleaf pine. SGAs have been designated a priority for restoration efforts within the conservation plan. (Protected Areas data adapted from Nelson, M. D. et al. 2010. Forest ownership in the conterminous United States: ForestOwn_v1 geospatial dataset. USDA Forest Service, Northern Research Station, Newtown Square, Pennsylvania. https://doi.org/10.2737/RDS-2010-0002. SGA/LIT boundaries data courtesy of The Nature Conservancy.)

As a follow-up to the development of the conservation plan, three U.S. cabinet-level departments—Agriculture, Defense, and Interior—signed a 2010 Memorandum of Understanding that established the Federal Coordinating Committee for Longleaf Pine. In recognition of the need to broaden the formal collaboration beyond federal agencies, the 33-member Longleaf Partnership Council was established in 2011 to facilitate implementation of the conservation plan through increased communication and further collaboration.

These efforts have resulted in more effective implementation of the conservation plan, additional funding opportunities, and increased collaboration—at both range-wide and local scales—among federal and state agencies, NGOs, and private landowners. Local collaborative teams representing the ownership-class interests have formed around the SGAs to lead on-the-ground implementation. These local implementation teams are supported by the Longleaf Stewardship Fund, a public–private grant program administered by the National Fish and Wildlife Foundation. From 2013 to 2015, the Longleaf Partnership Council has supported and documented the establishment of longleaf pine on >450,000 acres and the application of prescribed fire on 1.3 million acres of longleaf pine sites per year.

CURRENT STATUS OF LONGLEAF PINE

Current longleaf pine and longleaf pine/oak acreages represent a significant increase from the figures reported from the late 1990s, reflecting the dedicated work of many organizations and individuals. FIA data through 2010 estimated about 3.3 million acres in longleaf pine forests and another 985,600 acres in longleaf pine/oak forests for a total of 4.29 million acres (Oswalt et al. 2012). More recent queries of FIA data (through 2015) indicate a gain of about 204,000 acres in longleaf pine forests, which was offset by a concurrent loss of about 209,000 acres in longleaf pine/oak forests, for a total of 4.28 million acres (Miles 2016)—effectively unchanged since 2010. However, the Longleaf Partnership Council documented the planting of about 150,000 acres per year from 2013 to 2015 (Longleaf Partnership Council 2014, 2015, 2016), for a total of 450,000 acres, substantially more than the gains for the longleaf pine forest type shown in FIA data for this time period. This inconsistency may reflect FIA collection methods, which do not produce a true inventory but rather a sampling of a nationwide systematic grid, with each plot representing about 6000 acres (Oswalt et al. 2012). The wide distribution of these sample plots means that FIA data are most accurate at broad scales and do not always capture finer-scale changes. There are also potential time lags associated with sampling only a portion of the plots annually across each 5-year period. More important than the specific numbers are the valuable insights about overall trends that the FIA data provide from an established, statistically sound, long-term monitoring program. Clearly, any loss of longleaf pine acreage is a source of concern and a challenge for those working to restore these ecosystems.

Regardless of whether the losses reported represent a conversion to other land uses or change in forest type with hardwood proliferation in the absence of fire, any net gains of longleaf pine acreage are likely to be substantially less than the acreage gains of newly planted longleaf pine acres. Comparing FIA data from the 1990s with 2010, 5.6% of the existing longleaf pine forest acreage from the prior survey was converted to loblolly pine stands, with just over half of that as plantations, and 5.3% was lost to longleaf pine/oak forests (Oswalt et al. 2012). The transition of acreage from longleaf pine forests to longleaf pine/oak and natural loblolly pine suggests the influence of fire exclusion or insufficient fire intensity.

Striking changes have also occurred in the percentage of existing longleaf pine forests that originated from natural regeneration versus plantations. In 1985, the FIA estimated 319,000 acres of planted longleaf pine forest, or 9% of the total area of longleaf pine forests (Kelly and Bechtold 1990). The resurgence of interest in longleaf pine, and especially the implementation of government incentive programs, stimulated an unprecedented increase in longleaf pine plantation establishment over the next 25 years. By 2010, there were an estimated 967,000 acres of planted longleaf pine, or 29% of the total longleaf pine acreage (Oswalt et al. 2012).

The first of these incentive programs began in 1998 when the U.S. Department of Agriculture (USDA) designated planting of longleaf pine a National Conservation Priority Area under the Conservation Reserve Program, reflecting a growing appreciation of longleaf pine forests. This incentive program has been perhaps the most important catalyst for recent increases in longleaf pine acreage and is responsible for the establishment of about 398,000 acres of longleaf pine as of 2015 (D. Hoge, personal communication). Under this program, which is administered by the Farm Service Agency with technical assistance from the USDA Natural Resources Conservation Service, landowners retire acreage that is actively farmed and plant longleaf pine in exchange for establishment-cost reimbursements and yearly incentive payments that extend for a defined period. Initially this program prioritized farmlands with highly erodible soils, but later iterations broadened eligibility. Subsequently, other federal incentive programs made significant contributions to longleaf pine planting, resulting in the establishment of another 158,000 acres through 2015 (L. Jones, personal communication). Other landowner incentives, such as the U.S. Fish and Wildlife Service–Partners for Fish and Wildlife Program, that focus on wildlife habitat also made important contributions to longleaf pine establishment and stewardship.

DISTRIBUTION AND OWNERSHIP PATTERNS

The distribution of longleaf pine forests is quite variable across the historical range of the species; Figure 1.2 shows a summary of the percentage of the total acreage of the remaining longleaf pine forest found in each state, and Figure 1.3 shows 2015 county-level FIA data for distribution of both the longleaf pine forest type and the longleaf pine/oak forest type. Approximately 89% (2.9 million acres) of longleaf pine forest acreage is located east of the Mississippi River, with large concentrations in the Florida Panhandle, southern Alabama, Georgia, South Carolina, and southern Mississippi (Oswalt et al. 2012). Private landowners hold 62% of this acreage, and the remaining acreage is held by public land management agencies. Although this ownership distribution appears to emphasize private lands, the concentration of longleaf pine forests in public ownership (38%) is disproportionate to the total area (13%) of public lands in the Southeast (Oswalt et al. 2014),

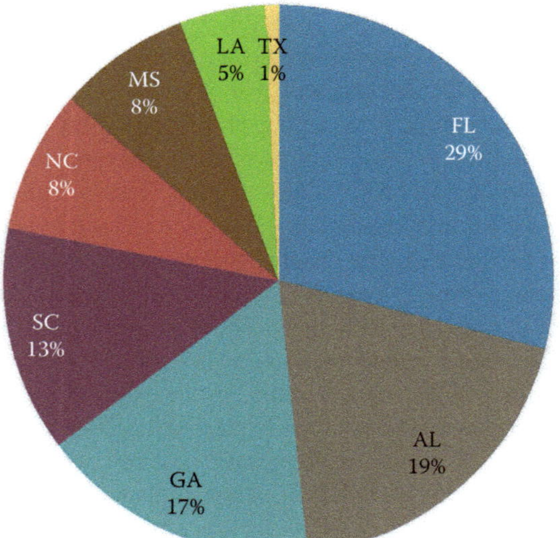

FIGURE 1.2 Percentage of longleaf pine forests and longleaf pine/oak forests by state from the 2015 U.S. Forest Service–Forest Inventory and Analysis database. (From Miles, P. D. Mon Dec 12 13:49:17 CST 2016. Forest Inventory EVALIDator web-application Version 1.6.0.03. St. Paul, MN: U.S. Department of Agriculture, Forest Service, Northern Research Station. Available at: http://apps.fs.fed.us/Evalidator/evalidator.jsp.)

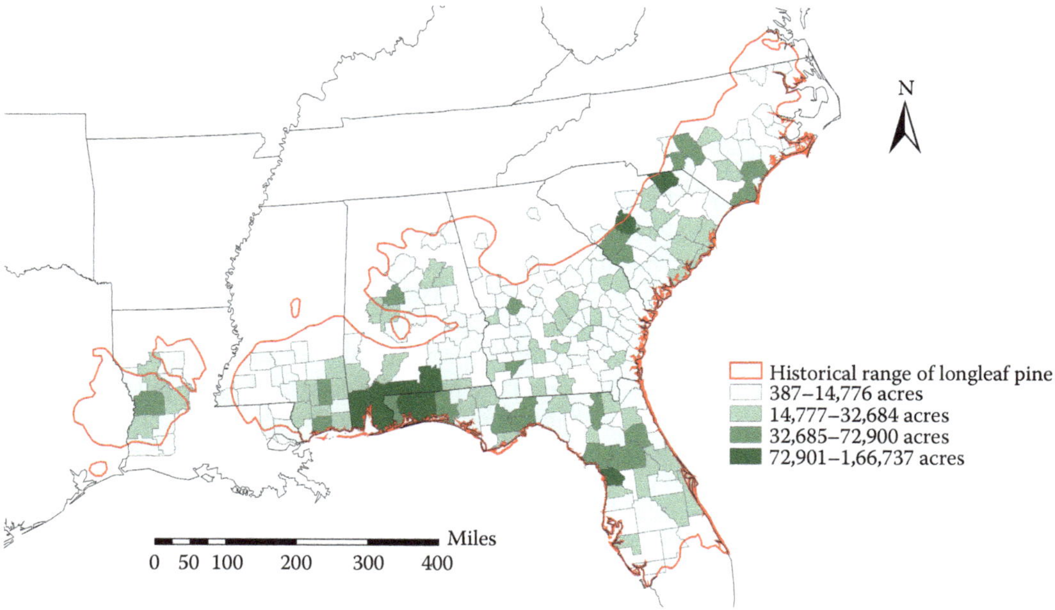

FIGURE 1.3 County-level distribution of longleaf pine and longleaf pine/oak forests from the 2015 U.S. Forest Service–Forest Inventory and Analysis database. (From Miles, P.D. Mon Dec 12 13:49:17 CST 2016. Forest Inventory EVALIDator web-application Version 1.6.0.03. St. Paul, MN: U.S. Department of Agriculture, Forest Service, Northern Research Station. Available at: http://apps.fs.fed.us/Evalidator/evalidator.jsp.)

underscoring the importance of public lands to longleaf pine conservation. Notably, public lands have higher than average longleaf pine volume per acre, and private lands have lower than average volumes (Oswalt et al. 2012).

Although longleaf pine acreage has increased significantly since the mid-1990s, its footprint (or spatial extent) continues to contract (Oswalt et al. 2012), particularly at the boundaries of its historical range. The age distribution of remaining longleaf pine stands also reflects its history of intensive harvesting, slow recovery, and the recent efforts to reverse its decline: 18% of all longleaf pine stands (longleaf pine forest type and longleaf pine/oak forest type) are <10 years old, 59% are <50 years old, and only 7% are >80 years old (Oswalt et al. 2012). FIA data from the mid-1990s show longleaf pine forests to be highly fragmented, especially on private lands, with about 75% of all longleaf pine stands <100 acres (Outcalt and Sheffield 1996). Moreover, the average forested parcel size continues to decrease, suggesting that fragmentation trends are continuing in the 21st century (see Chapter 3).

Data from FIA plots (Miles 2016) also provide insights into the distribution of longleaf pine by site type. Although they do not strictly align with the longleaf pine community classifications developed by Peet (2006), nor with the topoedaphic categories described below, broad comparisons can be drawn. Overall, across the range of the species, the ranked order of longleaf pine-dominated plots (those classified as longleaf forests) are mesic sites (47%), xeric sites (28%), flatwood sites (24%), and hydric sites (<1%). However, the variation in distribution patterns among states is considerable, a reflection of differences in the states' edaphic characteristics and land use trends. For example, longleaf pine plots were most commonly found on mesic sites in Alabama (88%) and Mississippi (92%), but were more common on flatwood sites in Florida (42%). Both South Carolina and North Carolina had a majority of plots located on xeric sites. Hydric longleaf pine sites were relatively rare and were reported only in Florida and Alabama. However, well-known hydric sites such as the Mississippi Sandhill Crane National Wildlife Refuge and the Green Swamp in North Carolina were not represented in the data—a function of the FIA sampling design and objectives and consequent difficulty of capturing all features at a finer scale.

FIRE-MAINTAINED LONGLEAF PINE VEGETATION

In addition to occupying a large historical range, longleaf pine occurred on a broad array of site types and had variable structures and compositions. Comprehensive classifications of longleaf pine ecosystems have identified patterns of vegetation composition that are associated with soil characteristics and physiography (Peet 2006). Broad-scale vegetation patterns also coincide with ecoregions that were identified by Omernik (1987) and Griffith et al. (2001). Overall, soil moisture and soil texture—coupled with fire frequency—are the primary influences on vegetation (Carr et al. 2009; Peet et al. 2014), with local composition varying in response to gradients of soil moisture and texture across relatively fine scales (Kirkman et al. 2001; Peet 2006).

The following sections describe the physical setting and vegetation structure of general longleaf pine vegetation types: sandhills and river dunes; dry upland forests and woodlands; mesic to wet-mesic upland forests, woodlands and savannas; pine flatwoods; and rocky pine woodlands (montane/Piedmont longleaf). With the exception of rocky pine woodlands, all occur in the Coastal Plain. More detailed discussions of geographic variation in vegetation composition can be found in Peet (2006) and Carr et al. (2009).

Sandhills and River Dunes

Physical Setting

Sandhill woodlands are ultraxeric communities that occur throughout the Coastal Plain on isolated upland ridges and knolls with deep, coarse, sandy soils; they are most abundant along the Fall Line (the physiographic boundary between the Coastal Plain and the Piedmont). River dunes are extremely deep sand ridges that are located parallel and to the east of major streams and are particularly well developed in Georgia.

Vegetation and Ecology

The extremely dry and infertile woodlands of these deep sandy sites often feature stunted trees and areas of bare ground or lichen cover. The patchiness of the vegetation and bare sand creates natural firebreaks in areas where fuel is absent. On very coarse sands, scattered longleaf pines are present with a subcanopy of pyrophytic oaks (*Quercus* spp.) that have thick bark, the ability to resprout from rhizomes, and other fire-tolerant traits (Cavender-Bares et al. 2004). Leaf litter accumulates beneath clusters of oaks or pine, forming islands of vegetation. Over time, these islands expand and grasses and forbs become established, ultimately producing enough fuel to carry fire. In periods of fire suppression, thick stands of midstory oaks can develop; however, without fire, the rate of hardwood encroachment is slower than it would be on more mesic sites (Bozeman 1971; Ware et al. 1993). Except in the most western and southern parts of the longleaf pine range, turkey oak (*Q. laevis*) is commonly associated with the coarse-textured sandy soils. The ground cover of these sandhill and river dune sites has the lowest species richness among the longleaf pine vegetation types.

Dry Upland Forests and Woodlands

Physical Setting

Dry (subxeric) upland woodlands occur on deep sandy soils in terrain that is nearly level to undulating, often along slightly more silty lower slopes of the sandy summits and ridges that are associated with ultraxeric sandhills and river dune communities (Wells and Shunk 1931). They also develop in shallow depressions within excessively drained sites where silt or clay has accumulated. In the upper Coastal Plain, dry upland woodlands develop on small knolls that form as caps of deep sand (>3 feet in depth) over loamy or clayey soils (Goebel et al. 2001). In the lower Coastal Plain, these woodlands also develop in level terrain that transitions into flatwoods (Peet 2006).

Vegetation and Ecology

With frequent prescribed fire, the vegetation of subxeric sites is characterized by a widely spaced longleaf pine canopy, few midstory species or shrubs, and a continuous ground cover that is dominated by grasses. With longer fire-return intervals, a midstory of pyrophytic oaks can become established including turkey oak, bluejack oak (*Q. incana*), sand laurel oak (*Q. laurifolia*), sand post oak (*Q. margaretta*), and southern red oak (*Q. falcata*). The distribution of oak species within sites and across landscapes depends on their individual tolerances to drought conditions and variable fire regimes (Monk 1968; Jacqmain et al. 1999; Cavender-Bares et al. 2004). Blackjack oak (*Q. marilandica*) and sand post oak may be prevalent in longleaf pine-dominated sites that have a clayey, sandstone, or ironstone horizon. Although the ground cover of dry woodland sites is less species rich than that of more mesic sites (Kirkman et al. 2001; Peet 2006), numerous legumes and forbs are usually present (Hainds et al. 1999). In these sites, soil moisture appears to be a limiting resource to ground cover seedling establishment (Iacona et al. 2010; Kirkman et al. 2016). For more detail, see Chapter 5.

Mesic to Wet-Mesic Upland Forests, Woodlands, and Savannas

Physical Setting

Mesic upland forests and woodlands, which occur on well-drained loamy sands over clay subsoils, were once a prevalent vegetation type throughout the upper Coastal Plain. Compared to the deep sandy soils of the dry upland woodlands, the loamy soils of mesic forests and woodlands retain more moisture and nutrients. They usually have a horizon with a significant accumulation of clay within three feet of the surface (Goebel et al. 2001). Depending on physiographic region and topographic characteristics of the site, these mesic sites can transition into wetter sites—supporting communities such as seepage herb bogs, shrub bogs, mesic slope forests, and savannas—or into flatwoods.

Vegetation and Ecology

The loamy sands support an open canopy of longleaf pine with few midstory and understory hardwoods. When frequently burned and relatively undisturbed, the grass-dominated ground cover has exceptionally high species richness. Maintaining forests with a sparse midstory requires prescribed fire with a return interval of 2–3 years (Mitchell, Hiers, et al. 2009). Other less fire-tolerant pine species can become established, either because of fire exclusion or moister conditions that restrict fire during seedling establishment. In the absence of frequent fire, fire-intolerant oaks and hickories (*Carya* spp.) can also become readily established. Most of these more fertile sites were converted to agricultural use before the 1900s (Williams 1989; Frost 2006). These communities can transition into more open savanna habitats with increased moisture, particularly along seepage slopes.

Pine Flatwoods

Physical Setting

Pine flatwoods are distributed throughout the outer Coastal Plain on deep, sandy, acidic soils (Stout and Marion 1993; Peet 2006). These low-relief sites occur on marine terraces where the water table is close to the soil surface. Thus, during wet winters, periodic saturation can occur, whereas during droughts the soil conditions are extremely dry. These soils have low nutrient availability, clay content, organic matter content, as well as cation exchange capacity (Abrahamson and Hartnett 1990). Often flatwoods stands have embedded wet depressions, oak domes, or small floodplain swamps.

Vegetation and Ecology

Species occurring in pine flatwoods are adapted to extremes in soil moisture. The scattered canopy is dominated by longleaf pine or mixtures of longleaf pine and slash pine. Canopy dominance can transition to slash pine or pond pine (*P. serotina*) depending on the position of the water table and the frequency of fire. The ground cover is dominated by grasses, runner oaks (*Q. pumila*), dwarf live oaks (*Q. minima*), saw palmetto (*Serenoa repens*), and other woody species such as gallberry (*Ilex glabra*) or numerous species of ericaceous shrubs. Frequent fire on drier flatwood sites tends to favor grasses. In the absence of fire, shrubs become more dominant, often forming nearly impenetrable thickets. With fire exclusion and the buildup of dense, flammable fuels, the risk of catastrophic wildfires increases significantly (Edmisten 1963). Large acreages of natural pine-dominated flatwood sites have been converted to industrial slash or loblolly pine plantations.

Rocky Pine Woodlands

Physical Setting

Rocky pine woodlands occur on the thin rocky soils of south to southwest-facing slopes in the Cumberland Plateau/Ridge and Valley, Blue Ridge, and Piedmont ecoregions of the Southeast. Although the original distribution of longleaf pine in woodlands above the Fall Line is not known, it appears to have been most widespread in Alabama and less frequent in Georgia and northward (Harper 1905, 1943; Peet 2006). Not surprisingly, given the climate of these ecoregions, rocky pine woodlands are occasionally exposed to ice storms and windthrow (Knight 2006).

Vegetation and Ecology

The canopy composition of remaining rocky pine woodlands is usually a mixture of longleaf pine, shortleaf pine (*P. echinata*), loblolly pine, blackjack oak, and other oak species. The current combination of hardwood species may reflect periods of longer fire-return intervals than occurred historically, although the historical fire-return interval has been debated (Edwards et al. 2013). Initial introductions of prescribed fire to stands with sparse ground cover have resulted in a substantial ground cover response, including the establishment of many species in common with frequently burned Coastal Plain upland longleaf pine stands (Currie et al. 2006; Cipollini et al. 2012).

SITE CONSIDERATIONS FOR RESTORATION AND MANAGEMENT

Practitioners who understand the state of longleaf pine forests, both historical and current, and the various site conditions that support them, will be more likely to develop realistic and appropriate goals for restoring and managing longleaf pine. However, in addition to ecological factors, social and economic constraints can also affect the ability to carry out sustainable management over extended periods of time at a given location.

Ecological Factors and Land Use History

The basic elements that define a site—topography and soil characteristics—and the initial level of site degradation strongly influence management and restoration practices. The degree of influence is mediated by ecosystem processes—such as competition, facilitation, nutrient cycling, persistent seed banks, fire behavior, and the functional roles of species—that vary across the gradient of sites. At any given site, consideration of this variation informs the choice of appropriate practices, both for the control of competing vegetation that limits the establishment of longleaf pine and other desired plant species, and for the maintenance of adequate fine fuels to conduct prescribed fires. In addition, the land use history of former longleaf pine sites spans a continuum of degradation that includes fire-excluded longleaf pine sites with remnant native ground cover, natural stands of longleaf pine with disturbed ground cover, plantations of off-site pine species in sites where native ground cover may or may not be present, and recently abandoned agricultural fields. These land use legacies also reflect environmental conditions such as altered soils or hardwood encroachment resulting from fire exclusion. Thus, understanding the basic ecosystem processes across a suite of site conditions is a fundamental step in integrating science and practice into restoration and management (see Chapters 4, 5, and 7).

Given that the establishment of a frequent fire regime is the most essential requirement for maintaining longleaf pine ecosystems across the full range of sites and initial conditions, successful management and restoration depends on an evaluation of fuel types, fuel loading, and the consequences of reintroducing fire (see Chapter 6). If the vegetation structure at a site is not conducive to reintroduction of prescribed fire, a variety of pretreatments is available for a range of site conditions (see Chapters 10 and 11). On fire-excluded mesic and subxeric sites, a series of carefully implemented burns is often necessary to reduce fuel loadings before beginning a maintenance fire regime. Similarly, on sites with a large midstory component (such as flatwoods), mechanical treatments are often required to reduce competition and heavy fuels to enable fine fuels development before longleaf pine seedlings can be planted. By contrast, in extremely dry sites (such as sandhills and river dunes, dry uplands), fuel levels may be too low to support the reintroduction of a prescribed fire regime, and fire-return intervals may need to be longer than they would be on wetter sites.

An interest in sustaining wildlife species uniquely associated with longleaf pine has driven much of the restoration interest because fire-maintained longleaf pine forests provide the structural characteristics that are essential for suitable habitat. Although there can be distinct differences in the details of vegetation structure between site types, broad structural attributes—mature, open-canopied overstory, little or no midstory vegetation, and grass-dominated but diverse herbaceous ground cover—are remarkably consistent across the range of longleaf pine-dominated ecological communities managed with frequent fire. However, land managers need guidance beyond these visual aesthetics, and the development of quantitative parameters for vegetation structure has been a focus of the longleaf pine restoration community in recent years (McIntyre 2012; Ware 2014; Nordman et al. 2016). Some general quantitative guidelines for desired structure and vegetation composition that correlate with healthy populations of longleaf pine-associated wildlife are as follows:

Canopy

- Canopy pine basal area: 30–80 ft^2/acre
- Pine canopy cover: 30%–65% canopy cover
- Canopy hardwood basal area: <20 ft^2/acre
- Stand age structure: basal area >20 ft^2/acre flat top, or >14 ft diameter at 4.5 ft above ground level (dbh)

Midstory

- Midstory overall cover: <20% cover woody midstory

Ground cover

- Overall native herbaceous cover: >40% cover
- Native warm-season grass cover: >25% cover
- Longleaf pine regeneration: >1% cover
- Invasive plant cover: <1% cover

Although specific structural and compositional factors will vary with site type, the greatest variance occurs at the dry and wet ends of the hydrologic gradient. Both xeric sites and wet flatwood and savanna sites may have lower canopy stocking (as expressed through basal area and canopy coverage) than longleaf pine communities that fall between these hydrologic extremes. Characteristics of shrub cover in the midstory stratum also vary between site types (Nordman et al. 2016) depending, in part, on site quality and fire history. Montane longleaf pine communities on rocky uplands also exhibit more variance in canopy composition, with shortleaf pine and various oak species commonly present, and even codominant, in the canopy.

These parameters are intended to be broad targets for desired structure and composition as managers manipulate habitat for certain wildlife species and monitor changes in their populations with changing forest structure. Management interventions (thinning, prescribed fire, herbicide treatments, etc.) will cause individual structural and compositional characteristics to move in and out of the desired ranges presented above. Natural disturbances, climatic variation, and altered hydrology will also influence vegetation structure and composition, and adequate monitoring and an adaptive management approach are essential to meeting goals and objectives for a given site (see Chapter 14). Even with this variability through time and between site types, the broad guidelines provide a useful tool for helping to define objectives that restore targeted wildlife communities along with the vegetation of longleaf pine forests.

Socioeconomic Factors

Beyond the consideration of ecological factors and on-the-ground management are the social and economic drivers that play a significant role in the real-world restoration of longleaf pine (see Chapter 3). Examples of these factors include types of ownership (public or private), changing land use patterns, and the role of regulations, incentives, and markets.

Although the type of ownership can affect objectives within the context of longleaf pine restoration, most longleaf pine restoration and management scenarios—whether for public or private owners—involve complex and nuanced mixtures of multiple values and objectives. As a broad generalization, public land management agencies and NGOs often place greater emphasis on ecosystem values such as at-risk species, water quality and quantity, or unique ecological communities. Although private landowners typically place more emphasis on economic returns than do public landowners, they may also assign equal value to recreation, aesthetics, and other noneconomic benefits. Regardless of how much emphasis a particular landowner or agency places on future economic returns, longleaf pine restoration requires substantial up-front investments of money and time, both for establishment and for ongoing management activities such as prescribed fire. Ownership class also influences management and restoration. For example, national forests and other federal holdings are subject to regulations and laws, such as the National Environmental Protection Act, that do not apply to private landowners. By the same token, private landowners face many challenges—such as heightened liability from prescribed fire or difficulties in obtaining burn permits—that do not apply to public lands.

Regardless of ownership, the role of local markets and mill infrastructure is an important consideration in the restoration process. Restoring a multiple age-class structure in longleaf pine forests typically requires ongoing intervention and removal of small-diameter trees. The locations and numbers of mills, as well as the types of materials they will accept, can determine what is economically possible during a spatially extensive or long-term longleaf pine restoration. For example, a fire-excluded forest with a canopy of longleaf pine (or other pines) frequently will have a well-developed midstory that requires mechanical treatments to reduce biomass and enable prescribed fire. Some locations in the longleaf pine range have local hardwood chip markets, allowing landowners to remove undesirable elements of vegetation structure without incurring costs. Without these markets, however, mechanical removal of the midstory would likely be cost prohibitive.

Finally, the changing landscape of the southeastern United States has impacts for restoration of longleaf pine ecosystems (see Chapter 3). Because of regional population growth that is among the highest in the United States, land use change and increased fragmentation is pervasive in the Southeast (Wear 2013). In many rapidly urbanizing areas where longleaf pine forests are present or that have the potential for restoration, forest management simply cannot compete with other land use alternatives when common economic metrics are used in land use evaluations. In addition, the rapidly growing population and accompanying infrastructure development bring new pressures to minimize smoke from prescribed fire (see Chapter 13), limiting options for managing longleaf pine and ultimately making restoration impractical.

SUMMARY

The restoration and management of longleaf pine ecosystems is a complex and long-term endeavor, requiring the practitioner to consider both past and current conditions when determining the desired future state for a given location. Longleaf pine historically was a canopy dominant in a variety of landscapes across its range, and although we often reference the "longleaf pine ecosystem," we do not have a single monolithic longleaf pine forest type to use as a target for all restoration trajectories. To address soil condition variations, topographic variations, land use history, and landscape considerations, successful restoration requires site-specific treatment regimes. Ecological and biological considerations often determine where and how restoration will be most effective, but other factors—such as the "social license" to conduct prescribed fire or the presence of local markets to offset the costs of harvesting operations—also have an influence. To incorporate all of these considerations (ecological, land use, social, and economic) requires an integrated and multidisciplinary approach to management and restoration—a requirement that is straightforward to grasp conceptually, but often difficult to implement. Exploring the relationships among these different elements is a principal objective of the chapters that follow.

ACKNOWLEDGMENT

Spatial data for Significant Geographic Area boundaries in Figure 1.1 were produced by Analie Barnett of The Nature Conservancy.

2 Biogeography

An Interweave of Climate, Fire, and Humans

Michael C. Stambaugh, J. Morgan Varner, and Stephen T. Jackson

CONTENTS

INTRODUCTION

Longleaf pine (*Pinus palustris*) is an icon of the southeastern United States and has been considered a foundation species in forests, woodlands, and savannas of the region (Schwarz 1907; Platt 1999). Longleaf pine is an avatar for the extensive pine-dominated, fire-dependent ecosystems (Figure 2.1) that provide habitats for thousands of species and have largely vanished from the landscape. Longleaf pine is one of the world's most resilient and fire-adapted trees (Keeley and Zedler 1998), widely perceived as the sole dominant in forests across a large area of the Southeast (Sargent 1884; Mohr 1896; Wahlenberg 1946). Longleaf pine was once a primary natural resource, providing high-quality timber, resins, and naval stores that fueled social changes and economic growth through the 19th and early 20th centuries.

Ecosystems dominated by longleaf pine are now among the most threatened in North America (Noss et al. 1995); fragmentation and decreasing populations have led to the recent designation of longleaf pine as an endangered species by the International Union for Conservation of Nature (Farjon 2013). Despite the historical importance and conservation interest of this species, its biogeography and ecological history have received surprisingly little scientific attention.

This chapter uses multiple lines of evidence to introduce the biogeography of longleaf pine from three perspectives: historical (from millions to hundreds of years ago), current (since the

FIGURE 2.1 Photographs depicting the variability in longleaf pine forest communities throughout its range: (a) A longleaf pine-bluestem (*Andropogon* spp.) forest at the western range margin—Angelina National Forest in Texas, (b) a closed-canopy longleaf pine and hardwood stand in the Kisatchie Hills of the western Gulf Coastal Plain—Kisatchie National Forest in Louisiana, (c) longleaf pine growing near its elevational maximum—Mountain Longleaf Pine National Wildlife Refuge in Alabama, (d) a Coastal Plain longleaf pine-wiregrass savanna at Ichauway—Joseph W. Jones Ecological Research Center in Georgia, (e) a longleaf pine woodland near the northern range limit—Zuni Pine Barrens in Virginia, (f) a longleaf pine savanna near the southern range limit—Platt Branch Wildlife and Environmental Area in Florida. (Photographs courtesy of [a] J. Sparks, [b] Michael Stambaugh, [c] Morgan Varner, [d] Michael Stambaugh, [e] Morgan Varner, and [f] Neil Pederson.)

(Continued)

(g)

(h)

FIGURE 2.1 (*Continued*) Photographs depicting the variability in longleaf pine forest communities throughout its range: (g) Longleaf pine growing along coastline at Choctawhatchee Bay—Eglin Air Force Base in Florida, and (h) longleaf pine growing on wet site—Triple N Ranch Wildlife Management Area in Florida. (Photographs courtesy of [g] Morgan Varner and [h] Amy Jenkins.)

early 20th century), and future. Biogeographic descriptions are important for several reasons: They can be a reference for broad-scale conservation, they raise awareness of challenges and opportunities, and they are often an integration of many information sources (Channell and Lomolino 2000).

Because the biogeography of any given region is dynamic through time, our goal was to cover the major drivers from global to local scales with a focus on the central themes of climate, fire, and human influences. We attempted to uncover new evidence of influences and trends from paleoecological studies through comparisons to modern-day longleaf pine ecology. As is common in most biogeographic studies, this information was often limited by resolution, disparate sources, and conflicting or ambiguous interpretations. Within these limitations, we tried to place the current status of longleaf pine into an appropriate context, focus attention on critical information gaps, and evaluate ensuing challenges likely to arise in our rapidly changing world.

BIOGEOGRAPHIC AND ENVIRONMENTAL HISTORIES

The biogeographic history of longleaf pine and other pine species of the Southeast is obscured by fragmentary fossil evidence and incomplete phylogenetic studies. Species differentiation among pines based on pollen morphology is difficult to impossible, and anatomically preserved needles and cones are scarce in sediments of the Southeast. Although phylogenetic studies provide general outlines of evolutionary history and relationships of *Pinus* (Figure 2.2), phylogenetic and phylogeographic studies specific to the distal clades (subsection and below) are lacking.

Longleaf pine is one of twelve species of *Pinus* subgenus *Pinus* in eastern North America, and one of eight species in *Pinus* section Trifoliae subsection Australes in the unglaciated southeastern United States (Figure 2.2). *Pinus* section Trifoliae occurs entirely in the Western Hemisphere, and subsection Australes is restricted to the southeastern United States, the Caribbean, and Mexico and adjacent Central America. This group has been documented as originating in the Western Hemisphere during the Paleogene (66–23 million years ago), but the timing and location of key divergences and originations is obscure (Millar 1998; Eckert and Hall 2006; Willyard et al. 2007).

During most of the Paleogene, the Coastal Plain of the Southeast was occupied largely by angiosperm forests. Although pine occasionally appears in the Paleogene pollen and leaf floras,

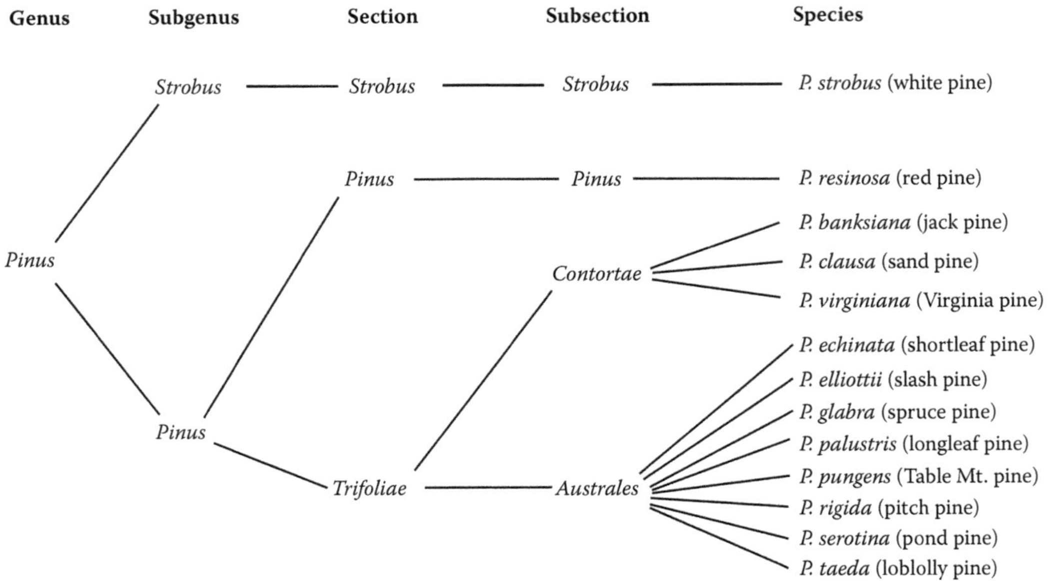

FIGURE 2.2 Taxonomy of eastern North American pine species. The lines are for heuristic purposes to show the taxonomic hierarchy; their length and angles impart no specific phylogenetic information. (Based upon classifications in Little, E. L., Jr. and W. B. Critchfield. *Subdivisions of the Genus Pinus. Miscellaneous Publication 1144*, USDA Forest Service, Washington, DC, 1969; Price, R. A. et al., *Ecology and Biogeography of Pinus*, Cambridge University Press, Cambridge, 1998; and Gernandt, D. S. et al., *Taxon*, 54, 29–42, 2005.)

subtropical and warm-temperate deciduous angiosperm trees largely dominated the assemblages (Dilcher 1973; Frederiksen 1980; Graham 1999; Dilcher 2000). Global cooling occurred through the Neogene (23–2.6 million years ago). Although pine is not well represented in the sparse Miocene floras of the Coastal Plain, pine-dominated forests were clearly established in Florida by the mid-Pliocene (about 3.5 million years ago) (Willard et al. 1993; Willard 1994), continuing into the late Pliocene and early Pleistocene (<2.6 million years ago) (Willard et al. 1993; Hansen et al. 2001). Although these records are from areas currently occupied by longleaf pine, slash pine (*P. elliottii*), and sand pine (*P. clausa*), the species represented in the pollen assemblages are unknown. Significantly, pollen of obligate upland nonwoody plants—ragweed (*Ambrosia* spp.), chenopods (*Chenopodium* spp.), and upland grasses (Poaceae)—are scarce in these assemblages, in contrast to the pine-dominated pollen assemblages from late Quaternary (<60,000 years ago) deposits in Florida (Hansen et al. 2001). The Pliocene assemblages indicate pine-dominated forests rather than savannas or woodlands.

The Quaternary, comprising the last 2.6 million years, was characterized by more than 20 glacial-interglacial cycles, each accompanied by dramatic changes in sea level and atmospheric carbon dioxide concentrations. Although terrestrial records in the Coastal Plain are scattered and discontinuous, these glacial-interglacial cycles are well recorded in marine sediments of the Gulf of Mexico (Joyce et al. 1990, 1993; Flower et al. 2004) as well as the subtropical Atlantic Ocean. Glacial-interglacial cycles were accompanied by high-magnitude variations in sea-surface temperature. For example, sea-surface temperatures in the Gulf of Mexico during the last glacial maximum 20,000 years ago were some 4°C lower than modern temperatures, declining rapidly during the last deglaciation (Flower et al. 2004; Williams et al. 2010). Although the Coastal Plain has been characterized as "climatically stable" throughout the late Cenozoic (Noss et al. 2015), geological and climatic records indicate that it, like the rest of the globe, has experienced substantial climatic change and variation throughout the Pleistocene and before.

The longest well-dated pollen chronology for the late Quaternary in the Coastal Plain is the 60,000-year record from Lake Tulane, on the southern Lake Wales ridge of Peninsular Florida (Grimm et al. 1993, 2006). Lake Tulane is farther south than the southernmost area of extensive longleaf pine forests; surrounding native vegetation is dominated by scrub oaks (*Quercus* spp.), as well as Ocala sand pine (*P. clausa* var. *clausa*) and South Florida slash pine (*P. elliottii* var. *densa*). The pollen record from 60,000 to 12,000 years ago shows a series of dramatic, rapid alternations between pine-dominated pollen assemblages and assemblages dominated by a mix of oaks, ragweed, and grasses (Grimm et al. 2006). These oscillations correspond temporally to the "Bond Cycles" recorded in ice-core and marine-sediment records across the North Atlantic region. The pine species that dominated during the peak pine phases of these cycles is not known, but the large, high-frequency oscillations illustrate the degree and rapidity of climate variation experienced during the late Quaternary, and firmly establishes the connectedness of the Coastal Plain to climate dynamics at hemispheric and global scales.

Several pollen records within the geographic range of longleaf pine span the past 10,000–20,000 years (Watts 1980; Watts and Stuiver 1980; Hussey 1993; Watts and Hansen 1994; Watts et al. 1996; Grimm et al. 2006), and macrofossil assemblages from several sites reveal late Quaternary biogeographic patterns and dynamics (Jackson et al. 1997, 2000). Forests and woodlands dominated by boreal and cool-temperate species—including white spruce (*Picea glauca*), jack pine (*P. banksiana*), and red pine (*P. resinosa*)—occurred at least as far south as central South Carolina, northern Georgia, and central and western Tennessee (Jackson et al. 2000; Liu, Andersen, et al. 2013). Forests in the Lower Mississippi Alluvial Valley were dominated by a now-extinct species of spruce (*Picea critchfieldii*), which also occurred in the Coastal Plain and the Appalachian foothills of Georgia (Jackson and Weng 1999; Jackson et al. 2000).

By contrast, spruce and northern pines were not found to occur in present-day Peninsular Florida during the last glacial period (Watts and Stuiver 1980; Watts and Hansen 1994; Jackson et al. 2000; Grimm et al. 2006). The dominant pines were clearly "southern" species (sand pine, the eight pine species that comprise the subsection Australes, or all nine species), but the dominant species and individual species distributions are unknown. Genetic (allozyme) studies of extant longleaf pine populations indicate a pattern of decreasing genetic diversity from west to east (Schmidtling and Hipkins 1998), a pattern not observed in other pine species of the Southeast (Schmidtling 2007). Schmidtling and Hipkins (1998) hypothesized that longleaf pine was restricted to coastal western Texas and the adjacent Mexican state of Tamaulipas during the last glacial period. However, the net decrease in heterozygosity from Texas to the Atlantic Coastal Plain is small (6%–8%), and longleaf pine populations might well have persisted in Peninsular Florida. Therefore, the hypothesis proposed by Schmidtling and Hipkins (1998) requires corroboration using other genetic markers and testing using paleoecological records.

Despite the scarcity of direct evidence to verify the whereabouts of longleaf pine populations during the last glacial maximum and the deglacial periods, pollen records provide minimum and maximum estimates of the antiquity of longleaf pine forests and savannas in the Southeast (Jackson 2012). Pine-dominated glacial-age forests in the Coastal Plain were replaced during the last deglaciation, between about 16,000 and 12,000 years ago, by forests that were dominated by oak, hickory (*Carya* spp.), beech (*Fagus* spp.), elm (*Ulmus* spp.), ash (*Fraxinus* spp.), hophornbeam/hornbeam (*Ostrya* spp./*Carpinus* spp.), and other hardwoods (Watts 1980; Watts and Stuiver 1980; Whitehead 1981; Hussey 1993; Watts and Hansen 1994; Grimm et al. 2006). Although many of the more mesic taxa declined in the early Holocene (10,000 years ago and later), oak-dominated forests persisted. Pine-dominated forests remain undocumented in the Coastal Plain during the late glacial and early Holocene periods.

All Holocene pollen records for the current range of longleaf pine show a transition from oak-dominated to pine-dominated assemblages, with pine persisting essentially unchanged until the land clearing that followed European settlement (Watts 1980; Watts and Stuiver 1980; Whitehead

1981; Hussey 1993; Watts and Hansen 1994). Comparing the timing of this transition at individual sites can provide an approximate beginning date for the establishment of longleaf pine ecosystems. However, the possibility of a transition from oak forests to pine forests dominated by another species—such as slash pine, sand pine, loblolly pine (*P. taeda*), and shortleaf pine (*P. echinata*)—followed by subsequent transition to longleaf pine dominance cannot be excluded. Unfortunately, the oak-to-pine transition is not well dated in the Coastal Plain; ongoing studies in Florida and South Carolina should refine the chronology (T. Krause and S.T. Jackson, unpublished data). The transition might have occurred as early as 8000 years ago in coastal South Carolina (Hussey 1993) and 5000–6000 years ago at other sites in Florida, Georgia, and South Carolina (Watts and Hansen 1994; Watts et al. 1996).

The underlying causes of the development of longleaf pine forests during the mid-Holocene remain obscure. It might represent a response to changes in Holocene climate, but evaluation is difficult until the timing of the transition is better documented and can be compared with independent paleoclimate records. Because long-term maintenance of longleaf pine as a dominant requires a high-frequency surface-fire regime, the transition might be the result of a shift in ignition frequency, fuel availability, or both. The development and maintenance of longleaf pine-dominated vegetation might have represented an interaction between climate change and influences from Native Americans. High-resolution pollen studies at sites in Florida indicate that longleaf pine forests persisted from the 16th through the 18th centuries—a time of disruption and transition in Native American communities across the Coastal Plain (Jones 2014). Either lightning was sufficiently frequent to continue igniting surface fires, or humans persisted in burning despite depopulation caused by European diseases and subsequent societal disruption and reorganization (Jones 2014).

MODERN BIOGEOGRAPHY OF LONGLEAF PINE

Climate

Climate is a major driver of the modern distribution of longleaf pine (Wahlenberg 1946). The climate of longleaf pine ecosystems is dominated by the influence of the Gulf of Mexico (Craul et al. 2005). The climate of the Southeast is humid-subtropical, lacking a distinct dry season (Peel et al. 2007), and relatively stable. Globally, humid-subtropical climates have undergone relatively little change through the 20th century (Chen and Chen 2013).

Data from 1981 to 2010 show that average annual precipitation within the range of longleaf pine is 134 cm with a range of 102–173 cm/year (Daly et al. 2008). The wettest area occurs along the eastern Gulf Coast, and the driest area is the eastern Piedmont. Seasonal precipitation is more bimodal along the Gulf Coast, transitioning to more singular wet and dry seasons into Peninsular Florida (Platt 1999).

Average annual temperature in the geographic range of longleaf is 18.1°C and ranges from 14 to 22°C. The warmest areas are located in Peninsular Florida; the coldest are in the Appalachian-Cumberland highland of Alabama and along the northern and inland margin of the Atlantic Coastal Plain. Throughout the range of longleaf pine, droughts during the last 500 years have been relatively short and infrequent compared to previous centuries (Cook et al. 2004). For about the last 1600 years, the frequency of decade-long droughts has increased northward from Florida to Georgia to North Carolina, but the frequency of single-year extreme droughts has increased from North Carolina southward to Florida (Figure 2.3). In some areas of the Southeast, climates are transitional between temperate and subtropical, and frosts are infrequent. The conditions associated with temperatures low enough to limit longleaf pine success are complicated. In general, seasonal low temperatures are important to the potential net primary production of pine species, with light and leaf area serving as primary limiting factors for tropical and subtropical pines (Richardson and Rundel 1998).

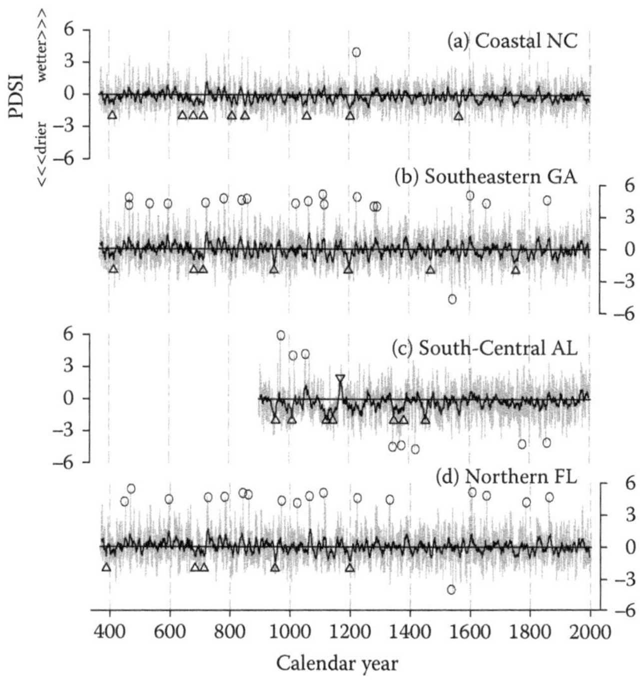

FIGURE 2.3 Reconstructed Palmer Drought Severity Index (PDSI)—with a value of +6 wettest and a value of −6 driest—at four locations across the range of longleaf pine: (a) Coastal North Carolina, (b) Southeastern Georgia, (c) South-Central Alabama, and (d) Northern Florida. Gray line represents annual indices, and black line is the 11-year moving average. Circles indicate years of extreme wetness (values > +4.0) and drought (values < −4.0). Arrows indicate periods of decade-long droughts and pluvials. (Reproduced from Cook, E. R. et al., 2004. North American Summer PDSI Reconstructions. World Data Center for Paleoclimatology Data Contribution Series #2004–045. NOAA/NGDC Paleoclimatology Program, Boulder, Colorado, USA, http://www.ncdc.noaa.gov/paleo/newpdsi.html.)

Damage and mortality from freezing temperatures and ice storms may be limiting factors for any northern or inland expansion of the longleaf pine range (Lipps 1966; Croker 1979), where heavy loads cause breakage of branches and foliage. Occurrence of freezing rain events varied from 0.1 to 2 days per year across the range of longleaf pine from 1949 to 2000 (Changnon and Karl 2003).

Physiography and Soils

The range of longleaf pine ecosystems before land clearance (Figure 2.4) is estimated to have covered >37 million ha (Frost 2006). Early accounts, from the explorations of Hernando de Soto (Clayton et al. 1995) and Cabeza de Vaca (Covey 1990) in the early 16th century to the botanical journeys of Bartram (1791) and Catesby (1771), depicted extensive longleaf pine-dominated landscapes. The extent and characteristics of longleaf pine ecosystems have been reconstructed from public land surveys conducted by the U.S. General Land Office in the 19th century. Narratives and datasets from southern (Predmore et al. 2007) and northern (Shankman and Wills 1995) Alabama and northern Florida (Delcourt and Delcourt 1977; Schwartz 1994) confirm that longleaf pine was dominant at the time of European settlement. Botanical and timber reports from the early 20th century further corroborate its extent and dominance (Reed 1905; Schwarz 1907; Harper 1913).

The distribution of longleaf pine spans a broad arc in the Gulf Coastal Plain, the Atlantic Coastal Plain, the Piedmont, and the Appalachian-Cumberland highlands (Figure 2.4). The western

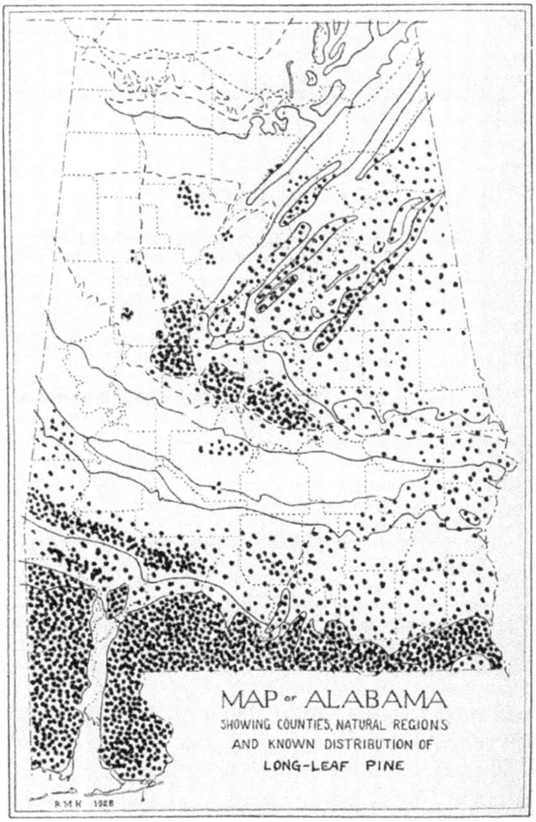

(a)

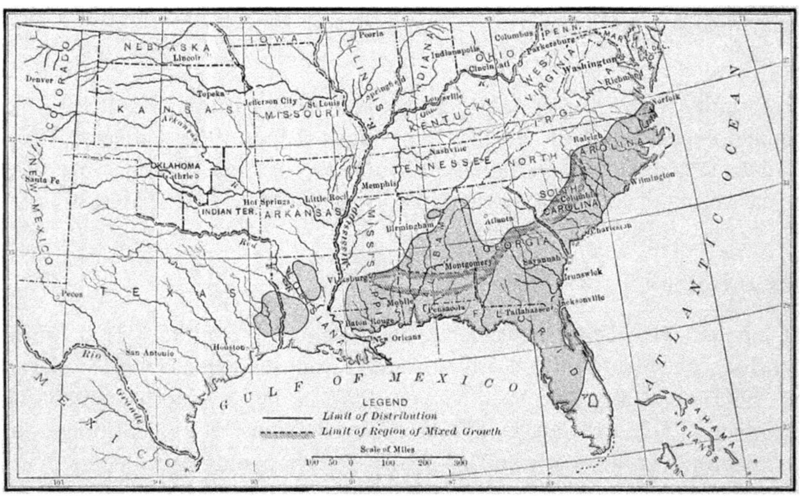

(b)

FIGURE 2.4 Longleaf pine: (a) Known distribution and approximate relative abundance in Alabama during the early 20th century. (From Harper, R. M. *Economic Botany of Alabama. Part 2. Geological Survey of Alabama, Monograph 9*, University of Alabama, Tuscaloosa, 1928.) (b) Range in the southeastern United States during the early 20th century. (From Schwarz, G. F., *The Longleaf Pine in Virgin Forest: A Silvical Study*, John Wiley & Sons, New York, 1907.) *(Continued)*

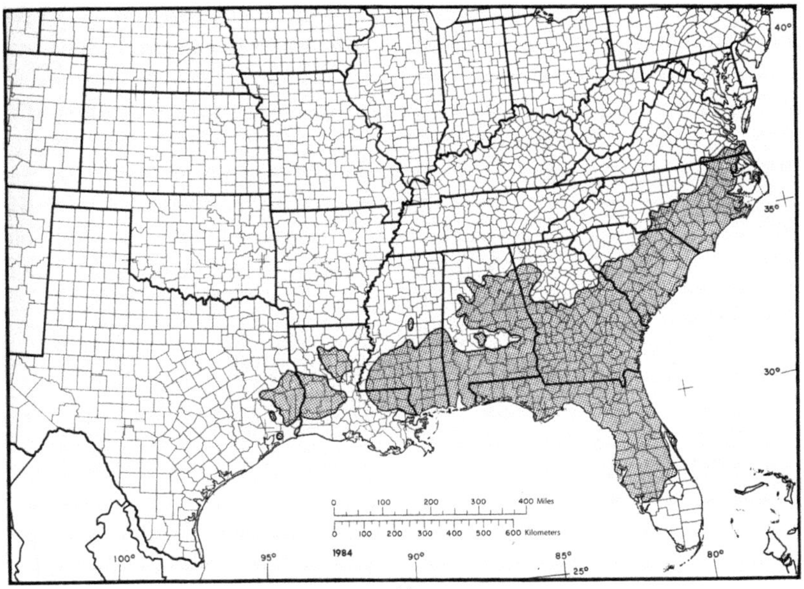

(c)

FIGURE 2.4 (*Continued*) Longleaf pine: (c) Range in the southeastern United States during the late 20th century. (From Little, E. L., Jr., *Atlas of United States Trees, Volume 1, Conifers and Important Hardwoods.* Miscellaneous Publication 1146, USDA Forest Service, Washington, DC, 1971.) Note that the largest differences between the two southeastern maps are attributed to changes in central Alabama.

range of longleaf pine (about 96° W longitude) is disjunct—with the Lower Mississippi Alluvial Valley (also known as "The Delta") serving as an important impediment and gap in the modern distribution. Fenneman (1938) classified the westernmost sites of eastern Texas and western Louisiana as the Coastal Prairie and Marsh, Pine Flats, Kisatchie Terraces, and Nacogdoches Terraces. To the east, longleaf pine dominates Coastal Plain sites in southeastern Louisiana, southern Mississippi, and adjacent southwestern Alabama—an area encompassing Fenneman's (1938) Pine Meadows and Southern Pine Hills; north of these areas are the rich sites of the east-to-west trending Red Hills. Inland interruptions of the range are provided by two large disjunct prairies: (1) the Jackson Prairie in south-central Mississippi, and (2) the Black Belt—or Blackland Prairies (Peacock and Schauwecker 2003)—with dark shrink-swell clays and alkaline soils that serve as an impediment to longleaf pine. Above the Black Belt, longleaf pine was once extensive in the Fall Line Hills extending eastward.

Inland, longleaf pine grows in the Piedmont Plateau and Appalachian-Cumberland highlands. In these areas, it grows up to 610 m above sea level and about 500–600 km inland (Figure 2.1). Its range is widespread in lowlands, on monadnocks, and on xeric ridges of the Piedmont Plateau and—to a lesser degree—in the Appalachian-Cumberland highlands (Reed 1905; Craul et al. 2005). Fenneman (1938) and Harper (1943) consider the northern stands to be part of the Blue Ridge, but some consider them to be part of the Southern Ridge and Valley (Bailey 1998).

Eastward along the Gulf Coastal Plain, longleaf pine once dominated over extensive plains of wiregrass (*Aristida stricta*). The East Gulf Coastal Plain of southern Georgia includes Fenneman's (1938) Brandywine and Coharrie Terraces, the broad Tifton Uplands, the wide Dougherty Plain, up to the Red Hills and Fall Line Sand Hills (Craul et al. 2005). Coastal wet flatwoods extend about 450 km south into Peninsular Florida where longleaf pine reaches its southernmost point at 27° N latitude (Figure 2.1). The longleaf pine range extends into southern Florida just north of Lake Okeechobee, overlapping with the range of South Florida slash pine. In the center of the peninsula along Fenneman's (1938) Sunderland Terrace, longleaf pine dominated mesic sites and xeric sand ridges, the latter intermingling with scrub ecosystems dominated by the serotinous Ocala sand pine and several oak species (Menges 1999).

Along the Atlantic Coastal Plain, longleaf pine grows along a diverse gradient that stretches inland from the coastline. Extensive wet flatwoods, which dominate the western coastline of Peninsular Florida, extend eastward across the state and then up the Atlantic Coast. In southeastern Georgia and South Carolina, longleaf pine dominates the Flatwoods, Wicomico Terrace, Sunderland Terrace, the Brandywine and Coharrie Terraces, and into the Fall Line Sand Hills (east to west). In North Carolina, it spans wet savannas in the Pamlico Terrace and spreads inland to the Wicomico Terrace, Sunderland, Brandywine and Coharrie Terraces, up to the Sand Hills. The wet savannas of the Atlantic Coastal Plain are among the most species-rich plant communities outside the tropics (Walker and Peet 1983). The Virginia Tidewater represents the northernmost outpost for longleaf pine, where it reaches 37° N latitude and comingles with shortleaf, loblolly, and pond pine (*P. serotina*) near wet spodic sites. The Virginia portion of the longleaf pine range contains many endemics that represent northern elements (Frost and Musselman 1987).

Wahlenberg (1946) compiled a thorough and thought-provoking treatment of longleaf pine's biogeography, plotting longleaf across a much wider swath of the Southeast (including a more northern range) than other authors. He suggested that, following harvest, wet longleaf pine sites along its southern range margins gave way to slash pine. In more northern sites, he suggested that loblolly pine and hardwoods repopulated harvested longleaf pine sites.

Longleaf pine can dominate a wide diversity of sites (Craul et al. 2005), including five soil orders (see Chapter 3). Sites include wet savannas (Entisols, Spodosols, and Inceptisols) of the Lower Coastal Plain, widespread mesic sites with clay argillic horizons (Ultisols) across the wider Coastal Plain and Piedmont, and xeric sandhill sites (Entisols) along the Fall Line boundary of the Piedmont and within ridges across the Coastal Plain. Along the loess bluffs on the eastern margin of the Lower Mississippi Alluvial Valley, longleaf pine grows in Alfisols. The rocky soils of the Appalachian-Cumberland highland are derived from schists, sandstones, amphibolite, and conglomerates with some serpentines (Craul et al. 2005). The uniting characteristics of longleaf pine sites are their tendency to have acidic and infertile soils; soil moisture is highly variable, from seasonally inundated wetlands to dry xeric ridges (Wahlenberg 1946).

Extant Longleaf Pine

With only a few exceptions, little information exists to describe the historical density of longleaf pine within its range. Harper (1928) published one of the earliest maps of longleaf pine density in Alabama (Figure 2.4). Wahlenberg (1946) plotted the distribution in 1935 with abundant notes on its relative dominance across its range.

By the end of the 20th century, longleaf pine was highly fragmented, covering approximately 1.2 million ha, or < 4% of its presettlement extent (Outcalt and Sheffield 1996). Loss of old-growth stands were even more dramatic, with fewer than 15 stands remaining across the range (Varner and Kush 2004). The causes and consequences of decline differed across the range, with areas subjected to extensive agricultural expansion, others converted to commercial loblolly pine and slash pine plantations, and still others subjected to exurbanization and fire exclusion. Each of these human-caused changes has fragmented or otherwise degraded remnant longleaf pine communities.

Longleaf pine has a native range that is more fragmented at its southern and northern extremes than most other pines (Stevens and Enquist 1999). Considering the dramatic decreases in extent since European settlement, the occurrence of longleaf pine within its native range has likely become even more fragmented.

TABLE 2.1
Longleaf Pine Growth Chronologies Based on Tree-Ring Widths: Chronologies Are Based on Crossdating Rather Than Actual Tree Age

Site	State	Measurement Period	Principal Investigators
Choccolocco Mountain	Alabama	1583–2006	Adam Bale
Flomaton Natural Area	Alabama	1814–1995	John Kush and others
Greenwood Plantation	Georgia	1661–2003	Troy Knight
Lavender Mountain	Georgia	1795–2003	Troy Knight
Sprewell Bluff	Georgia	1649–2002	Troy Knight
Jones Ecological Research Center[a]	Georgia	1802–1904	Neil Pederson and others
Kisatchie Hills	Louisiana	1587–2007	Michael Stambaugh and others
Jeffries Smokehouse	North Carolina	1608–1805	Aldos Barefoot
New Hill Beaver Tree Farm	North Carolina	1891–1994	Aldos Barefoot
Weymouth Woods State Park	North Carolina	1671–1979	Aldos Barefoot
Weymouth Woods	North Carolina	1690–2006	Jason Ortegren
Boyd Tract	North Carolina	1559–1982	Edward Cook and Scott St. George

Source: Data from International Tree-Ring Data Bank (ITRDB), National Oceanic and Atmospheric Administration, data.noaa.gov/dataset/international-tree-ring-data-bank-itrdb

[a] The Joseph W. Jones Ecological Research Center has developed additional chronologies of these samples stratified by site.

Growth and Survival

Under specific environmental and life-history conditions, longleaf pine can be long lived, with individual trees capable of surviving >500 years (K. Hiers and H. D. Grissino-Mayer, unpublished data). Its ability to produce preservation-promoting oleoresin allows wood to persist for centuries after a tree has died, enabling the retrieval of historical growth records from standing and down dead trees, stumps, aquatic wood, and historical structures such as dams and houses. Using this relict wood, dendrochronological methods using crossdating have extended longleaf pine growth chronologies well beyond current lifespans and have compensated for the scarcity of old trees in many locations (Table 2.1).

From tree-ring widths of living and long-dead trees, several dendroclimatological studies have concluded that longleaf pine radial growth increases with increased growing-season moisture (Lodewick 1930; Coile 1936; Schumacher and Day 1939; Zahner 1989; Meldahl et al. 1999; Henderson and Grissino-Mayer 2009). More recent work has shown that the seasonality of this growth-climate response is mixed across the longleaf pine range. In North Carolina, cool-wet springs are important for growth (Van De Gevel et al. 2007), but in Mississippi, the addition of warmer summer months is significant (Devall et al. 1991). At the northern extent of its range in Virginia, growth increases with increased February temperatures and precipitation levels (Bhuta et al. 2009). Analyses comparing climate-growth responses of earlywood and latewood have shown that latewood width is highly sensitive to current-year conditions across the longleaf pine range (Meldahl et al. 1999; Henderson and Grissino-Mayer 2009). Though less influential, prior-year climate conditions—such as September and October temperature—can also significantly affect current-year growth (Meldahl et al. 1999; Henderson and Grissino-Mayer 2009). These results show considerable range-wide variations in the seasonal patterns of climate conditions that are important to the growth of longleaf pine.

Limits to Dispersal and Dominance

Other southern pines share some longleaf pine traits, but none of them has the same combination of dispersal limitation and high resistance to stresses from fire, drought, and other disturbances

(Landers 1991). Longleaf pine is a classical "masting" species—its seed production occurs once every 4–7 years, with timing and abundance controlled by flower survival, tree size, and recent temperature and precipitation fluctuations (Pederson et al. 2000; Brockway et al. 2006). Seeds typically fall within 20 m of parent trees (Croker and Boyer 1975). After germination, seedlings undergo a delayed juvenile growth period (the so-called "grass stage") that stalls height growth for 1–20 years or longer. Once established, however, longleaf pine can tolerate intense fires, drought, and typically resist attacks by bark beetles much better than other co-occurring tree species (Boyer 1990; Landers 1991), but does not produce cones until around age 30 (Croker and Boyer 1975). These combined traits suggest that—compared to other pine species—longleaf pine was considerably more limited in its ability to spread across otherwise suitable landscapes (within its climate envelope) and, once established, was more resistant to changes and stresses.

HISTORICAL PYROGEOGRAPHY

The historical biogeography of longleaf pine reflects frequent fire with an evolutionary storyline that follows a rise to dominance and fall to near extinction. This trend was foreseen more than a century ago:

> It can be safely asserted that there is not and never has been a longleaf pine forest in the United States (a species that does not grow anywhere else) which did not show evidences of fire, such as charred bark near the bases of trees; and furthermore, that if it were possible to prevent forest fire absolutely the longleaf pine—our most useful tree—would soon become extinct. (Harper 1913)

From a global fire database covering the last 21,000 years, Power et al. (2008) identified millennial-scale influences on fire activity, including variations in insolation, atmospheric circulation patterns, and carbon dioxide concentrations. The period from 21,000 to 16,000 years ago was the lowest in fire activity since the last glacial maximum; this is consistent with the cooler and drier climate conditions and reduced carbon dioxide that might have led to an overall reduction in fuel production. The late-glacial period (15,000–12,000 years ago) had both increases and decreases in fire activity.

In the eastern United States, few sedimentary charcoal records precede the Holocene, and only two well-dated late Quaternary charcoal chronologies are available for the Southeast; new chronologies spanning the past 15,000–20,000 years are under development at sites in South Carolina and Peninsular Florida (T. Krause and S. T. Jackson, unpublished data). The Lake Tulane record in Peninsular Florida (Grimm et al. 1993; Watts and Hansen 1994) shows relatively constant fire activity through the Holocene. A record from Clear Pond in northeastern South Carolina shows an increase in charcoal that coincided with increased pine pollen 8000 years ago, but charcoal levels were very low starting 5000 years ago (Hussey 1993). This change in charcoal was unaccompanied by a change in the pollen record; pine pollen has stayed consistently high for the last 8000 years. The charcoal transition could represent a change in fire regime (for example, shifting from crown to surface or from mixed severity to low severity), perhaps accompanying a change in the dominant pine species; however, because sediment accumulation rates during this period were very low, the pattern could also be an artifact. A study of soil charcoal from the southern Appalachians showed an increase in fire activity starting about 1000 years ago, possibly related to human activity (Fesenmyer and Christensen 2010). Whether a corresponding change occurred in the Coastal Plain is unclear, but maize cultivation was adopted at about the same time, leading to major cultural changes (Hudson 1976; Smith 1989). Jones (2014) developed high-resolution charcoal chronologies spanning the past millennium from three Florida lakes in longleaf pine forests. All chronologies showed consistent charcoal deposition with no major patterns of fluctuation in the period before European settlement. Two of the sites showed modest decreases in charcoal during the past century, possibly the result of fire suppression.

Fire-dependent trees such as longleaf pine, having persisted in frequent fire regimes for hundreds to thousands of years, are often only able to persist within a relatively narrow range of fire frequency, intensity, and severity. Documentation of this fire tolerance range requires a finer temporal resolution than can be produced by charcoal studies and must rely on annually resolved tree rings and fire scars, field observations, or remote sensing. Although charcoal records of fire occurrence are among the longest available, their resolution is too coarse to adequately describe the relationship between fire effects and fire regime characteristics (such as frequency, severity, seasonality, extent, and type).

Generally, fire regimes capable of maintaining longleaf pine are characterized as frequent, low-severity surface fires. Estimates of historical fire frequencies in the range of longleaf pine include mean fire intervals (the number of years between fire events) ranging from 1 to 12 years (Frost 1998; Guyette et al. 2012), but commonly with mean fire intervals from 1 to 4 years in the period before European settlement. Hammocks (stands isolated from more flammable uplands) and wet sites probably burned less frequently (Harper 1911); however, this has not been verified, and the historical period might have had burning conditions and fuels that were significantly different from modern environments. Climate conditions suggest short historical mean fire intervals (1–4 years) existed across the Coastal Plain, Piedmont, and even in the areas of the southern Appalachians where mountain longleaf populations exist (Guyette et al. 2012). Modeling approaches that consider finer-scale influences of topography and vegetation also suggest comparable historical burning frequencies with variations associated with low-lying areas and wetting of soils and fuels (Keane et al. 2007).

In his hierarchy of the factors that govern longleaf pine's distribution, Wahlenberg (1946) ranked fire above all others. Along with soils and climate, Chapman (1932) cited fire as a criterion for determining where longleaf pine occurred and was dominant. He and his contemporaries recommended burning at a 2–3-year interval to control competing pines and hardwoods—a regime that fire management plans follow today (Hiers, Laine, et al. 2003). More frequent intervals tend to result in seedling mortality and a simplified structure of large, fire-resistant overstory pines in a woodland or savanna physiognomy. In the long term, the consequence can be reduced tree-growth potential (Boyer 1990) and altered spatial patterns of vegetation (Kirkman et al. 2013). Following decades to centuries of very frequent burning, woody vegetation would likely become increasingly relegated to marginal, sparsely vegetated sites such as shallow soils and rock outcrops, or to sites at wet extremes such as the margins of wetlands and streams.

During periods of longer fire-free intervals, more fire-sensitive species can survive to adult size, albeit species with typically thicker bark (Hammond et al. 2015). If fire exclusion extends beyond a couple of decades, the result will be losses of pineland herbaceous plants and deep accumulations of forest floor that can cause heavy tree mortality when fires do eventually return (Varner et al. 2005). These situations plague almost half of all remnant mature longleaf pine stands in the region (Outcalt 2000).

Fire Scar Records

Fire scars offer the longest and best records describing historical fire regime characteristics—such as fire frequency, seasonality, and severity—but little is known about the historical extent of fires. Observational data complement fire scar data, providing context for the inferences derived from longer fire scar records. However, fire scar data have potential biases. Trees might not show scarring evidence of all fires that occur, especially when fire intensity was too low to produce scars. The fire scarring "system" is a function of fire behavior, fuel conditions, and many properties of the recording tree. In addition, although fire scar records might have high temporal resolution, very few exist from the range of longleaf pine. Because record availability is limited, longleaf pine fire history datasets likely do not capture the full range of variability in historical fire regimes. Fire scar records in longleaf pine forests have been difficult to attain, primarily because of its resistance to fire scarring and the tendency of relict wood to be consumed by recurring frequent fires.

Studying longleaf pine populations in mountain areas where long records (≥400 years) are available can improve understanding of the historical fire regimes and fire ecology of the species. At two southern Appalachian sites, Bale (2009) successfully used fire scars on remnant longleaf pines to reconstruct historical fire regimes extending back to the middle of the 16th century. At these sites, mean fire intervals were 2.7–3.2 years prior to 1831—the date that marks the onset of major European settlement influences (Figure 2.5). Fire was more frequent in the 17th century as settlement effects increased (human population growth, increased ownership of livestock, and access to new land). Fires primarily occurred during the dormant season, plausibly between November and February. Based on the Bale (2009) collection, occurrence of low-severity fires was confirmed by the relatively low percentages of trees that were scarred during fire events. Perhaps most surprising are data showing that very frequent fire regimes, which were known to characterize Coastal Plain stands extended into more rugged and remote mountain locations despite surface roughness, natural fire breaks, limits to human access, reduced lightning ignition frequency, and other limitations on fire frequency.

Using methods comparable to Bale (2009), Stambaugh et al. (2011) documented the fire regime from a population of relict longleaf pines in the Kisatchie Hills of central Louisiana, where fires occurred every 3.2 years on average from 1650 to 1790. They found evidence that fire occurred twice within a 12-month period—likely a maximum fire frequency considering the limits of fuel reaccumulation rates and fuel flammability in the area. Fire severity—apparently low before European settlement—increased with the population growth and land development that accompanied the Louisiana Purchase. These fires occurred primarily during the dormant and late growing seasons.

Scattered fire history studies in the Atlantic and eastern Gulf Coastal Plain provide further evidence for frequent fire regimes. In a northern Florida savanna, Huffman (2006) found a mean fire interval of 3.2 years (1678–1868), with most fires occurring in the middle of the growing season. Other studies (Huffman et al. 2004; Henderson 2006) reported slightly longer fire intervals. Henderson (2006) reported on fire scar histories from smaller collections (≤19 trees) across the Atlantic and Gulf Coastal Plain. Periods of record varied by site, but most covered the 17th to 20th centuries. Average fire intervals ranged from 4.4 years at Big Thicket in Texas to 16.3 years at Eglin Air Force Base in Florida (Henderson 2006).

Taken together, fire scar records confirm the presence of long-term frequent fire in the longleaf pine ecosystems associated with the conditions first described in the documentary records of Bartram (1791) and Catesby (1771). The earliest documentary evidence described conditions of longleaf pine communities that coincided with records of long-term frequent fire regimes. Surprisingly, the fire scars of these regimes have undergone much less study than regimes in other areas of the midwestern and western United States—resulting partly from difficulties in locating old fire-scarred wood and partly from lack of effort. The potential for using fire scars to continue documentation of fire regimes throughout the range of longleaf pine is excellent, with several new efforts already underway. Demand for these data is high, fueled by their value in informing restoration and management efforts that use new applications developed to document spatiotemporal variability in historical fire regimes; reconstruct stand dynamics of frequent fire conifer forests; quantify departures from past conditions; analyze the biological legacies of past disturbance (such as herbaceous plant diversity); assess the relative roles of human and lightning ignitions; and model interactions among climate, fire, and tree growth.

Some clues to the historical dynamics of Southeastern fire regimes can be inferred by analyzing evidence from outside the region. For example, the frequency of fire in the Southeast likely equaled or exceeded most other adjacent regions based on fire environment characteristics such as frequency of potential ignitions, climate conditions, length of burning season, and type and production rate of fuels. Analysis of these characteristics suggests that fires could have been as frequent, or more frequent, than in the Midwest, where about 40 fire scar histories have documented an average fire-return interval of 3–10 years in the period before European settlement.

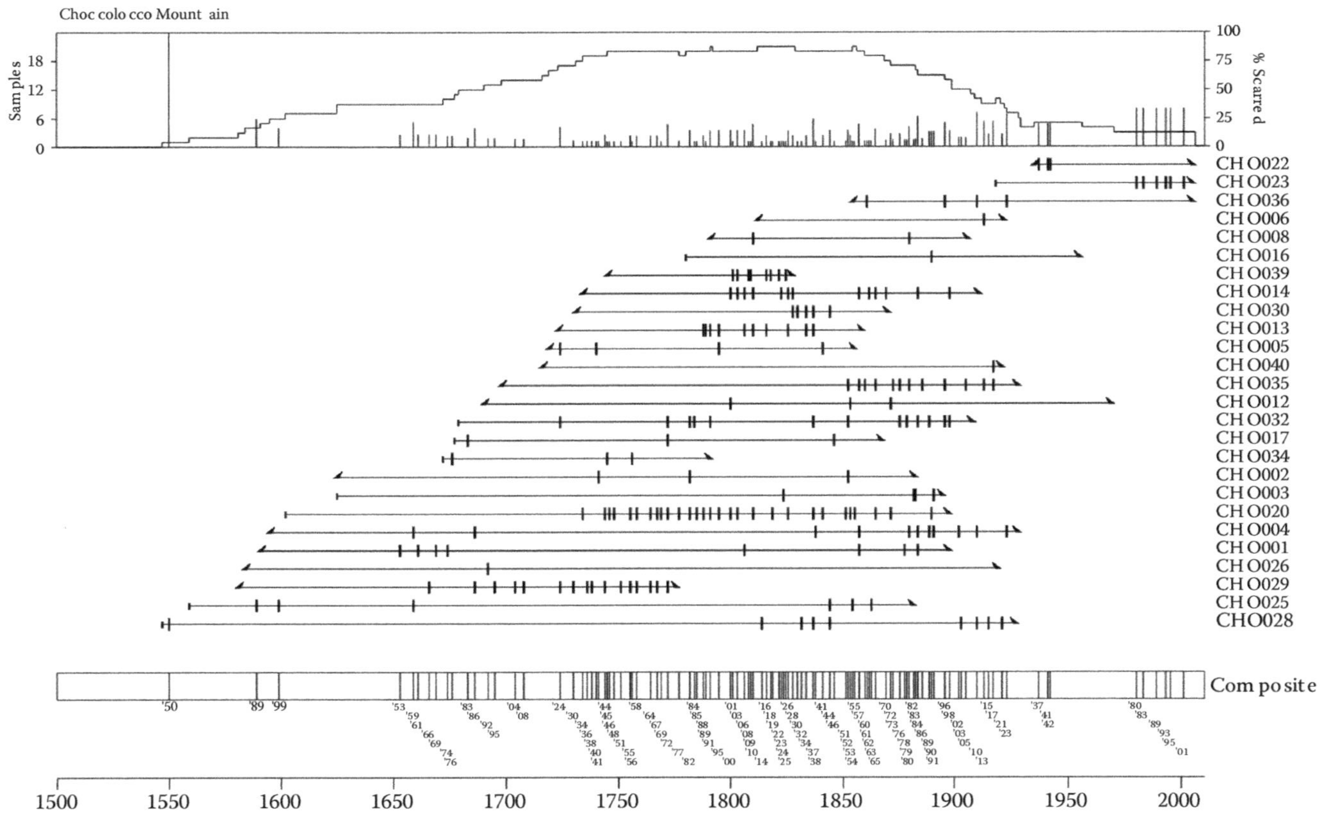

(a)

FIGURE 2.5 Longleaf pine fire scar history records from Choccolocco Mountain, Talledega National Forest, Alabama (1500–2000). (a) Choccolocco Mountain. The record has three parts: (1) An upper panel showing a line graph of sample scar depth with the percentages of trees scarred indicated by vertical bars; (2) a middle panel showing the periods of tree-ring record for individual trees as horizontal lines, each marked with bold interior vertical lines to indicate fire scar years, left-hand vertical ends to indicate pith years, left-hand diagonal ends to indicate inner ring year—rings missing to center, right-hand vertical ends to indicate bark years, and diagonal right-hand ends to indicate outer ring years; and (3) a lower panel showing a composite of all fire years at the site, with fire scar dates at the bottom representing the occurrence of fire events within the sites (about 1 km^2). *(Continued)*

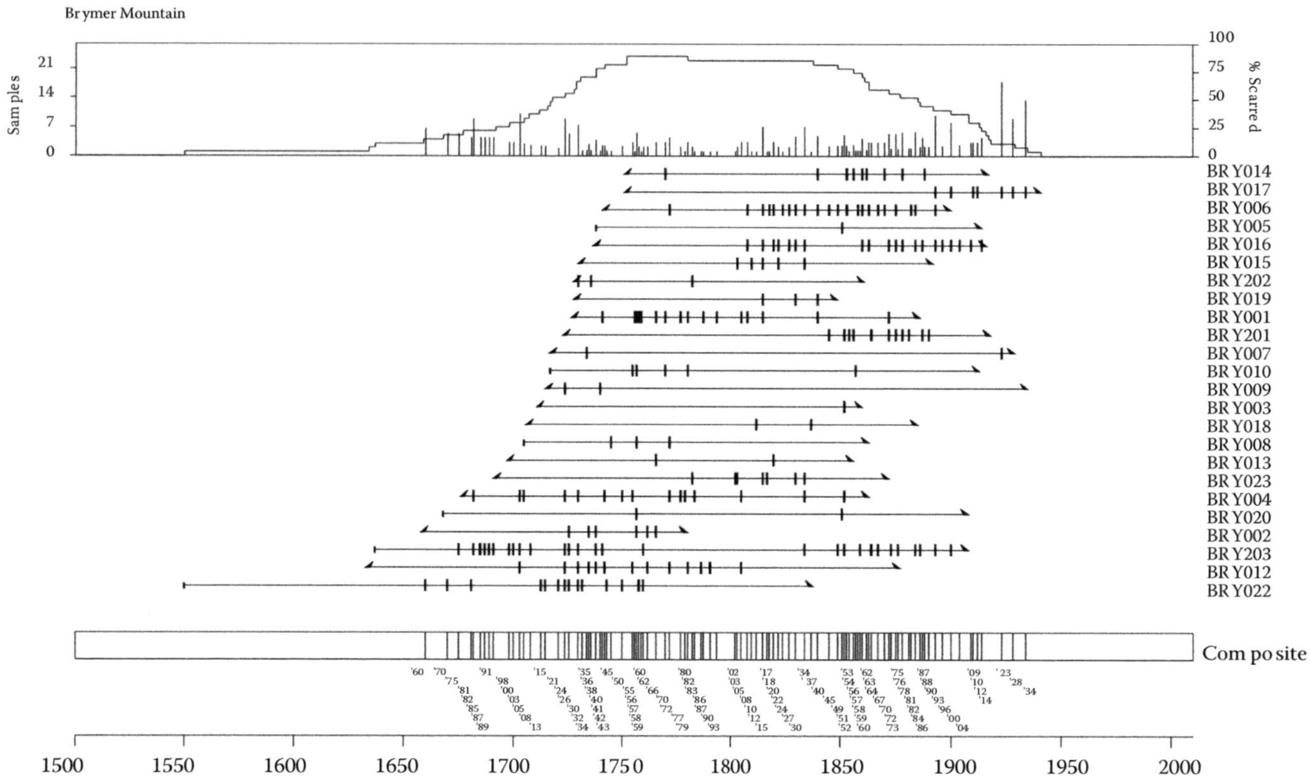

(b)

FIGURE 2.5 *(Continued)* Longleaf pine fire scar history records from Choccolocco Mountain, Talledega National Forest, Alabama (1500–2000). (b) Brymer Mountain. The record has three parts: (1) An upper panel showing a line graph of sample scar depth with the percentages of trees scarred indicated by vertical bars; (2) a middle panel showing the periods of tree-ring record for individual trees as horizontal lines, each marked with bold interior vertical lines to indicate fire scar years, left-hand vertical ends to indicate pith years, left-hand diagonal ends to indicate inner ring year—rings missing to center, right-hand vertical ends to indicate bark years, and diagonal right-hand ends to indicate outer ring years; and (3) a lower panel showing a composite of all fire years at the site, with fire scar dates at the bottom representing the occurrence of fire events within the sites (about 1 km^2). (Reproduced from Bale, A. M. 2009. Fire effects and litter accumulation dynamics in a montane longleaf pine ecosystem, MS thesis. University of Missouri, Columbia, Missouri.)

New environmental evidence based on changing patterns of fire regimes and human activity over time suggests that significant changes in fire frequency might have begun by about 1700 and that the changes followed a wave pattern (Marlon et al. 2008). A likely explanation is that fire activity in the northern hemisphere decreased sometime around the 17th century, then increased about 200 years later to levels that were well above the average for the preceding millennium. Compilations of eastern U.S. fire scar records confirm this pattern, but they suggest that the changes in fire frequency associated with European settlement—mirroring patterns of human migration—began earlier in the Southeast and Northeast than in more central U.S. areas (Hudson 1988; Stambaugh, unpublished data). Further, based on the pattern of fire activity documented by charcoal during the last two millennia (Marlon et al. 2008), fire frequencies of the 17th century might have been lower than the prior-millennium average, but those of the 19th century might have been higher. Ultimately, this calls into question the value of using fire scars from the past few centuries to generate fire regime data, and it emphasizes the importance of understanding not only the historical fire regime characteristics (frequencies, intensities, and seasonalities), but also those characteristics that are specific to longleaf pine success.

LANDSCAPE ECOLOGY AND DRIVERS OF FIRE IN LONGLEAF PINE ECOSYSTEMS

Modern fire records and models have advanced our understanding of southeastern fire regimes beyond what was possible using fire history reconstructions. From a global perspective, Archibald et al. (2013) showed that the modern fire regime of the Southeast is classified at an intermediate level for fire-return interval, fire energy release, and maximum fire size. From a national perspective, southeastern fire regimes are among the most frequent (Frost 1998; Keane et al. 2007; Guyette et al. 2012), likely only limited by the number of ignitions and weather conditions. Long-term fire records have been critical in demonstrating how historical fires differ from modern fires. For example, Knapp et al. (2009) showed that the seasonality of prescribed fire is sometimes opposite that of historical fires.

Lightning Patterns

Despite recurring debate among scientists and managers in the Southeast, measuring the degree to which historical fires were attributed to lightning versus human ignitions is impossible. Even when historical fire data, such as fire scars, can be attributed to the primary lightning season, the likelihood that human ignitions occurred in that same season is equally plausible. Some sources suggest that historical fires were predominantly of human origin based on population densities across the Southeast (Mann 2005), but others suggest that the number of lightning ignitions was sufficiently high to explain the observed fire frequency (Komarek 1964). Recently, Noss et al. (2015) more strongly asserted that frequent (1–3 years) fire regimes, particularly in the Coastal Plain, were primarily caused by lightning ignitions during dry periods and that this evolutionary force, not humans, was responsible for the high level of biodiversity in the subregion. Interestingly, areas of high plant endemism follow a spatial pattern that is similar to the spatial pattern of increased lightning-flash density (Sorrie and Weakley 2001).

Across the Lower Coastal Plain, lightning flash densities are concentrated in areas of Peninsular Florida and along the Gulf Coast. Lightning flash density and thunderstorm days per year generally decrease northward and westward (NOAA 2015). Less detailed information that describes patterns of lightning ignitions (not to be confused with lightning flashes) is available, particularly through modern remote sensing applications, but the prevailing pattern is for lightning-caused fires to decrease northward from Peninsular Florida eastward, even during drought conditions (Jurney et al. 2004). Based on state and federal wildfire records from 1916 to 1990, southeastern states have had 10–100 times more wildfires caused by humans than by lightning. These contemporary data, also common in other fire-prone areas, fail to acknowledge the possibility that the fire from a single lightning or human ignition might have spread over expansive areas and potentially burned for days or months, in many ways mirroring contemporary wildfires in the less fragmented western United States and other analogous fire-prone areas beyond the U.S. borders. Before human inhabitation of longleaf pine-dominated landscapes (about 12,500–20,000 years ago), lightning was the only plausible mechanism for fire (Komarek 1964).

Humans and Fire

Although humans have influenced southeastern ecosystems and fire regimes since about 14,500 years ago (Hudson 1976; Halligan et al. 2016), their populations likely grew large enough to have had a marked effect on biota during the last 1700 years. Evidence of plant cultivation in the eastern United States dates back to about 2000 years ago, and maybe as far back as 7000 years ago (Ford 1985). Whitehead and Sheehan (1985) reported evidence of maize (*Zea mays*) pollen in eastern Mississippi about 2400 years ago, coincident with a pollen boundary where oak decreased and pine increased; they attributed the change to agricultural practices and frequent burning. Earlier evidence (about 5000 years ago) of a transition to pine dominance is provided by Watts (1980), who surmised an effect caused by high water tables.

Intensive agricultural practices in the Southeast began as early as 1150 years ago. At the time of first European contact in the 15th century, Denevan (1992) estimated about 400,000 inhabitants, and Driver (1969) estimated 5–30 humans/100 km^2 outside the dense population centers of present day coastal Georgia and South Carolina. Using fire, humans imparted complex influences on the environment, with the spatial extent of those influences varying through time according to population characteristics (Delcourt and Delcourt 2004). The attribution of human influences to fire regimes is primarily based on changes in fire frequency, as evidenced in charcoal and fire-scar records that corresponded to changes in populations, cultures, and vegetation (Fowler and Konopik 2007).

More than 70 reasons have been documented for why humans historically burned vegetation and managed land; they include hunting, site preparation, cultivation of crops, improving growth and yield of plants, fireproofing areas, promotion of forage, cooking, felling trees, warfare, insect control, signaling, and clearing land for travel (Swanton 1979; Sitton 1995). Van Lear et al. (2005) provided information about the purposes and extent of Native American burning in the range of longleaf pine, and Jurney et al. (2004) reproduced several passages from 16th to 18th century expeditions that included reports of human ignitions in the Texas Coastal Plain and Mississippi Embayment. Pines were harvested for purposes that ranged from building structures and palisades by the Chickasaw and other peoples (Swanton 1979) to making "light-wood" torches for nighttime fire fishing (Lawson 1860). As a consequence, pine communities became less abundant in areas that surrounded villages than in areas that were farther away (Foster et al. 2004). Further evidence suggests the potential impact of Native Americans and European settlers in altering the seasonality of fires (Stewart 2002; Brose et al. 2013).

OUTLOOK FOR LONGLEAF PINE: CLIMATE, FIRE, AND HUMANS

The same factors that have dominated the prehistoric-to-present biogeography of longleaf pine will likely drive its success or failure in the future. Within site constraints, these drivers will almost certainly be climate, fire, and humans. Syntheses of projected climate change within the current longleaf pine range suggest that substantial alterations will occur by the end of this century (Liu, Goodrick, et al. 2013; Mitchell et al. 2014). Ensemble general circulation model projections suggest end-of-century temperature increases of 1.5°C–2.5°C within the current longleaf pine range (IPCC 2014). Precipitation projections are more spatially variable, with subtle increases (about 5%–15%) in the western part of the range and smaller increases (about 5%) in the eastern part. Increased growing-season moisture would likely cause increased tree growth (Meldahl et al. 1999), particularly on excessively dry sites or in the Piedmont and other areas where growing-season moisture is limited. Higher temperatures would also be expected to cause increased growth, especially if occurring during the winter season and in the northern parts of the range. Further growth increases from warming would be expected for longleaf pine if the annual number of freezing days and ice storm events were reduced.

Global climate change, which fundamentally is driven by the increases in atmospheric carbon dioxide, will likely have consequences that will be important for future longleaf pine ecosystems (see Chapter 15). In experiments with artificially elevated carbon dioxide (720 ppm, or double the level that occurred at the initiation of the study), the aboveground biomass of longleaf pine increased by 70%–100% in comparison to growth at ambient conditions (Davis et al. 2002; Runion et al. 2006). Co-occurring sand post oak (*Q. margaretta*) initially increased and then decreased over time, as did the dominant wiregrass (Davis et al. 2002; Runion et al. 2006). How these changes in plant productivity (and therefore fuels) will influence future longleaf pine ecosystems and their disturbance regimes is unclear. How future warmer, wetter, and transient climates interact with elevated carbon dioxide is also unclear, but will undoubtedly be important for long-term conservation of these ecosystems.

Future projections of temperature and precipitation changes, although seemingly subtle considering the wide variations throughout the geologic past, are projected to have weighty implications for the future distribution of longleaf pine. Using three general circulation models under two-fold and three-fold carbon dioxide scenarios, Iverson et al. (2008) predicted that the longleaf pine range would expand northward (Figure 2.6). In the western part of the range, importance values (a species' relative density and basal area) are predicted to increase and expand westward (more broadly into Texas) and northward into southern and central Arkansas and northern Mississippi. In the central part of its range, longleaf pine is predicted to increase and expand considerably northward in Alabama, Georgia, and Tennessee. The eastern boundary is more complicated, but the prediction is for contraction along the coast in response to rising sea level and expansion westward and northward (eastward in Virginia and perhaps as far as Delaware). Based on these projections, range contraction is expected in small areas of southern Mississippi and Alabama, along the Fall Line in South Carolina, and along the North Carolina coast—all areas where warming is expected to exceed the perceived tolerance of the species.

The projections for future longleaf pine distribution do not reflect future impediments to dispersal and migration or the compounding influences of soils, fire, land use, and other factors that are critical for longleaf pine ecosystems. The potential for northward migration out of southern Mississippi and Alabama is particularly concerning considering the large acreages involved and extensive efforts these states have already invested in longleaf pine conservation and management. For range-wide conservation efforts, consideration of future climate-forcing and local conditions will be needed for sites that are transitioning both away from and toward longleaf pine suitability. For conservation and investment efforts to succeed in areas that are projected to become part of the future longleaf pine range, they will need to consider the availability of land as well as the suitability of climate, site, and fire conditions. Assisted migration, for example, the planting of longleaf pine in sites where it was formerly absent but likely more competitive under predicted future climates, has been widely discussed and debated (McLachlan et al. 2007).

As mentioned above, projections of tree ranges fail to incorporate fire, likely because the use of fire depends as much on societal values, policies, regulations, and capabilities (Mitchell et al. 2014; Kobziar et al. 2015) as on the properties of combustion (see Chapter 13). The continued use of fire will be critical to distinguishing whether longleaf pine will be present as a scattering of single trees or as a functioning ecosystem. From a climatological perspective, projections of future fire regimes in southeastern pine ecosystems suggest that the length of fire seasons—when fuels are sufficiently dry to sustain fires—will increase by 1–3 months (Liu, Goodrick, et al. 2013). Changes in precipitation, more so than temperature, are expected to be a critical determinant of future drought. Measurements from the Keetch-Byram Drought Index, the drought index used to monitor fire weather (Keetch and Byram 1968), are projected to be significantly higher (indicating drier conditions) in the summer and autumn (Liu, Prestemon, et al. 2013).

Prescribed fire is a primary land management tool in the Southeast (Wade and Lunsford 1989; Melvin 2012). Predicting how prescribed fires might change under future scenarios is difficult given current and future impediments (Melvin 2012, 2015; Kobziar et al. 2015). Projected drying (Liu, Goodrick, et al. 2013) will presumably increase the number of potential days that fires can spread, but

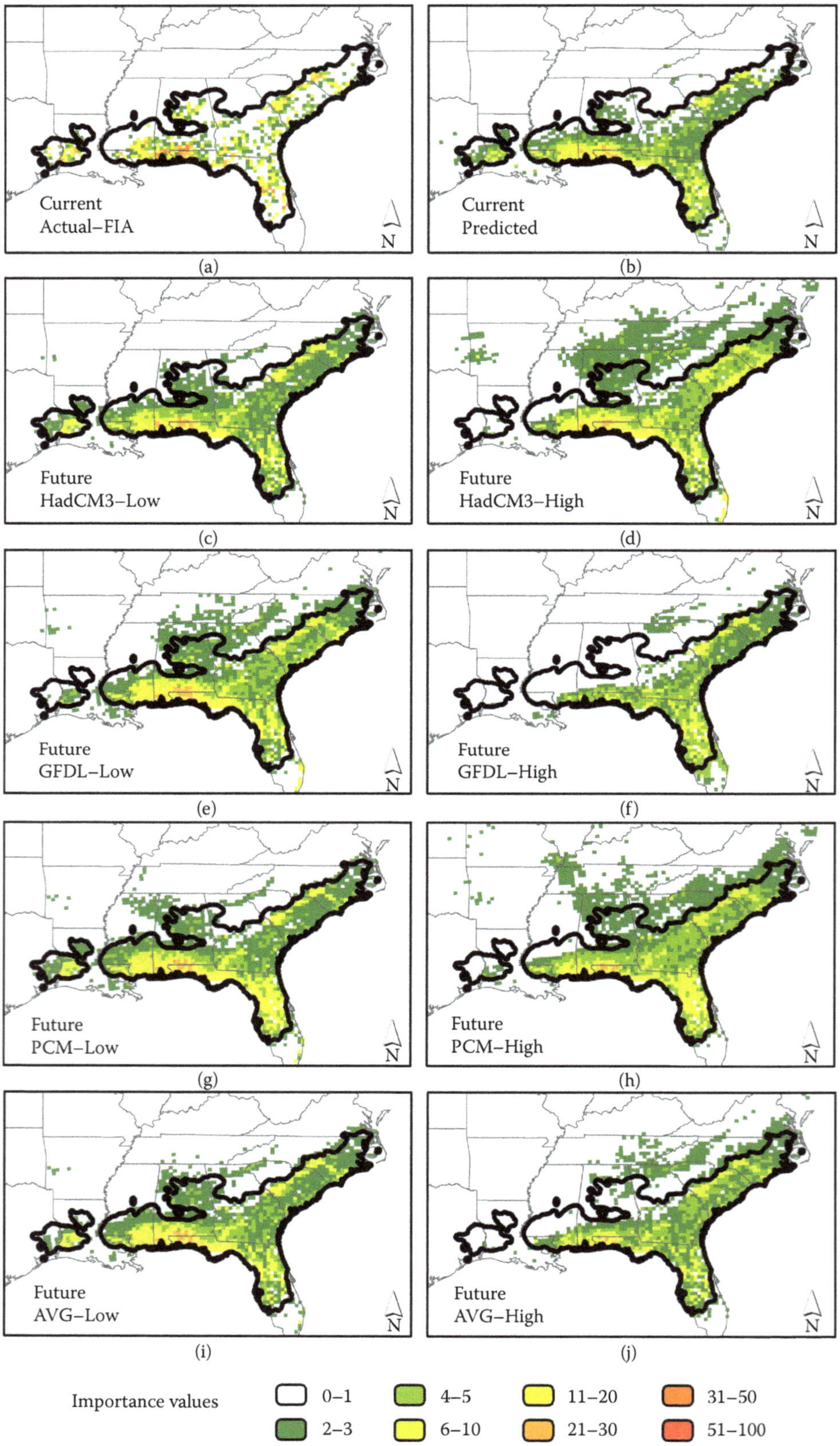

Current
Actual–FIA
N
(a)
Current
Predicted
N
(b)
Future
HadCM3–Low
N
(c)
Future
HadCM3–High
N
(d)
Future
GFDL–Low
N
(e)
Future
GFDL–High
N
(f)
Future
PCM–Low
N
(g)
Future
PCM–High
N
(h)
Future
AVG–Low
N
(i)
Future
AVG–High
N
(j)
Importance values
0–1
2–3
4–5
6–10
11–20
21–30
31–50
51–100

FIGURE 2.6 The current and projected future suitable habitat estimates—with importance values—for longleaf pine: (a) actual current estimate based on the U.S. Forest Service–Forest Inventory and Analysis database, (b) modeled current estimate, (c) low and (d) high projections based on emission scenarios from the Hadley CM3 model (From Pope, V. D., *Climate Dynamics*, 16, 123–146, 2000), (e) low and (f) high projections based on emission scenarios from the Geophysical Fluid Dynamics Laboratory CM2.1 model (From Delworth, T. L. et al., *Journal of Climate*, 19, 643–674, 2006), (g) low and (h) high projections based on emission scenarios from the Parallel Climate Model (From Washington, W. M. et al., *Climate Dynamics*, 16, 755–774, 2000), and (i) low and (j) high projections based averages from all three emission scenarios. Note that the selected emission scenarios encompass most 21st century projections, which range from a doubling to a tripling of preindustrial atmospheric carbon dioxide concentrations. (Data for scenarios from Prasad, A. M. et al., 2007. A climate change atlas for 134 forest tree species of the eastern United States [database]. USDA Forest Service, Northern Research Station, Delaware, Ohio, USA. (http://www.nrs.fs.fed.us/atlas/tree) and Iverson, L. R. et al., *Forest Ecology and Management*, 254, 390–406, 2008.)

might also limit those days given that prescribed fires in the region tend to be ignited under wet-dry conditions (Ryan et al. 2013). Because fire is fundamentally a chemical reaction, future changes in temperature and precipitation will alter the physical (fuel moisture, production) and chemical (reaction rate) fire environment. Based on the response of fire activity to climates across the United States during the past few centuries, Guyette et al. (2014) estimated that the probability of fire will increase throughout the Southeast, with even larger increases likely to the north (such as areas in the southern Appalachians) and smaller increases toward the western margin of the longleaf pine range. The potential longleaf pine range shift predicted by general circulation models might further complicate predictions. If it extends northward and inland—in Alabama, Georgia, South Carolina, North Carolina, and Virginia, and then into Tennessee—the future longleaf pine range could overlap with large, smoke-sensitive metropolitan areas such as Atlanta, Greenville–Spartanburg, Raleigh–Durham, and Virginia Beach–Hampton Roads. Prioritizing where and when prescribed burning should be applied within landscapes (Hiers, Laine, et al. 2003) will be required to manage fire effectively in a complex future (see Chapter 13).

SUMMARY

Currently, longleaf pine occurs simultaneously as a widespread dominant in open forests and savannas, and as a codominant or subdominant with many other pines and hardwoods across a wide variety of habitats. Monotypic stands, formerly widespread across the Coastal Plain, require a high-frequency surface-fire regime for survival, but longleaf pine is capable of maintaining populations in several other settings. The longleaf pine ecosystem is a relatively recent phenomenon, developing within the last 5000–8000 years, perhaps resulting from climatic change, human influences, or both. The southeastern Coastal Plain has experienced innumerable climatic and ecological changes in the past and will undergo additional changes in the future, driven in large part by the direct and indirect actions of humans. In the past few centuries, most of the original longleaf pine forest has been replaced by farming, establishment of short-rotation pine plantations, and urban/suburban development. The species now exists in highly fragmented populations scattered across the Southeast. Decreases in fire, owing in part to concerns over risks to health and property, result in continuing disappearance of longleaf pine forests and fragmentation of longleaf pine populations. To be effective, conservation planning and practice will need to incorporate the influences of climate, fire, and human activities—past, present, and future—on this "signature" species of the Southeast.

ACKNOWLEDGMENTS

We thank the editors and John Kush, Seth Bigelow, Benjamin Knapp and Kevin Hiers for providing thoughtful review comments on previous versions of this chapter.

3 The Social and Economic Drivers of the Southeastern Forest Landscape

R. Kevin McIntyre, Barrett B. McCall, and David N. Wear

CONTENTS

INTRODUCTION

The last quarter century has witnessed an unprecedented resurgence of interest in the management of longleaf pine (*Pinus palustris*) forests, a phenomenon that has been coupled with increased understanding of the ecology, management, and restoration of these ecosystems. As interest in longleaf pine becomes more mainstream among landowners and the general public, future opportunities for restoration will be strongly influenced by the context, both economic and social, in which they occur.

In this chapter, we examine the economic and social factors that affected the development of professional forest management in the southeastern United States in the decades following the harvest of the primary longleaf pine forests. We trace the development of new markets and new approaches to managing pine forests for fiber in the second half of the 20th century, which were transformational for the forested landscape of the Southeast, resulting in huge impacts on patterns of ownership, forest conditions, and expectations of land owners for their forests. We next review the literature on longleaf pine microeconomics that emerged in response to a heightened interest in restoration and management that began in the late 1980s. Finally, we address the three important changes that are affecting the forested landscape of the Southeast and its management in the

21st century: (1) a wholesale divestiture of forest industry lands, (2) the acceleration of land use change (especially urban development), and (3) the changing demographics of family forest owners. These dynamics present both challenges and opportunities for restoration and management of longleaf pine across the region.

HISTORICAL CONTEXT

The decline of longleaf pine ecosystems—from about 92 million acres prior to European settlement (Frost 2006) to about 3 million acres by the end of the 20th century (Outcalt and Sheffield 1996)—has been well documented. The rapid and ubiquitous harvesting of southeastern virgin pine forests during the early 20th century was the continuation of the boom and bust of regional "lumbering" rather than an application of forestry (Earley 2004; Boyd 2015). Most logging enterprises treated the land as an expendable commodity; when one area was depleted, the company and mill (and often the entire town) simply moved on to a new area of virgin forest. The rapid exploitation of longleaf pine forests, with little to no planning for the regeneration of cutover lands, led to a denuded landscape in many areas that could not support the continuation of the lumber industry as it was practiced at the time (Williams 1989; Boyd 2015). Estimates of cutover land in the "yellow pine belt" were as high as 100 million acres (Vance 1932).

Although there were some early successes with efforts to naturally regenerate longleaf pine or plant longleaf pine, these ultimately proved to be the exception rather than the rule. More commonly, cutover longleaf pine sites were captured by loblolly (*Pinus taeda*) or slash pine (*Pinus elliottii*), both of which are much more prolific and dependable seeders. Fire protection efforts that became widespread in the 1920s (Boyd 2015) thwarted longleaf pine reproduction and instead favored growth of newly established loblolly and slash pine, both of which require a fire-free interval for several years after germination. Although foresters were beginning to understand the complexities of longleaf pine and how to manage for the species (Wahlenberg 1946; Croker 1968), societal objectives for forests were evolving, opening a new chapter in the history of southeastern forestry.

Perhaps the most fundamental challenge to establishing widespread sustainable forest management in the region was the lack of confidence that managing second-growth forest stands could be profitable. Absent a well-defined production technology for regenerating and rapidly regrowing forests, and—given the availability of relatively inexpensive old-growth softwood timber in other regions—landowners simply viewed the economics of forest management as inferior to other capital investments. Indeed, the relatively low—though rising—timber prices and the hurdles of the high-opportunity costs of choosing forestry over alternative investments delayed any kind of long-term forest management practices on U.S. private lands for many decades (Hough 1878; Graves 1919).

Although the inability to reliably regenerate longleaf pine continued to frustrate those who took an interest in rehabilitating the forestlands of the Southeast, the ease with which loblolly and slash pine adapted to former longleaf pine sites and the rapid growth of these species did not escape notice. This phenomenon coincided with the growth of the nascent U.S. forestry profession, a discipline firmly rooted in the European tradition that viewed forests as a commodity that could be profitably regenerated, managed, and harvested at maturity (Chapman 1942). The rise of forestry as a profession, coupled with technologies developed in the late 1800s for producing pulp and making paper from wood fiber, gave birth to a new market that would profoundly change the way forests were managed in the Southeast (Boyd 2015). Further development of the paper-making technology improved the process of delignifying wood fiber from pine and other resinous tree species to produce kraft paper, ultimately enabling the production of newsprint and higher quality paper from southeastern pines. As the viability of paper production became recognized as a profitable business, the number of pulp mills in the Southeast began to grow, reaching 36 by 1935.

In 1932, production of southern yellow pine lumber was only three billion board feet, the lowest annual figure for the entire 20th century (Carter et al. 2015). After World War II, a growing supply of wood from second-growth forests and higher demand for lumber from the housing market led to new investments in modern sawmills (Wheeler 1969). Timber volume of southeastern forests grew dramatically, increasing by almost 70% from 1953 to 1977 (Oswalt et al. 2014). Although a fiber production model dominated in many areas, solid wood products remained an important component of southeastern forest outputs, with production tied closely to economic cycles and housing starts (Wear et al. 2013).

Growth of the pulp and paper industry resumed after World War II; by 1955, 73 southeastern mills were using local pine as feedstock. The opportunities presented by the GI Bill also contributed to unprecedented growth in the forestry profession, with university forestry program enrollment increasing nationally from about 1100 in 1944 to 7000 in 1946 (Gray 1988). From 1950 to 1961, five new university forestry programs were added to the eight operating in the Southeast. As the forestry profession grew, these newly trained professionals populated the ranks of private companies and public agencies, especially state forestry agencies. With state forestry agencies fully staffed, the new models of plantation forestry were more broadly disseminated and wildfire control efforts were much more successful, neither of which benefitted longleaf pine.

Early recognition that U.S. forest resources were not inexhaustible and that poor forestry practices had undesirable impacts on water resources resulted in legislation such as the Organic Act of 1897, the Weeks Act of 1911, and the Clark-McNary Act of 1924. In the Southeast, the cutover longleaf pine forests served as case studies for identifying and resolving these problems. Both private industry and the U.S. Forest Service successfully experimented with broad-scale plantings of pine in the 1920s; and the Civilian Conservation Corps planted >1.5 million acres across the region in the 1930s (Fox et al. 2007).

After World War II, the growth of the forest products industry and the economic importance of a dependable supply of fiber and sawtimber prompted further concerns about sustainable management of southeastern forests. Despite the natural regeneration of loblolly and slash pines on large acreages of cutover longleaf pine sites, estimates of degraded agricultural land and cutover longleaf pine in need of reforestation ranged from 13 to 29 million acres (Wakely 1954; Wahlenberg 1960). The planting of seedlings expanded rapidly, driven by both the interests of the burgeoning pulp and paper industry and the desire to rehabilitate degraded farm land through federal incentives such as the Soil Bank Program. Significant investment in nursery propagation of pines was underway, with the federal government, all the southeastern states, and many private forestry companies operating nurseries to supply seedlings for pine plantations.

The 1950s marked the beginning of a highly productive period for forestry research and development, largely the product of an unprecedented level of cooperative projects involving the U.S. Forest Service, land grant universities, and private industry. Some of the most important undertakings were tree improvement programs for both volume and form of loblolly and slash pine, most notably through the Texas Forest Service, the University of Florida, and North Carolina State University. Progeny from first-generation seed orchards were 20% more productive than progeny from wild-collected seed (Todd et al. 1995), with second-generation orchards projected to increase volume growth by 14%–23% over first-generation orchards (Li et al. 1997). Formal research programs also began to build upon the early efforts of practicing foresters to improve the performance of pine plantations through refined mechanical site preparation techniques, chemical control of competition, and fertilization to accelerate growth (Fox et al. 2007). Concurrent with these operational advances, more sophisticated growth and yield models for slash and loblolly pine were developed and refined, allowing the forest industry to predict the financial performance of their pine plantations with a higher degree of accuracy.

The advances in forest management resulting from these alliances among government, academia, and private industry ushered in what has been called the "Golden Age of Industrial Forestry," with the annual rate of plantation establishment reaching >2 million acres in the 1980s and early 1990s (Carter et al. 2015). As loblolly and slash pine plantation acreages increased under these new models of forest management, declines continued for the longleaf pine forests that remained after the initial cut of the late 1800s and early 1900s. By the 1980s, only 3.8 million acres remained of the 12 million acres that were reported in 1955 (Kelly and Bechtold 1990), and decreases continued until longleaf pine forests (defined as Forest Inventory and Analysis longleaf pine forest type) reached a low of about 3 million acres in the 1990s (Outcalt and Sheffield 1996). Additionally, there were as many as 600,000 acres of longleaf pine/oak forest type remaining at this time (Frost 2006).

FOREST OWNERSHIP IN THE SOUTHEAST

Before European settlement, total forest area in the Southeast was estimated at 354 million acres (Williams 1989). By the 1920s, forested acreage reached a low of about 220 million acres, but recovered to 231 million acres (Oswalt et al. 2014) by the end of the century. Contributing to this recovery were incentives and other efforts to encourage reforestation, combined with a comparative advantage of forestry over crop production on a portion of the landscape. The area of land used for urban or other developed uses has steadily grown, partially offset by a simultaneous transition of marginal cropland and pastureland to forest cover (Wear 2013). By 2012, forestland in the Southeast stood at about 245 million acres. Timberland, defined as acres capable of producing >20 cubic feet of industrial timber annually, accounted for roughly 210 million of that acreage (Oswalt et al. 2014).

The abbreviated history of the development of forestry and the forest landscape in the 20th century described above paints a rather simplistic picture. In reality, complex and dynamic public and private ownership patterns and approaches to forest management characterize the forest landscape of the Southeast.

PUBLIC OWNERSHIP

Publicly held forests are estimated at approximately 32.7 million acres (slightly >13% of all forestland), with 27.4 million acres classified as timberland (Oswalt et al. 2014). Federal timberland ownership is approximately 18 million acres; approximately 12.7 million of those acres are managed by the U.S. Forest Service, with most of the remainder managed by the Department of Defense, U.S. Fish and Wildlife Service, and National Park Service. Another 9.4 million acres are controlled by other government organizations such as states or counties (Oswalt et al. 2014). In 2010, longleaf pine occupied approximately 1.63 million acres of public land, or 38% of the total remaining longleaf pine acreage (Oswalt et al. 2012).

In addition to being the largest single steward of public timberland in the region, the U.S. Forest Service also manages the majority of public lands that have existing and potential longleaf pine acreage. Much of the area that is now national forest was once cutover timberland and highly eroded cropland (Shands and Healy 1977). Legislative efforts to reclaim these areas began in 1907 with the establishment of the first national forest, and reclamation efforts ultimately established national forests in every state in the region. Historically, management of national forests was broadly defined in the Organic Act of 1897, which outlined a mission of providing favorable conditions for water flow and furnishing a continuous supply of timber (Glasser 2005).

Because of the degraded conditions that characterized many southeastern national forests, early management focused on restoration and replanting (Williams 2003). As second-growth stands matured and U.S. demands for timber grew, harvesting timber from these forests became more common in the 1960s (Carter et al. 2015). The overall approach to forest management tended to follow what was considered to be the state of the art for modern forestry at the time; on harvested longleaf pine sites this often meant replanting using a plantation model for loblolly or slash pine (Earley 2004).

Public response to the increasingly visible and aggressive management of national forests throughout the United States resulted in controversies and legal rulings in the early 1970s and enactment of the National Forest Management Act of 1976. What followed was a significant policy shift in which a broader range of multiple uses—including ecological values, water quality and quantity, wildlife, and recreation—were put on equal footing with timber production as objectives for national forests (Carter et al. 2015). The new legislation also provided for extensive public input into decisions about national forest management as well as specifically encouraging forestry on private lands. Public controversy over old-growth logging on national forests in the Pacific Northwest and the federal listing of the northern spotted owl (*Strix occidentalis caurina*) under the Endangered Species Act led to a reduction in timber harvesting on all national forests and shifted harvesting pressure to the Southeast, particularly on private lands (Wear 2014). Similar to the situation in the Pacific Northwest, status of the red-cockaded woodpecker (*Picoides borealis*) drove many federal land management policies in the Southeast, particularly with respect to management of longleaf pines—a trend that continues today. U.S. Forest Service data from 2015 Forest Inventory and Analysis (FIA) surveys indicate approximately 690,000 acres of longleaf pine on national forestlands (Miles 2016). As of 2008, land and resource management plans for national forests in the region projected a long-term restoration goal of about 1.5 million acres for the forest type (ALRI 2009).

State ownership of timberland more than tripled from approximately 2 million acres in 1953 to 7 million acres in 2012, with >2 million acres of the increase occurring in Florida (Oswalt et al. 2014). Because Florida historically had an aggressive conservation land acquisition program and the largest concentration of longleaf pine, some portion of the total forest acreage increase presumably includes longleaf pine. Most state-managed lands are administered through state forestry or wildlife agencies; relatively little consistent information is available about the composition of these forests or how they are managed. FIA estimates from 2015 for eight of the states within the longleaf pine range suggest that approximately 456,000 acres of state land were occupied by longleaf pine (Miles 2016).

Total public timberland in the Southeast grew from 17.4 million acres in 1953 to 27.4 million acres in 2012, an increase of approximately 10 million acres, which included 3.3 million acres on federal land and 5 million acres on state land. During that same time period, total private timberlands decreased by about 4.5 million acres (Oswalt et al. 2014).

Private Land Ownership

The vast majority of the timberland in the Southeast has historically been controlled by the private sector. Private ownership held >91% of the acreage in 1953, and 87% remained in private ownership through 2012 (Oswalt et al. 2014). Private timberlands were historically divided into two broad categories: those owned by vertically integrated forest product companies (with the same company owning both the supply source and the processing facility), and those held by all other forest owners (historically referred to as "nonindustrial private owners"). Of the 182 million acres held in private timberland ownership in 2012, about a third was classified as corporate or industrial, and two-thirds was held by noncorporate or "family forest" owners (Oswalt et al. 2014). In 2010 about 2.7 million acres, or 62%, of longleaf pine acreage was on privately owned land (Oswalt et al. 2012).

In the last half of the 20th century, forest industry acreage steadily increased and peaked at 38 million acres in 1989. The late 1990s marked the beginning of sweeping changes, with most industrial timberlands divested into a diverse group of corporate ownership structures, primarily timber investment management organizations (TIMOs) and real estate investment trusts (REITs). By 2010, <7 million acres remained in traditional industrial ownership (Zhang et al. 2012), prompting a redefinition of ownership categories. Forest industry acres are now included in the "corporate" category, and the former nonindustrial private forest owners are now known as "family forests." Overall, the corporate category grew from 1953 to 2012, now standing at >61 million acres. During the same period, acreage of family forests (the former nonindustrial private category) decreased by 34 million acres to about 121 million acres (Oswalt et al. 2014).

Family Forest Owners

As the most significant group of forest owners in the region in both numbers and acreage, family forest owners have played a major role in shaping the southeastern landscape. Historically, forest management by this group, as measured by productivity, has been considered lacking when compared to industrial, public, and other ownerships, all of which typically employ forestry professionals. For example, a 1952 study found that productivity of family forests was 65% as high as productivity on forest industry land and 75% as high as productivity on national forests (Paley 1952). Efforts to address this gap included federal incentives, university forestry extension programs, and state forestry agency landowner assistance. In addition to these government efforts, many forestry industry companies began to collaborate on landowner assistance programs. For example, the Southern Pulpwood Conservation Association was founded by leaders in the pulp and paper industry to develop forest management guidance tools, establish "pilot forests" to demonstrate these concepts, improve access to technical assistance, and promote forestry to the general public (Fickle 2001). In the late 1950s, individual forest products companies further developed this effort by providing various levels of assistance within the wood-procurement areas of their mills, primarily to ensure a sustained supply of feedstock (Carter et al. 2015). The late decades of the 20th century also saw the rise of consulting foresters that offered their services to owners who had sufficient acreage to require occasional professional services but not enough to hire a full-time forester. Other efforts to encourage better stewardship and more active management included establishment of organizations such as the American Tree Farm System and the Forest Landowners Association.

Despite these efforts, family forest productivity as measured simply by the amount of timber harvested still lagged behind other ownerships. However, an alternate perspective from social science research concluded that the management styles of family and industry forest owners were both consistent with value optimization even though their management outcomes differed (Newman and Wear 1993). The most significant difference was that family forest owners placed a higher value on standing forests—consistent with recreation and other nontimber values—and required higher returns to motivate them to harvest their timber. These findings, which are encouraging for the design of policies to reestablish longleaf pine on family forestlands, suggest that many private land owners are likely to respond to well-designed incentives that help achieve their values and objectives for their land.

Forest Industry and Other Corporate Owners

The early 1990s saw a rapid expansion of international import/export markets, supported by evolving technologies and policies developed to support the North American Free Trade Agreement and other instruments that had been established to promote more connected global economies. Larger U.S. forest products companies were likely to have a well-integrated national presence, and some companies, such as International Paper and Westvaco, had begun to establish an international presence; but for the most part, forest industry in the Southeast was American owned and focused on regional operations (Carter et al. 2015). Changes occurred rapidly in the 1990s, when mergers and acquisitions created fewer but larger companies. At about the same time, a strong American dollar began driving exports down and increasing imports, including pulp and paper products. This forced the forest products industry to become much more competitive and much more globally focused, and ultimately contributed to a downturn in the U.S. domestic forest products business (Ince et al. 2007). Southeastern pulpwood prices peaked in 1997, decreased by 50% by 2002, and increased slightly after 2002. Since 1998, pulp and paper demands in developed countries have steadily declined, especially in response to expanding electronic media, causing U.S. paper production capacity to shrink (Wear et al. 2016).

Concurrent with these developments, institutional investors began to appreciate the potential of timberland as a component of their portfolios. The 1974 Employee Retirement Income Security Act encouraged diversification of pension plans beyond fixed income securities, but institutional investors also began to recognize that timberland assets are counter-cyclical to other investment vehicles and therefore could serve as a hedge against market volatility (Zinkhan 1988, 1992; Binkley et al. 1996). As investment analysts began to more fully understand the forest products industry as a whole, various pressures prompted these companies to contemplate a change in structure. These included the significant debt that many companies had taken on in the wave of consolidation that had occurred in the 1990s; the poor fit of standing timber values in standard accounting protocols, which negatively impacted overall balance sheets; and the more favorable tax treatment that alternate corporate structures offered compared to traditional C-corporations, in which income is taxed at both corporate and owner levels (Clutter et al. 2005; Binkley 2007). Furthermore, imports of raw materials were increasing and many family forest owners had begun providing fiber to mills, primarily because of silvicultural advances that had increased the productivity of their forests.

All this meant that maintaining timber supply for mills was no longer the concern that it had been historically, and many companies felt comfortable moving away from the vertically integrated model that had driven their land ownership. Perhaps most fundamentally, because coordinating the economic operations of timberlands with wood products production involves substantial transaction costs, these companies anticipated economic gains from separating the two functions (disintegration of many nonforest vertically integrated industries that occurred at the same time were based on these same economic fundamentals).

From 1998 to 2008, forest product companies divested about 75% of their lands. Most of this land was purchased by institutional investors such as pension funds and university endowments through TIMOs or by shareholders in REITs. Some forest products companies, such as Weyerhaeuser, restructured themselves as timberland REITs. TIMOs do not actually own forestlands, but rather acquire, manage, and sell them for their investors, often under term-limited investment periods. REITs differ in that they allow investors to pool capital for participation in real estate ownership, with shares in real estate that are either publicly traded or privately held.

This divestiture of forest industry land has been termed the largest U.S. land ownership transfer of the last century (Butler and Wear 2013). Institutional investment in timberland grew from approximately $2 billion in 1990 to over $40 billion by 2006. Today, TIMOs and REITs represent a fluid, but significant, ownership of large blocks of timberland within the range of longleaf pine.

RENEWAL OF INTEREST IN LONGLEAF PINE

The 1960s saw the beginning of heightened awareness of environmental issues in the United States. Landmark environmental legislation passed in the early 1970s included the Clean Air Act, the Clean Water Act, the National Environmental Policy Act, and of particular importance to the longleaf pine ecosystem, the Endangered Species Act. This legislation granted the U.S. Fish and Wildlife Service regulatory authority over habitat for listed species and prohibited federal agencies from engaging in activities or funding activities that might degrade habitat for listed species. The listing of the red-cockaded woodpecker as an endangered species effectively put longleaf pine ecosystems in the spotlight to a degree that had not been seen for decades.

Meanwhile, in the years after World War II, researchers and forest managers had begun to understand the basic ecology of longleaf pine and how to manage it successfully (Croker 1968). Knowledge of the fundamental aspects of the ecosystem, such as the role of fire, episodic masting and regeneration, competitive interactions, and wildlife-habitat relations, contributed to better-informed management and restoration programs. Research on seedling production, notably the development of containerized seedlings (Barnett and Brissette 1986), made longleaf pine a viable choice for afforestation or reforestation from the standpoint of seedling survival and growth. Federal policies dramatically slowed the conversion of second-growth longleaf pine to plantations on national forestlands, and many resource professionals in the U.S. Forest Service embraced the value, as well as the challenges, of managing for longleaf pine, prescribed fire, and the red-cockaded woodpecker.

The decline of longleaf pine ecosystems and their associated wildlife also became a subject of interest for academics, nongovernmental organizations (NGOs), and private landowners. Longleaf pine became an area of increased focus for research organizations such as the U.S. Forest Service—Southern Research Station, Tall Timbers Research Station, the Joseph W. Jones Ecological Research Center, as well as individual researchers in university natural resource programs across the Southeast, with important publications highlighting the conservation value and conservation status of the species (Frost 1993; Landers et al. 1995; Means 1996). Longleaf pine ecosystems and their associated rare species became conservation targets for The Nature Conservancy and other NGOs, and for state agency natural heritage and nongame wildlife programs. In the mid-1990s, a group of longleaf pine proponents came together to establish The Longleaf Alliance, an education and outreach organization originally housed at Auburn University. The Safe Harbor program for red-cockaded woodpeckers, an innovative U.S. Fish and Wildlife Service program that removes many disincentives for managing for high-quality longleaf pine on private lands, was established in 1995.

As momentum and interest in longleaf pine continued to grow, a collaborative group of 22 federal agencies, state agencies, and conservation NGOs formed in 2007 to develop a range-wide conservation plan for longleaf pine ecosystems (see Chapter 1). The plan was released as America's Longleaf Restoration Initiative (2009). The broad goal of this plan is to increase longleaf pine acreage to 8 million acres while improving the condition of existing longleaf pine forests and creating higher-quality habitat for longleaf pine-associated wildlife species. Since the plan's release, a diverse group has coalesced to form the Longleaf Partnership Council, whose primary mission is to facilitate implementation of the plan through enhanced communication and collaboration. Longleaf pine is an increasing focus of foresters, game managers, nongame wildlife biologists, botanists, academics, public natural resource agencies, NGOs, private landowners, and those concerned with and affected by threatened and endangered species. The membership of the Longleaf Partnership Council is representative of the diversity of this group and reflects the broad range of motivations, goals, and objectives of its members.

Although distinguishing—in a value-neutral way—between the reestablishment of longleaf pine as a tree species on a given site and the holistic restoration of the longleaf pine ecosystem is important, the two goals are not mutually exclusive; rather, reestablishment can be viewed as a prerequisite for restoration. However, one cannot simply equate reestablishment of longleaf pine trees with restoration of the ecosystem. Ecosystem restoration encompasses the full suite of structural, functional, and compositional elements found in reference sites. Reestablishment can buy opportunities for ecosystem restoration, but expecting that all acres of reestablished longleaf pine will be managed solely for ecosystem values is neither reasonable nor realistic. The rangewide conservation plan acknowledges this by targeting a 3-million acre subset of the 8-million acre goal for ecosystem management, or what is termed "maintenance class" condition (Ware 2014), with the remainder either managed for more utilitarian purposes or classified as developmental. However, framing this as a binary choice would be a mistake; for many ownerships, some mix of economic, ecological, and aesthetic goals will drive decisions.

As knowledge and experience have grown, it is clear that longleaf pine ecosystem restoration is a long-term developmental process rather than a discreet intervention. Beyond establishment, restoration will require ongoing inputs such as prescribed fire and thinning, and intermediate treatments such as midstory control or ground cover restoration. Perhaps most challenging is the protracted time scale over which longleaf pine ecosystem development unfolds and the long-term commitment required to achieve the desired structure and function. These longer time scales also present challenges for the economic performance of longleaf pine compared to other southeastern pine species.

LONGLEAF PINE ECONOMICS RESEARCH

For many private landowners, particularly those with large parcels, economic considerations play a role in their motivations for owning forestland. Economic considerations also factor into their decisions about forest management, including the basic question of whether to grow longleaf pine as opposed to other pine species. The modern literature on the economics of longleaf pine is limited compared to other southeastern pine species, reflecting the commercial dominance of loblolly and slash pine and the substantial support of the forest industry to meet applied information needs for these species. Beginning in the 1980s, research interest in longleaf pine management and economics began to grow; the literature subsequently reflects the heightened interest in the species and provides insight into the economics of longleaf pine compared to other southern pine species. The following section broadly summarizes trends and significant points from the emerging longleaf pine economics literature.

The use of capital budgeting techniques in the forest industry became more common in the 1970s; before then, relatively simple techniques such as payback analyses (the length of time required to recover capital investment) were often used. Increasingly, methods and concepts from the mainstream economics and business literature were used to incorporate the time value of money into forest investment analyses, including net present value (NPV), soil expectation value (SEV), internal rate of return (IRR), and other valuations (Bailes and Wendell 1979; Cubbage and Redmond 1985). Discount rates are annual rates of compound interest that are used to account for the time value of cash flows (Bullard et al. 2002). The discount rates chosen for forest investment analyses are often determined by the weighted cost of capital or the rate of return that could be earned from an alternative investment of similar length and risk.

NPV, the difference in the present (discounted) value of future cash inflows and future cash outflows, is one of the most common metrics used to analyze forest investments. For these investments, NPV is the difference between discounted costs, such as plantation establishment costs incurred early in the rotation, and discounted returns, which are typically profits from the sale of timber during and at the end of the rotation. A project is deemed acceptable if, for a chosen discount rate, NPV is a positive number; higher values are preferable to lower values. This metric does not indicate the relative scale of a project, meaning that a given figure for NPV does not distinguish between a large and a small investment, thus potentially obscuring the relative value. Also, NPV is typically calculated with a single value at the completion of the specific project, thus potentially confounding comparisons of projects with significantly different time frames. SEV, which is sometimes called land expectation value (LEV), is the net present value, per unit area, of the projected costs and revenues from an infinite series of identical even-aged forest rotations, starting initially from bare land. This allows comparisons of investments with different time horizons.

The internal rate of return (IRR) is the discount rate at which the present value of the costs is equal to the present value of the revenues, or the point at which NPV equals zero. It represents the actual rate of return on the investment, equivalent to a rate of compound interest that could be earned if the funds were invested elsewhere. This metric has the advantage of expressing results transparently in a common format, allowing comparisons of projects that are dissimilar in size or type. Although all of these metrics are employed in analyses and NPV is regarded by many as theoretically superior, IRR is often preferred as a quantitative decision criterion by the forest industry (Cubbage and Redmond 1985; Hogaboam and Shook 2004).

Historically, the prevailing perception was that economic returns from longleaf pine were categorically inferior to faster growing pine species. The range of approaches found in the literature for examining longleaf pine economics includes analyses of economic returns from longleaf pine alone, comparisons of economic returns from longleaf pine versus loblolly and/or slash pine, and comparative studies of volume growth of longleaf and other pine species without extending that analysis to capital budgeting analyses. A subcategory of this research addressed the economic aspects of managing for the red-cockaded woodpecker. For example, specific parameters of stand structure that have been defined in the red-cockaded woodpecker recovery plan require lower stocking rates and diameter distribution guidelines that inherently produce lower timber volumes and thus have negative impacts on economic returns (Lancia et al. 1989; Glenn 2012).

Most publications about the economics of longleaf pine management are found in the "gray" literature: conference proceedings, white papers, brochures, case studies, theses, and other reports that have not been subjected to rigorous peer review. The primary outlet for peer-reviewed forestry publications in the South was the *Southern Journal of Applied Forestry* (no longer published), but other outlets include a variety of forestry and economics journals. Taken as a whole, an important characteristic of these publications is the lack of consistency among the variables that drive the analyses. Differences in management regimes, discount rates, stumpage-value assumptions, and product output make direct comparisons across multiple analyses difficult. Some of these differences are attributable to subjective decisions on the part of the investigators; others involve changes in variables, such as stumpage values, during the 25 years that these analyses span.

However, these studies shared one consistent theme: the impact that the time value of money has on the economic returns for management of longleaf pine over longer rotations. A basic premise of financial analyses that use discounting, such as NPV, is that income earlier in the analysis cycle has higher value than income later in the cycle. Despite considerable gains in the development of longleaf pine establishment techniques that reduce time in the grass stage and accelerate early growth, the species still exhibits slower early growth than loblolly or slash pine in most situations (Schmidtling 1987; Cram et al. 2010). This is partly attributed to the inherent growth characteristics of the individual species, but also a reflection of decades-long tree improvement programs for slash and loblolly and the greater response of these species to fertilization and other silvicultural inputs (Bailian et al. 1999; Borders and Bailey 2001; Dickens, Moorhead, Morris, et al. 2012). Thus, the comparatively slower early growth of longleaf pine interacts with the time value of money and places it at an economic disadvantage compared to other pine species when analytical techniques that involve discounting are used, especially in short-term analyses. The essence of this disadvantage is that the other pines usually reach a valuable final harvest sooner than longleaf pine; thus, most economic analyses that involve direct comparisons of economic performance of longleaf and loblolly or slash pine invariably show longleaf pine to be inferior. However, many of these analyses also assume that products harvested from longleaf pine and other pine species are equivalent. Particularly when compared to loblolly pine, longleaf pine is preferred for pole production (Croker and Boyer 1975), produces dimensional lumber that has higher tensile strength, and has a higher weight-to-volume ratio (Meier 2016) as well as higher specific gravity (Jackson 1968).

To address the time value of money, several studies incorporated income pulses early in the analysis period (Hamilton 1998; Mills and Stiff 2008; Johnson 2011). For example, income from pine straw harvesting or incentive programs was shown to have very significant positive impacts on discounted metrics such as NPV and LEV as well as narrowing or eliminating disparities between the economic performance of longleaf pine versus other southern pine species. Incentive programs can offer early income in the form of establishment cost share or payments for given practices under a range of contractual periods (most typically within the first 10–15 years after establishment). Straw harvesting can begin at canopy closure, usually at age 6–8 depending on planting density and site productivity, and can continue for a decade or so until the first thinning. Many longleaf pine incentive programs prohibit straw raking, citing concerns about impacts on wildlife habitat and ground cover condition, illustrating that these two sources of early income are often mutually exclusive.

The discount rate (or rates) chosen for past analyses has varied widely, with many as low as 4%. Although a 4% discount rate would be appropriate for current analyses, using such a low rate in the 1980s and 1990s (and even early 2000s) does not accurately reflect the actual cost of capital during that time. Other confounding factors are variability in site indices used in different analyses and differences in management regimes. Finally, the disparity in stumpage prices and product mixes among studies was substantial. Assumptions about stumpage price are typically not explicit and are often referenced simply as being consistent with then-current market trends. Although the analyses that compare longleaf pine and loblolly pine rarely distinguish product classes and values between species, a few incorporate a reasonable percentage of poles in the longleaf pine harvest (Cubbage and Hodges 1989; Glenn 2012) and reflect the price differential in timber income. One analysis suggests that when combined with straw income, the higher rate of pole production yields equal or superior results for longleaf compared to loblolly in a range of scenarios (Mills and Stiff 2013). Another (Busby et al. 1993) modeled an unusually high rate of pole production (90% of all stems that met size/length requirements), perhaps skewing results favorably for longleaf pine. Yanquoi (1992) offered one of the better examples of a sensitivity analysis that highlights the significance of these variables and their interactions.

General timber-value trends for economic metrics are summarized in Table 3.1. For longleaf pine analyses that used both high and low discount rates, results are presented for the lower discount rates. Given the disparity in methodology among analyses, meaningful quantitative analysis or statistical exploration (such as meta-analysis) is not feasible.

The results for NPV of longleaf pine ranged from −$476 to + $766 per acre in analyses that used a 4%–5% discount rate. Within that range, NPV most commonly ran from about $150 to $500. Depending on management intensity, average results for loblolly pine (Siry 2002) ranged from $411 to $1082 at a 6% discount rate. This was considerably higher than longleaf pine, especially considering the discount rate used was 100–200 basis points higher. All other factors being equal, higher discount rates result in lower performance in discounted capital budgeting analyses because the time value of money has greater impact.

SEV for longleaf pine ranged from −$497 to + $967 per acre at a 4%–5% discount rate, with most results running from $200 to $600. At a 6% discount rate, SEV of loblolly pine ranged from $411 to $1411 per acre (Siry 2002).

IRR for longleaf pine ranged from 0%–10.1%, with most results in the 3%–7% range. IRR for loblolly (Siry 2002) in 2000 ranged from 9.6%–13%. This is consistent with later analyses (Cubbage et al. 2007) that suggest an IRR of 7.7%–12.5% for loblolly, reflecting recent overall decreases for pulpwood prices (Wear et al. 2013).

Rates of returns can be increased by >60% with annual income from pine straw in loblolly, slash, and longleaf pine forests (Dickens, Moorhead, Bargeron, et al. 2012). The inclusion of pine straw raking in several of the longleaf pine analyses demonstrates that this income source can improve

TABLE 3.1
Comparison of Averages for Commercial Value Measures for Longleaf Pine and Loblolly Pine Using Low and High Discount Rates (DR)

	Longleaf Pine		Loblolly Pine	
	Low DR	High DR	Low DR	High DR
Net Present Value[a] (US$/acre)	150	500	411	1082
Soil Expectation Value[b] (US$/acre)	200	600	500	1411
Internal Rate of Return[c] (%)	3	7	9.6	13

Note that analyses concentrated on timber only, rather than other commercial and ecological values of the two species.
[a] The difference between the present—or discounted—value of future cash inflows and future cash outflows.
[b] The net present value of the projected costs and revenues from an infinite series of identical even-aged forest rotations, starting initially from bare land.
[c] The discount rate at which the present value of the costs is equal to the present value of the revenues, or when net present value equals zero.

economic returns (Roise et al. 1991; Mills and Stiff 2008; Johnson 2011; Glenn 2012) for two primary reasons: (1) as outlined above, the ability to derive income from straw early in the investment cycle has significant impacts on metrics that use discounting, and (2) the income from straw over the life of a stand can equal or exceed income from timber. Recent multiyear averages range from $50 to $150 per acre per raking, although revenues as high as $300–$400 per acre have been reported in high-quality longleaf pine stands (Dickens, Moorhead, Bergeron, et al. 2012). Longleaf pine comparisons of LEV for timber-only and timber-plus-straw showed increases of as much as 300% when straw was included (Glenn 2012), with a more common range of 15%–90% (Mills and Stiff 2008). NPV increases ranged from 26%–350% (Mills and Stiff 2008; Johnson 2011).

Some analyses of straw-harvesting economic impacts in longleaf pine forests suggest either implicitly or explicitly that timber harvesting plus straw raking can be fairly competitive with timber harvesting alone in loblolly pine forests (Table 3.2). Interestingly, very few publications that compare longleaf pine and other pine species also incorporated a scenario in which loblolly or slash pine straw is harvested for income. One line of research examined straw management and revenues (Dickens et al. 2011, 2014; Dickens, Moorhead, Bergeron, et al. 2012), comparing straw raking and timber harvesting combinations for loblolly, slash, and longleaf pine. The studies suggested that although longleaf pine straw is clearly viewed as the superior product and commands wholesale prices that are nearly 100% higher than loblolly pine straw, loblolly pine forests produce 30% more bales per acre. Slash pine forests fall in between the two, both in price and in productivity. An informal survey (McIntyre, unpublished data) of wholesale straw suppliers suggests that slash pine is the largest seller by volume because of abundant supply and higher quality compared to loblolly, and that longleaf pine sales are constrained by supply. Analyses that include straw in economic comparisons would benefit from examining the potential for straw harvesting from all species of southeastern pines. A comparison of various loblolly, slash, and longleaf pine straw/timber scenarios found longleaf pine to have slightly higher SEV (+2.6%) than slash pine and much higher SEV (+12.3%) than loblolly pine at 33-year rotations, but it still greatly underperformed both species at 24-year rotations (Dickens et al. 2014).

One of the drivers for the range-wide efforts to restore longleaf pine is the availability of habitat for the suite of wildlife species that are associated with longleaf pine, from federally listed nongame species such as the red-cockaded woodpecker to more common game species such as the northern bobwhite (*Colinus virginianus*). Although longleaf pine-associated species may differ in the details of their individual requirements for optimal habitat, as a group they share many general characteristics of preferred habitat including a relatively open canopy, lower stocking rate, and grass-dominated ground cover that is maintained by frequent fire. All of these characteristics represent opportunity costs compared to longleaf pine forests managed for maximum timber production.

Two studies addressed this issue by examining opportunity costs within the context of red-cockaded woodpecker management. Lancia et al. (1989) found opportunity costs of $125–$250 per acre in SEV in areas that were managed for red-cockaded woodpecker foraging habitat as outlined in the 1985 recovery plan for the species. Comparing SEV of longleaf pine managed for maximum timber value with longleaf pine managed for red-cockaded woodpecker habitat, Glenn (2012) found opportunity costs ranging from $322 to $439 per acre (depending on site index and level of straw harvesting).

The management strategies in both of these studies essentially drove SEV into negative territory, the exception being two of the more aggressive straw harvesting regimes that resulted in positive SEV (Glenn 2012). Red-cockaded woodpeckers arguably represent the high end of the range of opportunity costs for wildlife habitat because of their preference for older trees and lower stocking, as defined in the habitat recovery plan for the species. Although the guidelines for red-cockaded woodpeckers differ significantly from the regimes in forests that are managed for fiber production, landowners who manage for sawtimber products would likely not view them as such a significant departure from their regimes. Managing for the broader community of longleaf pine-associated wildlife certainly involves some opportunity costs, but these costs are likely to be less than those incurred when managing

TABLE 3.2
Significant Economic Studies for Longleaf Pine and Other Southeastern U.S. Pines, Comparing Net Present Value (NPV)[a], Internal Rate of Return (IRR)[b], and Soil Expectation Value/Land Expectation Value (SEV/LEV)[c]

Investigator(s)	Scenario	Age (yr)	DR (%)	IRR (%)	NPV (Low DR)	NPV (High DR)	SEV/LEV (Low DR)	SEV/LEV (High DR)
					(US$/acre: Numbers in Parentheses Indicate Negative Values)			
Cubbage and Hodges (1989)	Longleaf natural regeneration	45	4.0	6.0	146	–	170	–
	Longleaf natural regeneration	80	4.0	6.8	443	–	460	–
	Longleaf artificial regeneration, 1 thinning	40	4.0	5.2	165	–	208	–
	Longleaf artificial regeneration, no thinning	40	4.0	6.2	408	–	516	–
	Longleaf artificial regeneration, 2 thinnings	50	4.0	6.3	538	–	626	–
Kessler and Straka (1991)	Longleaf natural regeneration and supplemental artificial regeneration, no thinning, with straw raking	37	4.0	–	709	–	627	–
Roise et al. (1991)	Longleaf natural regeneration, shelterwood, with straw raking	60	4.0	–	–	–	1198	–
	Longleaf natural regeneration, shelterwood, with straw raking	80	4.0	–	–	–	1227	–
	Longleaf natural regeneration, shelterwood, with straw raking	100	4.0	–	–	–	1268	–
	Longleaf natural regeneration, shelterwood, with straw raking	120	4.0	–	–	–	1298	–
Yanquoi (1992)	Longleaf natural regeneration, site index of 70	40–80	4.0	–	–	–	261–424	–
Alavalapati et al. (2002)	Longleaf, 42-year rotation	42	5.0	–	–	–	515	–
	Slash, 30-year rotation	30	5.0	–	–	–	1146	–

(Continued)

TABLE 3.2 (*Continued*)
Significant Economic Studies for Longleaf Pine and Other Southeastern U.S. Pines, Comparing Net Present Value (NPV)[a], Internal Rate of Return (IRR)[b], and Soil Expectation Value/Land Expectation Value (SEV/LEV)[c]

Investigator(s)	Scenario	Age (yr)	DR (%)	IRR (%)	NPV (Low DR)	NPV (High DR)	SEV/LEV (Low DR)	SEV/LEV (High DR)
					(US$/acre: Numbers in Parentheses Indicate Negative Values)			
Teeter and Somers (2005)	Longleaf artificial regeneration, site index of 70	50	5.0, 7.0	–	(20)–100	(60)–(160)	–	–
	Longleaf artificial regeneration, site index of 80,	50	5.0, 7.0	–	150–250	(60)–25	–	–
	Longleaf artificial regeneration, site index of 90,	50	5.0, 7.0	–	375–475	60–175	–	–
Cubbage et al. (2007)	Longleaf natural regeneration, southeastern United States	80	8.0	4.3	–	(165)	–	–
	Loblolly artificial regeneration, southeastern United States	30	8.0	9.5	–	133	–	–
	Pine spp. artificial regeneration, South America	18–22	8.0	10.5–16.9	–	460–748	–	–
	Eucalyptus artificial regeneration, South America	7–16	8.0	12.8–22.9	–	1486	–	–
Mills and Stiff (2008)	Loblolly artificial regeneration, low site index	Variable	5.0, 7.0	–	502	183	611	213
	Longleaf artificial regeneration, low site index	Variable	5.0, 7.0	–	245	49	308	53
	Longleaf artificial regeneration, with straw raking, low site index	Variable	5.0, 7.0	–	548	175	593	178
	Loblolly artificial regeneration, high site index	Variable	5.0, 7.0	–	895	470	1180	583
	Longleaf artificial regeneration, high site index	Variable	5.0, 7.0	–	766	385	967	457
	Longleaf artificial regeneration, with straw raking, high site index	Variable	5.0, 7.0	–	968	503	1109	597

(Continued)

TABLE 3.2 (*Continued*)
Significant Economic Studies for Longleaf Pine and Other Southeastern U.S. Pines, Comparing Net Present Value (NPV)[a], Internal Rate of Return (IRR)[b], and Soil Expectation Value/Land Expectation Value (SEV/LEV)[c]

Investigator(s)	Scenario	Age (yr)	DR (%)	IRR (%)	NPV (Low DR)	NPV (High DR)	SEV/LEV (Low DR)	SEV/LEV (High DR)
					(US$/acre: Numbers in Parentheses Indicate Negative Values)			
McIntyre et al. (2010)	Longleaf low intensity selection harvest, land value included in analysis	Ongoing	6.0	3.21	–	–	–	–
	Longleaf medium intensity selection harvest, land value included in analysis	Ongoing	6.0	3.29	–	–	–	–
	Longleaf high intensity selection harvest, land value included in analysis	Ongoing	6.0	3.48	–	–	–	–
Johnson (2011)	Longleaf artificial regeneration, timber harvesting, no straw raking	45	4.5, 6.0	6.0	197	(3)	–	–
	Longleaf artificial regeneration, timber harvesting with straw raking	45	4.5, 6.0	11.0	894	482	–	–
	Longleaf artificial regeneration, 50% cost share, timber harvesting, no straw raking	45	4.5, 6.0	7.0	282	81	–	–
	Longleaf artificial regeneration, Conservation Reserve Program incentives, timber harvesting, no straw raking	45	4.5, 6.0	29.0	787	545	–	–
	Longleaf artificial regeneration, Conservation Reserve Program incentives, timber harvesting, straw raking at year 15	45	4.5, 6.0	30.0	1317	883	–	–
Glenn (2012)	Longleaf, no straw raking	Variable	4.0	0.0–4.6	(476)–113	–	(497)–142	–
	Longleaf with conservative straw raking	Variable	4.0	2.3–5.4	(209)–192	–	(218)–386	–

(*Continued*)

TABLE 3.2 (*Continued*)
Significant Economic Studies for Longleaf Pine and Other Southeastern U.S. Pines, Comparing Net Present Value (NPV)[a], Internal Rate of Return (IRR)[b], and Soil Expectation Value/Land Expectation Value (SEV/LEV)[c]

Investigator(s)	Scenario	Age (yr)	DR (%)	IRR (%)	NPV (Low DR)	NPV (High DR)	SEV/LEV (Low DR)	SEV/LEV (High DR)
					(US$/acre: Numbers in Parentheses Indicate Negative Values)			
	Longleaf with moderate straw raking	Variable	4.0	3.6–6.1	(52)–463	–	(54)–584	–
The following studies are shown to provide comparisons of financial returns for loblolly pine.								
Sedjo (2001)	Loblolly, southwide average	*	*	12.5	–	–	–	–
Siry (2002)	Loblolly with five levels management intensity	25–30	6.0	9.6–13.0	–	416–1082	–	504–1411
Dickens et al. (2006)	Loblolly	24	8.0	6.2–11.9	–	–	–	–

DR: Discount rates;
–: Value not calculated; *: Information not available;
[a] The difference between the present—or discounted—value of future cash inflows and future cash outflows.
[b] The discount rate at which the present value of the costs is equal to the present value of the revenues, or when net present value equals zero.
[c] The net present value of the projected costs and revenues from an infinite series of identical even-aged forest rotations, starting initially from bare land.

primarily for red-cockaded woodpecker. A study that focused on management of mature longleaf pine for a broader suite of wildlife found IRR ranging from 3.21%–3.48% (McIntyre et al. 2010). This study incorporated income from a hunting lease and included land costs as an initial investment expense. IRR analyses for forestry do not often include land costs; sensitivity analyses (Cubbage et al. 2007) found that IRR for pine management increases by about 50%–60% on average when land costs are not included. Using this as a correction factor suggests that IRR results from McIntyre et al. (2010) would range from 5%–5.4% without land costs, comparable to other IRR analyses for longleaf pine.

Taken as a whole, these studies clearly suggest that, on many sites, longleaf pine lags in growth and thus economic returns (for timber only) compared to other southeastern pine species when using shorter time scales and discounted cash flow analyses. However, many of these analyses used older growth and yield models that did not incorporate the faster growth of containerized seedlings or the effects of modern longleaf pine establishment techniques. Also, few studies attempted to incorporate the economic advantages of long-term management, such as the minimal stand reestablishment costs afforded by natural regeneration under multiple age-class management strategies. Although yet-to-be-developed analytical approaches, such as updated growth and yield models, would likely cast longleaf pine in a better light, there are existing considerations that support longleaf pine as a viable economic choice for some land ownerships.

Several studies did not actually employ discounted capital budgeting analyses but rather focused on growth rates of longleaf pine and other southeastern pine species. These studies showed that volume growth of longleaf pine can compete with volume growth of loblolly pine on xeric sandy soils and other low productivity sites, suggesting that differences in economic returns may not be significant under these circumstances (Outcalt 1993; Cram et al. 2010). This has significant economic implications for large-acreage landowners who are evaluating fine-scale variations in site quality across their timberland portfolios. Further, several studies suggest that longleaf pine "catches up" to other southeastern pine species on many sites between the ages of 20 and 25 years (Schmidtling 1987; Boyer 1996; Harris et al. 2000).

Another aspect of longleaf pine that bears on this discussion is its reputation for high-quality wood products. Compared to other southeastern pine species, longleaf pine is known to produce higher percentages of poles (Williston et al. 1989) and wood that has higher tensile strength, specific gravity (Markwardt and Wilson 1935), and weight-to-volume ratio. These attributes should, in theory, result in higher economic returns for longleaf pine. Pole values are reflected in some of the analyses, and pole product classes have held their value in recent years. However, with few exceptions, longleaf pine sawtimber is generally lumped with other southeastern pine species with respect to stumpage prices. Although its heavier weight-to-volume ratio arguably captures some of this value, the broader attributes of longleaf pine are typically not valued in the marketplace. The degree to which this is captured in longleaf pine economics studies is unclear.

Research has also evaluated characteristics of longleaf pine that could provide greater diversification of a forest management portfolio to improve resilience for impacts from hurricanes, wildfires, and pine beetle outbreaks (Croker and Boyer 1975; Hodges et al. 1979; Strom et al. 2002; Johnsen et al. 2009). However, quantitative analyses of management risks have yet to be fully incorporated into economic comparisons of longleaf pine and other southeastern pine species. At the stand level, lower overall financial risks (a general finding from the studies listed earlier in this paragraph) should increase the relative return of longleaf pine management compared to other forest management choices. At a broader landscape level, considerations of risk management might suggest the increased use of longleaf pine in a portfolio of forest management strategies as a hedge against biophysical and financial risk, similar to risk spreading in a portfolio of financial assets (Dixit and Pindyck 1994). Beyond a consideration of risk, the different pattern of revenue streams across time and product classes (sawtimber versus pulpwood) for longleaf and loblolly pine suggests a countercyclical potential for a portfolio of forest investments that include both species (Wear et al. 2013).

Consistent with a basic principle of discounting in which cash flow earlier in an analytical cycle has higher value, the literature clearly suggests that supplemental income early in a forest management cycle, such as straw raking, establishment cost share, or incentive payments, can often make longleaf pine economically competitive. Finally, it should be recognized that the inclusion of broader management objectives in addition to economic returns, such as wildlife habitat and legacy values, implies some degree of opportunity cost. Some landowners may be willing to assume these opportunity costs, but for others, compensatory mechanisms such as incentives may be required to offset opportunity costs.

THE CHANGING LANDSCAPE—FUTURE PROJECTIONS FOR SOUTHEASTERN FORESTS

Wear and Greis (2013) developed forecasts of land use for southeastern forests using a set of scenarios or "cornerstone futures" as a framework for predicting change from 1997 to 2060. Their primary driving variables were growth (economic and population) and timber and crop prices (high or low to define increasing or decreasing demand) within an econometric model of land use change. Their historical land use categories (pasture, crops, forest, range, or urban) and area data were based on the Natural Resources Inventory (USDA NRCS 2001).

Urbanization is projected to continue driving regional land use patterns as rapid population growth continues, with forecasts ranging from 30 to 43 million additional acres of land developed for urban use by 2060. As a component of the overall acreage lost to urbanization, forestland is expected to decrease by 11–23 million acres. The highest collective losses are expected in the Piedmont; but forest loss in the Coastal Plain would also be significant, with most development expected to occur near the Atlantic Ocean and Gulf of Mexico, and with Florida projected to lose 34% of its existing forest cover (Klepzig et al. 2014).

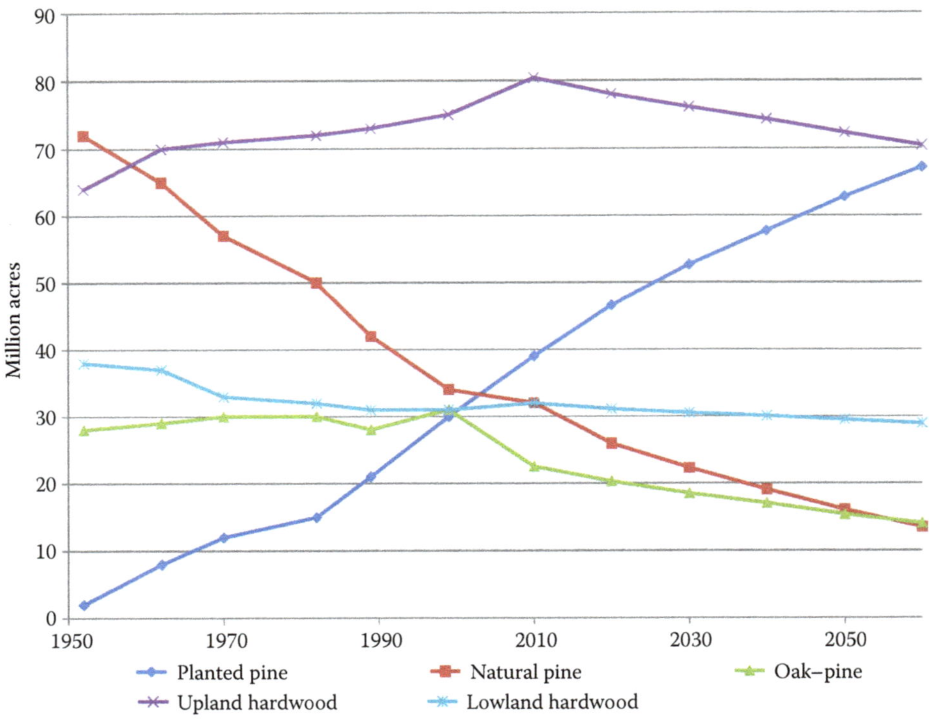

FIGURE 3.1 Forecasted forest area by forest management type, 1952–2060, for southeastern U.S. forests under the assumption of high urbanization, high timber prices, and more tree planting. (Redrawn from Huggett, R. et al., *Forecasts of Forest Conditions*, USDA Forest Service, Southern Research Station, Asheville, North Carolina, 73–101, 2013.)

Forest condition is also predicted to change over this time frame. Huggett et al. (2013) considered five broad categories of southeastern forests (Figure 3.1): planted pine (all species), natural pine, mixed pine and oak (*Quercus* spp.), upland hardwood, and lowland hardwood. They predicted that only planted pine will increase in area, from its current levels of 39 million acres (19% of southeastern forests) to somewhere between 47–67 million acres (24%–36% of southeastern forests), depending on the scenario. Declines in area of natural pine are expected to continue their historical trajectory and are inversely related to gains in planted pine, with losses projected to range from 7.6 to 18.0 million acres. Models also predicted overall forest carbon to increase slightly until 2020/2030 and then begin to decrease. Market projections also indicated a continued concentration of timber production in the Southeast, specifically in the Coastal Plain, which is consistent with the prediction of expanding areas of planted pine.

THE PLACE OF LONGLEAF PINE IN THE LANDSCAPE IN THE 21ST CENTURY

Current Status of Longleaf Pine

The current estimate of longleaf pine acreage stands at 4.28 million acres and is based on FIA data (Miles 2016). These data include two forest types: longleaf pine and longleaf pine/oak. The longleaf pine forest type is defined as those stands with greater than 50% stocking of pine in which longleaf pine is the dominant pine. The longleaf pine/oak forest type is defined as stands in which pine accounts for 25%–50% of the stocking and longleaf pine is the dominant pine species (Oswalt et al. 2012). Longleaf pine forests that remain today are not evenly distributed across the historical range of the species. About 89%, or 3.8 million acres, is located in an area of the Coastal Plain that stretches from the Mississippi River eastward to the Atlantic Ocean, with large concentrations in the Florida Panhandle, southern Alabama, Georgia, and Mississippi (Oswalt et al. 2012). Regionwide,

62% of the existing longleaf pine is controlled by private landowners, with the remaining acreage managed by public agencies. This concentration of longleaf pine forest type (38%) on public lands (Oswalt et al. 2012) is disproportionate to the total forested acreage in the Southeast (13%) that is occupied by public lands (Oswalt et al. 2014), thus demonstrating the importance of public lands to longleaf pine conservation.

Given the aggressive goals for restoration of longleaf pine forests and the rapid changes occurring on the southeastern landscape, significant questions emerge about how, where, and within what context these goals will be accomplished. The complex set of variables that interact to drive these considerations include land use changes, smoke management in the wildland-urban interface, the opportunity costs associated with longleaf pine, and the stability of land ownership.

Opportunities for Restoration on Public Lands

The fact that public lands contain 38% of the remaining longleaf pine on just 13% of the Southeast's forested land base highlights the critical role that public lands have played, and will continue to play, in the conservation and management of longleaf pine ecosystems. Going forward, they will also play an important role in the effort to increase longleaf pine acreage. Longleaf pine restoration on public lands represents perhaps the most valuable investment of available funding because of the low risk of conversion to other land uses or forest types, institutional capacity and resources to implement management activities, less concern about the economic opportunity costs of longleaf pine, and a commitment to long-term management, thus the potential for full realization of ecological restoration goals.

The ecological values of longleaf pine align well with the primary objectives of many public lands, including wildlife values, biodiversity conservation, carbon storage, and other ecosystem services. The advantages of prioritizing restoration on public lands include a much higher likelihood of longleaf pine persistence into the future, opportunities for expansion of significant core areas where longleaf pine dominates across functional landscapes, and consistent management planning over long time scales.

Perhaps the most important opportunity for longleaf pine restoration on public lands will occur on national forests, which comprise about 4 million acres within the historical range of longleaf pine (USGS 2016). In a review of Southeastern national forests land management plans, the U.S. Forest Service identified approximately 808,143 acres of existing longleaf pine and a target of 1,492,374 desired acres. Data for existing acreage differed from FIA sampling estimates because they originated from actual stand-level inventories (ALRI 2009). The other significant federal land holders in the Southeast are the U.S. Fish and Wildlife Service (national wildlife refuges) and the Department of Defense (military installations). The opportunities for national wildlife refuges are limited, primarily because of their historical emphasis on wetlands and the relatively low acreage of uplands in their land portfolio, but the U.S. Fish and Wildlife Service is actively working on longleaf pine restoration where appropriate. Although military installations have some of the best remaining examples of longleaf pine ecosystems because of their history of frequent fire, land requirements for the military mission may preclude significant expansion of longleaf pine acreage on these lands.

National forests regularly update management plans for their lands, with revised guidance set forth in the 2012 planning rule. The U.S. Forest Service has been a leader in the public sector for longleaf restoration efforts, including the development of the range-wide conservation plan. As southeastern national forests embark on revisions of their land management plans, it is anticipated that longleaf pine will become a higher priority on national forests within its historical range. In the interim, until forest plan revisions are undertaken in coming years, amendments to existing plans could be a tool for increased longleaf pine restoration. Prioritization of longleaf pine restoration on all appropriate national forest sites could result in significant increases in desired acres of longleaf pine. For example, the recently completed plan for the Francis Marion National Forest in South Carolina approximately doubles the goal for acres of upland longleaf and wet pine savanna and flatwoods ecosystems compared to the previous plan. Additionally, annual targets for prescribed fire were increased substantially and greater emphasis will be placed on growing-season fire for ecological objectives (USDA FS 2016). As additional national forests begin plan revisions, similar increases may be forthcoming given the evolving focus of national forest management and the new national planning rule. The U.S. Forest Service has a history of leadership in longleaf pine restoration, an existing capacity for forest management, a strong prescribed fire program, and a mandate for multiuse management that includes ecological values. With the largest portfolio of public lands within the historical range of longleaf pine, the opportunity to focus restoration efforts on sites with such a high degree of ownership stability that allows for management of longleaf pine over the long term is tremendous.

State-owned properties such as wildlife management areas and state forests are also good candidates for restoration of longleaf pine, but acreage varies widely among states. For example, for all of the states (excluding Texas) where FIA data report longleaf pine, Florida has as much state-owned land and almost twice as much longleaf pine acreage as all of the other states combined (Miles 2016). Across the region, longleaf pine on state-owned public lands comprise approximately 456,000 acres. As with national forests, the overall management purposes for these lands typically align closely with the attributes of longleaf pine ecosystems.

Assuming some level of increase across the board on national forests and some level of contribution from state lands and other federal lands, a forecast of ≥1 million acres of potential longleaf pine restoration on public lands would not be unreasonable. However, public agencies face a range of challenges, including budgetary limitations and constraints mandated by legislation such as the National Environmental Protection Act. From 2013 to 2015, annual longleaf pine establishment on all public lands averaged just 26,000 acres (Longleaf Partnership Council 2014, 2015, 2016). Thus, acceleration of longleaf pine restoration on public lands will be a critical component of meeting the goals of the regional restoration initiative.

Opportunities and Challenges for Restoration on Private Lands

Although public agencies support longleaf pine restoration and are more than willing to meet challenges and implement restoration on the lands that they manage, most of the acreage gains required to achieve the 8-million acre goal of the regional restoration initiative will need to occur on private lands—continuing a trend of acreage gains on private lands, particularly those held by family forest owners. From 2013 to 2015, approximately 450,000 acres of longleaf pine establishment were documented, with >75% occurring on private lands (Longleaf Partnership Council 2014, 2015, 2016); about 50% of these private land acres were established under incentive programs, for which corporate landowners are typically ineligible. This suggests that at least half of the increase in longleaf acreage occurred on family forests.

The multiple values of longleaf pine, which include aesthetics, wildlife habitat, moderate timber income, and legacy benefits, align well with the objectives of many family forest landowners. The increase in longleaf pine establishment from 2013 to 2015 built on the planting of >900,000 acres of longleaf pine plantations from 1985 to 2010 (Oswalt et al. 2012). However, these numbers for newly established longleaf do not tell the whole story of longleaf pine acreage dynamics. Based on preliminary examination of changes in FIA data from 2010 (Oswalt et al. 2012) to 2015 (Miles 2016), longleaf pine acreage losses appear to have negated any gains (see Chapter 1), with most of those losses presumably from older, established stands on private lands. These data illustrate some of the challenges involved in restoring longleaf pine on private lands.

Restoration on Family Forestlands

Individuals or families own two-thirds of the private forestlands in the Southeast, or about 132 million acres (Butler and Wear 2013). Insights into the motivations and dynamics of this diverse group can be taken from social science surveys conducted by the U.S. Forest Service since the late 1970s, initially through its stewardship program (Birch 1997) and more recently through its woodland owners survey (Butler 2008, 2016). Through the years, a consistent theme of these surveys has been that family forest owners value their forests for a variety of reasons, many of which are independent of economics (Figure 3.2). This evidence that family forest owners typically rank timber income lower in their priorities than other forest amenities is particularly encouraging for longleaf pine. Although timber income may be an important component of their suite of overall values, these surveys suggest that discounted economic metrics are not the primary drivers of their decision processes.

One of the most significant opportunities for longleaf pine on family forestlands is the ability of this group of landowners to access federal incentive programs. These programs offer cost-share support for longleaf pine establishment and for specific management activities such as prescribed fire. Other programs have also offered yearly incentive payments for a defined period, usually 10–15 years, to maintain longleaf pine. Many of the longleaf pine establishment acres are a direct result of USDA incentive programs offered through Natural Resources Conservation Service and Farm Service Agency, such as the Conservation Reserve Program, the Longleaf Pine Initiative, and Working Lands for Wildlife, as well as Department of Interior programs such as Partners for Fish and Wildlife. These programs have resulted in approximately 556,000 acres of longleaf pine establishment since the late 1990s (D. Hoge, personal communication; L. Jones, personal communication). Payments through these programs provide income early in the discounting cycle and are often large enough to offset opportunity costs of choosing longleaf pine over faster growing pine species.

Although there has been past success and opportunities remain, longleaf pine restoration on family forests also faces challenges, many stemming from a rapidly changing southeastern landscape.

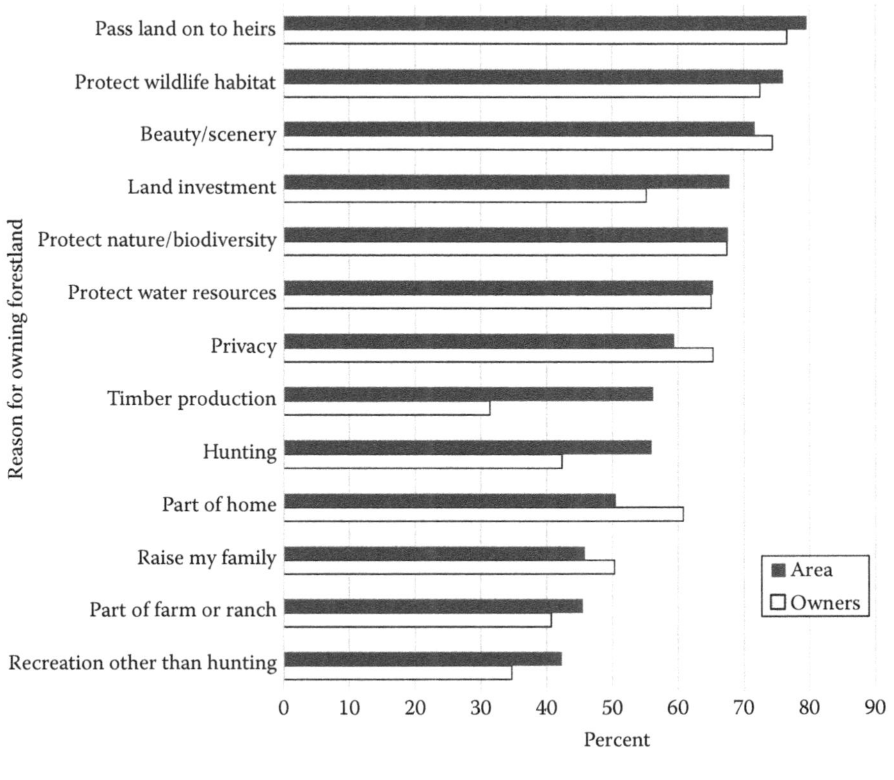

FIGURE 3.2 Survey data on motivations of U.S. Forest Service Southern Region family forest owners, 2011–2013 (excludes West Texas and West Oklahoma). (Modified from Butler, B. J. et al., *USDA Forest Service National Woodland Owner Survey: National, Regional, and State Statistics for Family Forest and Woodland Ownerships with 10+ Acres, 2011–2013*. Resource Bulletin NRS-99, USDA Forest Service, Northern Research Station, Newton Square, Pennsylvania, 2016.)

The average size of the family forest parcel is 29 acres (Butler and Wear 2013). FIA survey data suggest that parcelization is continuing (Figure 3.3). During the period covered by these surveys, the percentage of forests in parcels of ≤100 acres almost doubled, and tracts of ≥1,000 acres were reduced by half, reflecting both divestiture of forest industry lands as well as transfer of family forestlands. Decreasing parcel size of forestland in the Southeast is well documented and recognized as an issue for viability of ongoing forest management (Hatcher et al. 2013).

Small-parcel forest holdings have long been a source of concern in the forestry community for two primary reasons: first, economies of scale make implementation of certain forest management practices, such as prescribed fire and thinning, difficult on smaller tracts; and second, smaller tracts tend to have higher rates of parcelization. Addressing some of the management issues by aggregating tracts or establishing cooperative arrangements with neighboring properties has been suggested, but neither of these options has been widely used in the United States (Cubbage 1983; Kittredge 2005). As shown in Figure 3.4, approximately 86% of individuals who own family forests own parcels of <100 acres, but almost two-thirds of the actual acreage is in parcels ≥100 (Butler et al. 2016). Although direct causal mechanisms can be difficult to verify and may be highly variable from one situation to the next, the size of a forest holding is a good proxy variable for predicting several elements of forest management. As the size of forest holdings increases, land tenure increases, the probability that the owner will actively manage timber increases, participation in cost-share programs increases, the percentage of owners with a management plan increases, and the tendency to seek professional advice increases (Hatcher et al. 2013).

A primary scale-related issue associated with smaller-acreage family forests is the practicality and economic viability of operational management. Many of the costs associated with timber harvesting—such as equipment, labor, and transportation—are fixed and essentially independent of tract size or timber volume. Studies suggested that the minimum tract size for economically viable timber management ranges from 40 to 50 acres (Row 1973, 1978; Lazarus and Schaible 2015) to 80 acres (Wikstrom and Alley 1967). Wikstrom and Alley (1967) also suggested that prescribed burning of tracts <25 acres is prohibitively expensive and that the minimum average costs cannot be realized until tract size reaches 125 acres. Unless a smaller stand can be bundled with a neighboring property to reach a size that attracts a contractor for thinning or burning, executing operations on

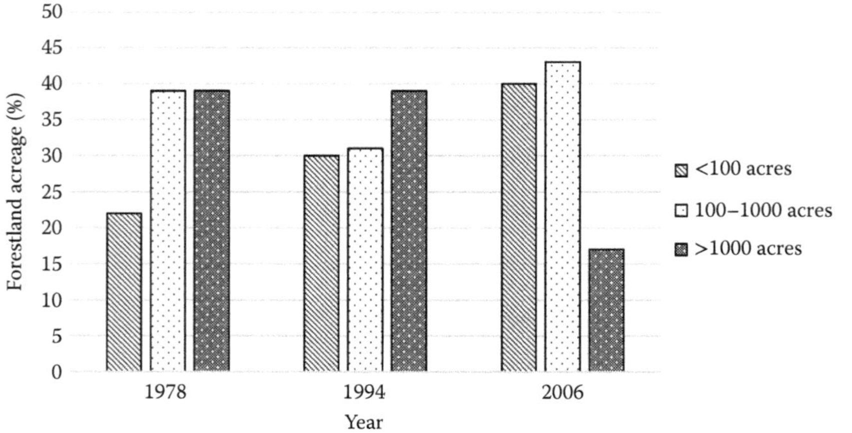

FIGURE 3.3 Parcel-size breakdown for forests in the southeastern United States, 1978–2006. (From Birch, T. W. et al., *The Private Forest-Land Owners of the United States*. Resource Bulletin WO-1, USDA Forest Service, Washington, DC, 1982; Birch, T. W., *Private Forest-Land Owners of the Southern United States, 1994*. Resource Bulletin NE-138. USDA Forest Service, Northeastern Forest Experiment Station, Pennsylvania, 1997; Butler, B. J., *Family Forest Owners of the United States, 2006*, General Technical Report NRS-27, USDA Forest Service, Northern Research Station, Newtown Square, Pennsylvania, 2008.)

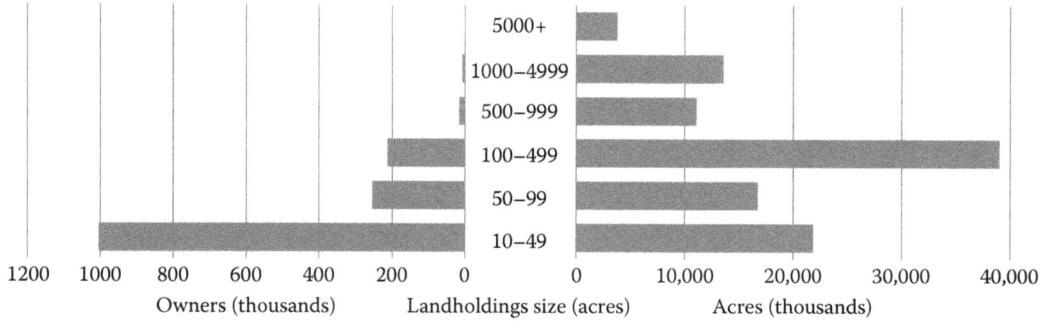

FIGURE 3.4 U.S. Forest Service Southern Region family forest ownership survey data, 2011–2013, comparing parcel size and number of land owners (excludes West Texas and West Oklahoma). (Modified from Butler, B. J. et al., *USDA Forest Service National Woodland Owner Survey: National, Regional, and State Statistics for Family Forest and Woodland Ownerships with 10+ acres, 2011–2013,* Resource Bulletin NRS-99, USDA Forest Service, Northern Research Station, Newton Square, Pennsylvania, 2016.)

smaller stands can be difficult. These studies illustrate the urgency of current parcelization trends caused by continuing ownership turnover and parcel fragmentation (Sampson and DeCoster 2000), but they also show the difficulty of deciding which landowners to target for limited technical assistance capacity and resources.

Family forest owners are also an aging demographic, with approximately 46% of their acres held by individuals ≥65 years of age and another 30% held by individuals aged 55–64 (Butler 2016). Although these demographics imply a somewhat uncertain future for these lands, especially in terms of intergeneration transfer, continuity of management, and overall interest in forest ownership, there is little data or research to document positive or negative outcomes.

Restoration on Corporate Forests

Estimates of corporate ownership in the Southeast, which now includes the forest industry sector, range from 54 million acres (Zhang et al. 2012) to 66 million acres (Butler and Wear 2013). Analyses of 2010 FIA data indicate that just <7 million acres of industry land remain, a reduction of about 80% from the peak of 38 million acres (Zhang et al. 2012). Acreage of TIMOs and REITs range from 16.5 million acres (Zhang et al. 2012) to 19.9 million acres (Butler and Wear 2013), respectively. Note that Zhang et al. (2012) does not include data from Mississippi and Louisiana in its estimates.

Opportunities for longleaf pine restoration by corporate ownerships have traditionally been viewed as problematic and many challenges remain for these ownerships. However, some new opportunities are unfolding. The overriding objective for many corporations is to generate returns for their investors. This is particularly true for TIMOs and REITs, which are legally required to maximize returns for investors (Ravenel et al. 2002). If opportunity costs for longleaf pine are negated or mediated in some way and investors are "made whole," longleaf pine could be a viable option.

Many of the industrial-land sales that involved TIMOs were quite large, with diverse portfolios of lands offering a variety of locations, site-quality levels, and other attributes. Given that their objectives are much broader than simply supplying timely and low-cost fiber to a mill, TIMOs often take a more nuanced approach to assessing the suite of opportunities for utilization of their lands. For example, it is not unusual for some portion of a TIMO portfolio to be targeted for residential development in areas where development returns would outperform continued forest management. These decisions are often based on spatial attributes that make such tracts appealing for development, such as proximity to existing development, transportation corridors, or natural amenities such as coastal zones, water bodies, or scenic areas. Acreage that is less suited for development is typically retained in forest management, particularly higher quality sites. On lower quality sites dedicated to forest management, such as xeric sites or low-productivity soils, more detailed analysis could reveal opportunities where longleaf pine is economically competitive. For example, the research discussed earlier relating to comparable growth/yield/productivity of longleaf pine and other southern pine species on xeric dry sites suggests that longleaf pine may be an economically viable management option on those sites. However, even if the site characteristics make longleaf pine competitive, some opportunity costs associated with reforestation must be addressed for corporate ownerships, such as the cost of longleaf pine seedlings, which is approximately 35%–75% greater than comparable quality loblolly or slash pine (IFCO 2016; Meeks 2016). Additional longleaf pine acreage could be established if eligibility for incentive programs that target the species were broadened to include corporate landowners. Corporate ownerships have traditionally been ineligible for incentives because current incentive programs have limits on adjusted gross income and acreage enrolled. Cash flow early in an investment period, which is typical for most incentive programs, can result in economic returns from longleaf pine that are competitive—even on higher-quality sites—with pine species that are not eligible for incentives (Mills and Stiff 2008; Johnson 2011).

Corporate owners with large forest holdings are likely better positioned to factor the risk mitigation values of a mixed portfolio of silvicultural approaches into their investment choices. In addition to risk mitigation from natural disturbances, different approaches to longleaf pine management can help diversify the mix of forest products, timing of product flow, and liquidity of forest products. These owners also are well suited to play a role in the development of longleaf pine markets, such as premium pine straw and high-quality solid wood. Many areas with aggregations of existing longleaf pine share a nexus with concentrations of corporate forestland, offering important opportunities to enhance landscape-level conservation of longleaf pine ecosystems (Figure 3.5).

Partnerships of corporate ownerships with NGOs and natural resource agencies offer potential for increasing longleaf pine acreage. Differences in longleaf pine productivity, wildlife-oriented stocking rates and management regimes, prescribed fire, and other opportunity costs could be compensated at a negotiated rate through incentive programs to incorporate these conservation values while allowing continued use of the lands as working forests. Some TIMOs have begun exploring opportunities to bridge the opportunity-cost gap that divides longleaf pine and other pine species; one such project is the Coastal Headwaters project in the Florida Panhandle (The Conservation Fund 2016). This project would offer permanent protection from development and convert the existing loblolly pine to longleaf pine under a long-rotation management system that balances wildlife habitat structure with timber income. Agencies and NGOs are increasingly using less-than-fee-simple strategies such as this on private lands as a means to achieve conservation goals in more cost-effective ways.

A portion of the corporate ownership category is made up of large family ownerships, a group that includes trusts and partnerships with legally defined fiduciary responsibilities. These owners often grapple with multiple objectives including the economic, recreational, and conservation uses of their land. For some, economic objectives are not only the optimization of return but also other considerations such as taxes and multigenerational preservation of wealth. This combination of objectives often translates into longer management rotations with a diversity of species and forest conditions; longleaf pine management would likely have a role in such a land portfolio. Tax benefits of working-forest conservation easements can be an additional motivating factor for this ownership.

Corporate owners that hold large acreages also bring the advantages of existing capacity for professional forest management and the economies of scale inherent with large aggregates of land. Stated simply, working with a few larger-acreage landowners is much more efficient than working with many small-acreage landowners. Additionally, parcelization and family-forest ownership trends suggest that land use and management on larger tracts are more stable than with smaller

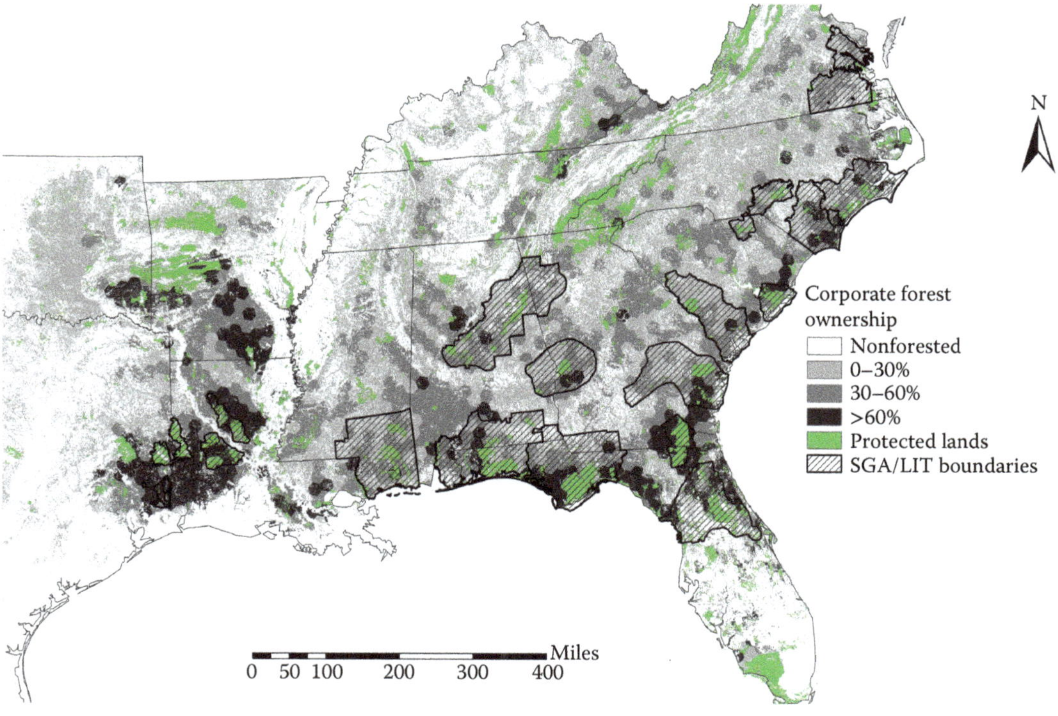

FIGURE 3.5 Corporate forest ownership, 2008, in Southeastern United States and 2016 Significant Geographic Areas (SGAs) and Local Implementation Team (LIT) boundaries. (From Nelson, M. D. et al, *Forest ownership in the conterminous United States: ForestOwn_v1 geospatial dataset*, USDA Forest Service, Northern Research Station, Newtown Square, Pennsylvania, 2010, https://doi.org/10.2737/RDS-2010-0002. SGA/LIT boundaries data courtesy of The Nature Conservancy.)

parcels (Hatcher et al. 2013), which is consistent with the longer time scales required for longleaf pine forest development.

In contrast to family forest owners, the challenges for longleaf pine restoration on corporate land are typically driven more by economic considerations. Corporate ownership structures, predominantly TIMOs and REITs but also including family trusts and other structures, have fiduciary and legal responsibilities to investors and shareholders and must be able to demonstrate that any decision has economic parity with other options. For this reason, strategies that mitigate opportunity costs of longleaf pine, such as establishment cost share, incentive programs, or purchase of conservation easements for working forests, would likely be a prerequisite for engaging this group of landowners.

Other economic challenges that discourage corporate owners from managing longleaf pine include the lack of well-developed longleaf pine growth and yield models and the relatively nascent state of tree-improvement programs for the species, both of which have been developed for loblolly and slash by substantial and well-endowed cooperative research programs between industry, universities, and government agencies. As a result, sophisticated investors have come to expect more accurate predictability for product yield and financial performance than is currently available for longleaf pine. Similarly, although tree-improvement programs have begun for longleaf pine, they lag far behind programs for loblolly and slash pine, thus further handicapping longleaf pine in the eyes of many corporate owners. Finally, many institutional investments through TIMOs consist of closed comingled funds with limited investment periods of 5–20 years (Zhang et al. 2012). These shorter time horizons may not be well suited for the longer time scales required for longleaf pine stand development.

Spatial Considerations

Just as longleaf pine may not be the most appropriate fit for the management goals of all land ownerships, there are spatial considerations that factor into the choice of the most appropriate locations for longleaf pine restoration. The regional longleaf pine conservation initiative has identified priority landscapes based on criteria such as known concentrations of existing or potential longleaf pine sites in protected status, occurrences of at-risk species, and pressures for land use change (ALRI 2009). Rather than engaging in random or opportunistic efforts throughout the historical range, the goal was to encourage spatially focused efforts to facilitate restoration of functional landscapes and viable populations of longleaf pine-associated wildlife. Within these broadly defined focus areas, additional analyses are needed to prioritize limited conservation resources.

For any conservation effort to be successful, some threshold questions must be addressed to identify the location, amount, and condition of the conservation target. Although the FIA program provides a good understanding of longleaf pine extent and condition at a coarse scale (Oswalt et al. 2012), spatially explicit information at finer scales is relatively limited. As a foundational element for future longleaf pine restoration, analyses are needed to identify appropriate longleaf pine sites based on soils, hydrology, and other biophysical characteristics as well as sites (existing and potential) that can serve as hubs and connectivity corridors for facilitating wildlife movement and sustaining species of interest (Hoctor 2013). The design of hubs and corridors should take into consideration the broader landscape-level matrix of forestland management and the potential role that traditional production-oriented forests could play in connectivity and buffering of high-priority longleaf pine sites.

Perhaps most importantly, spatial prioritization would need to include the compatibility of longleaf pine management with existing and projected land use as a proxy for the likelihood that an investment in longleaf pine restoration will produce long-term sustainable results (Figure 3.6). Some areas of otherwise-suitable longleaf pine restoration sites will undoubtedly be subjected to

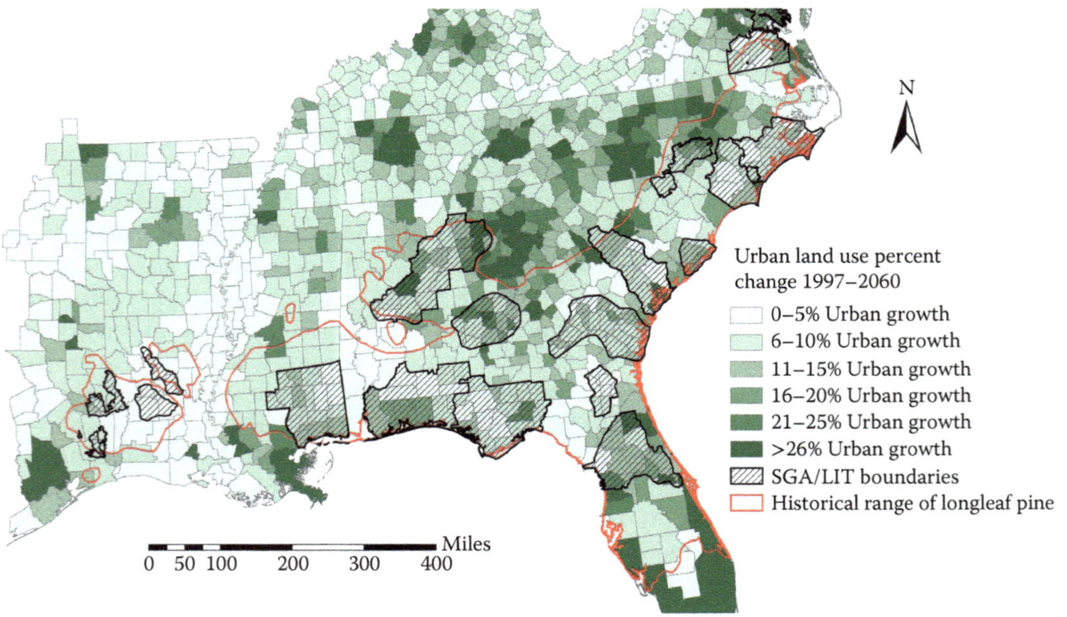

FIGURE 3.6 Southeastern U.S. urban land use projections, 1997–2060, juxtaposed against the historical range of longleaf pine and 2016 Significant Geographic Areas (SGAs) and Local Implementation Team (LIT) boundaries. (Modified from Wear, D. N., *Forecasts of Land Uses,* USDA Forest Service, Southern Research Station, Asheville, North Carolina, 45–71, 2013. SGA/LIT boundaries data courtesy of The Nature Conservancy.)

greater intensity of urban development, thus increasing the difficulty of continued forest management, particularly for longleaf pine, on land that has the potential for more lucrative uses (Wear and Newman 2004). Urban development also brings to the forefront social considerations such as conflicts with prescribed fire and smoke management in the wildland-urban interface (Wimberly et al. 2006). Of particular concern for longleaf pine are projections of forest loss and urban growth in Peninsular Florida, the western Gulf Coast, and the southern Atlantic Coast (Klepzig et al. 2014).

The long time scales at which longleaf pine functions and the developmental period it needs to reach the desired structure also affect spatial prioritization, particularly for sites that emphasize management for ecological values. Restoration and management of this ecosystem represents a substantial investment in personnel, funding, and time—suggesting that the highest priority sites for restoration should be on lands that are likely to remain committed to a consistent management trajectory over the long term. The highest priorities would logically be land that is protected from parcelization or development by enabling legislation (public land) or legal restrictions (easements on private land). Other candidates for prioritization would be larger parcels, particularly those farther removed from projected urbanization. Larger parcels would be more likely to persist on the landscape, less likely to be subdivided, and more likely to have professional, active forest management (Hatcher et al. 2013). Parcels that are distant from urban development are less likely to experience increased valuation of bare land driven by development pressure, a primary factor in change from forested land use (Wear and Newman 2004).

Another opportunity for prioritizing longleaf pine restoration efforts is existing stands of degraded longleaf pine. These are typically former longleaf pine-dominated sites that are no longer assigned by FIA to the longleaf pine (or longleaf pine/oak) forest type because of past management practices (or lack thereof) that resulted in insufficient stocking of the dominant species. Although they may have some remaining longleaf pine overstory and elements of associated ground cover communities, they are often compromised by encroaching mesic oaks or off-site pines—typically the result of inadequate fire management. These sites, particularly those with more substantial levels of longleaf pine stocking (20%–50%) can be restored relatively quickly and far more economically than sites requiring afforestation. This approach can offer the opportunity for landowners to derive some economic returns from restoration treatments that remove merchantable hardwoods or off-site pine from the stand, thereby reducing the overall cost of restoration. Preliminary estimates suggest that 1.24–1.82 million acres of degraded longleaf pine could be considered for this type of treatment (Guldin et al. 2016). The locations of these sites and their spatial contexts are critical information needs in conservation planning for longleaf pine restoration.

SUMMARY

The near demise of the once-vast longleaf pine forests that dominated the southeastern Coastal Plain is not unique within the historical context of U.S. development. In all regions, native ecosystems were pressed into the service of society as the land was settled and altered for economic gain. What is different, however, is that longleaf pine forests were perhaps discounted too quickly for their value to society, both economically and ecologically. As human populations grow and bring accompanying pressures for land use change, society will continue to expect more from the same acres of forestland. If forest loss continues, and perhaps accelerates into the future, it will be critical to be more thoughtful and creative in apportioning objectives and priorities among the remaining acres of forested land use. Although some forestlands will still be dedicated solely to economic purposes with others dedicated to ecological purposes, society will increasingly need to look for opportunities for concurrent gains in a broader suite of forest benefits. Longleaf pine could be uniquely suited for those multiple purpose acres.

The literature clearly suggests that longleaf pine can offer moderate, but not maximum, economic returns from timber products compared to other southeastern pines. On less productive sites, longleaf pine growth can compete with growth of other pine species, making it more economically attractive. The weaker economic performance of longleaf pine on many sites is typically not an issue for public agencies or conservation NGOs, and for many private landowners the opportunity costs inherent in longleaf pine may be adequately offset by other amenities such as aesthetics or wildlife habitat. Smaller-acreage family forest owners also have options to participate in incentive programs that can offset these opportunity costs. For corporate owners and other large-acreage landowners in the private sector, policy changes will be needed to broaden eligibility for existing incentives or new programs developed that compensate them for opportunity costs so that longleaf pine is a viable option. With about twice the acreage of private U.S. forestland owned in tracts ≥100 acres compared to tracts <100 acres, policy aimed at encouraging protection, care, and production for forestlands could more actively target the individuals and businesses that comprise this group (Larson 2004; Butler 2008).

The longleaf pine "market share" has room for further growth. Research and development for longleaf pine similar to that conducted on loblolly and slash pine over the last 40 years could improve economic returns from longleaf pine management. Tree-improvement programs have the potential to improve growth, form, and wood quality. A better understanding of longleaf pine growth and yield using data from stands established using modern silvicultural advances, such as containerized seedlings and competition control, would improve forecasting tools and would provide documentation of increases in productivity and economic returns from these silvicultural advances. If artificially regenerated, plantation-grown longleaf pine is of similar quality to logs from naturally regenerated forests over the long term, markets could be expanded for high-quality solid wood products as acreage of these size-class stands increases. Examples of these products include machine stress-rated lumber, heart pine flooring and cabinetry, and pole products. Nontimber products, primarily pine straw, can offer income pulses early in an analysis period that would mitigate opportunity costs from the slower growth that characterizes longleaf pine; preferably, these measures would be implemented in ways that reduce impacts to ecosystem function (Bailey 2015).

Longleaf pine offers great potential for adapting to a changing climate. Preliminary research indicates that longleaf pine may be significantly more water-efficient than other southeastern pines (Vose et al. 2011), suggesting that it could be a good hedge for the more frequent and severe droughts predicted for parts of the southeastern Coastal Plain (see Chapter 15). Longleaf pine is also more resilient to tropical cyclones, wildfire, and forest pests (Croker and Boyer 1975; Hodges et al. 1979; Strom et al. 2002; Johnsen et al. 2009), which are all expected to have greater impact on southeastern forests. Quantification of risk reduction could help support the economic case for longleaf pine as well as serving as a basis for targeted efforts; for example, incentives could be developed that encourage longleaf pine planting in hurricane-prone areas such as the Gulf Coast.

As discussed in this chapter, spatial prioritization of longleaf pine restoration that includes ecological, economic, and social criteria is critical. For issues such as capitalizing on soils that make longleaf pine growth competitive with other pine species, minimizing competition from other land uses, designing spatial arrangements that support healthy populations of vulnerable wildlife species and facilitate their movement across the landscape—location matters. Although longleaf pine restoration can bring benefits wherever it occurs, care must be taken to leverage scarce resources and direct restoration efforts to areas where these forests will be managed over the long term.

Forests provide many benefits to society, including clean air, clean water, carbon sequestration, wildlife habitat, and economic assets such as jobs, tax revenues, and capital investments. The logic and rationale for public support of public forests is clear. These same societal benefits are also produced on privately owned forests, but the costs of maintaining those forests are borne by the private sector. These benefits, or ecosystem services, could be monetized in future policies that address carbon storage, maintenance of water quantity and quality, and wildlife conservation as a means of maintaining private longleaf pine forests on the landscape. Defining the ecosystem service benefits of forests in the marketplace could facilitate a layering of these values with timber values, thereby ensuring more accurate valuation of forests.

Longleaf pine is not likely to ever dominate the forests of the southeastern Coastal Plain as it once did. However, the multiple values of longleaf pine forests have the potential to meet the objectives of a significantly larger land base than they currently occupy. New understandings of science and management can bolster the case for longleaf pine and justify expansion of its current footprint. Increasing longleaf pine acres on the landscape—and importantly, keeping those acres on the landscape long enough for the full suite of values to be realized—will require careful consideration of the dynamic economic and social contexts within which longleaf pine management and restoration occurs.

ACKNOWLEDGMENTS

The authors would like to acknowledge the helpful reviews and input from Jim Guldin and James Vose. Special thanks to Chris Oswalt for assistance with updated FIA data. Jessica McCorvey provided editorial assistance with figures, tables, and proofreading. Spatial data for Significant Geographic Area boundaries in Figures 3.5 and 3.6 were produced by Analie Barnett of The Nature Conservancy.

Part II

The Ecological Basis for Restoration

4 Regeneration Dynamics, Competition, and Seedling Response

Steven B. Jack and Stephen D. Pecot

CONTENTS

INTRODUCTION

Competition for available resources—light, water, nutrients, physical space to grow—is a primary driver of plant population dynamics. In the highly competitive forest environment, plant survival and growth ultimately come down to an ability to tolerate low resource levels. The relative manner in which plant species respond to biotic and abiotic stressors, either positively or negatively, whether they survive and grow or eventually die, leads to rankings of species' tolerance to a specific resource limitation.

In tree species, tolerance is most commonly associated with species' survival and growth responses to low light availability, and the term is often used interchangeably in forest management with the more ecologically precise term, "shade tolerance." Although various definitions can be found in the literature, for the purposes of this chapter we define shade tolerance as the ability of a plant to survive at low light levels (Valladares and Niinemets 2008).

Longleaf pine (*Pinus palustris*) has long been characterized as intolerant to competition in general (Wahlenberg 1946; Boyer 1985; Harrington 2006) and to shade specifically (Boyer 1990). Much of the historical ecological literature on longleaf pine regeneration focused primarily on the period from germination to the emergence of seedlings from the grass stage (Figure 4.1)—a growth form with little or no stem elongation aboveground and extensive root growth belowground (Pessin 1934; Boyer 1990; Brockway et al. 2006). Depending on conditions (such as soil, climate, and microsite), seedlings may remain in the grass stage from 2 to >15 years (Wahlenberg 1946; Boyer 1990),

(a)

(b)

FIGURE 4.1 In an 85- to 100-year-old, naturally regenerated longleaf pine forest in southwestern Georgia: (a) A patch of longleaf pine regeneration showing well-established grass stage seedlings and some seedlings beginning to emerge from the grass stage, and (b) natural regeneration "dome" of longleaf pine (single age cohort) in a canopy gap, demonstrating the bell-shaped curve of seedling and sapling growth response over time. (Photographs courtesy of Richard T. Bryant.)

frequently with low observed levels of mortality even though competition for light can have a large impact during this period (Pessin 1939; Wahlenberg 1946; Boyer and Peterson 1983; Brockway, Outcalt, Guldin, et al. 2005). Because the goal of these early studies was to improve longleaf pine silvicultural practices, their major (if not the only) focus was on light limitation from the overstory canopy, and the most common recommendation was to implement even-aged silvicultural systems. Subsequently, the typical suggested management application was to increase light availability and increase survival and growth of seedlings by modifying the overstory structure, generally by creating large reductions in the canopy cover.

This line of research, however, was at odds with the structural variation found in many natural longleaf pine forests that have been managed primarily with prescribed fire and, in some instances, conservative harvesting treatments. In these forests a heterogeneous canopy develops over time with significantly more size and spatial variation than typically results from applying the silvicultural methods that prevailed across the Southeast in the 20th century. The regeneration cohorts (groups of seedlings or saplings that are all the same age) seen in natural forests frequently occur in canopy gaps that are considerably smaller than those that forest managers would consider operational (R. McIntyre et al. 2008). If plant tolerance was primarily a light-driven biological strategy in these forests, one would not expect longleaf pine seedlings to survive unless gaps are large enough to maximize light availability and sustain growth beyond the grass stage. Further, many natural stands have been managed over decades with multiaged silvicultural practices that rely on individual-tree selection, a continuous canopy, and few large canopy gaps; according to substantial empirical and anecdotal evidence, these forests are self-sustaining populations with—over time—adequate recruitment of new age cohorts (Way 2011).

The inconsistencies between the published understanding of longleaf pine seedling biology and observations in natural forests have stimulated a considerable body of research in the last 25 years. Clearly, plant tolerance incorporates a myriad of competitive strategies including responses to water, soil nutrients, herbivory, fire, disease, and other plants (Boyer 1990; Brockway and Outcalt 1998; Palik et al. 2002; Harrington 2006). To reflect these multiple factors, the research focus has broadened to include other disciplines such as fire ecology, agronomy, hydrology, and climatology. Understanding the cumulative effect of these potential competitive interactions on the establishment and endurance of longleaf pine seedlings could help to explain the underlying variation observed in natural longleaf pine ecosystems.

In this chapter we cover the current body of knowledge with respect to seedling and sapling development in natural longleaf pine forests. We discuss the role that multiple factors (such as light, available soil nutrients, water, neighboring plants, and fire) have on the establishment, survival, and growth of longleaf pines as they germinate, become established as advanced regeneration, and reach the sapling size class. We describe how these competitive effects and responses change through time and give special attention to the role of a complex fire landscape as well as the management required to sustain it.

From this review, we can say with certainty that longleaf pine regeneration processes present an exciting and, on occasion, quite challenging paradox with respect to plant-plant relationships and their role within the larger context of the ecosystem.

QUANTIFYING NATURAL DISTURBANCE PATTERNS

Much of the recent research on the importance of belowground competition to longleaf pine seedling survival began with an effort to quantify disturbance patterns in mature (aged 70–85 years), naturally regenerated forests at Ichauway, the property in southwestern Georgia that is home to the Joseph W. Jones Ecological Research Center. The forests of Ichauway were a byproduct of extensive logging operations that had occurred in the region from the late 19th century until the 1920s (see Chapter 3). The relatively few mature trees that remained after logging provided the sole seed source for the present forest (Palik and Pederson 1996; Pederson et al. 2008). During the decades that the property was managed as a quail preserve (1930s through 1980s), the forest component was primarily managed and maintained by applying frequent prescribed fire. Infrequent low-volume timber harvesting occurred from the 1920s to the 1980s, but no timber harvests were conducted from the late 1980s to 1998 (Palik and Pederson 1996). Natural regeneration of varying age classes was scattered across the property. For these reasons, the forests at Ichauway served as an excellent base for understanding natural longleaf pine disturbance and regeneration dynamics.

Palik and Pederson (1996) used dendrochronological and redundancy analysis techniques to estimate the frequency of different causes of natural disturbance that occurred in these forests. They found that multiple agents act on tree distribution; the most common were lightning strikes that killed a single tree or a few trees, followed by suppression (due to competition between trees), and windthrow. Their results fit with the long-observed paradigm of natural disturbances in forests of the Southeast (Wahlenberg 1946; Komarek 1968; Boyer 1990), ranging from relatively frequent fine-scale events such as lightning strikes, to very rare landscape-scale events such as hurricanes (Croker 1987). They demonstrated that different types of disturbance tend to affect different tree sizes, with lightning and windthrow acting upon the middle to upper tree diameters and suppression removing smaller diameter trees (Palik and Pederson 1996). Lightning, the most common mortality agent in these forests (Platt, Evans, and Rathbun 1988; Palik and Pederson 1996), tends to strike in the same locations over multiple years and has a high probability of striking the trees on gap edges (Outcalt 2008). A tornado, hurricane, or other rare and broad-scale meteorological event typically creates canopy openings that resemble a large group selection or clearcut (Bengston et al. 1993; Stanturf et al. 2007; Johnsen et al. 2009); these events are much more common in coastal areas. Using a modeling approach, Gilliam et al. (2006) compared the influences of tropical storms and fire on longleaf pine forest structure. They found that tropical storms directly affect the structure and distribution of the canopy trees (via wind-caused damage and mortality) and indirectly affect the ground cover by changing shading and the distribution of fuels. In contrast, fire directly influences the juvenile longleaf pine and indirectly influences the ground cover plants through selective mortality.

The results of Palik and Pederson (1996) and Gilliam et al. (2006) are complementary and indicate that disturbance frequency and intensity both influence the structure of longleaf pine forests—with these factors varying according to location and stand history. Stochastic natural disturbance events, distributed unequally over time and space, produce the large size and age variations that are observed in many mature longleaf pine forests. The results from Palik and Pederson (1996) are important because they suggest that longleaf pine regeneration may not require large canopy gaps to become established and that management and restoration efforts are likely focusing too heavily on specific silvicultural approaches that create large openings. Not well known at the time of the study was (1) the extent to which resource availability would affect seedling survival and establishment, and (2) what, if any, lower threshold of canopy gap size would allow resource competition to prevent recruitment.

THE LIGHT ENVIRONMENT IN LONGLEAF PINE FORESTS

Because longleaf pine is regarded as very shade intolerant compared to other eastern conifer species (Daniel et al. 1979), describing the light environment below the canopy is important for understanding the competitive relationship between the overstory trees and longleaf pine regeneration. Natural longleaf pine forests are often described as woodlands (Chapman 1905; Reed 1905; Chapman 1932; Wahlenberg 1946) or as having open canopy structure (see Chapter 1), especially when compared to other forest types (Wright and Bailey 1982). In fact, whether natural longleaf pine forests achieve a state of true canopy closure—as typically defined in the literature (Smith 1986; Barnes et al. 1998)—is somewhat debatable. Although broadly accurate, the woodland and open canopy structure descriptions fail to convey the fine-scale light environment that is important for understanding regeneration dynamics.

At fine scales, the light environment below a natural longleaf pine canopy is highly variable, both spatially and temporally (McGuire et al. 2001; Battaglia et al. 2002, 2003; Pecot et al. 2005).

Canopy trees in multiaged longleaf pine forests generally occur in a spatially heterogeneous arrangement (Platt, Evans, and Rathbun 1988; Kush and Meldahl 2000). In addition, the distribution of foliage within crowns of individual longleaf pine trees is highly aggregated, with foliage grouped at the ends of branches (Sheffield et al. 2003; Kirkman, Brown, et al. 2007). These two structural factors combine to make a heterogeneous ground-level light environment that constantly changes both within a day and from 1 day to the next (Battaglia et al. 2002). Quantifying this variation can help in predicting how seedlings and saplings will respond to different levels of canopy cover (Brockway and Outcalt 1998, 2015; McGuire et al. 2001; Battaglia et al. 2003; Palik et al. 2003; Pecot et al. 2007) and in understanding longleaf pine regeneration dynamics.

In addition to the overstory, other vegetation layers impact the light environment of seedlings. Longleaf pine ecosystems that are treated with frequent low-intensity fires generally have (1) a persistent but not well-developed midstory layer, and (2) a well-developed ground cover composed of herbaceous forbs, grasses, and woody species. The midstory may have scattered numerous hardwood trees (mostly deciduous) and areas of sapling-size longleaf pine regeneration. Even in the presence of frequent fire, midstory hardwood trees can significantly alter ground-level light environments, if only at fine scales. Whereas the heterogeneous longleaf pine canopy produces microsites of high light availability—known as sunflecks and sunpatches—that shift throughout a day (Smith and Berry 2013), the density and arrangement of leaves in a dense hardwood layer produces relatively uniform and complete shading during the growing season. This heavier shading can have negative impacts on seedling establishment, survival, and growth (described in later sections) if not controlled by prescribed fire or other treatments. On some sites, other species can form dense vegetative layers above the ground cover (see Chapter 1) and create significant shade, although not typically reaching true "midstory" stature. Examples of these species include saw palmetto (*Serenoa repens*), gallberry (*Ilex glabra*), and ericaceous shrubs such as staggerbushes (*Lyonia* spp.).

Although the herbaceous and grass species found in the ground cover of a longleaf pine forest can produce ephemeral shading for germinants and grass-stage seedlings, they are generally stronger competitors for soil moisture and nutrients than for light (Mitchell et al. 1999; Ford et al. 2008; Iacona et al. 2012). Most ground-level sites in these open forests have measurements of light quantity and quality that greatly exceed the levels below which negative physiological effects are produced (Pecot et al. 2005; Iacona et al. 2012; Samuelson and Stokes 2012).

An active prescribed burning program usually top-kills most of the midstory hardwood stems and volatilizes most of the herbaceous and grass biomass in the ground cover (Hiers et al. 2009; Loudermilk et al. 2011, 2012; Wenk et al. 2011; Ellair and Platt 2013). When fire is infrequent or absent, however, a midstory composed of generally fire-intolerant hardwood species can encroach and quickly overtop established longleaf pine seedlings (Walker 1954) as quickly as 3 years after the last burn (Boyer 1988), greatly reducing available light (Wright and Bailey 1982; Boyer 1990). In contrast, some hardwood species such as sand post oak (*Quercus margaretta*) and turkey oak (*Q. laevis*) can persist under a frequent fire regime for very long periods, with some even living >200 years (Greenberg and Simons 1999; Knight 2004; Varner and Pederson 2004). These pyrophytic hardwoods, however, are typically present at low densities and do not greatly affect the light environment for longleaf pine regeneration (Hiers et al. 2014; Loudermilk et al. 2016).

This characterization of a shifting light environment at the ground level is generally applicable to naturally regenerated, multiaged longleaf pine forests. Light availability in intensively managed even-aged plantation forests, however, can be quite different because of their more uniform tree spacing and the larger leaf areas that are usually present. Subsequently, plant responses observed in these systems often do not readily apply in natural settings (Harrington et al. 2003; Harrington 2006). After canopy closure, a longleaf pine plantation tends to have much less light reaching the ground level (Harrington and Edwards 1999), which can negatively impact longleaf pine seedlings and other plants. Commonly applied plantation thinning operations will begin to open the canopy and provide a more variable light regime that, over time, can approach the variation found in natural longleaf pine forests (Harrington 2011).

COMPETITION EFFECTS ON REGENERATION

Although our understanding of the below-canopy light environment in longleaf pine forests has improved dramatically, information about the many factors that affect establishment, survival, and early growth of longleaf pine seedlings is less comprehensive and sometimes perplexing (Pessin 1938; Allen 1954; Palik et al. 1997, 2003; Brockway and Outcalt 1998; McGuire et al. 2001; Rodrìguez-Trejo et al. 2003; Harrington 2006; Pecot et al. 2007). The ability of a seedling to become established is the culmination of complex interactions that change over time and through space. These multifactored interactions create microsites that affect the scale of natural regeneration and lead to heterogeneous patterns of early survival and growth for natural longleaf pine seedlings (Wahlenberg 1946; Boyer 1963; Croker and Boyer 1975).

Wahlenberg's (1946) classic volume on longleaf pine—which consolidated the knowledge accumulated since the 19th century by researchers, botanists, and interested individuals—was used by subsequent researchers as a touchstone for developing and enhancing silvicultural systems that benefit longleaf pine (Boyer 1963, 1993; Croker and Boyer 1975; Smith 1986). The major emphases in this body of work were the overstory canopy, the influence of canopy manipulations, and the factors that stimulate rapid emergence from the grass stage and vigorous stem growth in individual trees (e.g., Croker and Boyer 1975; Boyer 1993). This focus was understandable (see Chapter 3) considering the relatively rapid growth of loblolly (*P. taeda*) and slash (*P. elliottii*) pines, the decline or exclusion of fire on a broad scale in all U.S. forests (Frost 2006), and the paucity of knowledge about longleaf pine regeneration dynamics.

Seedling survival in canopy gaps, whether created naturally or through silvicultural treatments, was attributed to aboveground competition for light, with fire intensity considered to be a secondary factor (Croker and Boyer 1975). The observed limited presence of seedlings near mature trees was thought to be from two concurrent overriding factors: the decrease in light availability beneath the crowns of overstory trees (Battaglia et al. 2002; Gagnon et al. 2003) and the increase in fire intensity, caused by increased fuel loading, from the crown drip line toward the stem (Williamson and Black 1981; Grace and Platt 1995a; Brockway and Outcalt 1998; O'Brien et al. 2008). The net effect is that seedling survival and growth decrease from the middle of canopy gaps to the gap edge and are lowest underneath overstory trees, creating "domes" of natural longleaf pine regeneration; these domes are characterized by the tallest seedlings occupying the middle of the gap and the height of seedlings from the same cohort decreasing toward gap edges (Figure 4.1). Because light attenuation is uneven along the north-south axis of the gap, the tallest seedlings are usually just slightly northeast of the gap center (McGuire et al. 2001).

The observed dome structure fit well with the shade-intolerance paradigm and the preponderance of research results indicating that seedling survival decreases as overstory basal area increases. Hence the long-standing general recommendation that larger (>0.2 ha) canopy gaps are "required" for successful regeneration of longleaf pine. More recent research, however, has demonstrated that canopy trees can actually facilitate survival of longleaf pine seedlings, primarily by providing shade and reducing water demands during periods of drought (McGuire et al. 2001; Rodrìguez-Trejo et al. 2003; Pecot et al. 2007). This observed facilitative relationship, with more longleaf pine seedlings surviving under canopies where light is less available, created an apparent paradox in our understanding of seedling shade tolerance or intolerance (Wright et al. 2014), and was the impetus for several studies by multiple investigators. These studies examined both the survival and growth of longleaf pine seedlings in response to competition from other plants.

Factors Affecting Seedling Survival

The confusion surrounding the relative importance of competitive and facilitative relationships for the ability of seedlings to survive aboveground competition may have arisen largely because belowground competition—both within and among species—is much more important than was previously assumed. Several studies conducted at the Joseph W. Jones Ecological Research Center at Ichauway and nearby sites in the mid- to late 1990s (in addition to other studies covered in this chapter) were instrumental in first illuminating these relationships.

Survival Responses in Artificially Created Gaps

Based upon the results in Palik and Pederson (1996) and observations of initial patterns of natural regeneration in longleaf pine forests, three studies were installed, two on the Ichauway property and one on land owned by a publicly traded timber company 30 miles away. The three study sites were similar in age and species composition. A series of experiments measured survival and growth of planted seedlings in relation to canopy gaps created by mortality from fire and lightning (Palik et al. 1997) or by harvesting (McGuire et al. 2001; Palik et al. 2003). Palik et al. (1997) examined the spatial effects and responses within overstory gaps (each about 0.1–0.2 ha) that had been created by crown scorch and lightning strikes. McGuire et al. (2001) studied responses within gaps (each measuring 0.11–1.63 ha) that had been created by group-selection harvesting. The third study had a more complex experimental design that examined effects and responses within different overstory retention treatments: single-tree selection, small and large group selection, and an uncut control (Battaglia et al. 2002, 2003; Palik et al. 2003; Jones et al. 2003; Jack, Mitchell, et al. 2006; Pecot et al. 2007). The results of these three studies were among the first to question the conventional wisdom that longleaf pine regeneration processes "require" large openings for successful establishment.

The studies showed that longleaf pine seedling survival is not always correlated with overstory tree abundance (Palik et al. 2003) and that survival sometimes decreases with increased gap size and closer proximity to the gap center (McGuire et al. 2001). In fact, under certain climatic conditions (such as extended drought), the survival rate was shown to be higher under the intact canopy than in the gap centers. Similar patterns of facilitation have been known to occur in other plant systems (Freckleton and Watkinson 2001; Bruno et al. 2003; Freestone 2006; Butterfield 2009), but were only rarely observed for longleaf pine (Allen 1954) and were unexpected in this heretofore classified "shade-intolerant" species. However, other studies (Brockway and Outcalt 1998; Gagnon et al. 2003) conducted at about the same time suggested that facilitation is not a universally observed phenomenon for longleaf pine.

The series of studies also found that ground cover and midstory biomass increased with increasing gap size; the findings were even more dramatic for woody plant species, especially midstory hardwoods in gaps ≥0.2 ha (McGuire et al. 2001; Jack et al. 2006). The effect was strong enough to suggest that newly created gaps above a certain threshold size would likely convert to hardwood domes if the hardwood trees in the ground cover and midstory were not controlled (Figure 4.2).

Aboveground versus Belowground Competition Influences

The gap-scale results described above suggest that light is not the only factor involved in controlling seedling survival. Indeed, Jones et al. (2003) demonstrated that root production of pines and other woody plants had opposite responses to gap size: increasing the gap fraction decreased root biomass of pines but increased root biomass of other woody plants. In addition, the different mycorrhizal associations formed by the roots of pines and other woody plants also affect the balance of root biomass and resource acquisition (Wallander et al. 2001; Treseder 2004; Hendricks et al. 2016). Jones et al. (2003) suggested a complex interaction occurring belowground between roots of pines and other woody plants that is mediated, at least in part, by relative resource availability (see Chapter 7).

To better understand the effects of belowground competition on longleaf pine seedlings, additional plots were incorporated into the Palik et al. (2003) study design. As described in Pecot et al. (2007), the additional plots were situated along a gradient of canopy coverage, each plot was divided into two equal subplots, and gap fraction (Battaglia et al. 2002) was used as a surrogate for light availability. The protocol followed a factorial combination of treatments: (1) trenching around plot borders and installation of barriers to prevent root incursion; (2) removal of all ground cover plants, both herbaceous and woody, by clipping and spot herbicide application on

FIGURE 4.2 Demonstration of the effect of canopy gap size on hardwood midstory population, 3 years after prescribed burning and 4 years after overstory gap creation in a multiaged, naturally regenerated longleaf pine forest in southwestern Georgia. Large gaps impede fire continuity by minimizing needle cast from overstory trees, ultimately creating "hardwood domes" and "fire shadows" where prescribed fire is more difficult to apply. (Photograph courtesy of Joseph W. Jones Ecological Research Center.)

subplots; and (3) a combination of these two treatments. Container-grown longleaf pine seedlings were planted in each subplot and their survival and growth were monitored over three growing seasons. Trenching to isolate aboveground and belowground factors has been employed as an experimental method in longleaf pine forests and many other forest types (Fricke 1904; Cieslar 1909; Craib 1929; Korstian and Coile 1938; Harrington et al. 2003) and addresses concerns that Harrington (2006) raised about isolating the different components of aboveground and belowground competition.

Results indicated a positive relationship between overstory abundance and seedling survival, with survival decreasing from the intact canopy to the centers of gaps (Figure 4.3); fewer seedlings were observed in the centers of gaps regardless of ground cover removal or trenching treatments (Pecot et al. 2007). Isolating overstory belowground competition through trenching had no effect on survival, but survival decreased in subplots where the ground cover was removed (Figure 4.3). These results suggest the existence of facilitative effects by the overstory and the ground cover during severe droughts and other periods of high water stress.

Factors Affecting Seedling Growth

The discussion so far has centered on the survival of the seedling from germination to the grass stage, with mixed effects from levels of canopy cover (both facilitation and competition) and demonstrated effects of belowground competition. Although the evidence supporting the impact of overstory abundance on survival is inconsistent, seedling growth is known to be greatly

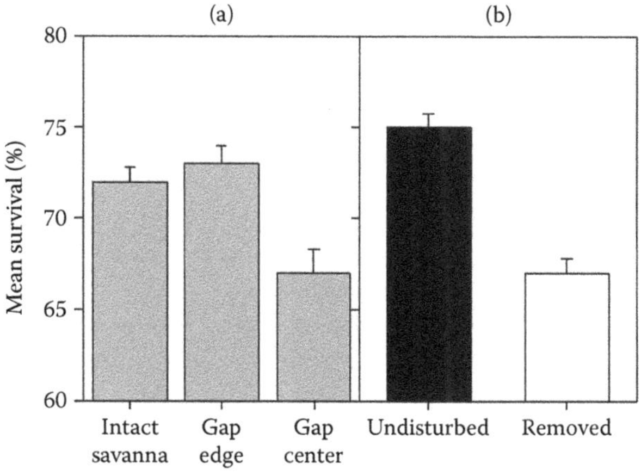

FIGURE 4.3 Longleaf pine seedling survival as influenced by: (a) Position relative to canopy gaps, and (b) presence or absence of understory (herbaceous ground cover and woody plants below midstory stratum) that was left undisturbed or removed by clipping and herbicide application. Lines above the bars show standard errors and indicate significant survival effects from location and understory. (Modified from Pecot, S. D. et al., *Canadian Journal of Forest Research*, 37, 634–648, 2007; © 2008 Canadian Science Publishing or its licensors. Reproduced with permission.)

affected by aboveground and belowground competition (Palik et al. 1997, 2003; Brockway and Outcalt 1998; McGuire et al. 2001; Pecot et al. 2007). The classic dome shape of increasing sapling height from the edges to the center of gaps (Figure 4.1) is the net result from years of competitive and facilitative seedling interactions with light availability, access to belowground resources, and fire intensity.

When a seedling emerges from the grass stage it begins the "rocket" or "bolt" stage with relatively rapid height growth (Boyer 1990; Brockway et al. 2006). Although some evidence indicates that the longleaf pine growth rate in the rocket stage is similar to other species at similar developmental stages (Wang et al. 2016), it is particularly dramatic after such an extended period of little or no stem elongation. When seedlings emerge from the grass stage, the tradeoff between survival and growth is most evident—the same factors that affect survival also mediate growth under the canopy, albeit with more predictability. Available light strongly and positively affects growth (Palik et al. 1997, 2003; Brockway and Outcalt 1998; McGuire et al. 2001; Gagnon et al. 2003; Kirkman and Mitchell 2006; Pecot et al. 2007), whether the seedlings were regenerated naturally or were planted under canopies of other pine species during restoration efforts (Harrington 2006; Kirkman, Mitchell, et al. 2007; Knapp et al. 2011; B. Knapp et al. 2013).

Belowground competition for soil nutrients occurs between the seedling and the community of other plants in the overstory and ground cover and also affects growth. For example, Pecot et al. (2007) demonstrated that seedling growth is linked to available light and soil nitrogen when competition from the ground cover is removed through herbicide, but is affected only by available light when the ground cover is left intact. The responses are intensified when belowground competition is removed by trenching (Table 4.1), reflecting increases in available soil nitrogen and moisture regardless of overstory basal area.

The Role of Fire in Mediating Competition

Given the importance of fire in the longleaf pine ecosystem (Greene 1931; Wahlenberg 1946; Waldrop et al. 1992; Landers et al. 1995; Gilliam and Platt 1999; Ryan et al. 2013), the large role that fire also plays in mediating competitive effects between longleaf pine seedlings and other vegetation

TABLE 4.1
Average Biomass (Grams) of Planted Longleaf Pine Seedlings Three Growing Seasons after Trenching and Ground Cover Removal: Experimental Treatments in 75- to 90-Year-Old, Naturally Regenerated Longleaf Pine Forests in Baker County, Georgia

	Not Trenched		Trenched	
Location	**Ground Cover Intact**	**Ground Cover Removed**	**Ground Cover Intact**	**Ground Cover Removed**
Under canopy	7.3	15.0	6.5	75.5
Gap edge	15.8	27.9	10.5	273.7
Gap center	16.4	97.7	27.2	460.4

Source: Recalculated from data in Pecot, S. D. et al., *Canadian Journal of Forest Research*, 37, 634–648, 2007; © 2008 Canadian Science Publishing or its licensors. Reproduced with permission.

is not surprising. Longleaf pine seedlings are not fireproof but they can survive low- to moderate-intensity fires quite well, with two exceptions: in the period after germination and before reaching a root collar diameter of 1 cm (Pessin 1934; Boyer 1990); and when height growth begins as seedlings emerge from the grass stage, exposing succulent shoots and buds that are vulnerable to lethal temperatures (Grelen 1983). Grass-stage longleaf pine seedlings are generally tolerant of fire, but the cumulative effects of competition and repeated fires can lead to seedling mortality.

Jack et al. (2010) examined these cumulative relationships by locating patches of naturally regenerated longleaf pine seedlings in wiregrass (*Aristida stricta*) dominated ground cover. In each patch, one of three fuel treatments was applied: (1) ambient pine litter (including only the amount naturally added by needle cast), (2) no pine litter (raking needles from the plots), and (3) twice the ambient pine litter (applying additional needles from the raked plots). The amount of fuel loading had a significant effect on seedling survival; in all treatments, however, mortality was concentrated on the smallest seedlings (Figure 4.4). Thus, the seedlings whose growth is most affected by competition are more likely to be killed by fire: although fire is the proximate cause of mortality, the contribution of competition within and between species is also decidedly important. These results confirm much earlier studies that were characterized by their operational focus on establishment of longleaf pine regeneration (Bruce and Bickford 1950; Bruce 1951) as well as more recent analyses of the seedling sizes that are most vulnerable to mortality from fire (Provencher, Herring, et al. 2001; Haywood 2002). Although fire risk is high when seedlings begin to grow taller, fire tolerance generally returns after the bud reaches a height above approximately 1 m (Pessin 1934; Maple 1975).

Because fuel loading affects seedling survival even in low-intensity fire, some argue that the lower survival of longleaf pine seedlings underneath the crowns of overstory trees can be partly attributed to the higher loading of fine fuels from needle cast (Croker and Boyer 1975; Platt and Rathbun 1993; Brockway and Outcalt 1998). Higher levels of pine fuel are known to occur around the base of individual overstory longleaf pine trees (Varner et al. 2009; O'Brien et al. 2010), and needle cast additions are largest beneath a canopy with low gap fraction (Brockway and Outcalt 1998; O'Brien et al. 2008). Although higher seedling mortality under an intact canopy has been documented in many studies (Croker and Boyer 1975), other studies concluded that seedlings can survive periodic fire with overstory-stocking densities ≤14 m^2/ha (Maple 1969). The interaction between increased fuel loading beneath the canopy and competitive relationships that reduce seedling growth and survival is likely strong (Grace and Platt 1995a; Avery et al. 2004). However, no direct relationship has been conclusively established because of the different seedling stages that were considered in the various studies that have been conducted (Avery et al. 2004), and longleaf pine seedlings can certainly be found surviving near mature trees (Pecot et al. 2007).

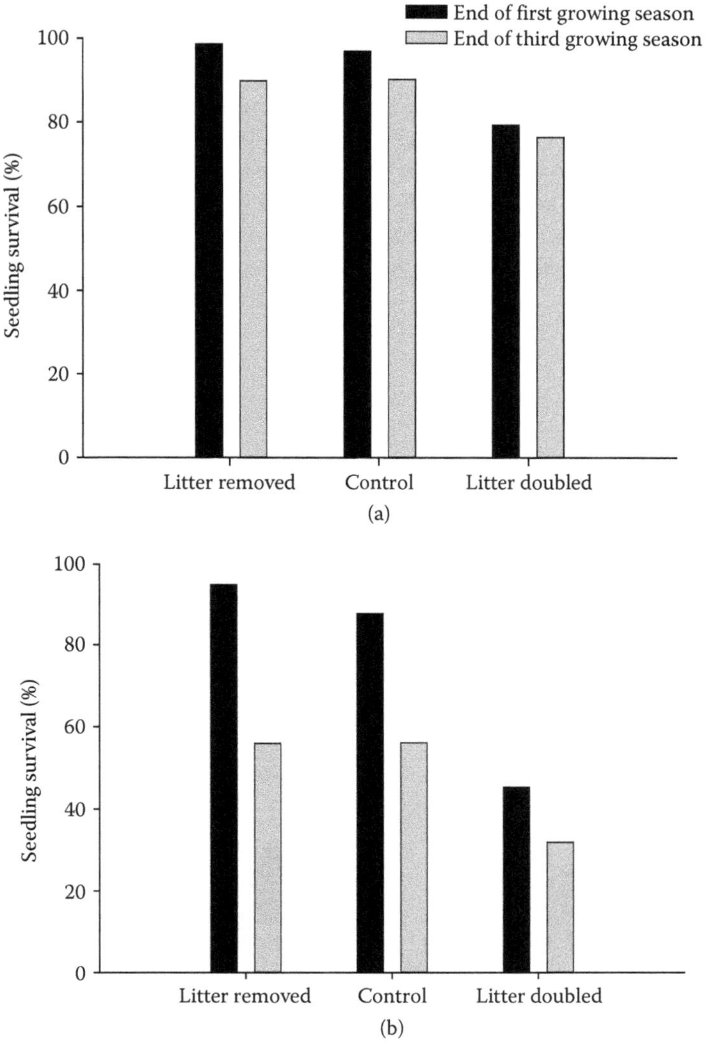

FIGURE 4.4 Effects of fuel-loading treatments on seedling survival one and three growing seasons after prescribed fire application for: (a) All seedling size classes—<0.2 m, 0.2–1.0 m, 1.0–2.0 m tall; and (b) only seedlings from the smallest size class—<0.2 m tall. (Redrawn from Jack, S. B. et al., *Fuel Loading and Fire Intensity—Effects on Longleaf Pine Seedling Survival*, USDA Forest Service, Southern Research Station, Asheville, NC, 2010.)

The influence of fire on hardwood density and stem size is well known (Wahlenberg 1946; Williamson and Black 1981; Waldrop et al. 1992; Glitzenstein et al. 1995; Ware et al. 1993; Brockway and Outcalt 2000; Kirkman, Coffey, et al. 2004; O'Brien et al. 2008; Mitchell, Hiers, et al. 2009; Haywood 2011). As discussed earlier in this chapter, the evidence that hardwood stems compete with longleaf pine seedlings for available resources—thereby affecting survival and growth—is also strong (Pecot et al. 2007). This means that fostering longleaf pine regeneration requires the use of frequent prescribed fire—with a fire-return interval of 1–4 years depending on site quality and local conditions—to maintain hardwood stems in a shrub state (Jacqmain et al. 1999). The exception to this general recommendation would be on extremely xeric sites where some level of hardwood canopy can decrease moisture stress and increase longleaf pine seedling survival (Loudermilk et al. 2016).

In the absence of fire, hardwood stems can rapidly grow in height from established root systems, quickly achieving midstory status (Williamson and Black 1981; Provencher, Herring, et al. 2001) and suppressing longleaf pine regeneration. Once they reach this size, hardwood stems can often tolerate fire intensities that are typical of prescribed fire regimes, and their leaves are also less flammable than pine needles (Pecot et al. 2007; O'Brien et al. 2008; Hiers et al. 2014), which can hinder the application of prescribed fire. Additional management actions, such as mechanical and chemical treatments, are frequently applied to reduce hardwood midstories (see Chapters 10 and 11), but fire is still the key to restoring conditions that are most suitable for longleaf pine regeneration and maintenance of a diverse ground cover community (Outcalt and Brockway 2010).

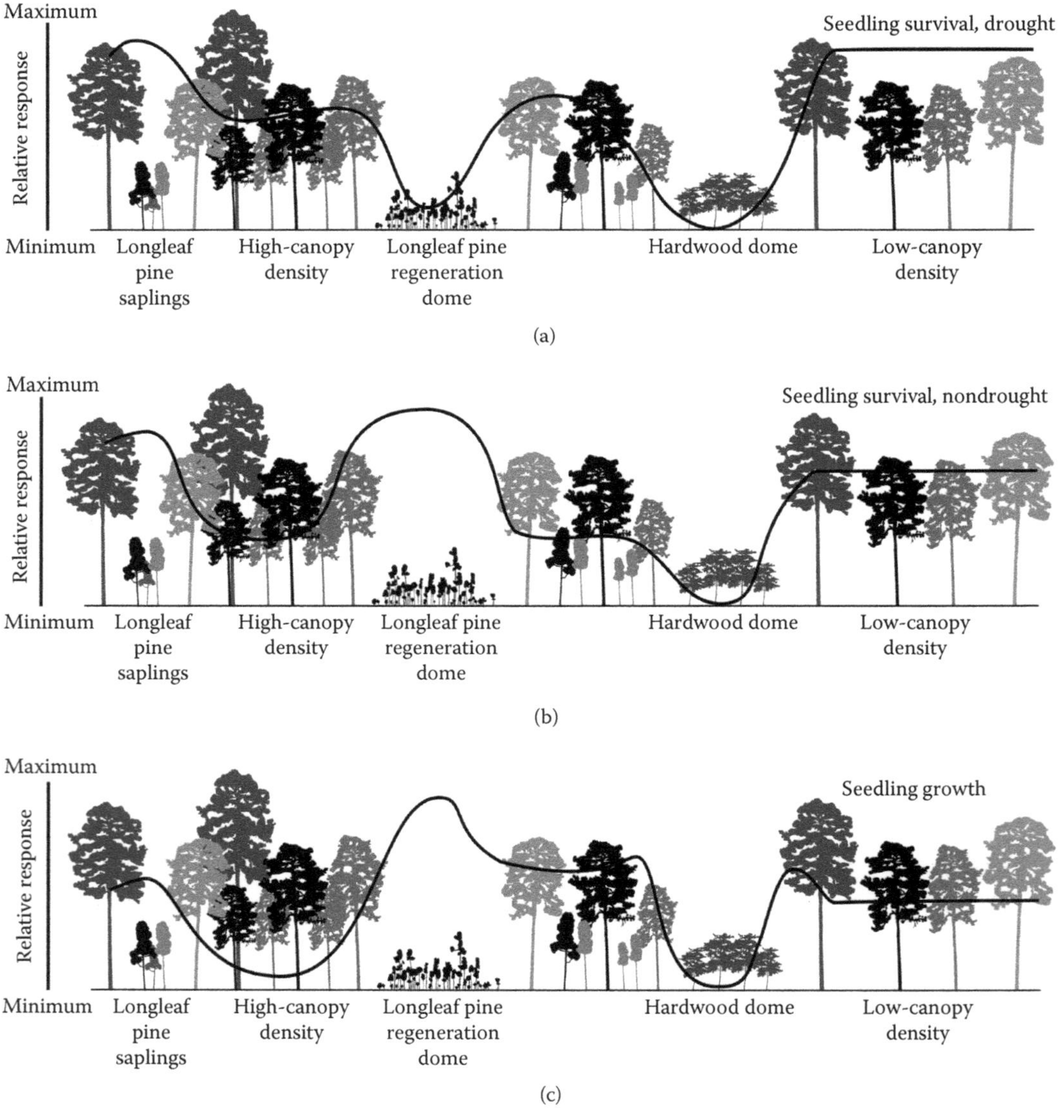

FIGURE 4.5 Stylized relationships for overstory, midstory, and ground cover competition effects on longleaf pine seedling: (a) Survival under drought conditions, (b) survival in nondrought conditions, (c) and growth. The relationships shown represent general responses to illustrate contrasting patterns and are not intended to be predictive.

Cumulative Effects on Survival and Growth

Over time, aboveground and belowground competition results in mixed patterns of longleaf pine seedling survival and growth in response to overstory density (Figure 4.5). As the size of canopy gaps increases, three important changes occur in ground cover and midstory plant communities that can affect seedling response. First, aboveground and belowground competition from the overstory are reduced, with increased light availability and decreased root competition (Jones et al. 2003). Second, although other woody plants also respond positively to increased light availability, they are more responsive than pines to reduction in belowground competition: roots of woody plants in the midstory and understory strata fill the belowground void created by the absence of the overstory trees, capturing nitrogen and other soil resources (see Chapter 7). Third, herbaceous ground cover biomass increases with increased light availability and soil nitrogen (Pecot et al. 2007). This sequence illustrates that the belowground "root gaps" can have as great an influence on longleaf pine seedlings as the aboveground canopy gaps.

This scenario demonstrates how the overstory and ground cover become uncoupled belowground in longleaf pine forests. It also provides significant credence to the applicability of Walter's two-layer hypothesis (Walter 1939) in natural longleaf pine ecosystems. A soil niche concept that has been used to explain the codominance of tree and grass species in savannas around the world (Walter 1971; Firbank and Watkinson 1985; Casper and Jackson 1997; Ward et al. 2013), the two-layer hypothesis proposes that the different rooting patterns of tree and grass species ensure that the two plant types get soil resources from different depths in the soil profile (see Chapter 7), and thus do not directly compete for the same pools of nutrients, water, and other soil resources. This phenomenon, along with their different physiological responses to environmental stimuli, allows the two groups to coexist in habitats where soil resources are periodically in limited supply.

The combination of resource compartmentalization and the associated differences in competitive interactions with varying abiotic conditions helps to explain the divergent survival and growth results reported for longleaf pine seedlings (Figure 4.5). In periods of drought or other stressful conditions, seedling survival generally decreases from the intact canopy into the center of a canopy gap, a response that can be attributed to facilitation by the overstory, the ground cover, or both (McGuire et al. 2001; Pecot et al. 2007; Knapp et al. 2015). Although counter to the published body of work about longleaf pine, seedling survival in these instances of facilitation would suggest that longleaf pine seedlings have some characteristics associated with shade tolerance, drought avoidance, or both. However, decreased overstory abundance is also often accompanied by increased abundance and size of midstory hardwood stems as well as nonwoody ground cover biomass; when droughts end and growing conditions improve, the plants in these strata can quickly shift to being strong competitors for available resources (Wright et al. 2015) and cause seedling survival to decrease within the gaps. Unlike seedling survival processes, the relationship between longleaf pine seedling growth and canopy density is straightforward, with growth increasing toward the gap centers and away from the intact matrix. This result is consistent with the large body of work that led to even-aged silvicultural recommendations for longleaf pine: seedling growth aligns with characteristics attributed to shade intolerance, even though seedling survival exhibits some responses that are characteristic of shade-tolerant species.

The relationships that link aboveground and belowground competition, fire, and gap size are also strong. In large gaps (usually >0.2 ha) where advanced longleaf pine regeneration is absent, a significant reduction in belowground competition will likely allow midstory hardwoods to dominate the site, with consequent decreases in pine fuel sources and fire intensity (Jack, Mitchell, et al. 2006). Ultimately, the proliferation of hardwood stems produces a "fire shadow" in which midstory hardwoods capture the gap and prevent the movement of a uniform fire front across the surrounding forest matrix (Mitchell et al. 2006; Mitchell, Hiers, et al. 2009). The general pattern of hardwood competitors dominating canopy gaps has been shown in multiple studies; however, differences among study sites (for example, uplands versus flatwoods) and their associated vegetation could lead to more complexities than those described here (B. Knapp et al. 2013; Brockway and Outcalt 2015).

These discussions show how the interactions among light, soil nutrients, and fire intensity can impact longleaf pine seedlings, both in their grass stage and as they begin to grow taller. However, seedling height growth is never uniform, reflecting the impacts of the long-term, complex interactions described above. Kirkman and Mitchell (2006) developed a three-level regeneration model to explain how survival and growth are affected by overstory stocking (using canopy gap fraction as a surrogate for basal area) and how this effect changes through time. In Level one (Figure 4.6), when gap fraction is <35%, seedling mortality after multiple years can reach 100% due to increased

(a)

(b)

FIGURE 4.6 Examples of two different longleaf pine stocking levels at Ichauway in southwest Georgia: (a) High stocking in which the canopy represents gap fractions of <35% and light reaching the ground cover is restricted, and (b) low stocking in which the canopy represents gap fractions of >65% and light at the ground level is not limited. (Photographs courtesy of Richard T. Bryant.)

TABLE 4.2
Distribution (by Percentage) of Total Forest Area into Each of Three Canopy-Coverage Classes, Indicating the Amount of Canopy Openness for Four Harvesting Treatments in 75 to 90-Year-Old, Naturally Regenerated Longleaf Pine Forests in Baker County, Georgia

	Percent of Total Area			
Canopy-Coverage Class (gap fraction)	Uncut Control	Single-Tree Selection	Small Group Selection	Large Group Selection
Level one (<35%)	3	0	2	0
Level two (35%–65%)	82	63	32	40
Level three (>65%)	15	37	66	60

These data illustrate how harvest selection rules affect the distribution of overstory canopy density, which influences the establishment of seedlings and the ability to conduct prescribed fire.

competition from the overstory and increased fire intensity. In Level two (about 35%–65% gap fraction), fire intensity decreases and light availability increases, with consequent improvements in seedling survival and biomass accretion (although not necessarily accompanied by the onset of height growth). Canopy conditions of Level two are also where facilitative effects of the overstory on seedling survival can be most apparent during drought conditions. At Level three (>65% gap fraction), the potential for seedling size and height growth would be highest (Figure 4.6), with reduced needle cast and increased hardwood midstory abundance reducing fire intensity to minimal levels; facilitation would not occur because the canopy is too open to ameliorate adverse conditions for the longleaf pine seedlings.

These cumulative effects provide important guidance for silvicultural manipulations of the canopy to promote successful establishment and maintenance of natural longleaf pine regeneration and provide fuels for prescribed fire. Table 4.2 shows the distribution of total area into each of the three canopy-coverage levels for sites that were harvested using single-tree selection, small and large group selection, and an uncut control area (Palik et al. 2003). The two group selection methods both have at least 60% of the area in Level three (gap fraction >65%); this level of canopy cover would produce the largest growth increase in longleaf pine seedlings, however, their survival would likely be lower during times of drought and the low canopy cover would reduce fuel levels, making prescribed fire more difficult to apply. In contrast, the single-tree selection method has >60% of the area in Level two. This level of canopy cover would provide adequate fuels for burning; longleaf pine seedling growth would be reduced, however, and facilitation could increase seedling survival during times of drought stress. These examples illustrate the balance that must be struck to manage the many competitive interactions that affect longleaf pine seedlings when implementing canopy manipulations to meet long-term regeneration goals.

BEYOND THE GRASS STAGE: SAPLING DYNAMICS

Compared to the accumulated data regarding seedling competition, survival, and growth, relatively little is known about similar dynamics in longleaf pine saplings (the stems that have grown out of the grass stage to a height >2 m but are not large or tall enough to be included in the canopy stratum), particularly for natural stands and localized cohorts in multiaged forests. Observations and empirical evidence from Platt, Evans, and Rathbun et al. (1988) and other studies, primarily at the stand level, indicate that competitive processes are active in the recruitment of saplings into overstory canopy positions. These processes, however, act over long time scales and have not been investigated at the individual tree level.

Some insights into sapling dynamics can be found from U.S. Forest Service studies that followed the development of natural longleaf pine regeneration in even-aged silvicultural systems. Boyer (1993) reported results for the development of longleaf pine regeneration under a range of residual basal area levels that were established to compare seed tree and shelterwood regeneration systems. Most of the treatment plots had little or no longleaf pine regeneration when the study was implemented. Thirty-four years after initial harvesting, Boyer (1993) found that the low and medium residual basal area plots had "reasonable" (≥4.5 m^2/ha) amounts of ingrowth from trees regenerated after harvesting, but the high residual basal area plots had very little ingrowth. The residual overstory on all plots, regardless of residual stocking levels, had reduced growth in the younger trees, and the diameter distributions for the two-aged stands were more similar to those expected for an uneven-aged stand (in that they exhibited a reverse-J-shaped curve). Similar results were reported by others (Farrar 1985; Croker 1990) for naturally regenerated longleaf pine saplings, strongly suggesting that—like seedlings—sapling growth is slowed by the presence of canopy trees.

Additional information is also available from studies that considered evenly spaced, plantation-grown longleaf pine. Boyer (1983) developed height-over-age curves for young longleaf pine plantations and examined growth for plots that had been established at a range of planting densities and with different site preparation treatments. The height-age curves varied significantly with different planting conditions (site preparation treatment) and initial densities; he attributed these height development patterns to different levels of competition. Haywood (2009, 2015) reported similar results for longleaf pine plantations, where treatments included herbicide applications for release and repeated prescribed fire in different seasons; the reduced competition resulting from herbicide treatments led to increased sapling growth (height and volume). Finally, Harrington (2011), Brooks and Jack (2016), and many others reported thinning responses for longleaf pine plantations, with increased growth of residual trees following density reduction. This result is a clear indication that longleaf pine trees in even-aged stands compete with one another for available resources.

The results of studies on plantations cannot be directly transferred to "patches" or clusters of sapling-sized, natural longleaf pine regeneration in multiaged stands. However, individual saplings clearly compete with each other, especially given the high stem densities that frequently occur in regeneration patches (Wahlenberg 1946; Boyer 1990). Their competitive relationships are complicated by the rate of emergence from the grass stage—individual trees emerge at different times, even within a single age cohort, resulting in a wide distribution of tree heights for many years. This differentiation of heights and crown classes carries over into the sapling stage as the trees continue to grow. The end result is that the dense patches of longleaf pine saplings (sometimes in localized densities equivalent to 7,000–12,000 stems/ha) rarely stagnate as occurs with other southern pine species (Boyer 1990; Croker 1990). The size differentiation of longleaf pine saplings, especially in conjunction with periodic fires, causes gradual mortality of smaller and less vigorous individuals over time.

We still have much to learn about the dynamics of longleaf pine trees that are beyond the germinant and grass stage and have initiated height growth. A better understanding of the interactions of sapling-sized trees, both with larger canopy trees and within cohort clusters, would likely improve the management of naturally regenerated multiaged longleaf pine forests and would help to refine prediction models for stand development and growth and yield. This is a fruitful area for future research.

STAND-LEVEL REGENERATION DYNAMICS

Until now, our focus has been on the responses of individual seedlings and saplings to competition and environmental factors, and generally at a fine scale (not more than a few meters around individual trees). Cumulatively these individual responses determine the stand-level regeneration dynamics that are important for management and restoration.

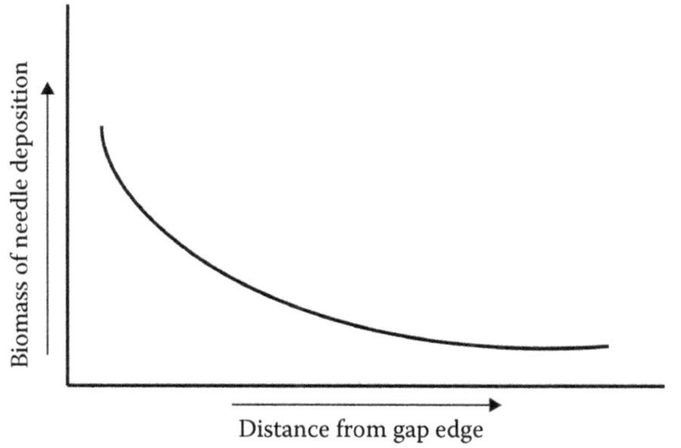

FIGURE 4.7 Schematic showing the relationship between decreasing needle deposition and the distance from the canopy edge to the center of a canopy gap.

When and where viable seed is produced are factors that also influence regeneration dynamics at the stand scale. Longleaf pine is a masting seed producer, with large seed crops produced only every 4–10 years (Platt, Evans, and Rathbun et al. 1988; Boyer 1990; Pederson et al. 2000; Haymes and Fox 2012). Although very large seed crops are produced periodically, some localized seed production also occurs each year, gradually providing a more continuous accumulation of regeneration cohorts throughout the forest (Platt, Evans, and Rathbun 1988; Landers et al. 1990). Especially after large seed crops, germinants can be found somewhat continuously across the forest floor, wherever good seed bed conditions are available. Given the previously described responses to competition and fire, however, the regeneration cohorts tend to accumulate in canopy gaps (Platt, Evans, and Rathbun 1988; Platt and Rathbun 1993; McGuire et al. 2001; Kirkman, Mitchell, et al. 2007) where more light and belowground resources are available. Over time, seedling densities within the cohort patches are reduced by competitive processes and the effects of fire. Because this self-thinning process is slow, attrition of the least vigorous individuals is gradual and high densities are often maintained for many years.

A forest sustained with natural regeneration in small cohort patches tends toward a multiaged condition with a mosaic of different cohorts—each with trees of varying sizes—unless a large disturbance, a management action, or fire exclusion disrupts these gradual stand dynamics. The multiaged, natural longleaf pine forest is thus maintained by a type of gap or patch dynamics (Pickett and White 1985; White et al. 1985; Yamamoto 2000) in which the periodicity of fire, variable seed production, and fine-scale canopy disturbances are tightly linked.

The balance between canopy gap size and its suitability for long-term regeneration success is delicate. Longleaf pine seeds are large and do not disperse far from their parent tree, typically 1–1.5 times the height of the tree (Boyer 1990). Thus, if a gap is too large, seed may fail to reach the central portion of the gap, and no regeneration will be captured. Likewise, needles shed from the trees at the gap edge are not dispersed very far into the gap (Figure 4.7), which can disrupt fire continuity within gaps (O'Brien et al. 2008). Because longleaf pine seedlings are susceptible to competition from other woody species, gaps that are too large (for example, >0.2 ha or >65% gap fraction) often do not have successful longleaf pine regeneration and may instead become fire-resistant hardwood domes that can gradually expand beyond the original gap (Jack, Mitchell, et al. 2006; Mitchell, Hiers, et al. 2009). Conversely, gaps that are too small (for example, <0.1 ha or <35% gap fraction) will have excessive competition and needle cast from the overstory longleaf pine; in this situation, the increased fire intensity and competition can result in seedling mortality. Seedlings that do survive in small gaps can persist for several years but will not begin height growth until a disturbance enlarges the gap (Boyer 1963; Brockway and Outcalt 1998; Brockway et al. 2006; Mitchell et al. 2006).

SUMMARY

Clearly, a tremendous amount of knowledge has been acquired about regeneration dynamics in natural longleaf pine forests. However, the establishment of regeneration in longleaf pine forests is never a "one-size-fits-all" endeavor. On the contrary, overstory gap sizes created through management need to reflect both the size distribution seen in natural longleaf pine forests and the state of the ground cover and midstory plant communities. In addition, regeneration dynamics almost certainly vary throughout the wide geographic distribution of longleaf pine. Addressing these issues will require further research to better understand differences in regeneration dynamics among site types (such as uplands versus flatwoods), along soil and hydrologic gradients, with evolving silvicultural methods, and in anticipation of predicted climate change scenarios. If anything has been learned, it is the need to use a range of canopy openings that have been selected in light of the individual site characteristics and the landowner's specific management and restoration objectives.

The divergence in seedling-survival results throughout the literature underscores the complexity of working with this species. That survival is inextricably linked to immediate growing conditions, whether mediated by resource availability or climate, is likely no surprise to any biologist or forester. However, a heretofore "shade-intolerant" plant exhibiting attributes of shade tolerance, at least for seedling survival, may warrant a reconsideration of how tolerance is defined for a particular species. Further studies can help to clarify definitions of tolerance for longleaf pine and other similar species that exhibit mixed responses to competition.

The more recent understanding of the competitive relationships that characterize longleaf pine seedling establishment and the translation of this new knowledge to the scale of forest stands has broadened our thinking about appropriate silvicultural systems for managing the species using natural regeneration, and specifically the application of uneven-aged silviculture (see Chapter 10). What, then, is "*se betsta læcedom*" (the best medicine) to use when creating canopy openings and regenerating these forests for restoration and long-term conservation goals? As is true for all silvicultural decisions, owner or manager objectives will help determine the appropriate choice of treatments. The silviculture of southern pines in the 20th century largely focused on wood and fiber production with a minor focus on other objectives and associated attributes (such as wildlife species, turpentine production, and soil erosion control). When the emphasis was on timber products, minimizing competitive interactions as much as possible and tracking the subsequent effects on seedling survival and growth were high priorities. In the 21st century, however, objectives are multifaceted and timber production is not always the forest manager's primary focus; within this new framework, managers can use competitive interactions to help achieve desired conditions rather than simply attempting to control most, if not all, competitive interactions that affect longleaf pine regeneration. The research presented in this chapter—combined with the documentation of applied knowledge from the small (but growing) cadre of practitioners managing natural longleaf pine forests—has the potential to establish a framework for successful, ecologically focused restoration and management of longleaf pine ecosystems.

ACKNOWLEDGMENTS

We thank Ben Knapp, Brian Palik, Seth Bigelow, Lindsay Boring, David Dyson, and Keville Larson for reviews of an early draft of this chapter. Jessica McCorvey and Micheal Simmons provided assistance with photographs and graphics. We also thank the many Joseph W. Jones Ecological Research Center colleagues, lab employees, and graduate students whose work provided the basis for this review. In particular we offer gratitude to Robert Mitchell (deceased), who informed and challenged our thinking on regeneration dynamics through countless discussions (and sometime arguments) and collaborations.

5 Mechanistic Controls of Community Assembly and Biodiversity

L. Katherine Kirkman and Jonathan A. Myers

CONTENTS

INTRODUCTION

Patterns of Diversity at Local and Regional Scales

The deceptively uniform bilayer structure of longleaf pine (*Pinus palustris*) ecosystems, featuring forests, woodlands, and savannas of canopy longleaf pines and grass-dominated ground cover, belies the exceptional floristic diversity associated with the herbaceous ground cover. Most notably, this ground cover harbors the highest levels of plant species richness in North America and globally rivals other species-rich herbaceous-dominated plant communities (Walker and Peet 1983; Peet and Allard 1993). Not only does high species richness occur at local scales (Table 5.1), but regionally, the species-rich ground cover is also characterized by many species endemic to the southeastern Coastal Plain. In addition, the entire North American Coastal Plain, which encompasses the former range of the longleaf pine, has been proposed as a global biodiversity hotspot (Noss et al. 2015).

Of the 1630 total vascular plant species endemic to the Coastal Plain Floristic Province identified by Sorrie and Weakley (2001), 1000 species are obligate associates of the longleaf pine ecosystem and many of these endemics are restricted to quite narrow subregions. Furthermore, fire-maintained longleaf pine communities provide habitat for numerous rare plant (Hardin and White 1989; Drew et al. 1998; Sorrie and Weakley 2006) and animal species (Earley 2004; Noss et al. 2015).

TABLE 5.1

Mean (and Maximum) Values for Species Richness (Number of Species) per Area in Longleaf Pine Communities of Florida (FL), Georgia (GA), Louisiana (LA), and North Carolina (NC)

Soil Moisture Regime	Community Type	State	Scale (m²) 0.01	0.10	1.00	10.00	100.00	1000.00	References
Wet	Wet savannas	NC	–	–	32.9	–	–	–	Walker and Peet (1983)[a]
			4.4	11.3	22.4	36.0	61.1	94.8	Noss (2013)
		FL	3.1	9.2	18.7 (46)	30.4	54.5	89.8	Noss (2013) and references therein
	Wet-mesic to wet flatwoods	NC	2.3	6.0	11.2	18.7	33.2	54.6	Noss (2013) and references therein
		FL	1.8	5.2	11.7	21.4	40.0	71.2	Noss (2013) and references therein
Mesic	Mesic savannas	NC	–	–	30.5 (50)	–	–	–	Walker and Peet (1983)[a]
		LA	1.9 (4)	–	–	–	–	–	Myers and Harms (2009a)
			–	–	30.8 (50)	–	–	–	Myers (unpublished data)
			–	–	–	–	103.0 (130)	–	Platt et al. (2006)
	Mesic to wet-mesic uplands	NC	2.7	7.5	15.7	27.1	51.7	81.4	Noss (2013) and references therein
		GA		7.9	18.7	33.0	57.9	–	Kirkman et al. (2016)
		FL	2.1	7.1	17.1	32.8	63.7	107.5	Noss (2013) and references therein
Dry	Dry savannas	NC	–	–	23.3	–	–	–	Walker and Peet (1983)[a]
	Sandhills	FL	–	–	18.2 (33)	–	–	–	K. Harms, S. Gagnon, and J. Myers (unpublished data)
		NC	0.5	1.3	3.2	6.6	12.9	22.5	Noss (2013) and references therein
		FL	1.1	3.2	9.5	20.9	42.6	74.1	Noss (2013) and references therein
	Subxeric uplands	NC	0.8	2.7	5.5	10.3	19.1	34.8	Noss (2013) and references therein
		GA	–	5.5	14.1	26.6	49.0	–	Kirkman et al. 2016
		FL	1.1	3.9	11.2	24.4	48.1	84.0	Noss (2013) and references therein

Source: Modified from Noss, R. F., *Forgotten Grasslands of the South: Natural History and Conservation*, Island Press, Washington, DC, 2013.

[a] Mean values of sites with annual and infrequent fire regimes.

Across the southeastern Coastal Plain, large species pools occur not only within individual sites, but they also contribute to a highly diverse regional species pool (Peet 2006). Based on region-wide comparisons of vegetation, Peet et al. (2014) describe a latitudinal increase in plant species richness from the Carolinas southward to the Florida Panhandle—a trend that is particularly apparent in subxeric and sandhill communities. They attributed this trend to a longer growing season and higher average mean temperature of more southerly areas, as well as the biogeographic history of the East Gulf Coastal Plain, which served as species refugia during periods of glaciation (Palmquist et al. 2014; Peet et al. 2014). Further, Carr et al. (2009) described a decrease in species richness of pine woodlands from the Florida Panhandle into Peninsular Florida despite similar soil moisture and fertility conditions. This pattern was probably caused by different biogeographic histories related to the timing of sediment deposition and sea level fluctuations and their effects on species pools (Myers 1990; Carr et al. 2009).

Purpose and Organization of this Chapter

Understanding the complex feedback mechanisms that create and maintain species diversity within the longleaf pine ecosystem is foundational to guiding conservation strategies for promoting biodiversity, particularly those strategies that guide reassembly and recovery processes during restoration of such species-rich communities. In this chapter we summarize the mechanisms that maintain high plant species diversity in the system; we review the ecological theories that seek to explain community assembly and patterns of diversity at different spatial scales in species-rich plant communities; and we examine the empirical evidence addressing these theories. In particular, we focus on field experiments where ecological processes have been manipulated in ways that help identify how plant community assembly occurs in natural or restored longleaf pine sites. Finally, we summarize what is known and not known about community assembly and diversity, and identify research gaps that would help to inform restoration approaches.

THEORETICAL FRAMEWORK FOR UNDERSTANDING COMMUNITY ASSEMBLY AND SPECIES DIVERSITY IN SPECIES-RICH COMMUNITIES

The Continuum from Niche-Assembly to Dispersal-Assembly Theories

Patterns of diversity reflect the interplay of processes at local and regional scales (Ricklefs 1987; Harrison and Cornell 2008). At local scales, the interactions among species and the responses of species to environmental conditions underlie classical coexistence and diversity models (Connell 1978; Tilman and Pacala 1993), which have been built upon extensively since their inception (Chesson 2000; Chase and Leibold 2003; Adler et al. 2007). Under this paradigm, species diversity patterns predominantly reflect a deterministic outcome of species responses to environmental conditions and species interactions, including interspecific competition, predation, and facilitation. In contrast, regional perspectives on community assembly emphasize the importance of the available species pool in determining the diversity of local communities (MacArthur and Wilson 1967; Ricklefs 1987). Under this perspective, patterns of local diversity are primarily determined by the size of the regional species pool, the colonization history, and the rate of immigration into communities. The processes that influence the size of this pool include those that operate at broader spatial and temporal scales, such as climate, dispersal, speciation, and extinction (Ricklefs 1987; Palmquist et al. 2014), as well as processes that operate at finer scales within and among local communities (Mittelbach and Schemske 2015).

Local- and regional-scale perspectives on community assembly can be organized along a continuum bounded at one extreme by niche-assembly theory and at the other extreme by dispersal-assembly theory (Hubbell 2001; Bell 2005; Gravel et al. 2006) (Figure 5.1), each embodying a large

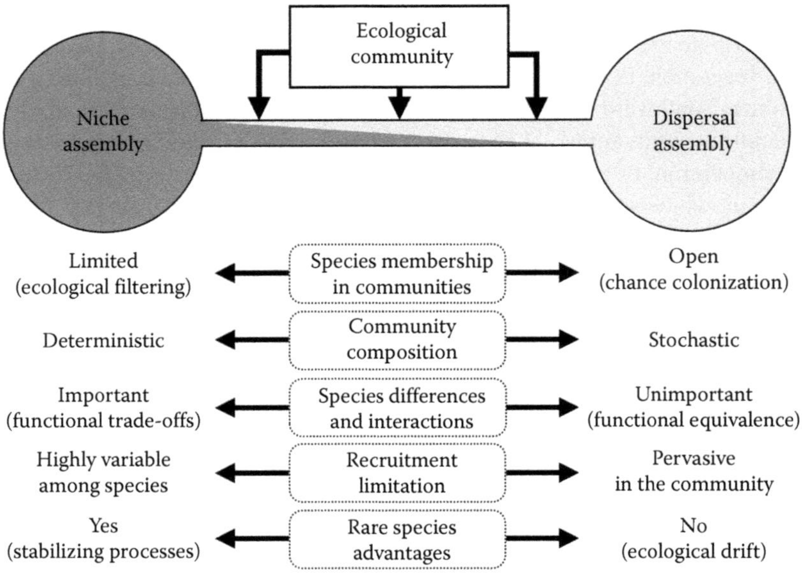

FIGURE 5.1 Schematic illustrating that community-assembly theory can be organized along a continuum, bounded at one end by niche-assembly theory and at the other end by dispersal-assembly theory. Deterministic community composition refers to membership in local communities that is nonrandom with respect to species identity, whereas stochastic community composition refers to species membership that is a random assemblage. An ecological community can be envisaged as falling anywhere along this continuum in space or time, depending on the presence and relative importance of mechanisms involved in community assembly. Note that not all of these mechanisms need to operate simultaneously at any given point on the continuum. (From Myers, J.A. 2010. Ecological assembly of high-diversity plant communities: dispersal, competition, and environmental filtering in longleaf pine savannas, PhD Dissertation. Louisiana State University, Baton Rouge, Louisiana.)

family of more detailed models and hypotheses (Chase and Leibold 2003; McGill et al. 2006; Chase and Myers 2011). These theories provide general—but contrasting—predictions about the principal mechanisms that control community assembly. Here, we focus on predictions based on these theories regarding local species richness (α-diversity) and spatial variation in community composition (β-diversity) (Anderson et al. 2011).

Patterns of Species Diversity in Dispersal- and Niche-Assembled Communities

According to dispersal-assembly theory, which was popularized in large part by Hubbell (2001), chance colonization, rates of immigration from the species pool, pervasive recruitment limitation, and/or stochastic births and deaths are the primary ecological mechanisms that influence community assembly (MacArthur and Wilson 1967; Sale 1977; van der Maarel and Sykes 1993; Hubbell 2001; Bell 2005). Hubbell's Unified Neutral Theory of Biodiversity and Biogeography (2001) represents an extreme version of the dispersal-assembly perspective because it assumes that all species within guilds are equivalent with respect to per capita rates of birth, death, and colonization; thus their abundances change randomly through time (ecological drift). Accordingly, local diversity is controlled by the size of the species pool (or metacommunity), the rate of immigration, and the degree of dispersal limitation within local communities.

Dispersal-assembly theory makes at least three predictions about species richness and variations in community composition across sites. First, local communities should have relatively "open-membership" assemblages—meaning that they are open to invasion by most if not all species in the regional species pool. Consequently, species richness should be positively related to the rate of immigration (dispersal) from the species pool (Chase et al. 2005). Second, the theory predicts that local communities with similar environmental conditions should have dissimilar species compositions (high β-diversity), reflecting dispersal limitation, historical contingencies resulting from variation in the arrival order of species into communities (priority effects), or stochastic colonization (Chase 2003; Fukami 2015). Third, in the extreme case of neutral models, where all individuals are assumed to be ecologically equivalent, β-diversity should not change systematically along environmental gradients (Chase and Myers 2011).

Niche-assembly theory, in contrast, proposes that membership in local communities is controlled by biotic and abiotic conditions that deterministically limit local species membership (Chesson 2000; Chase and Leibold 2003; Chase and Myers 2011) (Figure 5.1). In this case, local diversity is determined by species tolerances to local environmental conditions and the outcome of local interactions among species (Keddy 1992; Diaz et al. 1998). This theory makes at least three general predictions. First, local communities should be "limited-membership" assemblages in which α-diversity either saturates or declines as the rate of immigration increases. A decline in α-diversity is predicted when immigration introduces dominant competitors or predators into local communities (Mouquet and Loreau 2003), or when abiotic conditions impose environmental filters on species traits (Weiher et al. 2011). Second, local communities with similar environmental conditions are predicted to have similar species composition (low β-diversity) owing to deterministic filtering of the species pool (Chase 2003; Chase and Myers 2011). Third, β-diversity should change systematically along environmental gradients (Chase and Myers 2011).

Recruitment limitation is a key ingredient in both niche- and dispersal-assembly theory, but its mechanistic role differs between the two theories. Recruitment limitation can result from both the failure of propagules to arrive (for example, "seed limitation") and unsuitable conditions for establishment ("microsite limitation") (Nathan and Muller-Landau 2000). Seed limitation can be further decomposed into "dispersal limitation" and "source" or "fecundity limitation" (Clark et al. 1998). In dispersal-assembly theory, recruitment limitation is often assumed to be pervasive within communities. A fundamental consequence of community-wide recruitment limitation is the establishment of inferior competitors in sites where dominant competitors have failed to colonize, thereby reducing the rate of competitive exclusion (Hurtt and Pacala 1995). In niche-assembly models, recruitment limitation is often nonrandom with respect to species identity (deterministic). A classic example is the competition-colonization trade-off model, whereby populations of competitively dominant species have lower dispersal than populations of less competitive species (Tilman 1994). Evidence of recruitment limitation alone therefore cannot be used to disentangle the relative importance of niche-assembly and dispersal-assembly mechanisms (Adler et al. 2007; Clark 2009; Vellend et al. 2014), but evidence for recruitment limitation leaves open the possibility that both deterministic and stochastic processes contribute to community structure.

The relative roles of dispersal-assembly mechanisms and niche-assembly mechanisms are hypothesized to vary across disturbance and productivity gradients. Species pools are thought to be most important in limiting colonization and species richness in relatively unproductive habitats where competitive exclusion is slow and recruitment microsites are abundant (Grime 1979; Huston 1979, 2014; Huston and DeAngelis 1994; Pärtel et al. 2000; Zobel et al. 2000). However, on sites with higher productivity, competition can become more important in limiting establishment and species richness, regardless of the size of the species pool (Huston 1979; Foster 2001). Frequent fires and other disturbances can decrease species richness by increasing environmental filtering from the species pool (Pausas and Verdú 2008) or by causing stochastic local extinctions (Myers et al. 2015), but they can also increase species richness by decreasing interspecific competition (Burkle et al. 2015). The effect of disturbance on species richness can also vary with productivity (Huston 1979, 2014; Foster 2001). For example, disturbance can increase species richness in productive habitats by decreasing competition, but can also decrease species richness in unproductive habitats where low resource availability decreases birth rates or increases mortality. Interactions between disturbance and productivity can also scale up to influence patterns of β-diversity among communities (Burkle et al. 2015).

High-diversity plant communities have served a prominent role in theoretical and empirical studies of dispersal and niche assembly. Dispersal assembly is hypothesized to play a particularly important role in high-diversity communities that have many rare species and a large potential species pool. Rarity can contribute to dispersal assembly by increasing demographic fluctuations (Barot 2004); limiting pairwise species interactions, resulting in more diffuse competition (Grubb 1986; Hubbell and Foster 1986; Myers and Harms 2009b); and increasing recruitment limitation, thereby reducing the rate of deterministic competitive exclusion (Hurtt and Pacala 1995). Communities assembled from a large species pool are more likely to have higher variation in community composition (higher β-diversity) than communities assembled from a small species pool simply because of a greater probability of random sampling effects (Kraft et al. 2011; Myers et al. 2013) or a stronger influence of immigration history on community composition (Chase 2003).

EVIDENCE OF COMMUNITY ASSEMBLY MECHANISMS IN THE LONGLEAF PINE ECOSYSTEM

Early botanical descriptions as well as recent empirical studies suggest that three pervasive niche-based factors influence community assembly in the longleaf pine ecosystem: fire, soil moisture, and soil texture (as it controls soil fertility). Historical records describing fire effects on southeastern vegetation in regard to the origin of prairies and scrub barrens date back as early as the 1600s and later in the 1700s. Several references to the longleaf pine forest as a fire-maintained vegetation type were chronicled in the 1800s (Garren 1943, and references therein). Subsequent botanists and ecologists described the prevalence of fire in longleaf pine forests and the encroachment of hardwoods in the absence of fire (Harper 1914; Wells 1928; Wells and Shunk 1928; Chapman 1932; Heyward 1939). Others addressed the effects of soil texture and moisture on dominance patterns in longleaf pine and slash pine (*P. elliottii*) woodlands (Harper 1907, 1914, 1943; Gano 1917; Wells and Shunk 1928, 1931; Garren 1943). Although these studies clearly showed that fire is a structuring agent of native vegetation, they were met by considerable scientific opposition based on a perception of fire as a detrimental process that thwarts natural successional processes (Garren 1943; Vogl 1979).

Few studies in the early to mid-1900s specifically focused on the effects of fire on ground cover vegetation or species richness; rather, most studies of fire and pineland vegetation concentrated on the influence of fire on conifer growth, forage yield, or wildlife management (Greene 1935; Stoddard 1935; Wahlenberg et al. 1939). One study, which compared burned sites to unburned sites, identified a greater abundance and richness of legumes in burned sites (Lemon 1949). Although not specifically addressing species richness, later studies found that more species were present in the ground cover of longleaf pine stands when the time-since-fire decreased, and that groups of ground cover species (referred to as "fire followers") increased in abundance or productivity after fire (Lemon 1967). Vogl (1973) described the exceptionally high species richness of longleaf pine grasslands compared to other North American grasslands, and in a fire-exclusion study, reported that the absence of fire resulted in decreased abundance of grasses.

Evidence for Niche-Assembly Mechanisms

Fire and Environmental Conditions

Walker and Peet (1983) conducted one of the first studies on how species richness responds to fire frequency and soil moisture in longleaf pine communities. They sampled ground cover composition in 21 island-like grass-dominated sites that occurred on mineral soils within a peatland swamp matrix; the canopy ranged from treeless to 150 stems/ha, and site moisture conditions ranged from well-drained longleaf pine-dominated ridges to poorly drained depressions that were not dominated by longleaf pine-wiregrass (*Aristida stricta*). They reported the highest species richness observed in North American plant communities at fine scales (42 species/0.25 m^2, and 84 species/625 m^2). Species richness was highest in frequently burned sites that were near the middle of the soil moisture gradient. They attributed mechanisms of species co-occurrence to the elimination of woody competition and litter by fire and to phenological differences in biomass production by small- and large-stature species. Although their study gradient was not necessarily restricted to longleaf pine-wiregrass-dominated sites and captured a relatively small range of environmental conditions across the distribution of the longleaf pine ecosystem, they described strong variation in community composition across the environmental gradient. They suggested that the influence of fire is less pronounced at the extremes of the moisture gradient, where vegetation was sparser, presumably because the potential for competitive exclusion was lower.

Results consistent with these observations also indicated that environmental conditions interact with fire frequency to influence species richness. For example, compared to burned sites, flatwood sites that had been fire excluded for four decades had lower herbaceous species richness and higher hardwood abundance (Brockway and Lewis 1997). Evidence from several studies in longleaf pine stands indicated that fine-scale changes in environmental conditions along topographic gradients can contribute to changes in species richness, particularly when the changes are coupled with fire (Platt 1999; Glitzenstein et al. 2003; Carr et al. 2009; Palmquist et al. 2015; Kirkman et al. 2016). Glitzenstein et al. (2003) reported that species richness increased with decreasing fire intervals in wet-mesic sites occurring on Ultisols, as well as in longleaf pine flatwoods occurring on Spodosols. In flatwoods, the increase in species richness with greater fire frequency occurred only at finer scales. Also, a decline in species richness at fine scales was reported across a soil moisture gradient in a decade-long fire-exclusion experiment (Kirkman et al. 2016), as well as in longleaf pine sites that had experienced only a modest reduction in fire frequency (Palmquist et al. 2014). Collectively, these examples highlight the roles of fire history, current fire regimes, and various topographic conditions in maintaining species richness at multiple scales (Kirkman, Goebel, et al. 2004; Peet et al. 2014; Palmquist et al. 2015; Kirkman et al. 2016).

The effects of frequent fire on species richness are linked to alterations in forest structure that promote light penetration to the ground cover. Regardless of soil moisture, when the fire-return interval is lengthened, a midstory of fire-sensitive hardwoods will become rapidly established (Garren 1943; Monk 1968; Veno 1976; Myers 1990; Waldrop et al. 1992; Brockway and Lewis 1997; Haywood 2007; Kirkman et al. 2016). The buildup of leaf litter (herbaceous and hardwood) is also associated with fire exclusion, as is the consequent development of organic soil horizons (Varner et al. 2005; Hiers et al. 2007).

Although both conditions likely contribute to the suppression of ground cover diversity, the relative role of litter buildup versus shading from the midstory appears to depend on site conditions. Hiers et al. (2007) provide evidence that the influence of midstory encroachment on ground cover in xeric sites can differ fundamentally from more mesic sites because of differences in site productivity that govern the degree or rate of canopy closure. In sandhill sites with varying frequency of prescribed fires, they used structural equation modeling to examine the role of forest-floor conditions on species composition and found that forest-floor development (accumulations of litter, duff, and organic horizon development) was positively correlated with reduced fire frequency and was the primary factor explaining decreased richness of functional groups of ground cover species. In contrast, midstory density explained relatively little variation in richness (Hiers et al. 2007). Given that most ground cover species in the longleaf pine ecosystem are perennials that thrive in sunlight and resprout after fire, they suggested that the shade that is produced by litter reduces plant vigor and will inhibit species recovery if fire exclusion continues (see Chapter 6).

Additional supporting evidence came from an examination of ground cover richness in hardwood-encroached sandhill sites after the reintroduction of frequent fire with and without the removal of midstory oaks (Provencher, Herring, et al. 2001). With frequent prescribed burning for 15 years after midstory oak removal, species richness did not differ between the control and the treatment plots (Kirkman et al. 2013). Using structural equation modeling, Veldman et al. (2014)

similarly demonstrated that in extremely xeric sites frequent fire influences ground cover richness primarily by reducing the litter and duff depth, independent of tree abundance.

The fire-frequency interval that results in the highest ground cover species richness appears to vary with site fertility and production of fuels that can carry fire. In a long-term fire-frequency study conducted in mesic to subxeric sites, the highest level of species richness was associated with prescribed fires every 1–3 years (Glitzenstein et al. 2012). Further, no evidence of a decline in ground cover species richness due to overly frequent fire was observed, a finding that has been reported for other fire-influenced savanna and woodland systems (Glitzenstein et al. 2012, and references therein). Glitzenstein et al. (2012) attributed this phenomenon to the unique combination of characteristics of the longleaf pine ecosystem: an evolutionary history of very frequent low-intensity fire, the abundance of perennial herbaceous species that are adapted to survive and resprout after frequent fires, sufficiently productive climate and soil conditions to facilitate the competitive exclusion of fire-adapted species by fire-intolerant species during longer intervals between fires, and the absence of enhanced competitiveness of dominant C_4 grasses when frequently burned.

Soil moisture was the most important factor correlated with ground cover richness based on a large data set of fire-maintained forests and woodlands in the Southeast, regardless of scale (Carr et al. 2009; Peet et al. 2014). Other important factors included soil texture and cation availability. Across a natural edaphic gradient in frequently burned longleaf pine-wiregrass sites, Kirkman et al. (2001) examined the relationship of species richness to productivity as a function of nitrogen mineralization rates and soil moisture. Using the depth to the clayey layer and soil drainage classes, they sorted the sites into xeric, mesic, and wet-mesic conditions along the gradient. Their results showed that highest biomass and highest species richness occurred at the wet-mesic sites and that nitrogen availability was inversely correlated with richness and productivity; hence, percent soil moisture was found to be an important regulator of species richness and community production within the defined gradient of the study.

Competition

Many researchers have suggested that competition from bunchgrasses plays an important role in determining variation in species richness (Walker and Peet 1983; Keddy et al. 2006; Noss 2013; Peet et al. 2014); the assertion being that competitive effects of bunchgrasses are especially important in productive habitats, assuming that size asymmetries among species increase as productivity levels increase. However, little evidence exists to support a negative or unimodal relationship between increased ground cover productivity and species richness within frequently burned longleaf pine ecosystems. Instead, Kirkman et al. (2001) found that the proportional dominance of the ground cover by wiregrass persisted across a productivity and moisture gradient and that the relative abundance of wiregrass was not correlated with species richness. The monotonic relationship of species richness and productivity that they reported indicated that with frequent fire, competitive exclusion by wiregrass does not mechanistically structure variation in species richness along the environmental gradient—and they suggested that the growth form of wiregrass was perhaps a factor. The wiregrass growth form, which is nonrhizomatous (a perennial bunchgrass), results in distinct interstitial spacing between grass clumps. Across this gradient, water was likely a limiting resource to productivity, and stress due to water limitation was a major factor in seedling establishment in sandy soils (Kirkman et al. 2001).

Results from two experimental studies also support the idea that competition from dominant bunchgrasses has little influence on species richness in frequently burned communities. Roth et al. (2008) found that removal of bluestem (*Andropogon* spp.) in both wet and dry sites had no effect on local species richness 2 years after removal. At the same study site, Myers and Harms (2009a) also found no effect when slender bluestem (*Schizachyrium tenerum*) was removed, except at small neighborhood scales (0.1 m^2); they also found no effect on species rank-abundance distributions (Myers and Harms 2009a) or β-diversity (Myers, unpublished data). In contrast, they found a positive effect on species richness, but no effect on species rank-abundance distributions or β-diversity, when gallberry (*Ilex glabra*), a dominant clonal shrub, was removed. They suggested that their results were inconsistent with "peripheral-species" or "fugitive-species" concepts (Horn and MacArthur 1972; Keddy et al. 2006) in which local diversity is considered to reflect the escape of rare, small-stature species from dominant large-stature species in space or time. Further, other studies (Kirkman et al. 2001; Clark et al. 2008) suggest that the co-occurrence of a large number of infrequently occurring species may reflect the ability of small grasses and rosette-forming forbs to tolerate frequent fire and to survive between or under the dense canopy of larger grasses during the transient fire-free period (Walker and Peet 1983; Myers and Harms 2009a; Glitzenstein et al. 2012).

A niche-related factor that influences many legumes is the characteristically nitrogen-poor soils in this ecosystem, a condition exacerbated by frequent fire and volatilization of nitrogen with combustion (Wilson et al. 1999, 2002). In a nitrogen-enrichment study, Kirkman et al. (2016) found that legume richness decreased with the addition of nitrogen. They suggested that this was not necessarily a shift from belowground competition for nutrients to aboveground competition for light, but perhaps the response of legumes to the altered nitrogen-to-phosphorus ratio in the soil. In contrast to other plants, nitrogen-fixing legumes have a higher demand for phosphorus (Dixon and Wheeler 1983), meaning that they cannot compete effectively for very low levels of phosphorus when nitrogen ceases to be the most limiting resource.

Evidence for Dispersal-Assembly Mechanisms

Myers and Harms (2009a, 2011) proposed that the characteristically high number of rare species in longleaf pine ecosystems has important implications for community assembly processes. Rarity increases the importance of demographic stochasticity because it increases the probability of local extinctions. In individual species, rarity also limits the degree of interspecific interactions, such as competition, in turn diminishing the degree that assemblages are shaped by deterministic interactions. Furthermore, populations of rare species are likely to be recruitment limited through reduced dispersal and fecundity (Barot 2004).

Several lines of evidence support the hypothesis that dispersal assembly plays an important role in longleaf pine communities. In one ground cover study, Kirkman, Coffey, et al. (2004) examined the recovery of species richness from prior soil disturbance in a 64-year-old slash pine plantation that was located within close proximity to natural longleaf pine reference sites. They found high floristic overlap between reference and recovery sites. Additionally, similar species richness occurred at broader scales (≥10 m^2), indicating that substantial passive vegetation recovery (recovery without active reintroduction of species) had occurred. However, at finer scales, species richness was lower in the recovery site; this suggests that co-occurrence of a high number of species depends on local dispersal and that establishment at fine scales will likely take longer. They also observed that species with low dispersal potential (such as those that depend on gravity and ants) had a lower probability of occurrence than would be expected after 64 years relative to other species, even though source populations were present in the nearby longleaf pine-wiregrass communities (Kirkman, Coffey, et al. 2004). Collectively, these results suggest that dispersal limitation constrains establishment.

Similar correlative evidence suggesting dispersal and establishment limitation was provided in a comparison of paired plots from remnant longleaf pine sites and agricultural sites that had been abandoned in the mid-1900s (Brudvig et al. 2013; Grman et al. 2015). A broad-scale replicated experiment (Brudvig et al. 2009; Damschen and Brudvig 2012) demonstrated that landscape corridors between patches promoted species richness of animal-dispersed plants (including some native longleaf pine-associated species) compared to isolated patches; species richness of wind-dispersed plants increased in response to changes in patch shape created by corridors with higher edge-to-interior ratios (see Chapter 11).

Seed-addition experiments also have demonstrated that species richness is strongly limited by seed arrival from the species pool (Myers and Harms 2009a, b, 2011). Introducing seed of 38 ground cover species to experimental plots with and without two dominant functional groups (bunchgrasses and shrubs) increased species richness on all plots (Myers and Harms 2009a), and decreased β-diversity in shrub-dominated patches (Catano et al. 2017), indicating that high rates of immigration can homogenize community composition in the ground cover. In a companion experiment, Myers and Harms (2011) also found that seed addition of 31 ground cover species increased species richness irrespective of local fire intensity and soil moisture manipulations.

Potential for Interplay between Niche and Dispersal Processes

Experimental studies also indicate that dispersal-assembly processes interact with multiple niche-based processes to structure biodiversity in the ground cover. In particular, the effects of dispersal on community assembly are strongly influenced by two environmental filters—soil resource availability and fire intensity—through space and time (Table 5.2).

Soil Resource Availability and Dispersal

As an initial study within a long-term factorial experiment that included an examination of ground cover responses to resource manipulations (addition of water and nitrogen) across a soil moisture/texture gradient, Iacona et al. (2010) found that the mean number of species naturally recruited was increased by irrigation regardless of the soil moisture conditions of the site. In this short-term study, nitrogen addition did not affect species richness. To determine if establishment is limited by seed supply, Iacona et al. (2010) manipulated seed-addition densities of three species in subxeric and wet-mesic longleaf pine-wiregrass sites with and without water and nitrogen addition. Seedling recruitment increased at the mesic site but not at the subxeric site without additional water. They also found no differences in species richness or seedling density from seed rain samples (germinated in a greenhouse), suggesting that the local seed pools were similar in species richness (but not necessarily in species composition). Although seed limitation was observed across the naturally occurring moisture gradient, its relative importance to establishment depended on moisture availability. Together, these results suggest that water availability is an important driver of species richness in seedling recruits, which could explain the notably high levels of species richness that are observed at the mesic end of the naturally occurring moisture gradient (Figure 5.2). In subxeric sites, environmental conditions at recruitment microsites are rarely optimum for regeneration; at more mesic sites, a higher level of environmental variation (amplitude, frequency, or both) over time and space is more likely to result in optimum regeneration conditions. Seed limitation provides an upper bound on potential recruitment at both ends of the spectrum, but microsite limitation is the primary regulator of recruitment at the subxeric end.

Similarly, Myers and Harms (2011) found that soil moisture altered the relative importance of seed arrival in community assembly in a study that manipulated soil moisture (drought, water addition) and seed arrival (seed addition) of 31 ground cover species. Seed arrival had a weaker positive effect on total species richness in local communities with either drought or irrigation compared to control communities with natural soil moisture, further suggesting that spatial variation in soil moisture or temporal variation in rainfall interacts with dispersal to influence species richness. In contrast, local densities of individual seed-addition species were reduced under experimental drought conditions, but increased by watering. Despite these interactions, enhanced seed arrival increased total species richness regardless of soil moisture, suggesting that species-rich ground cover communities constitute relatively open-membership assemblages.

In contrast, in a 10-year irrigation experiment replicated across a natural soil moisture gradient in wiregrass-dominated communities, Kirkman et al. (2016) found that irrigation increased species richness as well as aboveground net primary production in subxeric sites. Irrigation had no influence on productivity in the mesic site, but increased species richness both for woody ground cover plants in mesic and subxeric sites and for legumes and forbs at the subxeric sites.

TABLE 5.2
Experimental Studies Conducted in Georgia and Louisiana That Tested the Influence of Niche-Assembly or Dispersal-Assembly Mechanisms on Species Richness in Longleaf Pine Ecosystems

		Experimental Manipulations			
Habitat	**Location**	**Niche Assembly**	**Dispersal Assembly**	**Effects on Species Diversity**	**References**
Subxeric uplands	Georgia	Nitrogen addition, water addition	Seed addition	No effect of nitrogen addition on species richness; positive effect of water on species richness; positive effect of seed addition plus water on recruitment	Iacona et al. (2010)
Mesic uplands	Georgia	Nitrogen addition, water addition	Seed addition	No effect of nitrogen addition on species richness; positive effect of water addition on species richness; positive effect of seed addition on recruitment	Iacona et al. (2010)
	Louisiana	Competition from dominant bunchgrasses[a]	None	No effect on species richness	Roth et al. (2008)
		Competition from dominant bunchgrasses[a]	Seed addition	No effect of competition on species richness, community composition, or species evenness; positive effect of seed addition on species richness but no effect of seed addition on β-diversity; no effect or weak interactive effects of competition and seed addition on species richness or β-diversity	Myers and Harms (2009a), J. Myers (unpublished data)
		Competition from dominant shrubs	Seed addition	Negative effect of competition on species richness; no effect of competition on β-diversity or species evenness; positive effect of seed addition on species richness; negative effect of seed addition on β-diversity; no interactive effects of competition and seed addition on species richness or β-diversity	Myers and Harms (2009a), J. Myers (unpublished data)
		Fuel addition (fire intensity)	Seed addition	Negative effect of fuel addition on species richness; no effect or weak negative effect of fuel addition on β-diversity[b]; positive effect of seed addition on speciesrichness; negative effect of seed addition on β-diversity; interactive effects of fuel addition and seed addition on species richness or β-diversity	Myers and Harms (2011), J. Myers (unpublished data)
		Water addition, water removal	Seed addition	Variable effects of water on species richness and community composition; positive effect of seed addition on species richness; interactive effects of water and seed addition on species richness	Myers and Harms (2011)
	Georgia	Facilitation by dominant bunchgrass[a]	Shading, seed addition	No facilitative effect of shading on recruitment	Iacona et al. (2012)

(Continued)

TABLE 5.2 (*Continued*)
Experimental Studies Conducted in Georgia and Louisiana That Tested the Influence of Niche-Assembly or Dispersal-Assembly Mechanisms on Species Richness in Longleaf Pine Ecosystems

		Experimental Manipulations			
Habitat	**Location**	**Niche Assembly**	**Dispersal Assembly**	**Effects on Species Diversity**	**References**
Subxeric uplands	Georgia	Facilitation by dominant bunchgrass[a]	Shading, seed addition	No facilitative effect of shading on recruitment	Iacona et al. (2012)
		Nitrogen addition, water addition, burn exclusion	None	Negative effect of nitrogen addition on species richness; positive effect of water addition on species richness; negative effect of fire exclusion on species richness	Kirkman et al. (2016)
Mesic uplands	Georgia	Nitrogen addition, water addition, prescribed fire	None	Negative effect of nitrogen addition on species richness; no effect of water addition on species richness; negative effect of fire exclusion on species richness	Kirkman et al. (2016)

Studies were conducted at the Joseph W. Jones Ecological Research Center at Ichauway in southwestern Georgia or at Camp Whispering Pines in southeastern Louisiana.

[a] Dominant bunchgrasses: broomsedge bluestem (Roth et al. 2008), slender bluestem (Myers and Harms 2011), and wiregrass (Iacona et al. 2012).

[b] No effect on β-diversity of all species; weak negative effect on β-diversity for seed-addition species.

Although these studies suggest that low soil moisture limits establishment and growth of ground cover species, their conclusions about the effects of irrigation on species richness were inconsistent. These differences could reflect a shift in the relative importance of different environmental filters among sites that vary in soils, water-holding capacity, or the composition of dominant functional groups (such as bunchgrasses). In addition, the contrasting effects of abiotic filters at different life-history stages (such as seed germination, seedling establishment, or adult performance) could, in part, explain some of the differences in results (Kirkman et al. 2016).

Spatial Heterogeneity in Fires and Interactions with Dispersal

Processes occurring at fine spatial scales, such as spatial heterogeneity in fire, influence patterns of community assembly (Thaxton and Platt 2006; Myers and Harms 2011; Gagnon et al. 2012; Wiggers et al. 2013, 2017). In burned landscapes, fine-scale fuel heterogeneity alters fire characteristics (see Chapter 6), which in turn can affect the availability of recruitment microsites, growth and survival of individual plants, and patterns of species composition and diversity (Thaxton and Platt 2006; Hiers et al. 2009; Wenk et al. 2011; Gagnon et al. 2012, 2015; Ellair and Platt 2013; Wiggers et al. 2013, 2017; O'Brien, Loudermilk, Hiers, et al. 2016).

Gagnon et al. (2012) reported differing responses (mortality and tussock size reduction) of two dominant bunchgrasses to fine-scale variation in fuel loads. In a later study, Gagnon et al. (2015) found that an increase in fuel loads resulted in greater duration of fire and soil heating, which in turn substantially reduced vegetation through mortality, damage to belowground perennating organs, and by reducing recruitment from the soil seed bank. They suggested that the localized short-term suppression of vegetation resulting from increased fuels provides episodic opportunities for recruitment. Additional evidence of the effects of fire heterogeneity was provided by Wiggers et al.

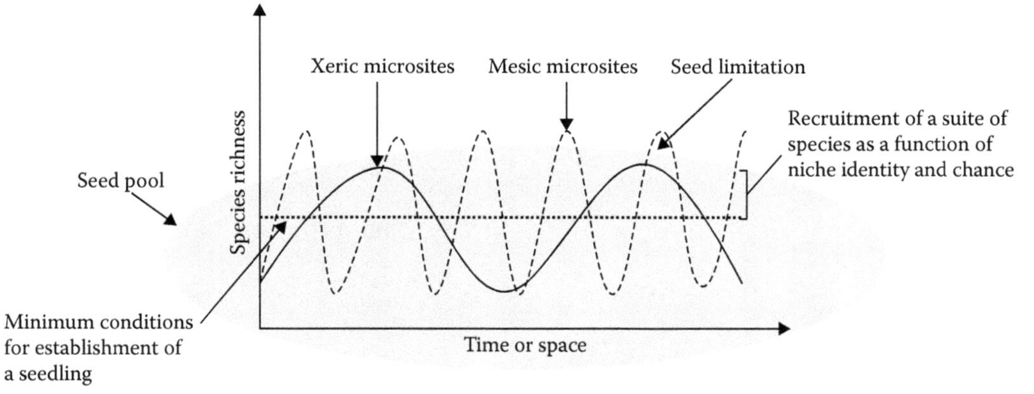

FIGURE 5.2 Conceptual model illustrating episodic recruitment depending on soil water availability. At xeric sites, environmental conditions at recruitment microsites are rarely optimum for regeneration; whereas, at more mesic sites, a higher amplitude and frequency of environmental variation over time and space could result in optimum regeneration conditions more commonly. At both locations, seed limitation provides an upper bound on potential recruitment, but at the xeric location, microsite limitation is the primary regulator of recruitment. (Modified from Iacona, G.D. 2008. Seedling recruitment as a driver of species richness in the understory of the longleaf pine savanna, M.S. thesis. University of Florida, Gainesville, Florida.)

(2013), who found that legume seed mortality increased in the vicinity of smoldering pine cones, but was not affected by increased pine needle fuel loads. They concluded that the prolonged duration of burning associated with coarse woody debris results in death of most seed or bud banks, but temperatures fall into a range that is favorable for germination of hard-coated seeds as depth and lateral distance from a pine cone increases (Wiggers et al. 2013, 2017). Moreover, Hiers et al. (2009) concluded that pine cones and other long-burning fuels are not distributed uniformly throughout longleaf pine stands (see Chapter 6), suggesting that recruitment into open microsites created by the combustion of these fuels exhibits similarly heterogeneous distribution, thus influencing fine-scale patterns of species distributions (Figure 5.3).

Myers and Harms (2011) suggested that local variation in fires may alter the relative importance of dispersal assembly. In a factorial field experiment that combined manipulations of fine fuel (addition of longleaf pine needles) and seed arrival (seed addition) of 31 ground cover species following burning, they found that postfire seed arrival had a stronger positive effect on species richness in local communities that burned under augmented fuel loads compared to those burned under control fuel loads. This finding suggests that fine-scale variation in fuel loads and fire intensity interacts with seed dispersal to create local variation in species richness. They also found a synergistic effect of seed arrival and local fire intensity on β-diversity, where composition was more similar among disturbed communities with high seed arrival compared to disturbed communities with low seed arrival (Catano et al. 2017). Collectively, these results suggest that fine-scale variation in fire intensity creates a mosaic of open microsites that promote the establishment of immigrating species—even within frequently burned landscapes.

SUMMARY

The maintenance of extraordinary numbers of ground cover species in the longleaf pine ecosystem appears to be mediated through a continuum of dispersal-related and environmental filtering factors. Collectively, these factors represent a unique complex of multivariate processes that regulate species richness in longleaf pine woodlands and forests. However, the interplay of various community-assembly processes as they occur at different spatial scales and how they are influenced by temporally fluctuating environmental conditions is largely unknown.

(a)

(b)

(c)

FIGURE 5.3 Fine-scale environmental heterogeneity from smoldering of coarse woody debris in which the prolonged duration of burning creates heterogeneity in the survival of regenerating plants and recruitment niche: (a) Flaming and smoldering snag, (b) a barren scar 6 weeks postfire, (c) recruitment of vegetation (resprout and germination) in the burn scar 6 months postfire. (Photographs courtesy of Richard T. Bryant.)

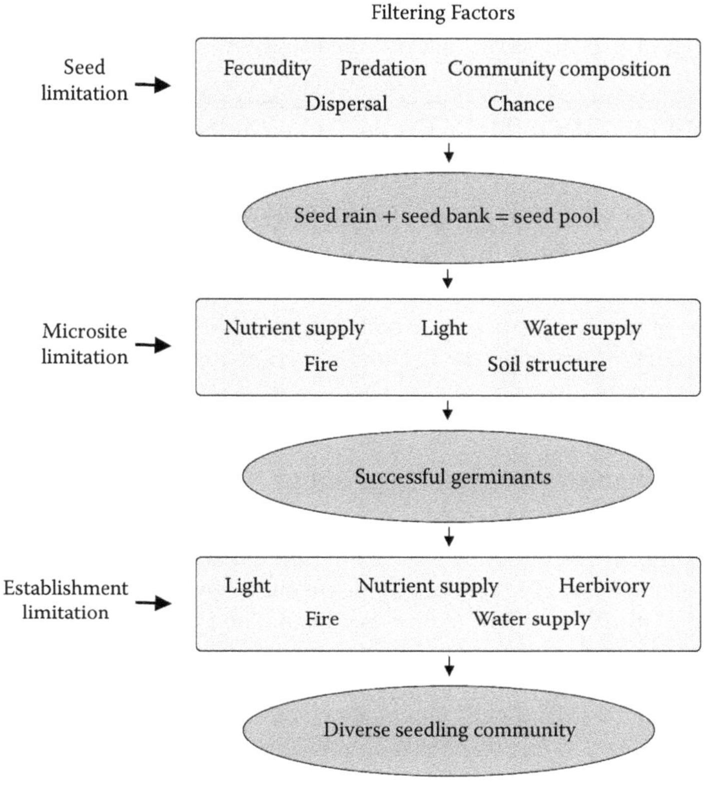

FIGURE 5.4 Conceptual model of seed availability and environmental filtering processes that regulate species richness in a frequently burned longleaf pine ecosystem.

The biogeographic and evolutionary processes that are associated with a history of frequent, low-intensity fires have resulted in large local and regional species pools, each composed of species with adaptations to survive burning and to resprout. This large species pool serves as a lottery-based seed supply that is sorted by various environmental filters imposed by regional habitats and gradients (Figure 5.4). Importantly, a key integrating mechanism that perpetuates high species richness of the longleaf pine ecosystem is an assemblage of dominant grasses that lack the trait of increased competitiveness with frequent fires or with increased resources in most sandy Coastal Plain soils. The abundance of species in frequently burned natural longleaf pine ecosystems at all scales, coupled with the infrequent occurrence of most species, suggests a strong influence of recruitment limitation at multiple levels including dispersal limitation, microsite limitation, and establishment limitation. The relative importance of these factors in regulating community assemblage across soil types has been addressed by only a few studies. At the local level, dispersal is coupled with the episodic availability of appropriate recruitment microsites. Although seed supply may be limited to some degree by dispersal, in some localities soil moisture availability is of paramount importance to species establishment. Interwoven with recruitment microsite availability and establishment is stochasticity associated with mortality of seeds and seedlings that results from fine-scale heterogeneity of fire behavior and intensity. Moreover, local-scale neighborhood interactions among individual plants associated with episodic availability of recruitment sites and seed arrival via seed rain or soil seed bank recruitment and the role of predation remain unexplored in most community types.

Finally, several gaps in empirical studies limit our understanding of diversity-maintenance mechanisms that are fundamental to linking theoretical, conservation, and restoration perspectives of the longleaf pine ecosystem. Future research contributions that would bridge these information gaps will need to focus on a broad range of questions (outlined below). The answers to these and other outstanding questions will play a central role in furthering our understanding of one of the most biologically diverse and relatively understudied ecosystems on the planet.

RESTORATION OF GROUND COVER COMMUNITIES: OUTSTANDING QUESTIONS AND EMPIRICAL GAPS

How Much Intersite Variation in Species Richness and Community Composition Originates from Species-Pool Variation across Landscapes?

Biogeographic and evolutionary processes that shape landscape species pools can strongly influence patterns of biodiversity across spatial scales (Ricklefs 1987; Harrison and Cornell 2008; Kraft et al. 2011; Myers et al. 2013). Cross-site comparisons of landscape-level influences on ground cover biodiversity will help address gaps in understanding the relative importance of landscape- and local-scale controls on community assembly and local biodiversity.

What Are the Mechanistic Roles of Dispersal in Ground Cover Community Assembly?

Dispersal can contribute to both nonrandom and random community assembly (Myers and Harms 2009a; Vellend et al. 2014) and can influence the outcome of restoration efforts (Kirkman, Coffey, et al. 2004). A few studies have examined the natural patterns of seed dispersal in species-rich longleaf pine ground cover communities (Mulligan et al. 2002; Stuble et al. 2010; Cumberland and Kirkman 2013; Chandler et al. 2016). Future studies comparing natural patterns of seed rain and intersite variations among species in seed dispersal and soil seed banking can be used to inform both theoretical models of community assembly and restoration of ground cover communities.

What Is the Relative Importance of Landscape-Scale Environmental Factors and Local-Scale Species Interactions as Determinants of Species Richness and Community Composition? How Do These Processes Vary among Community Types?

Landscape-scale factors such as topography, soil moisture, and soil nutrients strongly influence species diversity and composition in ground cover communities. However, the degree to which community assembly is influenced by landscape-scale factors, local-scale species interactions (such as competition, facilitation, or predation), and the interplay between them is largely unknown, especially in different soil moisture regimes and different community types (for example, uplands versus wet-mesic flatwoods). Although some experiments have examined the effects of competition among species on ground cover biodiversity (Roth et al. 2008; Myers and Harms 2009a), little is known about the effects of facilitation, pathogens, herbivory, seed predation, and other interactions among species.

What Is the Role of Ecological Drift in Creating Patterns of Biodiversity?

Theory suggests that ecological drift has an especially strong influence on the assembly of hyperdiverse communities, but empirical tests of ecological drift are currently lacking in longleaf pine ecosystems. Future experimental and observational studies can address key questions such as: (1) the extent to which ecological drift explains patterns of biodiversity; (2) the extent to which (and timing of) stochastic colonization and extinction dynamics contribute to community assembly; and (3) the extent to which fire frequency influences drift by altering local plant densities (community size).

Does the Relative Importance of Niche Assembly and Dispersal Assembly Vary among Different Functional Groups?

Within functionally diverse communities, the relative importance of niche- and dispersal-assembly processes can vary among different functional groups (such as perennial grasses, legume forbs, or nonlegume forbs). Studies that test assembly mechanisms both within and across functional groups can reveal the extent to which community-wide patterns are driven by differences or similarities in assembly mechanisms across guilds.

What Processes Determine Spatial and Temporal Patterns of Functional Diversity among Plants?

Community-assembly processes can be difficult to infer from patterns of species diversity alone, especially when communities contain large numbers of species with functionally redundant traits (Swenson et al. 2011). Little is known about patterns of functional-trait diversity in the ground cover communities and how they are shaped by different community assembly mechanisms.

ACKNOWLEDGMENTS

We acknowledge the input of Robert Mitchell (deceased), Kyle Harms, Paul Gagnon, and Heather Passmore in the numerous discussions that inspired the concepts presented in this chapter. Discussions by participants of the 10th Biennial Longleaf Alliance Conference workshop on heterogeneity in the longleaf pine ground cover were incorporated into our discussion of key unanswered questions. We also acknowledge funding provided by our respective affiliated institutions and the National Science Foundation (DEB 1144079, 1144084). We appreciate the constructive reviews of earlier drafts of this chapter by Kyle Palmquist, Robert Sutter, and Kyle Harms.

6 The Role of Fuels for Understanding Fire Behavior and Fire Effects

E. Louise Loudermilk, J. Kevin Hiers, and Joseph J. O'Brien

CONTENTS

INTRODUCTION

Fire ecology, which has emerged as a critical discipline, links the complex interactions that occur between fire regimes and ecosystems. The ecology of fuels, a first principle in fire ecology, identifies feedbacks between vegetation and fire behavior—a cyclic process that starts with fuels influencing fire behavior, which in turn governs patterns of postfire vegetative responses and the future production of fuels (Mitchell, Hiers, et al. 2009). Recent research has used this conceptual framework to understand the relationship between combustion science and ecology and to gain mechanistic understanding of fire effects (Johnson and Miyanishi 2001).

The purpose of this chapter is to synthesize research on forest fuels characterization—particularly as influenced by the overstory—and on the role that fuel heterogeneity plays in the feedback mechanisms between fire behavior and fire effects in the longleaf pine (*Pinus palustris*) ecosystems of the southeastern United States.

We begin by describing the state of the science in fire behavior within the context of ecosystem structure and function. We review advances in the modeling tools used to represent complex fuelbeds as wildland fuel cells (discrete patches of fuels). Further, we discuss model predictions of heterogeneity in fire behavior and fire effects that correspond to variations in wildland fuel cells. Our focus is on fire behavior and fire effects that influence ecosystem restoration, with an emphasis on introducing fire in sites where it has been long absent.

We also introduce an emerging concept in fire ecology that was developed in longleaf pine ecosystems: the "ecology of fuels." We review research on fuel and fuelbed characteristics including fuel accumulation rates, burning characteristics of various fuel types, and the effects of fuel moisture on fire behavior in longleaf pine ecosystems. We describe new multiscaled methods of measuring fire behavior and fuel variation that were developed to overcome the limitations of traditional approaches. We show how this emerging research can take advantage of next-generation fire-behavior prediction models to link interactions between fuel and fire behavior. We then focus on fire-fuel feedback interactions to describe an approach for fuel-based restoration of longleaf pine that transforms the traditional paradigm into one in which forest structure governs management. We conclude by discussing research directions that would advance fuel and fire ecology in longleaf pine ecosystems.

THE ECOLOGY OF FUELS

In longleaf pine and functionally analogous, frequently burned pine ecosystems, the ecology of fuels is a complex interplay among overstory structure, fuelbed heterogeneity, and ground cover vegetation (Figure 6.1) that collectively determines forest structure, composition, and functional

(a)

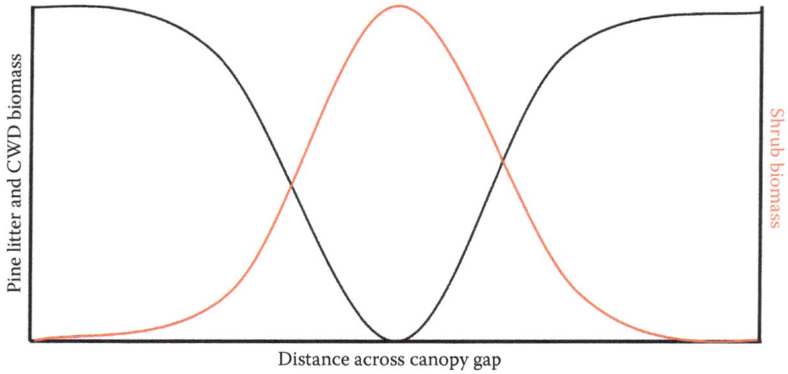

(b)

FIGURE 6.1 An illustration of the ecology of fuels as applied to overstory influences on fuel distribution in a longleaf pine sandhill habitat: (a) Cross-section from light detection and ranging three-dimensional point cloud showing encroachment or released growth of shrub in a canopy gap, and (b) changes in fuels shown directly (black line) as alterations in fuel from pine litter and coarse woody debris (CWD) distribution that were derived from the overstory and indirectly (red line) as subsequent shrub encroachment resulting from reduced fire intensity—less pine litter—and reduced belowground competition, notwithstanding interactions with grasses. Note that the biomass of each component depends on fire frequency, time-since-gap development, gap size, and habitat type; and that the dynamic conditions throughout the fuelbed create the fine-scale heterogeneity in wildland fuel cells. (From E. Rowell, unpublished data.)

processes. Moreover, these dynamics appear to vary categorically across the full range of edaphic conditions, ranging from mesic flatwoods to xeric sandhills (Wilson et al. 1999; Hiers et al. 2007). Research has focused on three aspects of fuels ecology regarding fire effects in longleaf pine ecosystems: (1) the importance of overstory canopy-derived fuels on fire behavior and resulting forest structure, (2) the mediation of biodiversity through fine-scale variations in fire behavior, and (3) the role of litter and forest floor in restoration of degraded ecosystems.

Because longleaf pine is the dominant species in the longleaf pine ecosystem, its influence on fire behavior and patterns of fire effects has been long recognized (Wahlenberg 1946; O'Brien et al. 2008; Mitchell, Hiers, et al. 2009). Overstory pines determine fuel distribution, both directly through fallen needles that support frequent fire, and indirectly through competitive interactions—largely belowground—with the less fire-tolerant hardwoods and ground cover plants that make up the complex fuelbed (Mitchell et al. 2006). The interaction between pine needles and bunchgrass produces important fine-scale variability in fire behavior, with needles adding significant residence time to fires that are propagated by grasses (Hiers et al. 2009; Loudermilk et al. 2014; Fill et al. 2016).

Fuel is the critical link between structure and function in these systems (Williamson and Black 1981; Rebertus et al. 1989; Glitzenstein et al. 1995). Pine needle litter distribution allows fire to spread across a heterogeneous ground cover (Hiers et al. 2009; Loudermilk et al. 2014). If this matrix is disrupted, through timber removal or pine straw raking for example, fire spread can be halted, leaving patches of unburned vegetation (O'Brien et al. 2008; Mitchell, Hiers, et al. 2009; Jack et al. 2010). If these fire-free patches coincide with canopy gaps, suppressed hardwoods will be released and grow to a fire-resistant height (Mitchell et al. 2006; O'Brien et al. 2008). Needle litter has long provided a link between timber and fire management; this relationship is epitomized in the Stoddard–Neel system of ecological forestry (see Chapter 10), which promotes maintenance of stand ecological integrity among its goals (Mitchell et al. 2006). Less understood is the feedback mechanism between needle litter and other ground cover fuels, particularly bunchgrasses; this mechanism also creates significant variations in fuels and fire behavior at fine scales (Hiers et al. 2009).

Wildland Fuel Cell Concept

The wildland fuel cell concept was developed to connect variations in fuels to a relevant scale of variations in fire behavior by aggregating fuels with similar characteristics—such as type, quantity, and spatial arrangement. The outcome is a linkage between the observed variability in fire behavior and variation in fuels—a critical component of the ecology-of-fuels feedback loop—and confirmation that coupled fuel characteristics (structure, type, and biomass) are correlated to observed variation in fire behavior. Thus, defining the heterogeneity in fuels and fire and identifying appropriate scales (Figure 6.2) provides a gateway both for understanding fire effects and for identifying mechanisms that control patterns of plant diversity (Hiers et al. 2009).

The wildland fuel cell concept was originally developed for the longleaf pine habitats of the Coastal Plain, where the scale of fuel and fire heterogeneity varies at about the 0.25-m scale. Similar scales of fuel and fire behavior variation also characterize the less productive xeric sandhill habitats (Loudermilk et al. 2014). The scale of heterogeneity, which likely would become coarser in flatwoods and other shrub-dominated habitats, has yet to be defined across the full range of variation in longleaf pine ecosystems.

Fuel and Fuelbed Characteristics of Longleaf Pine Sites

In longleaf pine sites with herbaceous ground cover, fires generally can burn successfully through a stand every 18 months (Glitzenstein et al. 2003; Reid et al. 2012). Ground cover growth commences immediately after a fire and when combined with pine needle litter, promotes a quick recovery of biomass and available fuels. On a site with 10.8 m^2/ha basal area of mature longleaf pine, needle

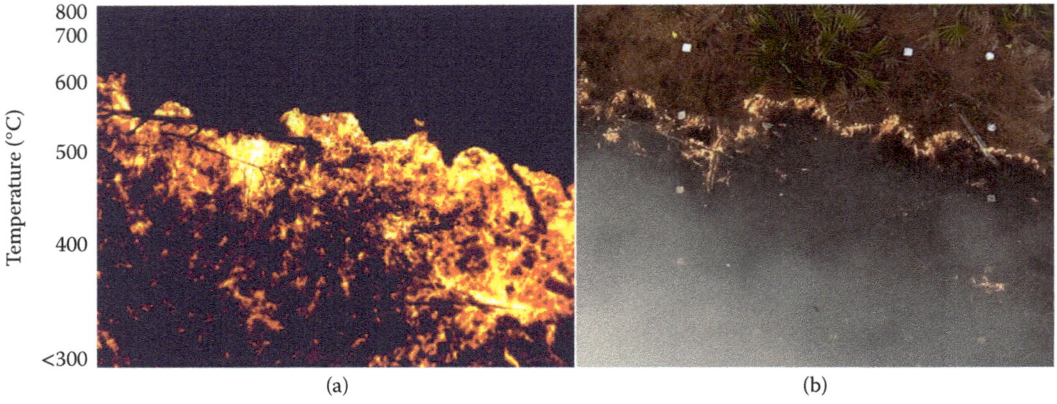

FIGURE 6.2 Depiction of a surface fire on a 4 × 4 m experimental plot captured by: (a) At-nadir forward looking infrared imagery, and (b) true-color photography showing the aluminum targets that were used for processing infrared imagery. Note the smoke penetration of the infrared image and the thermal signatures of flaming and smoldering combustion. (From O'Brien, J. J., E. L. Loudermilk, B. Hornsby, et al., *International Journal of Wildland Fire*, 25, 62–75, 2016.)

litter is produced at a rate of about 4884 kg/ha annually (Gresham 1982), providing about half of the fuel available to burn (Ferguson et al. 2002; Mitchell et al. 2006). Deciduous oaks (*Quercus* spp.) that are confined to the midstory by frequent fires can also contribute considerably to litter biomass, but their distribution and abundance can be quite variable (Wright 2013) and can be dramatically diminished by herbicide applications, mechanical tree removals, or other restoration efforts that affect oak abundance in a given longleaf pine area.

During burning, consumption of fine fuels ranges from about 50% to 70% (Goodrick et al. 2010), but can be as high as 92% (Ottmar, Hiers, et al. 2016). Coarse woody debris contributes about 10% of the fuel loading (Wright 2013). Small diameter (≤10-hour) woody fuels are relatively sparse because they are often completely consumed by frequent fires. Although large diameter (100- to 1000-hour) woody fuels are likely important in driving fire behavior and fire effects in small areas of a stand (Wiggers et al. 2013; Cannon et al. 2014; O'Brien, Loudermilk, Hiers, et al. 2016), they have little influence on stand-level fire behavior.

Fuel accumulation varies across longleaf pine sites depending on overstory basal area, habitat type, local climate, ground cover quality, and management regime. Larger basal area in the overstory results in more overstory-derived fuels (pine litter and woody fuels), but can also lead to smaller fuel loads that originated from ground cover (Mitchell et al. 1999; Gonzalez-Benecke et al. 2015). In general, wetter sites are more productive and therefore have larger postfire biomass accumulations, with differences across soil moisture gradients noticeable within a single year—for example, wiregrass (*Aristida stricta*) had twice the biomass in wet-mesic areas than in xeric areas at one site 1 year postfire (Kirkman et al. 2001). This suggests that maintaining frequent fire could be more important on mesic sites to prevent fuel accumulation and competition from species that grow faster in their more productive soils. Across the longleaf pine range, fuel biomass can change from a mix of herbaceous and woody fuels to predominantly woody fuels within just a few (3–4) years after a fire (Glitzenstein et al. 2003; Gonzalez-Benecke et al. 2015). In general, as the span of fire absence increases, fuel accumulation also increases, creating the potential for increased fire intensity (Ottmar et al. 2003). However, as the vegetation state changes from longleaf pine to a more oak-dominated forest—a phenomenon that can happen within just a few decades—grasses will be suppressed by litter accumulation, competition will degrade herbaceous vegetation, and pine recruitment will be inhibited (Hartnett and Krofta 1989; Reid et al. 2012). Also, for some oak species, the emerging oak canopy creates a denser fuelbed of oak litter that is less conducive to fire spread under typical prescribed fire conditions (Kane et al. 2008), creating a positive feedback in which the oaks are protected from fire damage and quickly grow to midstory height (Guerin 1993).

The interactions among fine fuels create a complex fuelbed matrix that contributes to fire spread and continuity (Hiers et al. 2009; Mitchell, Hiers, et al. 2009). Pine litter draped across shrubs, bunchgrasses, and other grasses in these communities creates a synergy among dry available fuels across the fuelbed (Fill et al. 2016) and an aerated medium that promotes continuous low-intensity fire spread.

Different individual fuel types create variations in burning characteristics within a fuelbed (Hiers et al. 2009). Longleaf pine needles are among the most flammable compared to other U.S. pine species (Fonda 2001), and their cones are among the longest smoldering (Fonda and Varner 2004)—making them an important fuel component for local fire behavior and fire effects (O'Brien, Loudermilk, Hiers, et al. 2016; Wiggers et al. 2013).

On longleaf pine sites, with consequent differences in flammability, oak variability can affect fire intensity and fire spread within and among those sites (Kane et al. 2008). In particular, turkey oak (*Q. laevis*) and post oak (*Q. stellata*) are pyrophytic, producing litter that is similar to longleaf pine litter in flammability, thus also facilitating fire spread. Other oaks such as live oak (*Q. virginiana*) and laurel oak (*Q. hemisphaerica*) have litter that burns poorly, impeding fire spread under their canopies. Many of the more pyrophytic oaks are critical components of longleaf pine systems, and their role in fire dynamics should be considered in restoration efforts (Hiers et al. 2014).

Flammability of vegetation can vary because different vegetative types have different fuel characteristics such as coarse woody debris, fine dead fuels, and height (Ottmar et al. 2007); different mixtures of species and guilds that affect the percentages of oak versus pine litter (Kane et al. 2008); different levels of fuel moisture content based on the amount of live versus dead moisture (Nelson and Hiers 2008); different leaf chemistry based on the amount of volatile content versus ash content (Gilliam 1988); or different physical traits that alter the arrangement and moisture of the fuel over time (Varner, Kane, et al. 2016). Historically, fire behavior modeling has often focused on fuel loading (the amount of fuel) and the relative size of fuel particles (Rothermel 1972; Andrews and Chase 1989); however, in the surface fire regime of longleaf pine ecosystems (Sandberg et al. 2007), fuel moisture dynamics (Reid et al. 2012), the interaction of flammability traits (Kane et al. 2008), and the heterogeneity that results could be more critical to fire behavior and fire effects. Suites of flammability traits have also been documented by species (Fonda 2001), and specifically for southeastern fuels (Reid and Robertson 2012), with categorization into litter syndromes that promote or facilitate high-intensity surface fires and those that dampen, diminish, or extinguish surface fires (Kane et al. 2008; Kreye et al. 2013; Mola et al. 2014). Such "suites of adapted traits" to survive and grow in longleaf pine landscapes result in increased diversity and forest structure, thereby perpetuating fire regimes through their influence on fire behavior—otherwise known as the ecology of fuels (Veldman et al. 2013; Hiers et al. 2014).

In the absence of fire, litter accumulates at a linear rate until a forest floor (organic horizon) develops and the process of decomposition and forest floor accumulation reaches an equilibrium (Olson 1963; Meentemeyer 1978). This equilibrium occurs after about 8–12 years postfire on longleaf pine sites (Olson 1963; Meentemeyer 1978) and perhaps longer on mixed pine sites (McNab et al. 1978). More importantly, the forest floor has a well-developed duff layer around adult pines, largely composed of decomposition-resistant pine bark litter (Varner et al. 2005). This duff layer can have considerable consequences for overstory mortality (particularly in larger trees) when fires are reintroduced in long-unburned sites. Fine roots that have grown into the duff layer are killed when the duff is consumed by fire, even in low-intensity prescribed fire conditions. This phenomenon is particularly prevalent in xeric sites, where water stress is heightened (Varner et al. 2007); for more detail see Chapter 7.

Fuel Moisture Characteristics

Fuel moisture content is a critically important factor governing fire behavior (Rothermel 1983). Both surface moisture and water in live and dead fuels alter fire behavior through several mechanisms: latent heat of evaporation, reduction in fire radiative power, and sensible heat flux. Nearly all models of fire behavior include fuel moisture content as an input variable (Matthews 2014).

Typically, fuel moisture is treated as a single characteristic, representing an average for 1-hour, 10-hour, and 100-hour dead fuel classes (Burgan and Rothermel 1984). Fuel moisture samples are often collected gravimetrically within a fuelbed. Expressed as the mass of water per unit mass of dry fuel, fuel moisture content can be >100%, particularly in live fuels. In the Southeast, researchers have collected fuel moisture content by four vegetative categories: herb, shrub, fine wood, and litter (Brenner 2002; Ferguson et al. 2002; Ottmar, Hudak, et al. 2016). Fine dead fuel moisture is less dynamic than live fuel moisture, which is quite variable (Heinsch et al. 2015) by species, season, and antecedent drought conditions. Although fire-behavior models typically homogenize the spatial variability of fuel moisture at the stand scale, such spatial variation can be important for fuel consumption and fire spread at certain scales (A. Smith et al. 2013).

Historically, the relationship between fuel moisture content and fire behavior has concentrated on moisture dampening curves at the stand scale, which terminate in the moisture of extinction for dead fuels, typically 25%–30% (Burgan and Rothermel 1984; A. Smith et al. 2013); the dead moisture of extinction is defined as "the characteristic moisture of dead fuels at which fire will not spread with a uniform front" (Burgan and Rothermel 1984). This definition implies that fire spreads heterogeneously and assumes that the moisture-of-extinction is nonuniform within the stand, but the role of spatial variability in altering fire spread is not understood (Viney 1991; Matthews 2014). The dynamics of live fuel moisture are also complex, interacting locally with fine dead fuel moisture within the fuelbed (Burgan and Rothermel 1984).

Spatial variability is probably responsible for the discrete patterns of fuel moisture within the fuel matrix that govern fire spread, both as fuels dry with exposure to sunlight and as they absorb moisture with increasing nighttime humidity. The observed patterns of moisture at various spatial scales are compatible with the tenets of the wildland fuel cell concept, with fuel moisture content within these cells undoubtedly contributing to observed differences in fire intensity that are documented at fine scales in longleaf pine (Hiers et al. 2009; Loudermilk et al. 2012).

At larger scales, fuel moisture represents a known source of error in estimating biomass consumption, but this issue can be overcome by using remote sensing to document spatial patterns of fuel moisture within forests. Chuvieco et al. (2002) analyzed multispectral satellite imagery and found that short-wave infrared bands are sensitive to water absorption. They also compared more traditional methods including the normalized difference vegetation index, which indirectly estimates fuel moisture via chlorophyll changes at a 30 × 30 m resolution. Another study (A. Smith et al. 2013) developed a spatial error correction for the effects of fuel moisture content on fire radiative power/energy. These landscape-scale approaches can be applied to wildland fuel cells at finer scales to understand the longleaf pine fuel moisture controls on fire behavior and energy release.

Ground fuels, such as duff or organic soils, are also critically responsive to soil moisture (Varner et al. 2005; Ferguson et al. 2002). For such fuels, fuel moisture conditions were found to significantly influence the patterns of fuel consumption that cause longleaf pine mortality after a smoldering fire (Varner et al. 2007; O'Brien et al. 2010). When compared to gravimetric destructive sampling, soil probes placed within organic fuels more closely mirrored the meteorological variables that caused trends in fuel moisture.

CHARACTERIZING SURFACE FUELS AND FIRE INTENSITY

Much of the focus of fire ecology has concentrated on investigating broad-scale patterns, at scales ranging from 10 m^2 to 10,000 ha (Hobbs and Atkins 1988; Turner et al. 1999; Finney 2001, 2003; Collins and Smith 2006). Usually the emphasis has been on understanding the mosaic of "green versus black" (burned versus unburned) areas across landscapes. Also, within-fire variations in intensity are usually not measured directly, but indirectly using coarse severity classes (Keeley 2009). Similarly, fuels classifications have focused on stand-level characterization (Ottmar et al. 2007; Ryan and Opperman 2013), even though variation of fuels within a stand can often exceed variation among stands (Brown and Bevins 1986). These kinds of classifications were developed to work with fire-prediction systems, which are based on semiempirical fire-spread models (Rothermel 1972; Burgan and Rothermel 1984) that assume fuel homogeneity. Furthermore, many of the techniques used to characterize fuels were developed to be effective in fuelbeds that are dominated by woody fuels (Brown 1974); fine fuels were usually collected in bulk and scaled up to the stand level.

In frequently burned ecosystems, low-intensity surface fires (Figure 6.2) often burn completely when fine fuels are continuous, leaving few unburned patches. Thus, understanding the variation in fire intensity within burned areas is especially critical to predicting fire effects (Hiers et al. 2009; Loudermilk et al. 2012). Furthermore, fine fuels are paramount in that they create the fuel continuity that carries fires across these landscapes. Capturing and understanding this fine-scale heterogeneity is important because this is where the most ecologically relevant fire effects occur (Rebertus et al. 1989; Mitchell et al. 2006; Thaxton and Platt 2006). Several researchers (Brewer et al. 1996; Thaxton and Platt 2006) acknowledged the potential importance of fine-scale variation in fuels, but the available tools for characterizing the combustion environment responsible for fire effects (including fine-scale patterns of fuels and fire behavior) remained inadequate to produce mechanistic connections. Before the early 2000s, attempts to measure fire were limited to indices of intensity derived from temperature-sensitive paints, evaporation of water, or thermocouples (Kennard et al. 2005). Similarly, fuels have been broadly and imperfectly categorized, with categories based on stand-level characterizations (Anderson 1982; Ottmar et al. 2003).

DeBano et al. (1998) showed that ignition properties, rates of spread, intensity, and other components of fire behavior are influenced by fuel loading and fuel depth (and thus density). Fuel properties, such as volume and loading, are drivers of models used to simulate fire behavior (Burgan and Rothermel 1984; Andrews and Queen 2001), and are important measurements for empirically understanding fire behavior and fire effects. Traditionally, measurements of surface fuelbed characteristics have been both direct and indirect. Common direct measurements are tallies of down woody fuels along planar transects (Brown 1974) coupled with destructive biomass sampling, also known as "clip plots" (Brown 1981). Indirect methods include visual cover estimates in plots or comparisons with photographs of known fuel loads or types (Ottmar et al. 2003; Keane and Dickinson 2007); although they provide estimates of characteristics—such as fuel load, bulk density, and packing ratios—that are useful for predicting fire behavior at the stand level (Burgan and Rothermel 1984; Reinhardt and Keane 1998; Andrews et al. 2004), such estimates are not suitable for calculating within-stand heterogeneity.

Furthermore, these methods have significant limitations. Direct sampling is labor intensive, often limiting sample size, particularly across large areas. Some techniques are not appropriate for all fuel types: for example, planar transects do not efficiently estimate grasses. Indirect measures can be subjective, resulting in biased estimates. Additionally, estimating volume for bulk density calculations relies on unrealistic simplifications; for example, shrub and grass volumes are calculated by assuming that the plants form simple geometric shapes, such as a spheroid or cylinder (Van Wagner 1968). Such traditional volume measurement techniques ignore complex plant architectural details that are important for characterizing leaf area and biomass (Loudermilk et al. 2009) and fire behavior (Loudermilk et al. 2012) at fine scales. These techniques were designed to estimate stand-level averages, which fail to capture the heterogeneity in fuels driving both fire behavior and fire effects in frequently burned ecosystems (Loudermilk et al. 2009; Mitchell, Hiers et al. 2009; Loudermilk et al. 2012; Wiggers et al. 2013, 2017; O'Brien, Loudermilk, Hiers et al. 2016).

Challenges of Surface Fuel and Fire Measurements

Johnson and Miyanishi (2001) observed that although "...the processes of combustion and heat transfer lie at the heart of fire ecology" very few studies actually quantify the energy released during a wildland fire—defined by the National Wildfire Coordinating Group (NWCG) (NWCG 2015) as any nonstructure fire (either prescribed burning or wildfire) that occurs in vegetation or natural fuels. As an example to illustrate measurement limitations, many studies have used pyrometers consisting of temperature-sensitive paints or waxes (Thaxton and Platt 2006; Davies et al. 2010; Brudvig et al. 2012) to characterize fire intensity. These point measurements are at best only qualitative approximations of fire temperature and are heavily influenced by their placement and construction (Iverson et al. 2004; Kennard et al. 2005). Thermocouples have also been extensively used to report fire temperatures, but these measurements are also limited by construction, placement, and probe energy balance (Yilmaz et al. 2008). They are primarily limited to measuring the convective energy fraction of fire at a single point, and their sensors have characteristics—such as convective cooling, thermal inertia, conduction along the sensor lead, and some radiant heating—that influence measurement accuracy.

The scale at which fuels vary in surface fire regimes makes measurements inherently difficult. Each fuel type found within the fuelbed matrix has a different set of properties such as fuel biomass, volume, bulk density, and surface-to-volume ratio (Fonda 2001; Ottmar et al. 2003; Fonda and Varner 2004; Kane et al. 2008) that influences fire behavior and fire effects (Loudermilk et al. 2012; O'Brien, Loudermilk, Hiers, et al. 2016) and that have consequences for longleaf pine restoration (Kirkman et al. 2013; Hiers et al. 2014). Many of these properties are related to fuel dimensions or structure.

Although biomass is the most difficult parameter to estimate using nondestructive means, considerable effort to derive biomass estimates for various fuel types in longleaf pine ecosystems has been made (Ottmar et al. 2003). The problem lies in connecting biomass, structure, and continuity across a fuelbed and throughout the forest matrix. For example, biomass estimates taken just outside plots that were measured for fire intensity or fire behavior did not necessarily represent the plots in question (Ottmar, Hudak, et al. 2016). This inconsistency was attributed to the spatial heterogeneity, random patchiness, and physical overlap of various fuel types—particularly low-growing shrubs and bunchgrasses—as well as heterogeneity in fuel consumption, wind gusts, and vegetative responses to fire.

With the possible exception of wiregrass growth patterns (Mulligan et al. 2002), longleaf pine litter and, to some degree, pine cone distribution, is likely the most predictable fuel type. Pine litter is produced regularly, with its abundance entirely dependent on site conditions and the size, density, and location of trees. The size and location of cone-producing trees can be used to predict cone distribution, but the timing of cone production is influenced by the highly variable and episodic masting events that are typical of longleaf pine (Boyer 1998). Time since last fire (needed to estimate fuel accumulation), site characteristics (such as land use and soil properties), and climate all contribute to the complexity of measuring and characterizing the fuelbed and should be considered in fuel measurement studies.

Characterizing wildland fuel cells or individual fuels across a fuelbed can provide insight into fuel-fire dynamics not apparent when focusing at the stand level. Because most ground cover plants are low growing, and with frequent fire coexist in small areas, fine-scale (<1 m) wildland fuel cells are quantifiable in longleaf pine systems (Hiers et al. 2009; Loudermilk et al. 2012; Bright et al. 2016). Connected across the fuelbed by ground cover plants mixed with leaf litter and coarse woody debris, wildland fuel cells can be categorized as bunchgrasses with perched pine litter, nonbunchgrasses, shrubs and perched pine litter, or pine cones and coarse woody debris (Hiers et al. 2009). Individual fuel types can be significant for influencing fire intensity and fire behavior, with pine litter serving as the main driver of fire spread in longleaf pine systems (O'Brien et al. 2008; Mitchell, Hiers, et al. 2009). Pine cones influence local fire intensity, potentially resulting in patches of plant mortality and affecting seed germination (Wiggers et al. 2013; O'Brien, Loudermilk, Hiers, et al. 2016). In addition, interactions between fuel patches and nonfuel patches create high-order nonlinear patterns of fire-atmosphere dynamics (Loudermilk et al. 2012) and produce "combustion legacies" within the burning environment. Combustion legacies are created when fuels that burn at one location alter the combustion environment for fuels either locally or at some distance. They can also result from patchy fuelbeds that contain areas of fuel and nonfuel. For example, sand mounds created by southeastern pocket gophers (*Geomys pinetis*) are common and can be numerous in many longleaf pine stands. These 30–50-cm circular mounds are devoid of vegetation and create fine-scale fuel-free patches that can dramatically alter the dynamics of fire as it spreads across the fuelbed (Figure 6.3). At low wind speeds (about 1.5–2 m/second), the mound can split a running head fire, and as the head fire passes the mound, it develops into two parallel flanking fires. Interactions between their convective plumes then pull the flanking fires together, resulting in a large patch of higher intensity fire immediately downwind of the mound. The head fire is then restored with the same geometry as the original head fire (O'Brien, unpublished data). Higher wind speeds would have a nonlinear impact on the fire-atmosphere and fuel dynamics (Figure 6.4): the fuel-free mound could have no influence on the head fire, as fire would flow over the mound and continue unimpeded (O'Brien, unpublished data).

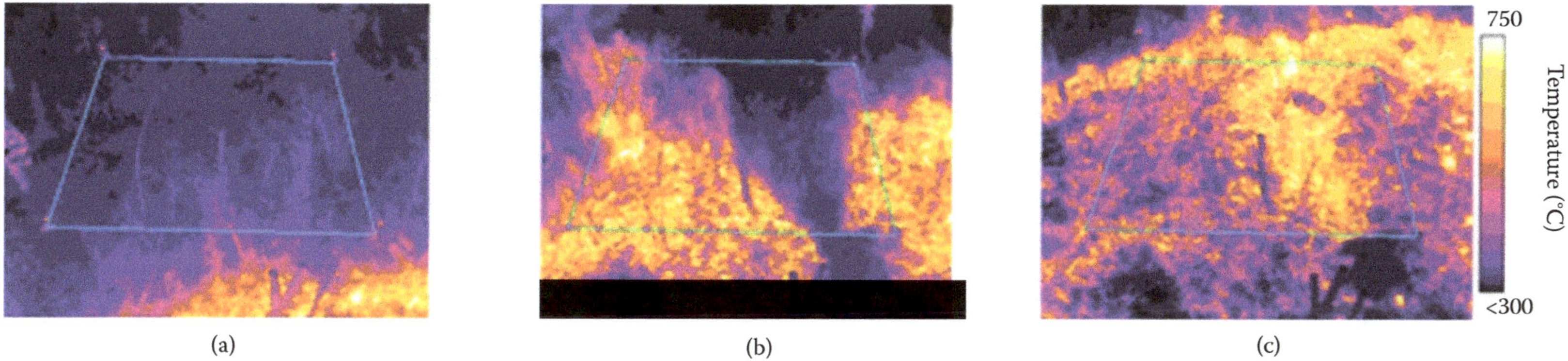

FIGURE 6.3 Three images, each collected every 30 seconds, of a head fire crossing a 4 × 4 m experimental plot in a longleaf pine stand, where wind speed was about 2 m/second: (a) Approaching fire in the lower-right corner, (b) mound of sand in lower-right corner breaking the head fire into two flanking fires, and (c) flanking fires recombining at a higher intensity and quickly reestablishing the head fire.

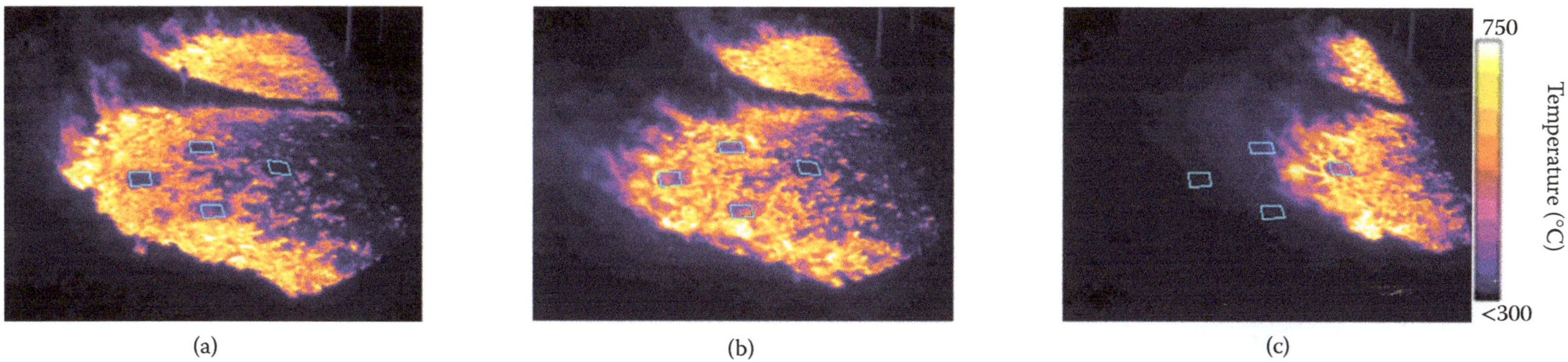

FIGURE 6.4 Images showing the progression of two 10 m lines of head fire that were ignited nearly simultaneously in longleaf pine fuels (upper fire line) and in four 1 m^2 plots that had been cleared of all fuel (lower fire line), with a 5 m/second wind coming from the right side of the images. Note that the wind speed was high enough to increase the intensity and rate of spread sufficiently to swamp any effect of fuel heterogeneity.

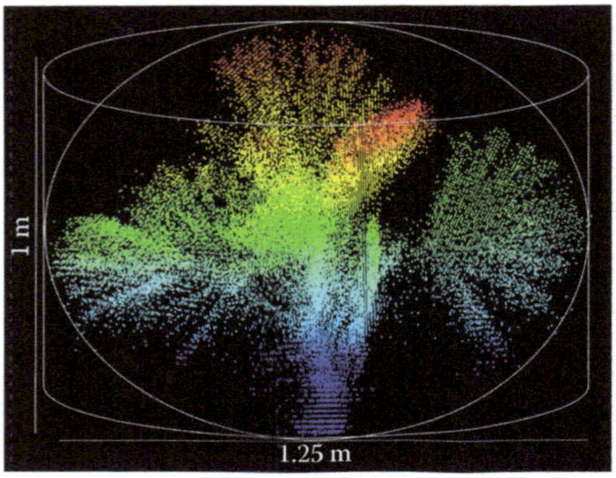

(a)

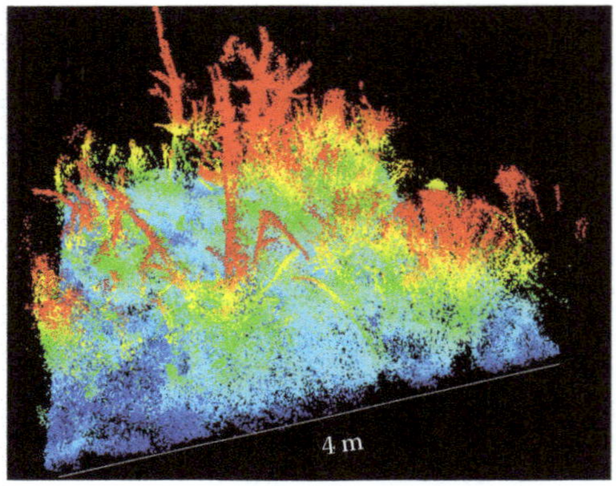

(b)

FIGURE 6.5 Output of three-dimensional point clouds from a terrestrial laser scanning for: (a) An individual saw palmetto shrub, and (b) a 4 × 4 m plot in a longleaf pine fuelbed, which has a maximum fuelbed height of 2 m. Note that fuel volume measurements of plants are usually based on a cylindrical or spheroid geometry. (From Loudermilk, E. L. et al., *International Journal of Wildland Fire*, 18, 676–685, 2009.)

Measuring and Characterizing Surface Fuels

Measuring fuels and fire at a scale necessary for meaningful interpretation requires nondestructive techniques, such as active remote sensing, that operate at high resolutions. Terrestrial laser scanning (TLS) is a remote sensing three-dimensional technique that has been used to capture structural heterogeneity of fuels in longleaf pine systems (Loudermilk et al. 2009, 2012; Rowell and Seielstad 2012; Rowell et al. 2015). TLS and other light detection and ranging (LiDAR) technologies provide a way to measure complex structures in the field with high accuracy and precision (Hopkinson et al. 2004). Unlike most LiDAR technologies, which are commonly used to quantify canopy structure across landscapes (Andersen et al. 2005; Hudak et al. 2009), TLS is positioned under the canopy to reduce the shadowing effects of overstory trees and can therefore provide fine-scale resolution (<1 m) data of the ground cover (Slatton et al. 2004). The high-density three-dimensional point data (>10,000 points/m^2) from TLS provide the precision needed to characterize the complex surface fuels within longleaf pine systems (Figure 6.5). Data extracted from TLS are in the form of fuel height distribution metrics (including average, maximum, variance, skewness, and kurtosis) and laser pulse intensity, which represents the combination of surface area and reflectance of individual fuel components. These values can be quantified in three-dimensional voxels or two-dimensional pixels (for example, 10 × 10-cm areas) across each measurement plot. This is useful for relating to other fuel characteristics, fire intensity measurements, or fire effects at similar scales.

TLS has also been useful for surface fuel characterization. Loudermilk et al. (2009) found that fine-scale volume estimates from TLS are strongly correlated to measurements of leaf area ($r^2 = 0.70$) and leaf biomass ($r^2 = 0.83$) for saw palmetto (*Serenoa repens*) and wax myrtle (*Myrica cerifera*)—two common shrub species in longleaf pine ecosystems. TLS measurements offered a significantly finer resolution and therefore were more precise than traditional methods of measuring volume (as a cylinder or spheroid), with discrepancies increasing as the size of plants increased. Results from other studies using TLS to estimate aboveground biomass and leaf area in colder and drier shrub-dominated systems (Olsoy et al. 2014; Greaves et al. 2015) can likely be applicable in longleaf pine areas.

Rowell and Seielstad (2012) used pre- and postburn TLS data to distinguish fuel types in a longleaf pine fuelbed. Their goal was to distribute field-collected estimates of biomass across various fuel types to provide a continuous representation of fuels and biomass. In a follow-up study, Rowell et al. (2015) developed fuel-height models from TLS scans of treeless 2-ha sites that were dominated by mixed grasses and shrubs and located in an area adjacent to longleaf pine stands in northwestern Florida. These data are valuable for employing TLS technology to examine the accuracy and bias of fine fuels measurements across a relatively large area at a fine resolution (about 2 cm). The study results showed that the scaled up (<1-m^2 resolution) TLS heights were comparable to field measurements. However, the coarse sampling methodology produced height distributions of the original TLS data that differed from the field height distributions. Ultimately, TLS provided a continuous spatially explicit representation of fine-scale fuel heights at a scale and data richness level that outperformed typical field methods. Measurements using TLS to estimate the influence of fire on surface fuels and consumption are increasing (Wang and Glenn 2009; Gupta et al. 2015).

Airborne laser scanning is often unable to provide quality estimates of ground cover vegetation, partly because its horizontal resolution is limited to only a few decimeters and partly because of canopy-obstruction issues (Slatton et al. 2004). To overcome these obstacles, researchers have used airborne laser scanning to map shrub percentage cover over large landscapes in varying canopy densities of ponderosa pine (*P. ponderosa*) or mixed conifer forests (Martinuzzi et al. 2009; Wing et al. 2012) in two western forests. This work has promising applications for longleaf pine stands, as many of the ponderosa stands were of similar stand and fuelbed construct.

Although these technologies are powerful and can provide invaluable information, some limitations of using TLS or LiDAR should be considered. The expense of instrumentation and amount of time required for data acquisition and processing are both excessive, as is the expertise needed to handle data collection and analysis (Dassot et al. 2011). Estimating important fuel within the three-dimensional TLS point cloud types is difficult—and is especially so for pine needle litter, which drapes across other living and dead fuels and is virtually indistinguishable from the soil and other litter. Multiple scans are used to reduce shadowing effects, however, these effects still occur and are more pronounced in denser vegetation. Even though TLSs have large-range capabilities (for example, ≤1500 m for Optech's ILRIS-36D), the area that can be measured is limited by the density and size of tree boles, obstruction from other vegetation, overall vegetation density, and the differences in point-density that occur across larger areas (with spatial bias occurring toward the plot edges that tend to be closer to the TLS instrumentation) (Rowell et al. 2015).

An alternative to TLS for characterizing surface fuels is photogrammetry techniques. Since the 1930s, these techniques have used overlapping (aerial) photographs to create three-dimensional "stereophotos" for use in timber cruising, detecting land use and land cover changes, and estimating tree and stand characteristics (Spurr 1960; Slama et al. 1980; Miller et al. 2000; Naesset 2002; Zagalikis et al. 2005). When LiDAR was introduced in the 1980s, and soon afterward became more affordable and accessible, it quickly took the place of photogrammetry. Photogrammetry has, however, advanced over that time (Miller et al. 2000; Zagalikis et al. 2005), and digital imagery and photogrammetric software or workstations have replaced hard copy photographs and stereoscopes. Furthermore, photogrammetry has recently become competitive with light detection and ranging technology, producing high-quality three-dimensional renderings of urban and forest structures for a fraction of the cost (Dandois and Ellis 2013). In addition, Bright et al. (2016) have developed a method for using photogrammetry to measure three-dimensional ground cover vegetation and coarse woody debris in a xeric longleaf pine ecosystem. The resulting photogrammetric height metrics (similar to TLS) and color photo values (red, green, and blue) can be used to characterize centimeter-scale fuelbed height distributions similar to 10-cm scale point-intercept sampling, and to predict plant functional groups at this same scale. This work shows that three-dimensional photogrammetric points can provide fine-scale measurements of ground cover fuels and plants that are comparable to those derived from TLS.

New ways of creating virtual fuels and fuelbeds have been introduced to circumvent the need for physically measuring the ground cover fuelbed. Rowell et al. (2016) worked with the same photographs for photogrammetry as those used by Bright et al. (2016) to create models of individual fuel types (such as leaf litter, shrubs, grasses, and pine cones) and then constructed fuelbeds of mixed fuel types (Figure 6.6). With this approach, metrics of interest (such as bulk density, total volume, and height) can be extracted from the fuelbed without shadowing effects, variable point density distributions, limits on fuel type identification, and other issues associated with TLS. There is promise in constructing fuelbeds across larger areas (stands or management units) with simple guidance from photographs, measurements from field sampling and literature searches, and other known

(a) (b)

FIGURE 6.6 Images of a fuelbed shown as (a) a three-dimensional synthetic model of fuelbed created from in-situ point-intercept fuel-type and height data, and (b) an at-nadir photograph. (From Rowell, E. and E. Loudermilk, unpublished data.)

variables such as time since last fire and fluctuations in cone production. In addition, this approach offers the potential to serve as a bridge connecting estimates of type and biomass from field sampling to the three-dimensional structural data from TLS. It can be used to produce continuous and consistent estimates of fuel types, biomass, and volume that would be valuable in efforts to estimate stand fuel loading and prepare spatial inputs to fire behavior models.

Measuring Fire

Capturing spatial fire behavior measurements is difficult and requires techniques beyond those that have been typically used in fire ecology studies. For many years, the available technology for measuring wildland fire intensity was limited to qualitative estimates, point measurements, and relative indices of intensity that do not lend themselves to comparisons across multiple studies (Kennard et al. 2005). These limitations hampered efforts to mechanistically link the energy released by fire to actual fire effects, especially for spatially disparate variables. Measurements of energy transfer (watts, joules), not just temperature, are essential for predicting and understanding both first- and second-order fire effects (Van Wagner 1971; Johnson and Miyanishi 2001; Dickinson and Ryan 2010).

Advances in infrared thermography have increased the possibilities of directly connecting fire behavior to fire effects (O'Brien, Loudermilk, Hiers, et al. 2016; O'Brien, Loudermilk, Hornsby et al. 2016). This well-established technique (Maldague 2001; Meléndez et al. 2010) is especially useful for measuring radiation emitted by surfaces that are heated by fire (Figure 6.7) and integrating the impact of

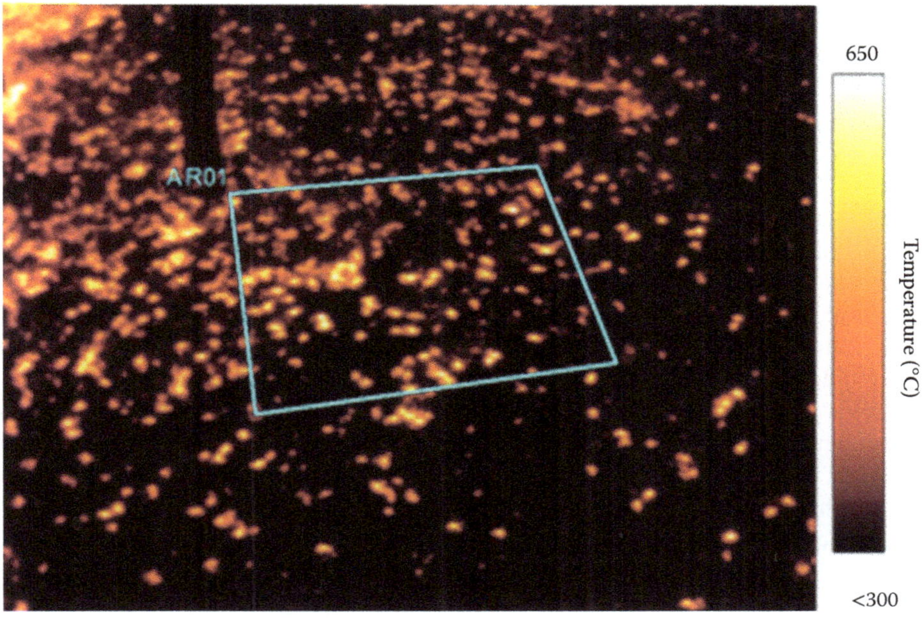

FIGURE 6.7 Infrared thermography capturing smoldering small coarse woody debris (pine cones and a few other 10-hour fuels) after the passing of a low-intensity fire through fuels under a longleaf pine (tree bole in upper left corner). The area is about 4 × 4 m, illustrating the fine-scale variability in fire intensity. (From Mitchell, R. J. et al., *Journal of Forestry*, 107, 391–397, 2009.)

radiative, convective, and conductive heating. Many available products offer high spatial and temporal resolutions (for example, 1 × 1-cm pixels collected at a frequency of 1 Hz), and some offer long-wave infrared radiation—a band especially useful for smoky environments (Rogalski and Chrzanowski 2002).

If exceptionally high-resolution information is not required, coarser-scale infrared instrument platforms are available to provide information on fire energy release and spread that can answer other relevant questions. For example, dual band radiometers (Kremens et al. 2012) can capture quantitative data on fire radiant heat release that can be used to examine the impact of forest structure on fire behavior and fire effects (Cannon et al. 2014). In addition, aerial and satellite platforms (Lentile et al. 2006; Dickinson et al. 2016; Hudak et al. 2016) can provide synoptic information on whole-fire heat release and fire spread.

Linking Fuels to Measurements of Fire

The advances in methods to measure multiscale fuels and fire described above provide a means to capture data on the mechanisms that control fire behavior in longleaf pine ecosystems. In addition, advances in statistical techniques facilitate the analysis of the resulting (potentially very large) data sets that have complex nonlinear, high-order relationships (Prasad et al. 2006; Seni and Elder 2010). In a southwestern Georgia longleaf pine woodland, Loudermilk et al. (2012) developed nonlinear correlations between fire behavior measurements recorded with infrared thermography and fine-scale fuels measured by TLS and field data. This work demonstrates the importance of coupling fuelbed height metrics, fuelbed continuity, and fuel types as driving influences on fire dynamics. They found that fire behavior is best predicted by characterizing fuelbed heterogeneity and continuity across multiple plots that have similar fire intensity, optimizing plot-to-plot variability in fuel characteristics and fire weather conditions; and their assessments of individual plots confirmed the significance of individual fuel types. These studies hint at the potential of such technologies and approaches to characterize fuels and fire in new ways, providing novel opportunities to advance fire-effects research in longleaf pine and associated ecosystems.

FUEL INFLUENCES ON FIRE EFFECTS DURING RESTORATION

Overstory Disturbance Impacts of Fuels

In the past two decades, efforts to restore longleaf pine communities have begun to move away from a simplistic focus of fire as a monolithic disturbance that primarily influences forest structure and toward a more holistic view of vegetation, fuels, and the variability of fire behavior—with fire in essence seen as many disturbances. This perspective has become a critical factor in the development of successful restoration projects, particularly on sites where fire regimes have been disrupted.

In many fire-excluded and poorly burned degraded stands, restoration has focused primarily on the removal of midstory stems of fire-intolerant hardwoods (Provencher, Herring, et al. 2001; Hiers et al. 2014). This strategy led to many unintended consequences for subsequent attempts to reintroduce fire, including overstory and old-growth mortality, escaped prescribed fire, and loss of species (Varner et al. 2005, 2007). Recognizing the necessity of first restoring an appropriate fuelbed has come slowly because doing so challenged the prevailing paradigm as to why ecosystems were degraded in the first place—that fire-exclusion creates an environment in which midstory plants encroach, blocking out light from the diverse ground cover flora of longleaf pine forests (Provencher, Herring, et al. 2001).

A more nuanced view resulted from a series of studies that quantified the role of duff consumption in prescribed fires that were applied during restoration projects (Varner et al. 2005, 2007; O'Brien et al. 2010). The researchers found that duff consumption not only compromises the integrity of the longleaf pine overstory through excessive mortality (Varner et al. 2005, 2007; O'Brien et al. 2010), but it also complicates future prescribed fires—in the short run by suddenly adding dead fuels (Varner et al. 2005), and in the long run by removing the overstory and thereby reducing needle cast (Mitchell et al. 2006; Mitchell, Hiers, et al. 2009). Just as critical, however, was the discovery that the accumulation of forest-floor matter plays a direct role in ground cover degradation and that it also inhibits ground cover recovery, at least in xeric sites (Hiers et al. 2007).

The prioritization of fuels management as a precursor to diversity recovery has become more widely accepted, but an overemphasis on forest structure—particularly on the presence of pyrophytic midstory hardwoods in xeric sites—continues to absorb resources while compromising restoration goals (Hiers et al. 2014). Taking a fuels-management approach to the restoration of fire regimes relegates forest structure to a secondary goal to be achieved by multiple burns over a longer time frame. Restoration objectives should be focused on the reduction of encroachments by semideciduous and evergreen hardwoods and shrubs to promote the reestablishment of fuelbeds that will propagate fire through stands (Hiers et al. 2007).

Plant diversity in longleaf pine communities is a direct product of frequent fire, both across and within stands. Any fuel disruption caused by overstory removal (hence, removal of pine needle sources), soil disturbances, or a combination of both, will affect plant recruitment patterns both directly and indirectly. Disturbances that disrupt soil profiles and alter soil structure and topography by compacting and churning mineral soils (such as entrenchments and vehicle tracks) can alter patterns of fire behavior, disrupt fire spread, and inevitably change plant recruitment patterns as well.

Incorporating Time into Fire-Based Restoration

Among the principles that guide ecological forestry is the need to incorporate time into strategies for recovering degraded ecosystems (Franklin et al. 1997, 2007). The fuels-driven strategy for ecosystem restoration is built on the old adage, "it took 50 years to degrade, it will take 50 years to restore." Despite the recent widespread trend of using herbicide and mechanical treatments, no surrogate is available to replace fire in ecosystem recovery (Menges and Gordon 2010; Outcalt and Brockway 2010). Often, the value of the "rush to restore with fire" approach is minimal because the recovery of fuels can take decades, particularly when considerable duff is present. An approach that incorporates time and patience is also likely to produce more lasting outcomes during periods of rapid ecological change, when variations of ecological conditions can drive restoration trajectories in unpredictable directions (Hiers et al. 2012; Loudermilk et al. 2016). Increasingly, understanding the role of time is also altering perceptions about appropriate restoration targets (Hiers et al. 2012; Kirkman et al. 2013). Longer term cycles can involve temporal dynamics that range from structural changes—such as those resulting from the massive 1996 longleaf pine seed crop—to compositional changes that are associated with recovery from long-term perturbations (Kirkman et al. 2013). Such long-term views on ecological variability will be critical in maintaining longleaf pine resiliency in a future of climate uncertainty (Loudermilk et al. 2016).

Fire Regime Effects

Long-term fire regime effects can dramatically alter fuelbed properties; these, in turn, mediate fire behavior through the compounded impacts of positive feedback loops that are associated with the ecology of fuels. Such effects can operate over several decades or more, and are particularly important for managed fire regimes because the edge effects from managed fire regimes can alter habitats within longleaf pine stands (Lashley et al. 2014). Fuelbeds in longleaf pine systems are dynamic in that their structure and continuity continuously change, both within a single fire cycle and over multiple fire events. Understanding these changes through time, in particular with changing fire-return intervals, is important for long-term restoration management. Applying fires as often as fuels will permit (often at 1–3-year intervals) maintains the highest levels of native plant diversity as well as maintains fuel levels, both of which experience significant changes at slightly longer (6–7-year) return intervals (Glitzenstein et al. 2003; Kirkman, Goebel, et al. 2004; Glitzenstein et al. 2012); for more details see Chapter 11. When fire-return intervals are regular over long periods, the effects can be seen directly in the relative proportion of fuel types—primarily grasses, forbs, and woody species (Glitzenstein et al. 2012). These changes are accompanied by alterations in fuel moisture, relative proportion of live and dead materials, and ultimately modifications of fire behavior. The changes in fire behavior, in turn, have a reinforcing effect on fire regimes over time (Glitzenstein et al., 2012).

IMPORTANT DIRECTIONS FOR FUTURE PROGRESS

The advancement in applications of fire ecology research to longleaf pine restoration requires the development of fire and ecosystem models that are increasingly mechanistic. Two examples of new mechanistic models are the ecosystem process models used to estimate carbon consequences of burn regimes (Martin et al. 2015), and the numerical models coupling fire-atmosphere dynamics (Linn et al. 2002) that examine the convective dynamics of wildland fires, including effects from aerial ignitions (Department of Defense Strategic Environmental Research and Development Program #RC-2643). At individual burn scales, model capacity to document convection-driven heat exchange is already challenging the conventional wisdom of a previous model-based management recommendation (NWCG 2015).

Also needed for understanding the complexities of fire behavior and effects is the ability to examine the impacts of fires on soil properties, plant diversity, longleaf pine recruitment, and interactions with other disturbances (such as the effects of drought on fire-soil interactions or the effects of hurricanes on canopy-fire feedbacks). The importance of understanding the variability of disturbances on ecological patterns and processes is gaining in recognition. Such variability not only drives ecological trajectories of longleaf pine community composition (Kirkman et al. 2013), but also can be critical for other processes affected by fire (Lashley et al. 2014; Hiers et al. 2014).

In a recent review of soil moisture research, Matthews (2014) summarized field research and modeling efforts since 1991, building on a review from Viney (1991), and outlined several areas of research needed to rectify important information gaps about fuel moisture and fire behavior:

- Although soil moisture research has concentrated on pine fuels, the diverse effects of hardwood litter are also important and deserve greater attention (Kane et al. 2008; Varner, Arthur, et al. 2016).

- An open question remains about the interaction of soil moisture with surface fuel moisture and drying rates in fuelbeds, particularly with respect to the role of capillary action. Microtopography and microclimate at the wildland fuel cell scale could dramatically improve predictions of low-intensity surface-fire regimes, but not the role of soil moisture at that scale.
- The temperature of fuel and its relationship to fuel moisture content has been assumed to be a geometrical model function, but field work has shown that solar radiation is confounded with or depends on vegetation structure or micrometeorology.
- Spatial data reflecting patterns of fuel moisture content are needed to test existing model limitations experimentally and improve understanding of model functioning (Matthews 2014). Spatial variability is likely to be critical to fire behavior in low-intensity surface-fire regimes, such as those used in longleaf pine ecosystems. Incorporating fuel moisture content explicitly into wildland fuels cells common to longleaf pine systems would improve the ability to predict fire behavior and to connect fire behavior to postburn fire effects.

For this emerging knowledge to be incorporated into management decisions by forest managers and owners, the wildland fuel cells concept, remotely sensed data, and three-dimensional fuel models need to be translated into useful products (such as those used to map fuels). To integrate wildland fuel cells into management, additional research must identify the relevant scale of fuel variation that drives fire behavior and fire effects for models. This would require further development of the wildland fuel cell concept within various types of longleaf pine communities. Information is available on the scale of wildland fuel cells, how to characterize them, and how they link to surface fire behavior in longleaf pine sandhills and other upland pine habitats; but shrub-dominated fuels, such as those found in longleaf pine flatwoods, remain an area of uncertainty.

More information is needed on how wildland fuel cells function (either with each other or when influenced by fire-atmosphere dynamics) to affect fire intensity and fire movement throughout a stand. Also needed is testing of the wildland fuel cell concept in other fire-driven ecosystems—focusing on the scale at which fuels and fire interact to produce fire effects. The use of TLS, airborne laser scanning, and photogrammetry to capture fuel heterogeneity more thoroughly is becoming widespread, and offers a tremendous opportunity to characterize variability relevant to fire behavior and ecosystem response, but linking fuels at multiple scales requires additional work. New fuel measurement techniques and the development of three-dimensional modeling techniques will provide valuable improvements in the quality of the fuel characterizations that are used for fire effects research and modeling fire behavior.

The challenge remaining for landscape-scale application of fire research is the tension between expanding the number of acres treated with prescribed fire and minimizing impacts on air quality and other societal values (see Chapter 13). To help navigate this increasingly narrow management space will require better tools for modeling smoke, seasonal fire effects, and ecosystem responses to less frequent fire regimes. Last, ecosystem process models—and the incorporation of realistic fire disturbances into their framework—would provide insights into longleaf pine and associated ecosystems and predictions of how these ecosystems might respond to changes in climate, climate-fire dynamics, and management in an era of continued constraints from urbanization. The value of such predictions will ultimately rely on robust monitoring of long-term ecosystem trends as well as understanding new and unexpected feedback loops within the ecology of longleaf pine fuels.

ACKNOWLEDGMENTS

We thank Andy Hudak, Michael Stambaugh, Morgan Varner, and Eric Rowell for their thoughtful reviews on this chapter. We would like to add a special acknowledgement of our appreciation for the mentorship, friendship, and insight of the late Bob Mitchell. His legacy continues to influence conservation management and research in longleaf pine ecosystems.

7 Ecosystem Processes and Restoration

Lindsay R. Boring, Joseph J. Hendricks,
R. Scott Taylor, and Daniel Markewitz

CONTENTS

INTRODUCTION

Belowground processes are exceptionally important in regulating the productivity and nutrient availability of longleaf pine (*Pinus palustris*) ecosystems, given that these fire-dependent ecosystems lose large amounts of aboveground biomass and litter with burning and that they mostly occupy an area of the United States—the southeastern Coastal Plain—that is subject to extremes in water and nutrient availability. Compared to other forest types, these frequent-fire forests are more intricately influenced by overstory and ground cover species composition, belowground patterns of carbon (C) allocation and root turnover, nitrogen (N) fixation, and multiple limiting soil resources. When fire exclusion or conversion of land to other uses alters these sites, either will leave a legacy of significant changes in edaphic resource availability and biogeochemical cycles in addition to changes in species composition. Thus, the key compositional, structural, and functional attributes of longleaf pine ecosystems cannot be successfully restored without also reestablishing the conditions needed to: (1) foster the soil processes that are integral to long-term sustainability, and (2) build resilience to environmental stresses, both currently and in an uncertain climatic future (Millar et al. 2007; Hobbs et al. 2009; Bizzari et al. 2015; Hanberry et al. 2015).

Multiple limiting factors—such as water, nutrients, and sunlight—influence species composition, productivity, and other processes of longleaf pine ecosystems; the relative importance of these factors varies across an edaphic gradient (Kirkman et al. 2016; Starr et al. 2016). Coastal Plain sites occur on a soil moisture gradient that ranges from ultraxeric to hydric:

- Ultraxeric sandhill sites are characterized by excessively drained Entisols and dry Ultisols (Myers 1990; Peet 2006).
- Mesic terrace sites have moderate soil moisture content; they occur on soils that are somewhat excessively drained to somewhat poorly drained Ultisols (Goebel et al. 2001; Peet 2006).

- Most hydric sites are characterized by soils that are poorly drained Ultisols; soils on flatwood hydric sites are Spodosols (Abrahamson and Hartnett 1990; Peet 2006).

To illustrate the relationship between soils and ecosystem types, previous publications on the distribution of longleaf pine ecosystem components have included coordinate plane graphs of surface-soil moisture versus silt (Peet 2006) and of hydrologic regime versus fire frequency (Goebel et al. 2001). Figure 7.1 shows a conceptual diagram of the soil moisture gradient, using an equilateral triangle with ultraxeric and hydric soils at the base angles and mesic soils at the vertex angle. The correlation between soil moisture and aboveground net primary production (ANPP) is well documented within the context of other soil physical and chemical properties and with respect to species richness, diversity, and composition of longleaf pine ecosystems (Myers 1990; Mitchell et al. 1999; Wilson et al. 1999, 2002; Kirkman et al. 2001, 2016; Goebel et al. 2001; Peet 2006). The relationships among soil moisture, vegetation, and key belowground processes are especially important for longleaf pine ecosystems because of their influence on efforts to restore sites that have been degraded by agriculture, fire exclusion, or conversion to intensively managed pine plantations (Palik et al. 2000; Kirkman et al. 2013).

In this chapter, we summarize the influences of fire regimes on biogeochemical cycles as well as the effects of fire exclusion and agricultural land use on soil processes across soil types within the longleaf pine range. We also discuss the changes in ecosystem attributes (particularly C and N cycles) that are associated with degraded longleaf pine sites, and the implications of restoration and fire reintroduction on fire-excluded sites. Most of the longleaf pine ecosystem studies cited in this chapter spanned the soil moisture gradient for longleaf pine ecosystems on sites that included Fort Benning Army Base and the Joseph W. Jones Ecological Research Center at Ichauway in southwestern Georgia, Eglin Air Force Base in the Florida Panhandle, and the Savannah River National Laboratory in southwestern South Carolina.

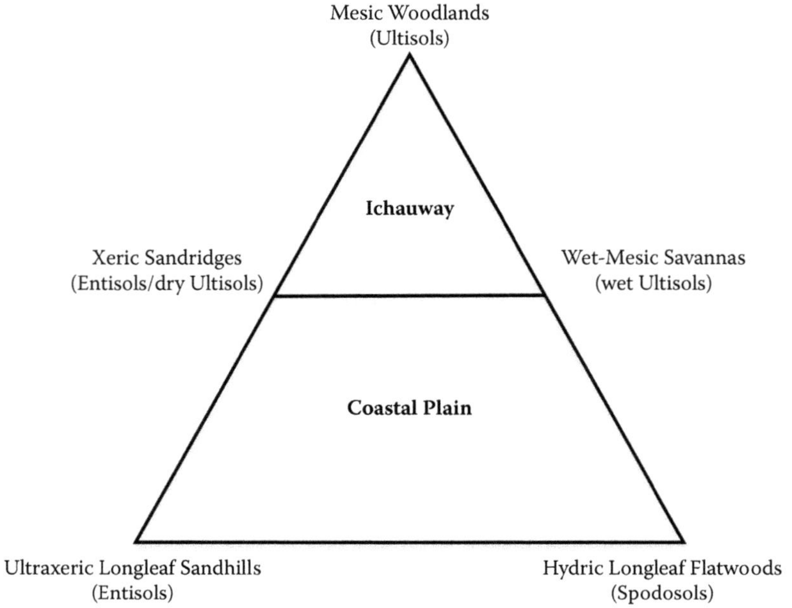

FIGURE 7.1 Schematic of longleaf pine grouping by soils along an edaphic gradient for the southeastern U.S. Coastal Plain and an upland area in southwestern Georgia (Joseph W. Jones Ecological Research Center at Ichauway).

KEY BASELINE LONGLEAF PINE ECOSYSTEM AND SOIL PROCESSES

Aboveground Productivity

Across a gradient of longleaf pine-wiregrass (*Aristida stricta*) site types at Ichauway, ANPP positively correlated with moisture availability (Figure 7.2). Although annual rates of ANPP varied more than two-fold from xeric (3.7 Mg/ha) to wet-mesic (7.5 Mg/ha) sites, the proportional allocation of ANPP within the various sites was consistent across all site types, averaging 43% to ground cover production, 32% to overstory foliage production, and 25% to overstory woody stem and branch production. The large proportional allocation of ANPP to photosynthetic tissues (with as much as 75% comprised of overstory foliage and ground cover tissues) has a functional significance, as the litter of these nonwoody tissues is the primary fuel source in these fire-dependent ecosystems (Mitchell et al. 1999, 2006; Boring et al. 2004).

ANPP was positively correlated with soil moisture availability but negatively correlated with N availability across the gradient (Table 7.1), even though N has always been a presumed limiting nutrient to production in longleaf pine-wiregrass ecosystems (Pecot et al. 2007; Ford et al. 2008; Hendricks et al. 2016). The cumulative N mineralization rate (annually ranging from 3.5 kg/ha for wet-mesic to 11.8 kg/ha for xeric sites), ranks among the lowest estimates reported for North American forests (Wilson et al. 1999). The inverse relationship between ANPP and N mineralization might be attributed to the open overstory canopy and ground cover structure on low-productivity xeric sites, which would allow more light penetration to the forest floor; the resulting higher soil temperature increases could stimulate microbial activity. Additionally, unlike other site types, xeric sites are codominated by fire-tolerant oaks (*Quercus* spp.). The deposition of more oak leaf litter, with its higher substrate quality and faster decomposition rates than pine needle litter, may have contributed to higher N mineralization rates (Wilson et al. 1999; Hendricks et al. 2002). Finally, soil disturbances by fossorial animals, such as gopher tortoises (*Gopherus polyphemus*) and pocket gophers (*Geomys pinetis*), which are more common at the intermediate to xeric end of the moisture gradient—can also stimulate N mineralization (Mitchell et al. 2006).

A 10-year edaphic resource manipulation study in wet-mesic and xeric site types has provided insights into the controls on ANPP in longleaf pine-wiregrass forests (Kirkman et al. 2016). In wet-mesic sites, neither irrigation nor N fertilization had a significant effect on midstory/overstory,

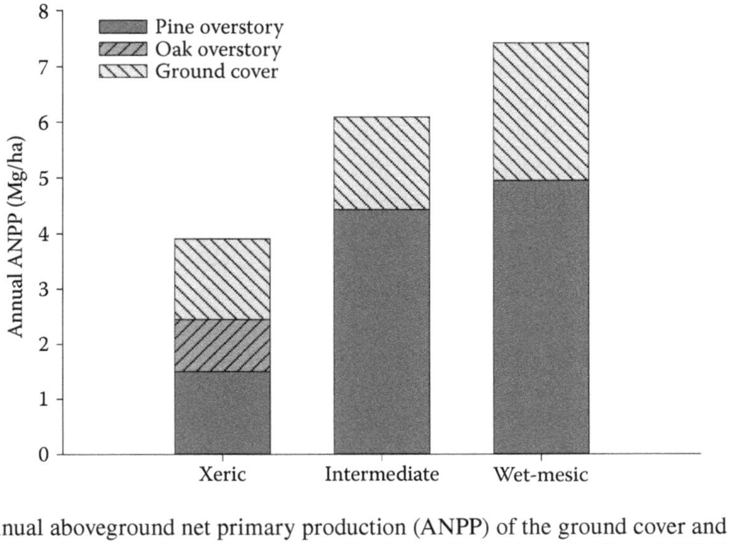

FIGURE 7.2 Annual aboveground net primary production (ANPP) of the ground cover and overstory (pine and oak) across an environmental gradient at an upland area located on the Joseph W. Jones Ecological Research Center at Ichauway in southwestern Georgia. (Adapted from Mitchell, R. J. et al., *Canadian Journal of Forest Research*, 29, 743–751, 1999; © 2008 Canadian Science Publishing or its licensors. With permission.)

TABLE 7.1
Average (±1 SE) Aboveground Net Primary Production (ANPP), Net Nitrogen Mineralization (NNM), and Other Selected Physical and Chemical Properties of Surface Soil (0–10 cm) from a Fire-Maintained Forest Ecosystem with a 2-Year Fire-Return Interval

Site Type	Annual ANPP (Mg/ha)	Annual NNM (kg/ha)	Annual Moisture (%)	Annual Temperature (°C)	Carbon (%)	Nitrogen (%)
Xeric	3.7 (0.2) C	11.8 (0.6) A	4.4 (0.3) C	22.2 (0.6) A	0.80 (0.07)	0.03 (0.01)
Intermediate	6.9 (0.2) B	6.8 (1.4) B	9.0 (0.9) B	22.1 (0.6) A	1.21 (0.06)	0.04 (0.01)
Wet-mesic	7.5 (0.1) A	3.5 (0.4) C	12.4 (1.1) A	20.1 (0.1) B	1.26 (0.08)	0.06 (0.02)

Source: Reprinted from Wilson, C.A. et al., *Soil Biology and Biochemistry*, 34, Soil nitrogen dynamics in a fire-maintained forest ecosystem: results over a 3-year burn interval, 679–689, Copyright (2002), with permission from Elsevier. Data are from the Joseph W. Jones Ecological Research Center at Ichauway in southwestern Georgia. Values within columns with different letters (A, B, or C) denote significant differences within columns at $p < 0.05$.

understory hardwoods/ground cover, or total ANPP, suggesting the importance of nonedaphic resource (such as light) controls on productivity. By contrast, in the xeric sites, irrigation significantly increased ANPP in all layers and N fertilization stimulated a marginal ANPP increase in understory hardwoods and ground cover. Collectively, these results point to the potential for multiple resource limitations of ANPP across site types, and the potential variation in resource controls among the plant life forms within an ecosystem.

Belowground Productivity

That belowground C and nutrient allocation could be particularly important in longleaf pine-wiregrass ecosystems (Auld 1987; Jacqmain et al. 1999; Hendricks et al. 2006, 2016) is suggested by the species' initial grass stage growth habit, whereby height growth is limited in favor of extensive root system development (Boyer 1963; Landers et al. 1995; Mitchell et al. 1999, 2006).

A substantial proportion of belowground allocation, used to develop and support the feeder root system, is important for acquiring limiting edaphic resources in sites where soils are coarse textured and infertile (Mitchell et al. 1999; Wilson et al. 1999, 2002). Belowground minirhizotron assessments at Ichauway revealed that fine root production was positively correlated with soil moisture and ANPP (Hendricks et al. 2006). In addition, fine root production estimates were significant relative to aboveground (overstory and ground cover combined) production estimates. Comparisons of annual production estimates in the hydric sites (4.6 Mg/ha for fine roots versus 6.5 Mg/ha for foliage) and xeric sites (2.3 Mg/ha for fine roots versus 3.5 Mg/ha for foliage) showed that fine roots amounted to 70% of the foliage production on hydric sites and 66% of the foliage production in xeric sites. In addition, mycorrhizal fungal mycelia extending from the fine roots into the soil matrix could account for substantial C and nutrient allocation; the annual mycorrhizal fungal mycelia production estimate of 2.8 Mg/ha was slightly more than half of the needle production estimate of 5.4 Mg/ha in a 25-year-old longleaf pine plantation (Hendricks et al. 2016). These results suggest that fine root and associated mycorrhizal fungal components account for as much or more C allocation and production as the foliage component of longleaf pine-wiregrass ecosystems.

Allocation to storage in coarse roots may be particularly important in longleaf pine-wiregrass ecosystems that are sustained by regular burning (Sword and Haywood 1999; Langley et al. 2002; Guo et al. 2004; Clarke et al. 2010). Carbohydrate reserves stored in coarse roots are protected from combustion and from consumption by herbivores. Although the importance of these reserves in rapidly regenerating photosynthetic leaf area after burn events has long been recognized (Landers et al. 1995), their role in maintaining an extensive postburn feeder root system (fine roots and mycorrhizal fungi) has also been demonstrated. For example, in foliar scorching experiments, defoliation of mature longleaf pine plantations caused a significant reduction in the stored C reserves of coarse roots, but did not affect the production, mortality, or nonstructural carbohydrate concentrations of the fine roots (Guo et al. 2004, 2008; Aubrey et al. 2012), or ecosystem soil respiration, measured as carbon dioxide efflux (Clinton et al. 2011; Aubrey et al. 2012). Likewise, foliar scorching had no significant impact on the standing biomass, production, or turnover of longleaf pine ectomycorrhizal fungal mycelia (Sims et al. 2007; Hendricks et al. 2016). These results suggest that stored C in coarse roots compensates for the loss of photosynthate production, thereby maintaining a level of fine root and mycorrhizal fungal dynamics that is comparable to unscorched stands (Guo et al. 2004; Varner et al. 2009; Aubrey et al. 2012). Although many studies have demonstrated that fine root and mycorrhizal fungal dynamics are directly correlated to current photosynthate production, decreasing dramatically after defoliation (Ekblad et al. 2013), the absence of a foliage scorching effect on the feeder root system and soil respiration in longleaf pine stands is consistent with other plant species that are adapted to regular foliar disturbances such as fire, grazing, and consumption by insects (Wallace 1987; Eom et al. 1999; Kosola et al. 2001; Langley et al. 2002; Vargas et al. 2009). Collectively, these results support the hypothesis that tree species in frequently disturbed ecosystems allocate proportionately more C to storage than species in less frequently disturbed ecosystems, and are consequently more resistant and resilient to disturbances that affect photosynthate supply (Guo et al. 2004; Aubrey et al. 2012).

Mitchell et al. (1999), Wilson et al. (1999), and Ford et al. (2008) found that the production of the root system in longleaf pine-wiregrass ecosystems is likely limited by the availability of water, N, and possibly phosphorus (P), but the dominant plant life forms in these ecosystems may respond differently to these edaphic resources (Pecot et al. 2007; Ford et al. 2008). Walter (1971) proposed a two-layer model for savannas in which the two dominant vegetation forms (grasses and trees) occupy different rooting zones and consequently depend on different water sources along the soil profile. Assuming that the deeply rooted woody species have access to the more persistent subsurface water table and that the shallow rooted grass species are limited to more ephemeral precipitation inputs near the surface, the two life forms may compete less for water and more for nutrients. However, if the water table drops to a level that is lower than the rooting depth of the woody species, more intense competition for water will result, eclipsing the competitive interactions for nutrients. This model is generally applicable for longleaf pine-wiregrass ecosystems in which the roots of wiregrass and other nonwoody ground cover plants predominantly occur in the upper 30 cm of the soil profile (Saterson and Vitousek 1984) and pines and hardwoods are more deeply rooted (Jacqmain et al. 1999; Addington et al. 2006; Hendricks et al. 2006). Consistent with this model, a soil trenching study found that longleaf pine roots competed more intensely with deeply rooted hardwood species than with nonwoody ground cover species (Pecot et al. 2007). In addition, Ford et al. (2008) reported that the minimum annual water table depth was more strongly coupled to the ANPP of longleaf pine than wiregrass across xeric and mesic site types, and that 95% of the variation in ANPP could be accounted for by differences in water table depth (Figure 7.3) and N additions. This means that belowground competition among plant life forms may vary based on the relative abundance of limiting soil resources.

Belowground competition for water and nutrients can have important implications for longleaf pine-wiregrass ecosystem restoration efforts. Because longleaf pine is shade intolerant, restoration research has focused largely on aboveground canopy gaps to enhance light availability for seedling establishment and growth (Palik et al. 1997; McGuire et al. 2001). However, belowground gaps formed after an aboveground disturbance may be critical for the development of the coarse and

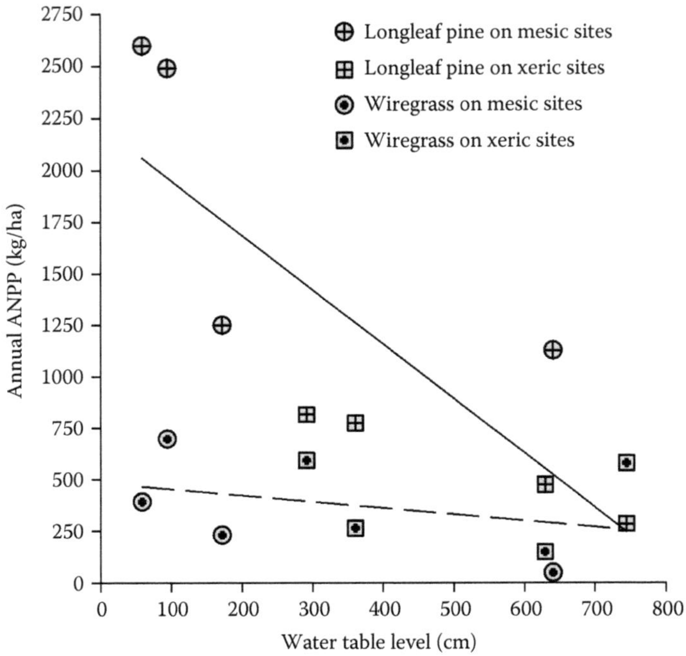

FIGURE 7.3 Annual aboveground net primary production (ANPP) of longleaf pine (solid line) and wiregrass (dashed line) biomass with respect to belowground water availability, expressed as the distance of the water table from the surface, at sites located on the Joseph W. Jones Ecological Research Center at Ichauway in southwestern Georgia. (Adapted from Ford, C. R. et al., *Canadian Journal of Forest Research*, 38, 2118–2127, 2008; © 2008 Canadian Science Publishing or its licensors. With permission.)

feeder root systems (McGuire et al. 2001; Jones et al. 2003; Pecot et al. 2007). Figure 7.4 shows that longleaf pines and associated ground cover species are highly responsive to root gaps, where a reduction in the production of longleaf pine roots results in the proliferation of nonpine roots (and vice versa). Consequently, root gaps are typically more ephemeral than canopy gaps, a belowground dynamic that could dictate the success of restoration efforts.

Fire Influences

As demonstrated in Figure 7.5, regular burning (such as a 1–3-year return interval) influences the species composition, vegetative structure, resource availability, and net primary productivity (NPP) of longleaf pine-wiregrass ecosystems, which, in turn, influence the quantity and quality of fuels for subsequent burn events (Boring et al. 1991, 2004; Hendricks et al. 2002). Central to this complex feedback system is the foliar litter and ground cover that together serve as the primary fuel source for fires and contain potentially large pools of C and nutrients whose form and availability can be altered by burning. Regular burning removes both the ground cover layer and the organic layer on the mineral soil surface. However, soon after a burn event, ground cover vegetation, often dominated by wiregrass and other warm-season grasses, reestablishes a dense crown structure of erect tillers that suspends longleaf pine needles and other litter above the soil surface (Figure 7.6). Hendricks et al. (2002) found that elevated litter can account for more than half of the total litter standing stock in ground cover 1–2 years after a burn event. In addition to having a substantial impact on fuel loading and fire behavior, litter accumulation in ground cover can impose multiple resource limitations by reducing light penetration to vegetation, altering evapotranspiration rates, and immobilizing C and other nutrients (Seastedt et al. 1992; Hendricks et al. 2002; Mitchell et al. 2006).

Litter elevated in wiregrass crowns decomposes slowly because of its reduced contact with decomposer populations and by its altered microclimatic conditions (Hendricks et al. 2002). Decay

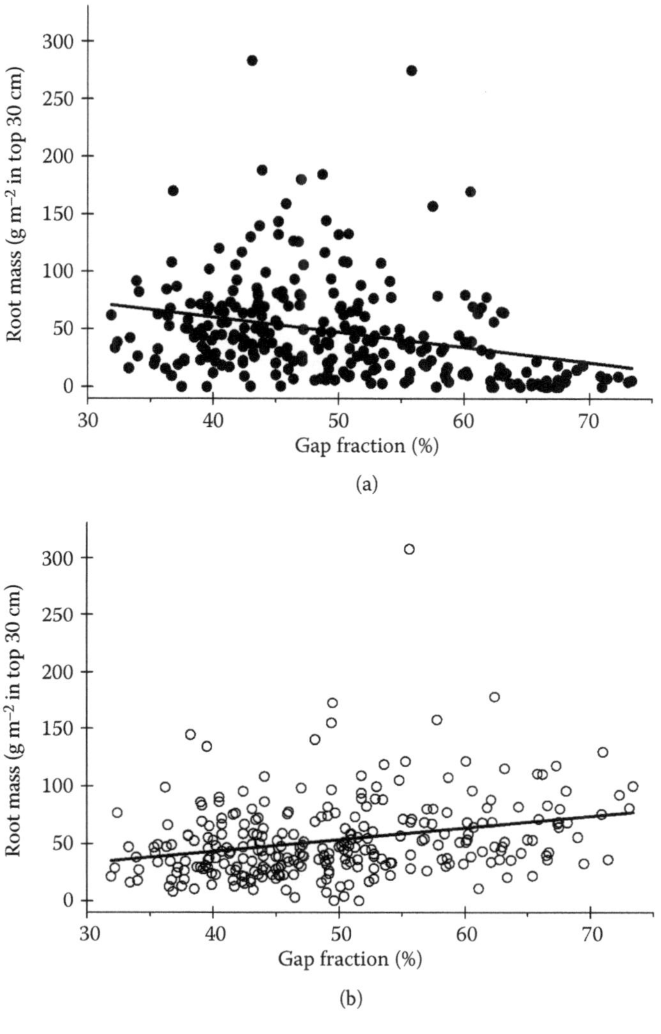

FIGURE 7.4 Effects of (a) longleaf pine and (b) nonpine root mass reductions resulting from disturbance—and consequently increased gap fraction—over an 11-month period at a site located on the Joseph W. Jones Ecological Research Center at Ichauway in southwestern Georgia. (With kind permission from Springer Science + Business Media: Jones, R.H. et. al., *Oecologia*, Controls of fine root dynamics across a gradient of gap sizes in a pine woodland, 134, 132–143, 2003; © Springer-Verlag 2002.)

rates of elevated longleaf pine needles and wiregrass tillers was about 50% lower than the same tissue types measured on the soil surface, which is among the lowest reported rates for all forest ecosystems (Hendricks et al. 2002). This suggests that the ground cover, consisting of the accrued dead vegetative matter and perhaps living biomass depending on the season of the burn, constitutes a relatively large pool of C subject to volatilization and loss from the ecosystem. Although prescribed burning of litter and ground cover may occasionally result in overstory canopy scorch, most other C pools in the ecosystem are not directly reduced. Some of this C is retained as partially combusted organic matter, or "black carbon." Black C mixed into the mineral soil primarily by soil fauna may constitute as much as 8% of the soil C pool to a 50-cm depth (Ike 2010).

Fire has a major impact on the cycling of N and P, nutrients that are potentially colimiting to NPP in longleaf pine woodlands (Walker and Peet 1983; Pecot et al. 2007; Ford et al. 2008) and are among the most immobile elements in the litter (Hough 1982; Gilliam 1988; Hendricks et al. 2002). P, which has a relatively high volatilization temperature, is generally converted to more readily available inorganic

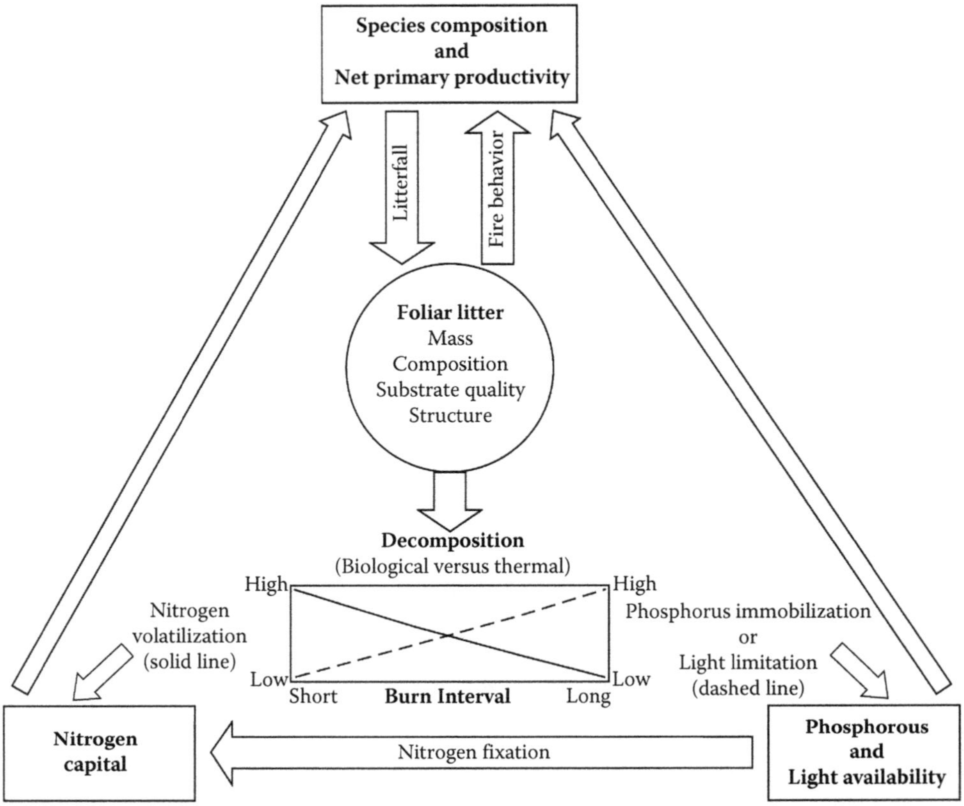

FIGURE 7.5 Diagram illustrating the influences of litter decomposition on resource availability, net primary productivity, and species composition of a longleaf pine-wiregrass ecosystem. (Adapted from Hendricks, J. J. et al., *Canadian Journal of Forest Research*, 32, 928–941, 2002; © 2008 Canadian Science Publishers or its licensors. With permission.)

forms and conserved in the ash during a typical longleaf pine burn event (Boring et al. 2004). In contrast, N has a relatively low volatilization temperature (Raison et al. 1985; Ojima et al. 1994; Gillon et al. 1995). N in litter that has not been converted to ammonium and transferred to the soil through biological mineralization is converted to molecular N_2 and transferred to the atmosphere by thermal mineralization during the burn event, thereby reducing the N capital of the ecosystem (Boring et al. 1991, 2004). However, the brief pulse of both N and P immediately after the burn may be critical to postfire nutrient dynamics and vegetation regrowth (Boring et al. 2004; Dean et al. 2015).

Although frequent low-intensity fires are clearly essential for the restoration and management of longleaf pine-wiregrass ecosystems, evaluating the effect of various burn regimes on the N capital of these inherently nutrient-poor ecosystems is instructive. In a study that measured the effects of burn season, ground cover types, and fuel loading, Boring et al. (2004) found that N losses were substantial for some treatments. For example, a growing-season burn of wiregrass with about 5 Mg/ha of pine litter resulted in an N loss of more than 30 kg/ha—but they concluded that frequent dormant season or even intermittent dormant and growing season burning would produce only modest N impacts. It would also not degrade the N balance of longleaf pine ecosystems due to moderate N inputs from N_2-fixation and atmospheric deposition.

Fire-Legume-Nitrogen Dynamics

Populations of fire-adapted herbaceous legumes (Figure 7.7), many of which are capable of symbiotic N_2-fixation (Boring et al. 1991; Hainds et al. 1999; Hendricks and Boring 1999; Hiers, Mitchell, et al. 2003; Hiers and Mitchell 2007; Cathey et al. 2010), can offset the deleterious effects of burning

FIGURE 7.6 The role of wiregrass in the interception and elevation of longleaf pine needles above the soil surface. (Photograph courtesy of Joseph W. Jones Ecological Research Center at Ichauway.)

FIGURE 7.7 Legume abundance within wiregrass-dominated ground cover. (Photograph courtesy of Stephen Golladay.)

on the N capital of longleaf pine-wiregrass ecosystems. Regular burning confers competitive advantages to these fire-adapted species. Their characteristically hard seed coats are often scarified when exposed to fire, stimulating germination (Cushwa et al. 1969, 1970). The establishment and growth of germinating legumes are likely enhanced by the increased light, water, and nutrient availability in the postburn landscape (Fox and Fox 1986; Waldrop et al. 1987). Additionally, established perennial legumes store large carbohydrate and nutrient reserves in their coarse roots, a trait that allows vigorous resprouting after fire-induced top-kill (Auld 1987; Pate et al. 1990). Ash residues are a source of mineralized nutrients, such as P and calcium (Ca), which can increase soil pH and enhance N_2-fixation rates (Sanginga et al. 1995; Marschner 2012). Finally, even if the abundance of legumes is reduced by competitive interactions or long intervals without burning (or both), their hard seed coats allow their seed to remain in the ecosystem through a persistent soil seed bank (Kaeser and Kirkman 2012); for more detail, see Chapter 11.

Numerous studies have documented the common occurrence of native and naturalized herbaceous legumes in frequently burned pine and mixed pine-hardwood ecosystems of the Southeast (Hainds et al. 1999; Hendricks and Boring 1999; Lajeunesse et al. 2006). Hainds et al. (1999) reported that herbaceous legumes were diverse (at least 37 confirmed distinct species, mostly perennial), ubiquitous (presence confirmed in 94% of the 85 2-m^2 sample plots), and abundant (peak density of almost 120,000 stems/ha) across a complex edaphic resource availability and aboveground productivity gradient. With the exception of the few species that occupy only extremely wet or dry sites, legumes maintain high diversity, distribution, and abundance in frequently burned longleaf pine-wiregrass ecosystems across a broad environmental gradient.

Although legumes are common in regularly burned ecosystems, population size alone is not a reliable indicator of N_2-fixation activity (Hendricks and Boring 1999; Hiers, Mitchell, et al. 2003). Assessments with ^{15}N isotope dilution showed that the dominant herbaceous legumes native to longleaf pine-wiregrass ecosystems vary widely in their ability to fix atmospheric N_2 (Hiers, Mitchell, et al. 2003; Hiers and Mitchell 2007; Cathey et al. 2010), which is expressed as the percentage of N derived from the atmosphere ($\%N_{dfa}$). In a common garden study using three dominant species exposed to a factorial of shade (±) and burning (±) treatments, Hiers and Mitchell (2007) determined that the $\%N_{dfa}$ in control plots (exposed to ambient light and not subjected to burning of aboveground tissues) ranged from 44% for dollarleaf (*Rhynchosia reniformis*) to 82% for goat's rue (*Tephrosia virginiana*). Likewise, in a pot study, Cathey et al. (2010) determined that the $\%N_{dfa}$ under ambient light for nine dominant species ranged from 0% for Piedmont leather-root (*Orbexilum lupinellus*) to 55% for Atlantic pigeonwings (*Clitoria mariana*). Surprisingly, in both the Hiers and Mitchell (2007) and Cathey et al. (2010) studies, light availability did not have a significant main treatment effect on the $\%N_{dfa}$ for the dominant legume species. However, Hiers and Mitchell (2007) reported that light availability did interact with burning and species treatment effects to yield a significant curvilinear relationship between total biomass and $\%N_{dfa}$ ($R^2 = 0.56$, $p < 0.0001$) for all species and treatments combined. This indicated that relatively small increases in the biomass of smaller plants translates into relatively large increases in $\%N_{dfa}$ (and conversely, relatively large increases in the biomass of larger plants translates into relatively small increases in the $\%N_{dfa}$). This wide variation in $\%N_{dfa}$ among the dominant legumes appears to be driven by a complex interaction among species and environmental variables that affect photosynthate production and support the energetically expensive N_2-fixation process (Hiers, Mitchell, et al. 2003; Hiers and Mitchell 2007; Cathey et al. 2010).

The impact of legumes on the N capital of longleaf pine-wiregrass ecosystems also depends on the fate of fixed N, much of which is contained in aboveground tissues that are vulnerable to volatilization during regular burn events. The net balance between the biological and thermal mineralization of N in the tissues of legumes and other species is critical because a fire-induced decline in N can trigger a positive feedback system that progressively decreases the N capital of regularly burned ecosystems (Mooney and Gulmon 1982; Monleon et al. 1997; Aber and Melillo 2001; Carter and Foster 2004). As the N capital and mineralization rates of ecosystems decrease, the substrate qualities of litter tend to degrade (high lignin and low N concentrations), often resulting in relatively slow decomposition and N mineralization rates. Consequently, the litter accumulates N that is vulnerable to volatilization during subsequent burn events, thereby promoting the positive feedback system. This potential for burning regimes to decrease the N capital emphasizes the importance of understanding the complex interactions and feedbacks among edaphic resource availability, NPP, and disturbance regimes of longleaf pine-wiregrass ecosystems (Mitchell et al. 2006; Kirkman et al. 2016).

FIRE EXCLUSION AND RESTORATION

In longleaf pine woodlands, fire exclusion—either by suppression or by infrequent use of prescribed burning—has dramatic impacts on the structural components of the ecosystem (Christensen 1981; Gilliam and Platt 1999; Glitzenstein et al. 2003). It changes the structure of the forest canopy and forest floor, alters biogeochemical processes in vegetation and soil, and produces a long-term successional transformation to dominance by upland hardwood species. Fire exclusion favors shade-tolerant hardwood seedlings, shrubs, and vines, which can easily outcompete the exceptionally diverse native ground cover that is typical of longleaf pine woodlands (Hiers et al. 2007).

Without frequent combustion, leaf litter and coarse woody debris accumulate in longleaf pine forests (Hiers et al. 2007), quickly producing large organic pools of C and N on the forest floor. Also, the delayed use of fire reduces the pyric mineralization of P and cations such as Ca, potassium (K), and magnesium (Mg). Although N and C are no longer lost by combustion from the litter and surficial soil pools, litter immobilization of P and possibly other elements may result (Hendricks et al. 2002; Boring et al. 2004). Sudden decreases of long-term C storage may occur when these forests are subjected to inevitable wildfires (Varner et al. 2005; Hurteau and North 2009; O'Brien et al. 2010).

Restoration efforts in fire-excluded forests may require the mechanical removal of woody biomass, which often has no merchantable value and will likely be disposed of on-site. Removing this biomass often involves piling and burning (Korb et al. 2004; Phillips and Waldrop 2008). Studies suggest that increases in P and cations and changes to other soil chemical properties are associated with burning slash piles. Creech et al. (2012) described how the high temperatures (>1000°C) and the long durations (>24 hours) that are associated with burning slash piles resulted in decreased species richness, elimination of the soil seed bank, changes to soil chemical properties (including an increase in pH from weakly acidic to neutral or basic), significant increases in P levels, and increased N mineralization with time since slash pile burning. Although this study showed limited colonization by nonnative plants, other studies have reported increased recruitment of aggressive nonnatives after slash pile burning (Haskins and Gehring 2004; Korb et al. 2004), which can present challenges to restoration. In addition, the persistence and dominance of species of Asteraceae were reported by Chapuis-Lardy et al. (2006) and Creech et al. (2012) in areas where soil P increased. This vegetation dominance coupled with an absence of grasses, which did not respond to P as effectively as broad-leaved species (Halsted and Lynch 1996), may lead to reduced fuel continuity with consequent limitations on prescribed burning within the fire scar.

In contrast to mesic and xeric forests, fire-excluded longleaf pine forests on flatwood sites with Spodosols and fuels dominated by evergreen shrubs burn with higher severity and higher losses of C and N. Recovery of C and N pools in the soil is estimated to take >6 years (Lavoie et al. 2010). Lavoie et al. (2014) showed no differences in soil C and N between reference and fire-excluded plots in an ultraxeric sandhill prior to hardwood removal treatments, and the nutrient-poor soils showed no lasting effects on soil nutrients 15 years after restoration.

In fire-excluded longleaf pine ecosystems on xeric to mesic sites, development of a forest floor can increase soil organic C, but the effect is limited to surface soils (Table 7.2). This may be partially attributed to another outcome of fire exclusion—the loss of burrowing animals and invertebrates normally living in healthy longleaf pine ecosystems. Soil bioturbation by pocket gophers, gopher tortoises, and soil invertebrates (Figure 7.8) is an important process that incorporates both surficial organic and black C into lower soil depths in open longleaf pine woodlands (Kalisz and Stone 1984a; Simkin and Michener 2005). As young stands of longleaf pine mature into open

TABLE 7.2
Average Carbon Content for Soils from 0 to 5 cm Depth That Were Collected in Southwestern Georgia

Land Cover	Carbon Content	
	Percent	*kg/ha*
Longleaf pine forest	2.19 BC	15,132 AB
Other upland pine forest	2.11 BCD	11,527 BC
Upland pine/hardwood forest	2.10 BCDE	12,270 A
Upland hardwood forest	2.05 BCDE	11,964 A
Longleaf pine plantation	1.51 CDE	11,346 BC
Fallow agricultural site	1.38 DE	9479 CD
Old-field agricultural site	1.33 E	9157 CD
Functioning agricultural site	1.29 E	8898 CD

Source: Adapted from Ike, J. C., Spatial variability and land use change: Effects on total soil carbon contents in the coastal plain of Georgia, MS thesis. University of Georgia, Athens, Georgia. p. 89. Letters indicate significant differences among categories based on Duncan's Multiple Range test ($p < 0.05$).

FIGURE 7.8 Abundant pocket gopher mounds in a longleaf pine-wiregrass savanna, and mixing of soil with both organic litter and black carbon. (Photograph courtesy of R. Scott Taylor.)

savannas, the colonization of native soil vertebrates and invertebrates is necessary for mixing soil organic matter, black C, and N to deeper soil horizons.

Exclusion of frequent fire increases the rate of C accumulation, both in forest litter and in the woody biomass of the changing ground cover and overstory. However, this accumulation is potentially unsustainable because it also increases the risk of wildfire (Hurteau and North 2009; Mitchell, Hiers, et al. 2009). Using fire-suppression models to compare healthy forests managed with and without frequent prescribed fire and fuel reduction treatments, Hurteau et al. (2011) demonstrated that fire-suppressed Ponderosa pine (*P. ponderosa*) stored more C but had higher susceptibility to wildfires in the short term and experienced higher long-term C losses. Similar results are expected for longleaf pine forests.

Longleaf pine restoration practices vary depending on fire history, stand composition (including the longleaf or other pines in the overstory), and other factors. Restoring fire to long-unburned stands may require the mechanical removal of encroaching oaks and other successional species as well as clearance of large litter and duff accumulations resulting from the slow decomposition of longleaf pine needles and woody debris (Hendricks et al. 2002). Although the reintroduction of fire must be a major component of restoration, it must be applied judiciously to prevent deleterious effects to residual pines. Their fine roots, which colonize the deepening forest floor of fire-excluded forests, are susceptible to severe fires; damage or removal of these important feeder roots would kill the overstory pines or increase their susceptibility to bark beetle attacks (Varner et al. 2005; O'Brien et al. 2010).

LAND USE AND RESTORATION INFLUENCES ON ECOSYSTEM PROCESSES

The current interest in restoring longleaf pine on marginal agricultural fields and land previously planted in loblolly (*P. taeda*) and slash (*P. elliottii*) pine plantations presents specific challenges and provides an opportunity to study and quantify the ecological processes that influence these conversions. Decades of fertilizer use, surface soil erosion, and alterations in plant competition and dominance patterns dramatically change belowground processes (Richter and Markewitz 2001; Markewitz et al. 2002; Flinn and Marks 2007). The influences of subsistence and industrial agriculture on soils and ground cover vegetation for sites slated for restoration can present many problems.

The long-term negative effects of historical agricultural land use on soil quality are best exemplified by the significant and long-term decreases of C and N that have occurred in the Southeast. Moreover, one of the greatest concerns for successful longleaf pine restoration is the legacy of agricultural fertilization for increased P and base cations that remain in the soil for decades (Richter and Markewitz 2001; Markewitz et al. 2002). This past use of industrial and organic fertilizers, when combined with substantial surface soil erosion, has had significant impacts on the chemical and physical properties of southeastern soils that can influence the establishment and growth of young planted longleaf pine and native ground cover species (Figure 7.9). Markewitz et al. (2002) described soil and stand characteristics along a chronosequence of 1-year-old to 14-year-old longleaf plantations that had been established on former agricultural sites in Coastal Plain upland Ultisols. Compared to natural stands, these sites exhibited reduced pools of soil C and N for 14 years postagriculture, with steadily decreasing soil bulk density from 1 to 14 years after planting. The decrease in bulk density over time was attributed to root growth by young trees and to bioturbation changes in soil structure and porosity caused by burrowing animals and other soil fauna (Table 7.3). In contrast to studies of reforestation of agricultural lands to northern hardwood forests (Flinn and Marks 2007) where decreases in extractable P were observed, substantial reserves of soil P and cations remained from prior tillage and fertilization. These elevated nutrients could provide a competitive advantage for nonnative plants or invasive old-field species over longleaf pine seedlings and native warm-season grasses, the desired species for restoration of longleaf pine communities.

Ground cover recovery is a major factor in restoring the function of the forest floor and soils, primarily through the reestablishment of N_2-fixing legumes and native warm-season grass species. They are critical to the recovery of the depleted N and C cycles by providing major inputs to the surface soil horizons (Hendricks and Boring 1999; Hiers, Mitchell, et al. 2003; Boring et al. 2004;

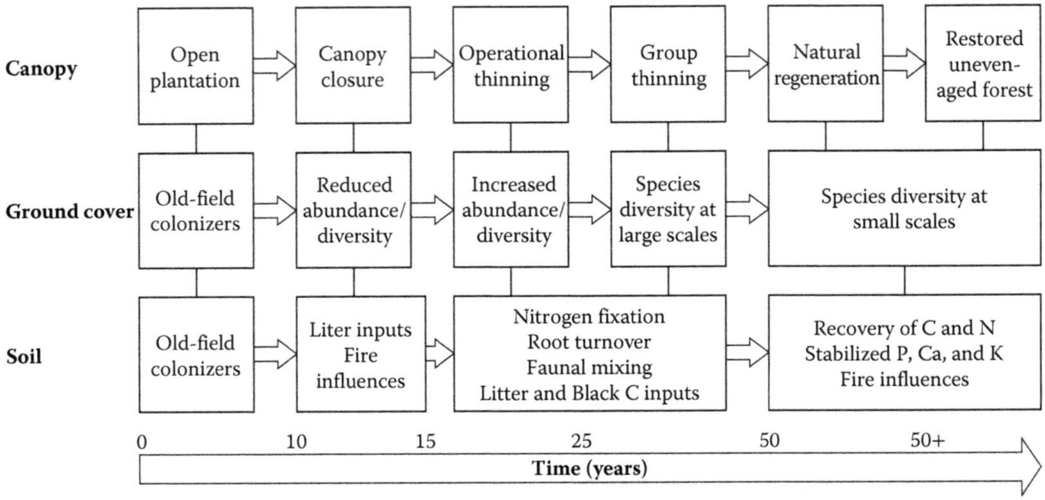

FIGURE 7.9 Conceptual diagram of restoration influences on canopy, ground cover, and soils of a developing longleaf pine forest with old-field conditions as the starting point and a restored uneven-aged forest as the result. C = carbon; N = nitrogen; P = phosphorus; Ca = calcium; and K = potassium.

TABLE 7.3
Average (±1 SE) Soil Measurements for Longleaf Pine (1 and 14 Years Old) Planted on Former Agricultural Sites and Never-Tilled Mature Forests in Southwestern Georgia (Joseph W. Jones Ecological Research Center at Ichauway)

Soil Depth (cm)	Stand Age (years)	Bulk Density (g/cm³)	Total Carbon (%)	Total Nitrogen (%)	Total Phosphorus (μg/g)	Potassium (exchange)
0–10	1	1.69 (0.07)	0.69 (0.05)	0.030 (0.006)	215 (10)	0.12 (0.02)
	14	1.50 (0.01)	0.75 (0.08)	0.030 (0.006)	236 (6)	0.08 (0.02)
	Reference[a]	1.22 (0.04)	2.19 (0.18)	0.060 (0.006)	112 (8)	0.06 (0.01)
Significant contrasts		A, B, C	A, C	A, C	A, C	B, C
10–20	1	1.81 (0.02)	0.53 (0.13)	0.013 (0.003)	209 (8)	0.09 (0.02)
	14	1.60 (0.07)	0.54 (0.02)	0.017 (0.003)	201 (8)	0.04 (0.01)
	Reference[a]	1.4 6 (0.04)	0.91 (0.06)	0.023 (0.003)	97 (7)	0.03 (0.01)
Significant contrasts		A, B, C	A, C	A, C	A, c	A, B, C
20–50	1	1.76 (0.03)	0.26 (0.02)	0.003 (0.003)	166 (3)	0.10 (0.01)
	14	1.56 (0.01)	0.28 (0.03)	0.003 (0.003)	149 (9)	0.04 (0.01)
	Reference[a]	1.54 (0.02)	0.44 (0.06)	0.013 (0.003)	97 (6)	0.02 (0.01)
Significant contrasts		A, B, C	A, C		A, C	

Source: Adapted from Markewitz, D. et al. Soil change and carbon storage in longleaf pine stands planted on marginal agricultural lands. *Ecological Applications,* 12, 1276-1285, 2002. © 2002 by the Ecological Society of America. With permission from John Wiley & Sons.

A = Never tilled land versus plantations, $p < 0.05$; B = A 1-year-old plantation versus a 14-year-old plantation, $p < 0.05$; C = A contrast for linearity with age, $p < 0.05$; c = A contrast for linearity with age, $p < 0.10$.

[a] Reference data from a mature stand.

Cathey et al. 2010). Markewitz et al. (2002) reported that dense young longleaf pine plantations demonstrated little recovery of C and N soil pools for at least 14 years after planting. They attributed the low N to depletion or absence of the ground cover that would otherwise support native N-fixing legumes and N_2-fixation. More recent studies of second-growth longleaf pine woodlands on old-field Ultisols indicate that soil and biomass pools require 50–75 years to attain the C and N storage of mature reference stands (Craft and Chiang 2002; Ike 2010; Gonzalez-Benecke et al. 2015).

Many other studies have documented the long-term soil impacts of agricultural land use legacies in the Southeast and Northeast (Richter and Markewitz 2001; McLauchlan 2006; Flinn and Marks 2007; Grossmann and Mladenoff 2008). Although they suggest that changes from erosion, compaction, and other physical disturbances are often dominant, other studies also implicate large reductions in total soil C content, significant increases in residual P, and variable changes in total N and base cation content. On average, surface soil C tends to decrease by half when forests are converted to agriculture (Davidson and Ackerman 1993; Brudvig et al. 2013). Bizzari et al. (2015) studied the collective effects of agricultural history and fire exclusion on reforestation or restoration of longleaf pine ecosystems. Compared to undisturbed frequently burned longleaf pine woodlands on upland Ultisols, the previously cultivated sites had agricultural legacies that persisted for up to 60 years, including decreased soil C, increased bulk density, and elevated P as deep as 30 cm.

One study used >800 forest monitoring plots to examine the effects of land use, forest management, and spatial variability on soil C (Ike 2010). On upland Kandiudults, average C concentration and content in surface horizons were highest in frequently burned reference longleaf pine woodlands, followed closely by frequently burned maturing old-field upland longleaf pine and >75-year-old mixed pine-hardwood forests. Actively farmed agricultural fields and fallow fields had the lowest C concentration (a 40% reduction), but young planted longleaf pine plantations had intermediate values (Table 7.2).

In the eroded and clayey surface soils of developing loblolly pine plantations, total C did not begin to increase until decades after pine establishment (Richter et al. 1999). Despite continuous C inputs, the low accumulation of C in the soil was likely attributable to low incorporation rates of decomposing organic matter, soil compaction, erosion, and low site productivity—characteristics of many southeastern soils that have been altered by former agricultural practices. Another consideration is that loblolly pine forests on highly compacted clay soils do not accommodate the same abundance of soil fauna that are found in Coastal Plain longleaf pine ecosystems, and are therefore unlikely to experience the same "mixing" of soil organic matter to deeper horizons.

Within 50–75 years post agriculture, fire-maintained longleaf pine with appropriate ground cover and root turnover, as well as soil mixing by indigenous fauna, will rebuild C and N pools to a soil depth of 50 cm (Ike 2010). Furthermore, the black C produced by frequent burning is also a significant component of the soil C pool in reference stands, comprising about 8% of the soil pool to a depth of 50 cm (Ike 2010). In Florida, Kalisz and Stone (1984a, b) described a distinct dark surface soil layer that had been highly enriched by charcoal from frequent fires, a rich herbaceous ground cover, and active soil mixing by fauna. The quantity of black C in these longleaf pine ecosystems does not match quantities measured in high-intensity burns of Ponderosa pine forests or warm-temperate grasslands; but it is an important component of total soil C and a source of long-term C storage in comparison to other fractions of soil organic matter in longleaf pine forests (Ansley et al. 2006; DeLuca et al. 2006).

FUTURE CHALLENGES FOR LONGLEAF PINE RESTORATION

Although many factors have contributed to the drastic reduction (>96%) of the longleaf pine ecosystem throughout its range, conversion to agricultural land use and elimination of frequent fire are the two primary driving forces (Frost 1993). The legacies of these land use changes include both short-term and long-term impacts on soil properties and other key ecosystem processes that will have varying influences on the success of longleaf pine restoration, depending on the edaphic conditions of the restoration site.

Belowground ecological processes are uniquely important in that they influence the multiple limiting soil resources of water and nutrients in fire-maintained longleaf pine ecosystems. These include biological fixation to replace lost N, soil mixing by soil faunal bioturbation, a high ratio of root productivity and belowground decomposition, the differential rooting depths of pines and warm-season grasses, pyric mineralization of P and cations, and a long-term soil sink for black C. All strongly affect the availability of soil resources, the resulting site-specific productivity, and plant species dominance. Future climate changes that are predicted will likely alter edaphic conditions and soil processes over space and time (Vose and Klepzig 2014; Hanberry et al. 2015). Mitigating climatic impacts will require new adaptive ecological restoration approaches that vary across edaphic gradients.

Ecosystem studies are conducted at increasingly larger landscapes and increasingly longer time scales to predict climatic influences on forest productivity and C fluxes (Becknell et al. 2015). Current sensor technologies (Figure 7.10) permit measurement of ecosystem processes at larger scales and decadal periods that should increasingly provide a clearer understanding of edaphic regulation of productivity, C, and water flux. Starr et al. (2016) compared the carbon dynamics of longleaf pine woodlands using eddy covariance data from longleaf pine stands located at the ends of a soil moisture gradient. The 7-year study included four prescribed burns, a severe drought period, and periods of normal precipitation. The combination of drought and fire over 3 years resulted in large C emissions into the atmosphere from both the xeric and the mesic sites (Whelan et al. 2013; Starr et al. 2015). Two years of normal precipitation caused both sites to recover quickly and begin functioning as significant C sinks. The rapid recovery of the xeric site from the prolonged drought indicates that longleaf pine ecosystems have a strong adaptation to disturbance and climatic fluctuations.

FIGURE 7.10 An eddy-flux tower at the Joseph W. Jones Ecological Research Center at Ichauway, used in measuring carbon and water budgets of longleaf pine ecosystems.

Sophisticated regional ecosystem models are currently guided by new remote sensing technologies and eddy covariance measurements (Becknell et al. 2015). Understanding ecosystem processes at broad regional scales will require a better consideration of belowground dynamics across ecological gradients for fire-dependent ecosystems such as longleaf pine, for the younger plantings that are used for ecological restoration, and for other forest types. Belowground ecological processes vary greatly across the southeastern edaphic gradient, and longleaf pine forests on different sites will likely have a range of responses to the diverse stresses and disturbances that will undoubtedly accompany climate change.

The potential for increased regional C sequestration by forested ecosystems has attracted much attention from policy makers. Given its extensive cover of pine forests, the Southeast is one of the best candidate regions for C storage and sequestration. Samuelson et al. (2014) stated that longleaf pine stands of similar age and structure accumulate C similarly to other pine ecosystems when proper assessments of C stocks are used. However, Remucal et al. (2013) found that standardized pine equations underestimated aboveground C in mature longleaf pine stands by 36%, perhaps related to the higher wood density and unique structure of longleaf pine. In comparison, the C sequestration capacity of traditional loblolly and slash pine forest production systems are likely to be reduced by predicted disturbances such as drought, insect infestations, hurricanes, and catastrophic fire (Mitchell et al. 2014). Thus, restoration of the relatively disease-, drought-, insect-, and stress-resistant longleaf pine forests could play a much larger role in the future of southeastern forest resource management and in the provisioning of multiple ecological services, especially wildlife and water (Millar et al. 2007). For more on this topic, see Chapter 15.

ACKNOWLEDGMENTS

The authors acknowledge the many years of teamwork contributed by technicians, postdoctoral researchers, and graduate students at the Joseph W. Jones Ecological Research Center, where a large amount of the cited research was conducted. They include Andres Baron, Jim Bradley, Mary Cobb, Dali Guo, Jason McGee, Stephen Pecot, Dwan Williams, Carlos Wilson, and Roger Winans. We thank Jack Ewel, Jennifer Knoepp, and Steven Brantley for helpful reviews of earlier drafts.

Most of all, we were fortunate to work with our valued colleague Robert Mitchell, who passed away prior to this synthetic endeavor. Although we sorely missed him as a coauthor, his many research contributions are a large portion of the work that we cited in this chapter.

8 Considering Herbivory and Predation in Forest Management

L. Mike Conner and Michael J. Cherry

CONTENTS

INTRODUCTION

Wildlife populations—particularly game species and rare, threatened, or endangered species—often become a consideration in the restoration or management of ecosystems, and are sometimes used as an indicator of a successful intervention (Siddig et al. 2016). In other projects, restoration targets are based on the perceived ability of a landscape to meet the needs of a species or suite of species (GCPO LCC 2013). Only rarely are the influences of wildlife on vegetation considered when restoring or managing an ecosystem.

A growing body of evidence suggests that predators play an important role in structuring food webs (Estes et al. 2011). Historically, the traditional perspective of population ecologists has been that predators reduce herbivores numerically, thereby releasing plant communities (Rosenzweig 1973). However, animals are not immobile organisms that are consumed at will by predators; rather, prey species actively attempt to avoid predation. These behavioral interactions can also ripple through food webs and cause trophic cascades, which occur when predators suppress the abundance or alter the behavior of their prey, thereby releasing the next lower trophic level from predation. Behaviorally mediated indirect interactions between predators and plants through herbivores can alter the structure of food webs in place of (or in addition to) the predator's numerical influence on herbivores (Schmitz et al. 2004). The behaviorally mediated trophic cascade hypothesis explains these interactions (Beckerman et al. 1997).

Although the effects of wildlife on plant communities within the longleaf pine (*Pinus palustris*) forests of the southeastern United States have not been subjected to much research attention, identifying the hypothetical links that may have existed within food webs before European settlement and comparing them to contemporary longleaf pine ecosystems is still possible. This can help to explain how animals previously influenced vegetation and how they may be influencing vegetation today. Although coarse, such an examination can serve as a theoretical underpinning for adaptive management studies that use wildlife to enhance management or facilitate restoration of a desired landscape condition.

Originally known as the "food cycle" (Elton 1927), the food web likely represents the greatest conceptual leap since the inception of the community ecology discipline, describing both the flow of energy through ecosystems and the cycling of nutrients within and among ecosystems. Importantly, food webs reveal the role of wildlife in the movement of energy and nutrients and can be used to evaluate the potential for species or guilds to affect other species through the processes of plant consumption, predation, and competition.

Food webs allow scientists to conceptualize the effects of a wildlife population on adjacent trophic levels and within their guild. The species composition within an ecosystem affects how energy and nutrients are moved through the system, which subsequently affects other components of the food web and has important implications for restoration. Because ecological management seeks to move an ecosystem from a current state or condition to a desired state of structural and functional attributes that are sustainable, consideration of food webs is useful for identifying desired future conditions and ensuring the functionality and sustainability of restored ecosystems.

A goal of ecological management is to strengthen the stability of a desired ecosystem condition (Hobbs and Norton 1996). This can be conceptualized as "deepening the cup" to avoid tipping points—the shifting of an ecosystem to an alternative stable state that requires substantial inputs for a return of the ecosystem to its former state (Lenton 2011). We suggest that the success of achieving any ecosystem management goal (such as restoring a forest) can be measured by the human subsidies that are required to maintain the ecosystem in the desired condition. Therefore, the incorporation of natural (nonhuman) processes that enhance system stability, while reducing the need for human intervention, is a top priority for ecological management. Greater inputs therefore indicate a reduced level of restoration; whereas, lesser inputs correspond with greater degrees of restoration. When taken to its logical conclusion, this concept suggests that complete restoration can only occur when a desired condition is achieved with no human inputs needed to maintain the condition. Admittedly, this level of restoration is not practical given contemporary landscapes, and some degree of long-term management will always be needed to maintain a desired condition. The goal, then, is to minimize these inputs.

In this chapter, we provide an overview of how wildlife communities within the range of longleaf pine have changed since European settlement. We also suggest that understanding food webs, particularly the direct and indirect effects of predation, can aid in identifying the structural and functional components of longleaf pine ecosystems that are useful in reducing the need for human inputs.

With a discussion that is based largely on vertebrate communities, we describe how wildlife may have contributed to the development of the historical longleaf pine community and how restoration of these communities, or surrogate species, can reduce the need for human inputs in the maintenance of desired conditions. We begin by briefly discussing a few key species that are now largely absent from the longleaf pine ecosystem. We then describe contemporary communities, distinguishing between native species and nonnative species that have more recently become established. We outline possible management actions by describing the potential for three-level trophic cascades in contemporary longleaf pine ecosystems and comparing those food webs to what existed before European settlement. Finally, we introduce the interaction among fire, browsing, and predation as a potential mechanism for influencing restoration and management decisions.

HISTORICAL WILDLIFE COMMUNITIES

Descriptions of wildlife communities within historical longleaf pine forests are limited to fossil records, accounts of explorers, and current records of species distributions. Thus, an exhaustive description of wildlife diversity in the longleaf pine ecosystem before European exploration and settlement is not possible. The focus of this chapter will be on highly interactive species—those species that are thought to play a disproportionate role within food webs by facilitating interactions among trophic levels. These species tend to be either large herbivores that have significant impacts on vegetation (Côté et al. 2004; Gordon et al. 2004; Bakker et al. 2006), or predators that serve as top-down forces and can contribute to trophic cascades (Ripple et al. 2001; Estes et al. 2011).

Herbivores

Historical records place bison (*Bison bison*) as far south into present-day Florida as Tampa and verify that they occurred within the range of longleaf pine as recently as 1772 (Rostlund 1960). Although little is known about their abundance in the Southeast before the early 1600s, their potential effects on the longleaf pine ecosystem warrant consideration from a restoration perspective. Historical accounts do not mention large herds of bison that were typical of the western Plains; rather, they tend to reference smaller groups. Still, bison were present in longleaf pine ecosystems and their foraging habits affected vegetation.

Among herbivores, diet is often characterized along a browser-grazer continuum that is strongly related to the ecology, morphology, and natural history of the organism (Fritz and Loison 2006). Species that exist at the browser end of the spectrum consume primarily woody plants, whereas those at the grazer end consume grasses and herbaceous plants. Modern plains bison are often considered grazers. However, a recent attempt to reconstruct paleodiets of bison from across their historic range revealed substantial variation in their diet (Rivals et al. 2007). Samples from Florida portray wood bison in pine savannas as mixed feeders that consumed both woody browse and grasses. This suggests that bison foraging may have influenced the ground cover structure and the grasses, forbs, and hardwood saplings within native open-pine forests. A large grazer, such as bison, in a longleaf pine savanna could have provided wiregrass (*Aristida stricta*) a competitive advantage over more palatable grasses—potentially influencing the composition of grasses, and ultimately, fuel attributes and fire ecology.

Elk (*Cervus elaphus*) were also present within the historical range of longleaf pine; however, their geographical range was largely restricted to the upper Coastal Plain (Laliberte and Ripple 2004). Unlike bison, historical accounts do not suggest that herd sizes in the eastern United States differed from their western counterparts or from restored contemporary eastern populations. Elk are also intermediate feeders along the grazer-browser continuum and consume a variety of vegetation types. However, when elk were reintroduced into the east they also consumed acorns (Lupardus et al. 2011) more frequently than their western counterparts (Collins and Urness 1983); this apparently learned behavior may have been common in historical eastern herds. Historical effects of elk within eastern open-pine forests are largely unknown, but may have been substantial, particularly if their browsing or mast consumption influenced the regeneration of fire-impeding oaks (*Quercus* spp.). Moreover, contemporary research in the western United States has suggested that elk profoundly influence vegetation through trophic cascades (Ripple and Larsen 2000; Ripple et al. 2001; Fortin et al. 2005).

Predators

As with large herbivores, several conspicuous large predators are now functionally absent from most of the historical range of longleaf pine. The loss of large predators has caused considerable reorganization of food webs in numerous ecosystems (Ripple et al. 2014). This "trophic downgrading of planet Earth" could substantially affect the dynamics of disease, wildfire, carbon sequestration, invasive species, and biogeochemical cycles (Estes et al. 2011).

Red wolves (*Canis rufus*) were once present throughout the Southeast (Hinton et al. 2013). In the 1970s, remaining red wolves were removed from the wild to support a captive breeding program, and captive-born wolves were reintroduced into a few highly isolated areas (Hinton et al. 2013).

Today, the total population of red wolves is most certainly <100 individuals. Presumably, red wolf predation on large herbivores such as bison and elk would have occurred because prey as large as adult moose are readily taken by gray wolves (*Canis lupus*), usually paired but sometimes as individuals (Thurber and Peterson 1993). Like gray wolves, red wolves frequently hunt in packs, suggesting that their predation on large herbivores was quite likely (Hinton and Chamberlain 2010). Red wolves are also thought to have preyed on raccoons (*Procyon lotor*) and other smaller mesocarnivores (McVey et al. 2013). Predation likely resulted in direct reduction of prey populations, but the effects of these reductions are largely unknown.

Although the historical impacts of red wolves on large herbivores are not known, their influence may have been similar to the modern-day effects of coyotes (*Canis latrans*) on white-tailed deer (*Odocoileus virginianus*). Although coyote predation is more common on young white-tailed deer than on adults (Cherry, Turner, et al. 2016), adult deer exhibit antipredator behaviors in response to coyotes (Cherry et al. 2015). Coyotes are smaller than red wolves and presumably less capable of preying on adult deer. The increased ability of red wolves to prey on adult deer may have resulted in more pronounced antipredator behaviors than are currently exhibited in response to coyotes, but this remains a mystery.

Although more abundant in the Southeast than red wolves, breeding populations of cougars (*Puma concolor*) are currently restricted to southern Florida (Laliberte and Ripple 2004); however, their historical range encompassed all of the historical range of longleaf pine. This remaining population of the subspecies is locally known as the Florida panther (*Puma concolor coryi*). Although their populations occur south of the longleaf pine range, they inhabit a similar frequent-fire conifer ecosystem dominated by slash pine (*P. elliottii*), and dispersing males periodically move through longleaf pine forests in Peninsular Florida. Cougars are a major predator of white-tailed deer; historically they likely had strong numeric and behavioral effects on deer but a lesser effect on elk and bison in longleaf pine ecosystems (Hernández and Laundré 2005).

Black bears (*Ursus americanus*) are considered game animals throughout most states within the geographic range of longleaf pine. The range of this species has declined drastically in the past two centuries. However, black bears still exist within remnant longleaf pine forests (Larkin et al. 2004), but their presence within the ecosystem as a whole is rare. Black bears are perhaps the prototypical large omnivore. Omnivory is generally considered destabilizing in communities (but see Fagan 1997); if so, the presence of large omnivores within longleaf pine forests may have had disproportionate effects on the system because a large omnivore can function as a large herbivore, as well as a large predator. Finally, although the black bear is generally not considered an important predator of vertebrates, it can be a significant predator of white-tailed deer fawns (Mathews and Porter 1988; Vreeland et al. 2004).

Although bison, elk, red wolf, cougar, and black bear are largely absent from within contemporary longleaf pine ecosystems, their populations remain in other areas. Thus, these species could be reintroduced in appropriate landscapes. Minimally, ecologists can assess the role that each plays within its habitat and can formulate hypotheses about the role that it may have played within the historical range of the longleaf pine ecosystem. Reintroductions can be costly and impractical in current landscapes; however, because they are technically possible, consideration should be given as to whether these species would serve an important function in longleaf pine restoration efforts.

Some species of potentially great ecological importance to the ecology of longleaf pine ecosystems are now globally extinct. For example, mastodons and giant tortoises were likely once very common throughout the range of the longleaf pine ecosystem. Indeed, their presence may have contributed or even driven the ecological processes responsible for the rise of the longleaf pine ecosystem (Noss 2013). Clearly, the ecological role of globally extinct species cannot be determined, nor can they be replaced. In such situations, surrogates must be found to provide these ecological services unless such services have been rendered superfluous by the attainment of an alternate stable state.

CURRENT WILDLIFE COMMUNITIES

Numerous longleaf pine-associated wildlife species garner special attention from a conservation perspective. However, examining most of them would not likely improve our understanding of trophic relationships and how those relationships can be used within a management or restoration perspective. Instead, we again turn to those species or communities that are highly interactive within the ecosystem.

NATIVE HERBIVORES

White-tailed deer are common throughout the geographic range of longleaf pine. In virtually all circumstances, white-tailed deer are the largest native herbivore remaining within longleaf pine-dominated forests, where their impacts on vegetation within longleaf pine and other open-pine ecosystems have been well documented (Cherry, Warren, et al. 2016). Unlike the extirpated bison and elk, white-tailed deer are considered browsers; therefore, their impacts on plant communities within the longleaf pine ecosystem are likely most evident on preferred browse species. The effects of overbrowsing by white-tailed deer are particularly noticeable, and overpopulation of the species has been linked to losses of both plant and animal diversity (Côté et al. 2004). Cherry, Warren, et al. 2016 suggest that predation risk alone is sufficient to alter white-tailed deer behavior to the point that vegetation communities are affected.

Eastern cottontails (*Sylvilagus floridanus*)—and to a lesser extent, marsh rabbits (*S. palustris*) and swamp rabbits (*S. aquaticus*)—have great potential to interact with adjacent trophic levels within longleaf pine ecosystems. Often an important component in the diets of coyotes and bobcats (*Lynx rufus*)—the two largest predators (Godbois et al. 2003; Cherry, Turner, et al. 2016) in current longleaf pine ecosystems—rabbits also affect plant communities (Del-Val and Crawley 2005). All three species are intermediate feeders, capable of both grazing and browsing. Unlike deer and elk, rabbits are coprophagic hind-gut fermenters, which allows for the extraction of more nutrition from lower-quality forage such as woody browse or graminoids.

Gopher tortoises (*Gopherus polyphemus*) also influence vegetation through herbivory and seed dispersal (Diemer 1986). They forage on a diverse array of plant species (>68 genera of plants from 26 families), but grasses are the major component of their diet (MacDonald and Mushinsky 1988). Gopher tortoises also affect vegetation by mechanically altering their surroundings through burrowing and mound building (Kaczor and Hartnett 1990). In addition, burrowing creates habitat for many other wildlife species that use gopher tortoise burrows, leading some to suggest that the gopher tortoise is an ecosystem engineer (Kinlaw and Grasmueck 2012). Nest and hatchling predation can limit gopher tortoise populations (L. Smith et al. 2013); as a result, predation may ultimately affect how these ecosystem engineers influence their environment.

Small mammal communities within longleaf pine forests are generally dominated by just a few species. Fire facilitates predation on all small mammals within the longleaf pine ecosystem by removing cover and making small mammals more susceptible to predation. However, these effects are much more pronounced in the cotton rat (*Sigmodon hispidus*), one of the most common small mammal species within longleaf pine forests (Morris, Hostetler, Conner, et al. 2011). Cotton rats have a notably high reproductive output, often reproduce at 5–6 weeks of age, have an average litter size of seven, and produce multiple litters per year. Within the geographic range of longleaf pine, this species historically reached densities of >200 individuals/ha (Cameron and Spencer 1981); this represents considerable herbivore biomass and could have had a profound impact on vegetation. Throughout the Southeast, cotton rats are also an important prey species for other mammals (Godbois et al. 2003; Conner et al. 2011; Cherry, Turner, et al. 2016) and predatory birds (Preston and Beane 1993). Within the longleaf pine ecosystem, the strong relationships among prescribed fire, cotton rats, and predators have implications for prey, predator, and plant communities.

Native Carnivores

Throughout most of the geographic range of longleaf pine, large carnivores, like large herbivores, have been extirpated, and a suite of mesocarnivores, raptors, and snakes are now the dominant predators. Current apex predators include a few relatively abundant species, which unlike their historical counterparts, are relatively small. The small body size of remaining native predators likely limits their impact on white-tailed deer and other large herbivores that occur in longleaf pine ecosystems.

Unlike most other southeastern mammalian predators, which are largely omnivorous, bobcats are entirely carnivorous. As a result, they can be considered a contemporary apex predator in the longleaf pine ecosystem. Bobcats can reach relatively high densities within open-pine forests (Miller and Speake 1978), particularly within systems that can support abundant prey. Bobcats prey on all of the primary mammalian herbivores in longleaf pine ecosystems, including white-tailed deer, rabbits, and smaller species. Although bobcats are known to prey on white-tailed deer fawns (Nelson et al. 2015) and adults (Labisky and Boulay 1998), they are generally considered a predator of smaller mammals and birds, particularly when larger predators are present (Thornton et al. 2004). In many southeastern pine-dominated forests, cotton rats make up the bulk of their diet (Miller and Speake 1978; Godbois et al. 2003).

Numerous species of native raptors are found throughout the Southeast, and all are considered predators of small (<4-pound) mammals, reptiles, and birds. Raptors are more likely to impact small mammalian prey than mammalian predators (Wiegert 1972). When mammalian predators are absent (Morris, Hostetler, Conner, et al. 2011) or when their numbers are significantly reduced (Ellis-Felege et al. 2012), raptors can compensate, keeping prey survival rates from increasing significantly. Similarly, the longleaf pine ecosystem supports several species of snakes that prey on amphibians, other reptiles, small mammals, and birds. Collectively, snakes can have substantial impacts on prey populations, exerting a compensatory effect when other predator populations are suppressed (Conner et al. 2011; Ellis-Felege et al. 2012).

Finally, a suite of medium-sized mammals—including raccoons, opossums (*Didelphis virginiana*), and gray foxes (*Urocyon cinereoargenteus*)—is largely omnivorous. These species generally feed on smaller prey (Neale and Sacks 2001) and can serve as significant dispersers of soft mast seeds (Wilson 1993). The global ecological consequences of their irruptive growth following the collapse of larger predators has garnered much attention worldwide (Crooks and Soulé 1999; Prugh et al. 2009; Ritchie and Johnson 2009), but such consequences have been largely unreported in longleaf pine ecosystems. Nevertheless, abundant populations of medium-sized mammals within longleaf pine ecosystems cause challenging management scenarios for the conservation of birds, reptiles, and amphibians.

Native Granivores

Granivores affect longleaf pine ecosystems primarily through seed dispersal. For example, blue jays (*Cyanocitta cristata*) often cache acorns far from their source (Darley-Hill and Johnson 1981). Gray squirrels (*Sciurus carolinensis*) and fox squirrels (*Sciurus niger*) also cache various seeds, potentially aiding in the regeneration of trees that produce hard mast. Ants are instrumental in dispersing many ground cover plant species. Although the role of native granivores in plant dispersal is generally acknowledged, their effect on vegetation communities within longleaf pine ecosystems has not been quantified and therefore is poorly understood.

Nonnative Animals

As is true for most regions worldwide, the Southeast is home to many species of nonnative wildlife. For the most part, these species are viewed negatively; however, some could fill functional roles vacated by locally or globally extinct species.

The coyote is a relatively recent addition to the Southeast; within much of the geographical range of longleaf pine, the species has been present for <100 years. Coyotes can strongly influence communities through intraguild interactions (Newsome and Ripple 2015) and by altering herbivore behavior (Cherry et al. 2015) and abundance (Chitwood et al. 2015). Coyotes are considered a pest species in most states and have been blamed for the demise of many game and nongame species. For example, they have been linked to reductions in white-tailed deer populations (Kilgo et al. 2010). However, coyotes can facilitate increased biodiversity in birds (Crooks and Soulé 1999) and mammals (Henke and Bryant 1999), and their intraguild interactions with red fox can result in increased nest success for waterfowl (Sovada et al. 1995). The coyote is a highly adaptable, opportunistic omnivore; as a result, it plays a prominent role in food webs of contemporary ecosystems throughout its range.

Wild pigs (*Sus scrofa*) are considered among the worst vertebrate pests worldwide and have become established across the continental United States. Nationally, damage from wild pigs costs >$1 billion annually, and the ecological damage by wild pigs can be catastrophic. In longleaf pine forests, wild pigs destroy pine seedlings (Lipscomb 1989) and native ground cover (Bratton 1975) through their rooting activities and affect other wildlife species through predation (Fordham et al. 2006), competition (Focardi et al. 2000), and habitat alteration (Mack and D'Antonio 1998). Longleaf pine ecosystems are likely especially susceptible to damage by wild pigs for two reasons: (1) they host numerous ground-nesting species of birds, reptiles, and amphibians; and (2) many fire-facilitating ground cover species (such as wiregrass) are sensitive to soil disturbance and could be harmed by rooting. Wild pigs can also induce indirect links in food webs; for example, on the Channel Islands off the California coast, wild pig populations served as an abundant food resource to support golden eagle (*Aquila chrysaetos*) populations, which in turn have caused the near extinction of three island fox (*Urocyon littoralis*) subspecies (Roemer et al. 2002). Although the indirect effects of wild pigs on food webs in longleaf pine ecosystems have not been reported, we suspect the impacts are substantial.

Prior to the mid-1850s, the range of nine-banded armadillos (*Dasypus novemcinctus*) in the continental United States was limited to southern Texas (Taulman and Robbins 1996). Armadillos now occur throughout most of the Southeast, the result of natural range expansion (Fitch et al. 1952) and accidental introductions (Talmage and Buchanan 1954). Although absent from longleaf pine ecosystems <200 years ago, its population has likely reached >50 million individuals in the Southeast today. Armadillos disturb soil in search of invertebrate prey (Hawthorne 1994) and also consume eggs of ground-nesting birds (Ellis-Felege et al. 2012) and reptiles (L. Smith et al. 2013); thus, their potential for altering contemporary food webs is great.

Red imported fire ants (*Solenopsis invicta*) are now common in the longleaf pine range, and recent research suggests that they can have detrimental effects on native vertebrates (Long, Conner, et al. 2015; Long, Knapp, et al. 2015), native ant communities, and plants (Stuble et al. 2010; Cumberland and Kirkman 2013). Fire ants were the primary nest predators of shrub-nesting songbirds (Conner et al. 2010); they reduced survival of cotton rats (Long, Conner, et al. 2015) and gopher tortoise hatchlings (Dziadzio, Long, et al. 2016); they decreased reproductive output of eastern fence lizards (*Sceloporus undulatus*) (Long 2015); they altered the foraging behavior of cotton rats and southern toads (*Anaxyrus terrestris*) (Long, Knapp, et al. 2015; Darracq et al. 2016); and they have been implicated in decreasing survival of white-tailed deer fawns (Allen et al. 1997).

FROM POPULATIONS TO FOOD WEBS

The major changes in the wildlife community of longleaf pine forests described above have implications for management and restoration. Many species are now locally or globally extinct; likewise, remaining longleaf pine ecosystems support populations that were not present only 200–300 years ago. Clearly, the loss of large herbivores and carnivores resulted in a dramatic reorganization of food webs within longleaf pine ecosystems.

By their very nature, food webs explain a portion of an organism's role in its environment. Even in the simplest of ecosystems, food webs can be incredibly complex. From a management perspective, however, using generalized conceptual food webs to identify links among species within a community can help managers understand the ecosystem, identify potential effects of management actions, and determine which management actions are appropriate for achieving the desired change. Minimally, acknowledging the presence of linkages permits: (1) formulation of hypotheses about the processes by which trophic interactions may have contributed to the structure of the ecosystem, (2) determination of whether these interactions are still at play within the ecosystem, and (3) identification of clues for restoring interactions that are no longer present.

Identifying linkages within food webs provides insight into what might happen if a particular population or community is absent, has become suppressed, or is becoming more abundant. Simplification of food webs into primary producers, primary consumers, and secondary consumers is useful in describing the fundamental elements of trophic cascades. Basic, three-tiered, terrestrial, trophic cascades—such as would be expected within the longleaf pine ecosystem—occur when secondary consumers (predators) exert top-down pressures on primary consumers (prey/herbivores) that result in changes to primary producers (plants). Today, examples of terrestrial trophic cascades are well documented in a wide variety of ecosystems (Estes et al. 2011).

Examination of two simplified food web models, one before European settlement and one contemporary, reveals obvious differences between the two periods, with top-down forces leading to different cascading effects. Most striking is the absence of large predators and herbivores in contemporary food webs. The loss of bison—and to lesser extent, elk—likely removed substantial grazing pressure from the ecosystem and resulted in a community shift toward the browser end of the grazer-browser continuum. The specialist apex predators (primarily cougar and red wolf) were replaced by the coyote, an opportunistic generalist predator that is less dependent on specific prey populations. Furthermore, the shift from cougars and red wolves to bobcats and coyotes likely altered the numeric effect of these predators on herbivore populations, shifting the bulk of mortality from adults to juveniles.

Even with these shifts, the behavioral effects of the predators on herbivores did not necessarily change from the earlier to the later period. When the risk of predation increases, many species in the deer family (including elk) increase their use of woody browse over a more palatable, but less safely accessible diet (Creel and Christianson 2009). A shift in diet selection toward browsing on woody, fire-impeding species could have important implications for the ecology of fuels in longleaf pine systems. For managers who are trying to create and manage open-canopied longleaf pine, the degree to which prior trophic interactions influence perceptions about desired management endpoints is particularly relevant. If the desired endpoints reflect historical top-down pressures that are now severely diminished or even absent, then our approach toward managing for these endpoints will need to consider how to mimic the influence of these pressures. To borrow from niche theory (Connell 1980), perceptions of the desired endpoint for a longleaf ecosystem might be based on the ghosts of trophic cascades past. Wildlife populations in the longleaf pine ecosystem have changed, as have food webs and the forces that contribute to tropic cascades. Attempts to manage toward a particular longleaf pine-dominated condition should consider whether we can reasonably reach management goals with the faunal conditions that currently exist on the landscape and determine what level of human inputs will be required.

The longleaf pine forest is now a fraction of its former range, and the wildlife communities within its current range are clearly quite different from those that contributed to the trophic interactions of the past. Therefore, helping managers move toward the desired condition for their forests requires an understanding of how contemporary wildlife communities interact with other environmental factors.

Large carnivores (Wallach et al. 2015) are declining globally (Ripple et al. 2014); with rare exceptions, they have been absent from the geographic range of longleaf pine for at least a century. In the absence of large carnivores, the smaller mesocarnivores became the new apex predators.

Because they are generally less effective predators of large prey, one would expect that predator pressure on large prey has decreased with consequent increases in their populations and foraging and decreases in their preferred forage. However, modern longleaf pine forests may be the exception to this expected outcome.

Throughout much of the Southeast, contemporary white-tailed deer populations are largely the result of restoration efforts that took place during the early to mid-1900s. During most of this period and throughout most of the region, natural predators of white-tailed deer were rare because coyotes were not yet well established. More recently, coyotes have become common; during the past 20 years coyotes have been linked to declines in fawn survival (Kilgo et al. 2012), recruitment (Gulsby et al. 2015), and population growth (Kilgo et al. 2010; Chitwood et al. 2015). On a longleaf pine-dominated study site in Georgia, fawn survival was among the lowest reported for the region (Nelson et al. 2015)—even though fawns were captured 3–4 days after birth, which would likely have resulted in an overestimation of their survival rates (Gilbert et al. 2014).

In response to high predation rates, adult female deer exhibit both proactive (Conner et al. 2016) and reactive (Cherry et al. 2015) behaviors. Conner et al. (2016) reported that during the fawn-rearing season, deer selected predator exclosures (four 40 ha fenced pens permeable to deer but excluding coyotes) over control plots. Cherry et al. (2015) experimentally demonstrated that female deer spent more time feeding at baited camera traps in predator exclosures than in control plots. These studies collectively suggest that does perceive predation risk to be both great and highly focused on their offspring, and that they respond with multiple antipredator defenses. Recent experimentation within a longleaf pine-dominated forest provided evidence that cascading effects may also result from coyote presence (Cherry, Warren, et al. 2016).

As described above, trophic cascades are the result of predation that limits herbivore populations and thereby reduces the impacts of herbivory on plant communities. However, Beckerman et al. (1997) suggested an additional mechanism whereby predators influence plant communities through behaviorally mediated trophic cascades. Indeed, the trophic cascade resulting from the coyote-deer interaction described above may be largely the result of antipredator behavior. Although healthy, adult female white-tailed deer can be preyed upon by coyotes, predation of fawns by coyotes and bobcats is much more common. Fawn predation can have a limiting effect on low-density white-tailed deer populations (Chitwood et al. 2015); but in most ecosystems, fawn predation alone would be unlikely to limit populations to the point where a trophic cascade results. However, the presence of coyotes does alter deer foraging behavior (Cherry et al. 2015) and space-use patterns (Conner et al. 2016), both of which affect vegetation communities.

Before European settlement, large herbivores were likely very important to the persistence of large carnivores. The mesocarnivores (primarily coyotes and bobcats) that serve as today's apex predators are unlikely to rely solely on a diet of white-tailed deer, which is the current largest wild herbivore in the longleaf pine forest. Instead, these more recent apex predators are more dependent on small mammals such as cotton rats and eastern cottontails (Godbois et al. 2003, Cherry, Turner, et al. 2016). Thus, for antipredator behavior in white-tailed deer to result in a vegetation response, the quantity of smaller prey must be sufficient to support mesocarnivore populations at levels that are great enough to affect white-tailed deer behaviors.

Fire and browsing interact to influence the balance of trees and grasses in numerous savanna ecosystems (Scholes and Archer 1997; Higgins et al. 2000; Barnes 2001; Sankaran et al. 2004). For example, Staver et al. (2009) found that both fire and browsing reduced tree growth, but only their combined effects limited tree density in an African savanna. Aboveground survival of oaks in frequently burned longleaf pine savannas is a function of seedling or sapling size at the time of the fire, local fuel characteristics, and fire conditions (Glitzenstein et al. 1995; Ellair and Platt 2013). Growth of oak saplings after a prescribed fire can be inhibited by white-tailed deer browsing (Adams and Rieske 2001), thus illustrating that herbivores can indirectly influence oak mortality. This is especially relevant in longleaf pine forests, woodlands, and savannas, where frequent fire limits midstory encroachment by oaks and other fire-impeding species (Glitzenstein et al. 1995; Kirkman, Coffey, et al. 2004).

Browsing by white-tailed deer strongly influences ecosystems (Côté et al. 2004). Because they browse on fire-impeding species (such as oaks), deer can decrease sapling survival following fire. Like other cervids, they increase their use of woody browse when predation risk increases in more desirable food source patches (Edwards 1983; Christianson and Creel 2008). In longleaf pine ecosystems, fawn-rearing does avoided recently burned areas, even when those areas offered higher nutritional content and digestibility; this counterintuitive behavior was attributed to predation risk (Lashley et al. 2015). Deer vigilance while foraging decreased with increasing time-since-fire, suggesting that unburned patches are perceived as safer than the surrounding open woodlands Cherry et al. (2017).

Although oak browse is common forage for deer in longleaf pine savannas, its quality is relatively poor (Lashley et al. 2015). The relationship among white-tailed deer foraging behavior, coyotes, prescribed fire, and vegetation response sets up a potentially complex scenario in longleaf pine ecosystems—with browse selection of hardwood midstory species by deer regulated by coyote predation, which in turn reduces the encroachment of nonpyric vegetation (Kilgo et al. 2010; Cherry et al. 2015). Cherry, Warren, et al. (2016) demonstrated this scenario by experimentally excluding coyotes; the result was an increase in the number of oak sprouts and a decline in preferred deer browse species. Because observed patterns in vegetation represented both positive and negative indirect interactions between predators and plants, they concluded that cascading effects are behaviorally mediated through shifts of diet selection (increased use of woody browse with predation risk) and space use (allocating time to areas relative to predation risk).

Although the herbivores and predators are smaller in contemporary longleaf pine ecosystems, the processes and outcomes could be more similar to historical food webs than previously presumed. Thus, the cascading effects present today could be similar to those that originally shaped the longleaf pine ecosystem. This suggests that the ghosts of cascades past are not really ghosts at all. In other words, current systems may have enough top-down pressures to facilitate the creation and maintenance of desired conditions within longleaf pine-dominated forests—maybe the cascading effects still persist and only the "players" have changed.

APPLICATION: RESTORATION AND MANAGEMENT

Antipredator behaviors are ubiquitous in nature (Lima and Dill 1990). Some of these behaviors result in changing herbivore diets (Edwards 1983; Beckerman et al. 1997), while others result in altered space use (Creel et al. 2005; Conner et al. 2016); both outcomes can alter the spatial distribution and species composition of vegetation. Research is only beginning to explore how these complex interactions influence local vegetation structure across large spatial scales. However, the general pattern is that herbivores increase their use of food items in or near safe areas when predation risk increases, and they are free to select their most preferred forage items—regardless of location—in the absence of predation risk.

Because few data are available for predicting how complex behaviors will influence vegetation structure, offering management recommendations that rely upon antipredator behaviors would be premature. However, monitoring the influence of antipredator behaviors on vegetation composition and structure within an adaptive management context would be appropriate and would lay the groundwork for interesting and potentially productive research. Adaptive management that includes monitoring of predator and prey populations as well as resource availability over time could ultimately reveal the degree to which predator-prey interactions influence vegetation.

Prey species must balance the acquisition of resources for their survival and reproduction against the risk of predation (Brown 1999)—they must eat without being eaten. Other general predictions extend from this simple theory; prey species avoid areas of high risk with low reward and they prefer areas of low risk with high reward, neither of which is common in nature. Instead, areas of high risk and high reward and areas of low risk and low reward create a continuum of net value from which a forager must select (Brown and Kolter 2004). Foraging decisions are complex and are influenced by a suite of intrinsic and extrinsic factors, but predation risk is primary among them. Predator abundance and diversity affects where and on what prey will forage, indirectly linking predators to vegetation in a process that has been undervalued until recently. Adaptive management that includes monitoring of predator and prey populations as well as resource availability over time can illuminate the degree to which predator-prey interaction influences vegetation.

The contemporary species assemblages in longleaf pine ecosystems and their associated interactions could serve as surrogates for those wildlife species that originally shaped the ecosystem. Capitalizing on these interactions as a component of management could reduce human inputs required or shorten the time needed to reach a given management goal. Minimally, managers would benefit by recognizing that these interactions exist and have the potential to affect the outcomes of restoration activities.

The question, then, is how well today's species can serve as surrogates for the species or processes that helped to create the ecosystems used to define the desired conditions for management activities. Cattle (*Bos taurus*) are commonly grazed in pine savannas. Whether the herbivory pressure from this domesticated ruminant is similar to the pressure from the bison that for thousands of years foraged beneath the longleaf pine under risk of predation from intact predator communities is a valid question. However, a more important question may be the likelihood of this surrogate to help us meet modern objectives such as hardwood control in pine-dominated uplands. Is the "downsized" trophic cascade involving the coyote–deer–plant community functionally similar to the historical version involving wolf/cougar/bear–bison/elk/deer–plant community interactions in terms of balancing management objectives and desired conditions? Are forest conditions governed by the same interplay between biotic and abiotic forces, but with smaller and more domesticated versions of beast and fire? Ultimately, the question is how restoration of large herbivores, predators, or both might influence conservation and our ability to maintain desired conditions.

Whether ecological management should include restoration of historical apex predators is largely a subjective decision based on societal objectives. Recent investigations have demonstrated that large predators play an important role in food webs and the availability of ecosystem services that benefit humans (Ostfeld and Holt 2004; Estes et al. 2011). The result has been a dramatic turnaround in public opinion toward coexistence with large mammalian carnivores. After centuries of effort to remove predators from the landscape, predator restoration is occurring in ecosystems widely distributed across North America. Cougar and wolf populations are expanding. Federal agencies now maintain active programs that promote the conservation and restoration of Florida panthers, gray wolves, red wolves, Mexican wolves (*Canis lupus baileyi*), grizzly bears (*Ursus arctos*), and Louisiana black bears (*Ursus americanus luteolus*).

Restoration of large predators could be an important management tool or an indicator of successful restoration in the longleaf pine ecosystem. For example, the loss of dingoes (*Canis lupus dingo*) in Australia has been linked to losses in plant biomass and diversity caused by large increases in herbivore populations, and to widespread losses of small- and medium-sized native mammals largely caused by increased red fox predation (Letnic et al. 2012). Similarly, in North America the restoration of wolves resulted in changes in herbivore abundance and behavior, ultimately linking wolves to plant community restoration.

These examples demonstrate the potential for predators to be harnessed as restorers of ecosystems (Ritchie et al. 2012). Unfortunately, given the current degree of fragmentation within the historical range of longleaf pine, few areas are large enough to support restoration of large carnivore populations.

Finally, humans are also an apex predator and perhaps should be considered a surrogate for large carnivores. Sport hunting, trapping, and lethal population manipulation provide direct interactions between management and wildlife populations within food webs. Sport hunting can be a very effective method of controlling large herbivore populations (Riley et al. 2003), resulting in reduced plant consumption and a change in vegetation composition. Likewise, predator control may reduce predator populations—if even for brief periods (Conner and Morris 2015)—leading to increased herbivore populations (Howze et al. 2009; Conner et al. 2016), increased herbivory, and associated vegetation response (Cherry, Turner, et al. 2016). Additionally, antipredator behavior has been observed in response to sport hunting of white-tailed deer (Little, Demarais, et al. 2014); these behaviors may potentially affect vegetation in mechanisms that are similar to the effects of nonhuman predation. However, hunting regulations generally restrict predation by humans to brief segments of the year, and antipredator behaviors in response to hunting would not be reinforced throughout the remainder of the year.

Effective management regimes enhance the ecosystem services that flow from forests. These ecosystem services often include sustaining biodiversity and restoring food web dynamics. Assessing how past conditions in food webs influenced the evolution of an ecosystem is difficult, particularly when some of the species were eliminated before any useful data could be collected. Nonetheless, contemplating the processes that contributed to the evolution of the ecosystem can provide insights that guide future management and that identify surrogates for extinct processes.

ACKNOWLEDGMENTS

We thank Marcus Lashley, John Kilgo, and Lora Smith for their comments, edits, and suggestions. This work was greatly influenced by discussions with Joel Brown, Andrea Darracq, Robert McCleery, Darren Miller, Gail Morris, Nova Silvy, Bob Warren, and numerous graduate students.

9 Geographically Isolated Wetlands
Embedded Habitats

Lora L. Smith, Amanda L. Subalusky, Carla L. Atkinson, and L. Katherine Kirkman

CONTENTS

INTRODUCTION

Geographically isolated wetlands (GIWs) are an integral component of the longleaf pine (*Pinus palustris*) ecosystem. These upland embedded wetlands occur throughout the range of longleaf pine, and can be locally abundant and variable in size, vegetation type, hydroperiod, and landscape position. Although they are not directly connected to surface drainages, these "geographically isolated" wetlands (Tiner 2003) do not function as discrete systems (Leibowitz 2003; Cohen et al. 2016). Instead, hydrological, biogeochemical, and biological processes connect them to surrounding terrestrial systems, streams, and other wetlands. This connectivity with adjacent environments is dynamic in both degree and importance, varying with recurring cycles of wetting and drying.

Within the longleaf pine ecosystem, frequent fire is a critical process that promotes habitat connectivity between embedded GIWs and the uplands around them; the frequency and intensity of fires spreading from upland sites into wetlands are influenced by wetland hydrology. Similarly, temporal fluctuations of soil saturation in upland-wetland ecotones create sites that are important for microbial processing of nutrients, and the infrequent connections that result from ephemeral streams or drainages in some GIWs have cumulative hydrologic impacts at watershed scales (Cohen et al. 2016). Collectively these physical characteristics and dynamic processes contribute to the role of GIWs in supporting the exceptionally rich biodiversity of the longleaf pine ecosystem.

The contribution of GIWs to the overall biodiversity of the longleaf pine ecosystem has been largely overlooked. Nearly 40% of the plants (L. Kirkman, unpublished data) and 30% of the amphibians attributed to the longleaf pine ecosystem (Moler and Franz 1987; Guyer and Bailey 1993) are associated with GIWs and the ecotones between GIWs and the surrounding longleaf pine forest. Moreover, many amphibians, some reptiles, and a few other specialized wildlife groups rely on both GIWs and longleaf pine forests at different life stages and they must traverse between the two habitats to complete their life cycle (Guyer and Bailey 1993; Means 2006). The magnitude of wildlife migrations into and out of GIWs is noteworthy, particularly for wetlands embedded within longleaf pine uplands (Gibbons et al. 2006) and likely represents a significant contribution to both aquatic and terrestrial food webs (Regester et al. 2006; Harper et al. 2015).

In this chapter, we describe the status of GIWs within the southeastern United States in light of the extensive loss of longleaf pine forests; we review the geology, hydrology, soils, vegetation, and biogeochemistry of GIWs within longleaf pine forests; and we describe the influence of these factors on patterns of biodiversity. We provide examples of the relevant types of habitat connectivity for amphibians, reptiles, and the other groups that are unique to GIWs in longleaf pine (wetland to upland, wetland to wetland, and among wetlands and other aquatic systems). We focus largely on amphibians because their complex habitat requirements present unique challenges for restoration of longleaf pine forests and embedded GIWs. We summarize the challenges of restoring GIW-dependent wildlife within altered landscapes, and present examples of several GIW restoration projects. Last, we identify management and conservation challenges presented by climate change predictions, and conclude with a discussion of future research needs.

BACKGROUND ON GIWs IN THE LONGLEAF PINE ECOSYSTEM

Historical Distribution and Current Status

Although GIWs are abundant across much of the southeastern Coastal Plain and throughout the range of longleaf pine (Tiner 2003), estimating their numbers is difficult because many are too small to be detected by traditional wetland mapping techniques, particularly during dry periods. Lane et al. (2012) used a U.S. Fish and Wildlife Service National Wetlands Inventory and a U.S. Geological Survey hydrography dataset to identify potential GIWs across eight southeastern and mid-Atlantic states, including Alabama, Florida, Georgia, South Carolina, and North Carolina, which are within the longleaf pine historical range. Although the scope of their estimate (>700,000 isolated wetlands across eight states) extends beyond the historical range of longleaf pine, it demonstrates the remarkable abundance of GIWs in the Southeast. On a smaller scale, but within the historical range of longleaf pine, Martin et al. (2012) estimated that approximately 12,000 GIWs occur within the 668,940 ha Dougherty Plain physiographic district of southwestern Georgia (a density of 1.7 wetlands/km^2). Most of these wetlands are small (84% are <4 ha, 54% are <1.2 ha); Semlitsch and Bodie (1998) reported similar percentages for the Savannah River Site on the Atlantic Coastal Plain of South Carolina (87% are <4 ha, 46% are <1.2 ha).

Documenting the current status and condition of GIWs within the range of longleaf pine is challenging for the same reasons. Dahl (1990) estimated that 23%–59% of all wetlands in the southeastern states have been lost since the 1780s, but did not specify the proportion of those wetlands that were GIWs. Regulations, including those that support the U.S. Clean Water Act, have slowed the loss of wetlands in general; however, because many GIWs are excluded from federal and state regulations, GIW losses and alterations often go unchecked. In some areas, GIWs remain on the landscape (Martin et al. 2012), but they have been altered dramatically by human land uses (Lane et al. 2012; Stuber et al. 2016) (Figure 9.1). In the five southern Atlantic states, Lane et al. (2012) estimated that by 2000, only half the GIWs in the longleaf pine range were in reference condition (Figure 9.2). Their results likely overestimated habitat quality, because the study used an index of landscape development intensity (Brown and Vivas 2005) to assess wetland condition—such indices rely on land use and land cover data and often underestimate the impacts of fire exclusion, groundwater withdrawals, construction of ditches and canals, and other landscape-level hydrological alterations (Stuber et al. 2016). Furthermore, landscape-level connectivity for those GIWs that remain may also have been lost. Stuber (2013) found that from 1948 to 2007, more than half the GIWs in the Dougherty Plain of Georgia lost contiguous forested corridors, likely affecting access to adjacent terrestrial habitat for semiaquatic species, as well as linkages to other aquatic systems (Figure 9.3).

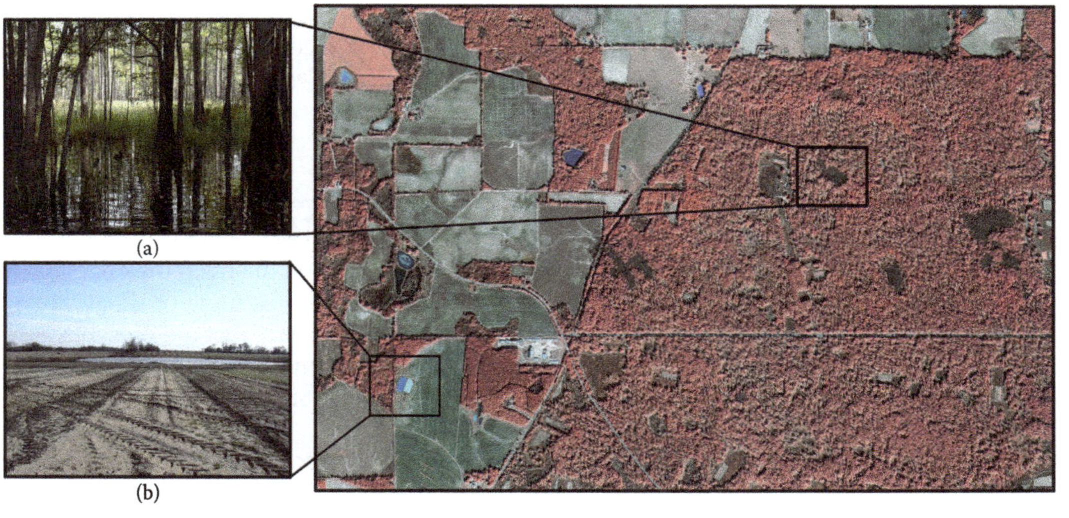

FIGURE 9.1 Geographically isolated wetlands (GIWs) within a longleaf pine forest and adjacent agricultural landscape in southwestern Georgia: (a) A forested GIW with an intact ecotone leading to the surrounding longleaf pine forest, and (b) a degraded GIW within an unplanted irrigated agricultural field. (Photographs courtesy of Aubrey Heupel Greene and Stephen Golladay.)

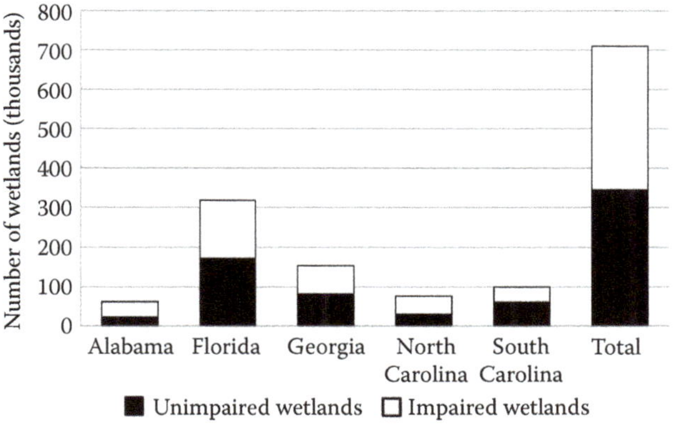

FIGURE 9.2 Estimated number of geographically isolated wetlands in the southeastern United States as of 2010. (Modified from Lane, C. R. et al., *Wetlands*, 32, 753–767, 2012. Note: Wetland conditions were determined using landscape development intensity coefficients applied to U.S. land cover data from Brown, M. T. and M. B. Vivas, *Environmental Monitoring and Assessment*, 101, 289–309, 2005.)

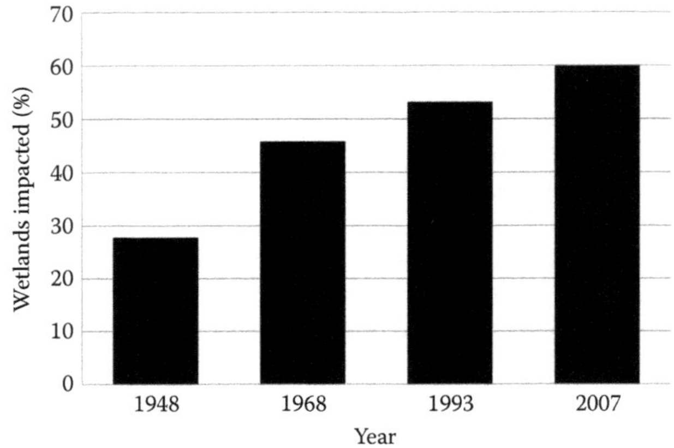

FIGURE 9.3 Percentage of geographically isolated wetlands impacted by human land uses in the Dougherty Plain of southwestern Georgia, 1948–2007. (Modified from Stuber, O. S. 2013. The relationship between land use and the ecological integrity of isolated wetlands in the Dougherty Plain, Georgia, USA, MS Thesis, University of Georgia, Athens, Georgia.)

GEOGRAPHICALLY ISOLATED WETLANDS: REGULATED OR NOT?

Congress enacted the Clean Water Act (CWA) of 1972 to restore and maintain the chemical, physical, and biological integrity of U.S. waters. In addition to including traditional navigable waters (waters that are subject to the ebb and flow of the tide or waters that are presently, have been, or could be used to transport interstate or foreign commerce), the CWA was originally interpreted broadly to include adjacent and upstream waters such as headwaters and wetlands that significantly affect the integrity of navigable waters.

Recent challenges to the CWA (see below) led to uncertainty about federal jurisdiction over the other waters and wetlands, particularly geographically isolated wetlands (GIWs), included in the broader interpretation. In a 2006 decision, the U.S. Supreme Court ruled that the U.S. Environmental Protection Agency (EPA) could not regulate wetlands unless a significant nexus (meaningful physical, chemical, or biological connection) with downstream waters could be demonstrated. Because connections between GIWs and navigable waters can be cryptic and difficult to demonstrate, determination of a significant nexus for jurisdictional purposes requires assessment on a case-by-case basis.

Major Supreme Court Decisions

U.S. v. Riverside Bayview (1985)

The Court deferred to the opinion of the U.S. Army Corps of Engineers that adjacent wetlands are inseparably bound up with waters to which they are adjacent. The Court took a broad, systemic view of the goal of the CWA in maintaining and improving water quality.

Solid Waste Agency of Northern Cook County v. U.S. Army Corps of Engineers (2001)

The Court held that the use of isolated, nonnavigable intrastate ponds by migratory birds was not by itself a sufficient basis for the exercise of federal regulatory authority under the CWA. Although the ruling did not call into question earlier decisions that upheld jurisdiction of waters adjacent to traditional navigable waters, it created uncertainty about jurisdiction of other waters and wetlands and introduced the concept of "significant nexus" to explain the Court's reading of the CWA as it applies to nonnavigable waters.

***Rapanos v. United States* (2006)**

The Court agreed that the term "waters of the U.S." encompasses some waters that are not navigable in the traditional sense. Justice Kennedy's opinion indicated that the critical factor in determining coverage of the CWA is whether a water has a significant nexus to downstream traditional navigable waters, such that the water is important to protecting the chemical, physical, or biological integrity of the navigable water.

Recent Developments

In 2015, the EPA released a report (U.S. EPA 2015c) that included a review and synthesis of the scientific evidence that streams and wetlands are connected to downstream waters. Based on the conclusions of this report and an EPA Science Advisory Board review of its conclusions, the agency published a new rule in 2015 (U.S. DOD 2015), which clarified that the CWA applies to all tributaries, headwater streams, and "adjacent" wetlands. Adjacent wetlands were defined as those that border, are contiguous with, or neighbor other U.S. waters. Although the new rule recognized that nonfloodplain wetlands and open waters without obvious hydrologic connections to a tributary stream (including many GIWs) provide functions that are important to downstream water quality and integrity, it maintained that these wetlands and open waters must be individually evaluated as to whether they fall under the jurisdiction of the CWA because generalizing such functions based solely on the available science was not possible.

Geology, Hydrology, Soils, and Vegetation

In the southeastern Coastal Plain, GIWs are depressional wetlands that were historically embedded within a matrix of frequently burned longleaf pine forests. The origins of these depressions depend on their topographic position and the underlying geology of the site. In some areas, they are formed when limestone, weathered clay, or ironstone dissolves, causing surface soils to collapse and settle (Hendricks and Goodwin 1952; Folkerts 1997). In other areas, particularly in the Carolinas, where GIWs are called "Carolina bays," they have a distinctive elliptical shape that is attributed to unidirectional wind on water in surface depressions (Soller and Mills 1991). Regardless of geologic origin, fine clays gradually accumulate in depressions, forming an impermeable lens that results in ponding of rainwater and restricts surface water and groundwater interactions (Hendricks and Goodwin 1952). Typically, GIWs fill in late fall and winter with seasonal wet cold fronts, and they dry down in late spring or summer as evapotranspiration exceeds rainfall. The exception is Peninsular Florida, where most rainfall occurs from June through November and water levels peak during September and October (Greenberg and Tanner 2004). Shallow groundwater can be transferred to and from GIWs, depending on the depth of the water table, subsurface topography, and the permeability of underlying clay layers (Hendricks 1954; Lide et al. 1995; Pyzoha et al. 2008).

Cycles of inundation and dry downs, coupled with frequent fires, shape the development of GIW vegetation (Kirkman et al. 2000). Prescribed fire moving through adjacent longleaf pine stands during dry seasons or during droughts will occasionally burn through wetlands. Fires discourage the establishment of hardwood trees that could alter plant communities and possibly influence the hydrological regime—thus illustrating how GIWs are affected by the timing of prescribed fire in surrounding forests.

In relatively undisturbed and frequently burned longleaf pine forests, GIW vegetation structure ranges from open-canopied, herbaceous-dominated communities to closed-canopied swamps or shrub bogs, depending on physiography, hydroperiod, and fire frequency (Figure 9.4). Herbaceous-dominated communities include grass-sedge marshes and cypress savannas, which occur primarily in wetlands that have sandy surficial soils over clay and experience frequent dry downs. The combination of fluctuating water levels and periodic fires maintains a sparse to absent canopy condition

(a)

(b)

FIGURE 9.4 Four major vegetation types ofvegetation types of geographically isolated wetlands embedded in the longleaf pine forests of the southeastern Coastal Plain: (a) A marsh dominated by grasses and sedges, (b) a cypress savanna with grass-sedge ground cover and a sparse canopy of pond-cypress (*Taxodium ascendens*). *(Continued)*

(c)

(d)

FIGURE 9.4 (*Continued*) Four major vegetation types ofvegetation types of geographically isolated wetlands embedded in the longleaf pine forests of the southeastern Coastal Plain: (c) A cypress-gum swamp with sparse ground cover and a closed canopy dominated by pond-cypress and swamp black gum (*Nyssa biflora*), and (d) a shrub-bog with evergreen shrubs and occasional pine or pond-cypress in the canopy. (Photographs courtesy of Carol Nourse, Hugh Nourse, and Aubrey Heupel Greene.)

(Kirkman et al. 2000; Menges and Marks 2008). Species-rich herbaceous plant communities are common, and scattered pond-cypress (*Taxodium ascendens*) occurs in cypress savannas. Floating species dominate in the deepest areas or across entire GIWs if hydroperiods are semipermanent. Emergent grasses and sedges often occur in more intermediate hydroperiods, and woody shrubs occur primarily along the wetland edge.

In comparison, closed-canopied forests of mixed cypress and gum (*Nyssa sylvatica* var. *biflora*) tend to develop in GIWs that have longer hydroperiods and, consequently, less frequent fires. The canopy is dominated by pond-cypress or swamp gum or a mixture of both. The herbaceous ground cover in cypress-gum swamps is often sparse because of both shading by the closed canopy and frequent inundation. In the upper Coastal Plain, these GIWs generally develop in interridge landscape positions or floodplain terraces (Kirkman et al. 2000; De Steven and Toner 2004), but in the lower Coastal Plain and Peninsular Florida, they are commonly embedded in wet-mesic pine flatwoods (Ewel 1998; Casey and Ewel 2006). The surface soils are usually highly acidic organic mucks and peats.

Shrub-bog vegetation is characterized by dense evergreen shrubs with occasional pond pine (*P. serotina*) or pond-cypress emerging into the canopy (Christensen et al. 1981; Sharitz and Gresham 1998). This vegetation develops in GIWs with peat or sandy peat soils, long hydroperiods (6–12 months), and fire-return intervals of 20–50 years (Christensen 1988; Casey and Ewel 2006). Shrub-bog GIWs most commonly occur in the lower Coastal Plain (Sharitz and Gresham 1998; Richardson 2003; Laliberte et al. 2007) or in Florida (Richardson and Gibbons 1993). Fires can be severe if accumulations of flammable evergreen fuels are heavy. When fires burn through deep layers of peat that exposes bare mineral soil, the resulting vegetation is a mixture of marsh and shrub-bog species (Richardson 2003; Casey and Ewel 2006). Overall plant species richness in shrub-bogs is much lower than that of other southeastern GIWs (Laliberte et al. 2007).

Biogeochemistry

Despite their small size, GIWs may play a large role as biogeochemical "hot spots" of ecosystem processes due the cycles of wetting and drying in these wetlands (McClain et al. 2003). Rates of litterfall in southeastern GIWs are among the highest reported for wetland ecosystems (Watt and Golladay 1999). Litterfall varies by wetland type, with much more litterfall in cypress-gum swamps (410–582 g/m^2) than in cypress savannas or marshes (67–290 g/m^2) (Watt and Golladay 1999; Craft and Casey 2000). Measured turnover rates of organic matter were >1 year (Watt and Golladay 1999), indicating net accumulation of organic matter, which can influence elemental cycling (Dunne et al. 2007; Yellick et al. 2016). However, accumulations of organic matter as peat in GIWs are typically low, suggesting there may be interannual variability in turnover rates, other controls on organic matter accumulation (such as fire), and/or efficient remineralization of organic matter (Watt and Golladay 1999). Indeed, GIWs have high rates of microbial metabolism, which both remineralizes organic matter and contributes to the wetland food web through a microbial loop (Mann and Wetzel 1996; Opsahl 2005).

As a result of abundant litterfall inputs, levels of dissolved organic carbon (DOC) in GIWs are relatively high, ranging from 10 to 40 mg/L in southwestern Georgia (Opsahl 2005). GIWs vary in DOC concentrations by vegetation type, with the highest levels in forested wetlands, followed by savannas and marshes. DOC might be expected to stimulate microbial metabolism; however, oxygen consumption, which can be an indicator of microbial metabolism, showed a reverse pattern, with marshes having the highest levels and forested wetlands having the lowest levels (Opsahl 2005). This apparent discrepancy could be attributed to variations in lability of organic material across wetland vegetation types, the degree of photochemical processes, or the amount of algal production. Across all wetland types, only a small portion of the DOC was highly bioavailable, and microbial oxygen consumption was limited by the bioavailability of organic carbon (Opsahl 2005).

Nutrient levels in GIWs are relatively low, and most nutrient inputs are from internal recycling of wetland plant material (Battle and Golladay 2001a). Studies in southwestern Georgia found that phosphorus was the primary limiting nutrient of primary production in GIWs (Craft and Chiang 2002) and that microbial metabolism was also a contributing factor (Opsahl 2005). Research suggests that GIWs are very efficient at recycling nutrients, which would be expected in sites with low nutrient availability (Watt and Golladay 1999). GIWs also may retain nutrients transported from the surrounding upland catchment via rainfall (Watt and Golladay 1999; Craft and Casey 2000). Research on GIWs within agricultural landscapes found higher levels of phosphorus in wetland soil and vegetation than in corresponding upland areas, suggesting that GIWs may play an important role in mitigating nutrient and sediment transport, particularly in disturbed landscapes (Battle et al. 2001; Dunne et al. 2007). However, this ecosystem service may be compromised by ditching, filling, or otherwise altering wetland hydrologic conditions (Whigham and Jordan 2003).

Biogeochemical cycling in GIWs is strongly influenced by hydrology, as their wetting and drying regimes produce conditions that support high levels of microbial processing (Battle and Golladay 2001a, 2007; Inkley et al. 2008). During dry down, organic matter tends to accumulate in GIWs and decomposition is slow. Once GIWs are inundated, accumulated reactants may be reactivated and/or mobilized by associated microbial assemblages, leading to "hot moments" of nutrient cycling and litter decomposition (Sørensen 1974; Orchard and Cook 1983; Moore 1990; Battle and Golladay 2001a; Craft and Chiang 2002; McClain et al. 2003). Although results of studies on the effects of wet-dry period frequency are mixed, most suggest that multiple inundation and dry-down periods accelerate decomposition of organic material and mineralization of nutrients (Yates and Day 1983; Glazebrook and Robertson 1999; Watt and Golladay 1999; Battle and Golladay 2001a). The degree of variability in hydrology across GIWs promotes heterogeneity in biogeochemical cycling across the longleaf pine landscape.

Variability in hydrology also influences the severity of fire effects within and around GIWs, but the interactions of fire and biogeochemical cycling are complex and not fully understood. Battle and Golladay (2003) reported that the effects of fire on water quality in GIWs vary with soil moisture or inundation conditions before and after the fire, and showed that pH, alkalinity, and dissolved inorganic carbon (DIC) increase when GIWs are only partially inundated, allowing fire to encroach. They attributed the increase in pH and alkalinity to carbonates and hydroxides leaching from ash, and the increase in DIC to increased levels of bicarbonate formed when carbon dioxide produced in the fire dissolves in water. When the GIWs are fully inundated and fire in the surrounding uplands is followed by heavy rainfall, DOC and ammonium increases, presumably from the transport of materials from surrounding areas.

Patterns of Biodiversity

In longleaf pine forests, the relative contribution of GIWs to biodiversity is disproportionate to their total area (Kirkman et al. 1999). Nearly 40% of the plants and 30% of the amphibians attributed to the longleaf pine ecosystem are associated with GIWs and the ecotone between GIWs and surrounding landscapes (Moler and Franz 1987; Guyer and Bailey 1993; L. Kirkman, unpublished data). Invertebrate species are also highly diverse (Leeper and Taylor 1998; Battle and Golladay 2001b, 2002; Kirkman et al. 2012), although their numbers are poorly documented compared to plants and amphibians. At a landscape level, the collective effects of large and small wetlands, wet and dry years, and long and short hydroperiods provide an exceptionally diverse and dynamic assemblage of environmental conditions (Sharitz 2003; Whigham and Jordan 2003), supporting an array of species that are adapted to the extreme conditions that accompany cycles of wetting and drying. These unique habitats support numerous endemic, threatened, and endangered species (Sutter and Kral 1994; Kirkman et al. 1999; Edwards and Weakly 2001; Dodd and Smith 2003; Sharitz 2003).

Plants

The exceptional plant species richness in herbaceous-dominated GIWs is attributable to variable fire regimes and the fire-maintained upland-wetland ecotones that are associated with the longleaf pine ecosystem (Kirkman et al. 1998; Kirkman and Mitchell 2006; Kaeser and Kirkman 2009). Within these ecotones, plant species richness can be extremely high, often >50 species/m^2 (Kirkman et al. 1998), with many areas supporting numerous regionally rare species.

In a survey of rare plants in GIWs of six southeastern states, Edwards and Weakly (2001) found nearly 200 species of concern, 69 of which were state listed as threatened. Most were perennials occurring in grass-sedge marsh and cypress-savanna habitats or in adjacent fire-maintained ecotones. Two federally listed endangered species, chaffseed (*Schwalbea americana*) and pondberry (*Lindera melissifolia*), particularly favor these upland-wetland ecotones (Kirkman et al. 1998; Norden and Kirkman 2004; Aleric and Kirkman 2005).

Amphibians

The diverse assemblage of amphibians in GIWs also reflects the diverse structural and hydrological conditions and the spectrum of landscape positions that these habitats occupy in the longleaf pine ecosystem (Guyer and Bailey 1993; Gibbons 2003; Means 2006; Smith et al. 2006; Liner et al. 2008; Kirkman et al. 2012). Many amphibian species breed only in GIWs that lack predatory fish (Hecnar and McClosky 1997). Of the >30 amphibian species in the longleaf pine ecosystem, 13 breed exclusively in GIWs (Moler and Franz 1987). Those that depend on GIWs for breeding habitat tend to be long-lived; their high reproductive output compensates for missed breeding opportunities during droughts. One species, the eastern spadefoot (*Scaphiopus holbrookii*) can breed year round, taking advantage of the short-hydroperiod wetlands that are filled by extreme rain events. In a single breeding event, a female can produce more than 4000 eggs, and larvae develop into juveniles in only 3–4 weeks (Wright 1932; Greenberg and Tanner 2004). Other species, such as the southern leopard frog (*Lithobates sphenocephalus*) and marbled salamander (*Ambystoma opacum*), have longer larval development periods (3–6 months) but also form large breeding aggregations that produce large numbers of eggs and offspring (Wright 1932; Gibbons et al. 2006). Although episodic, the biomass of amphibians produced in GIWs can be considerable (Pechmann et al. 1989; Semlitsch et al. 1996). For example, at a single Carolina bay on the Savannah River Site in South Carolina, 24 species were reported, and the biomass of the >360,000 juveniles produced in a single breeding season was >1400 kg (Gibbons et al. 2006).

Another adaptation to the variable hydroperiods of GIWs is facultative paedomorphosis, a phenomenon that occurs in two southeastern salamander species (Dodd 1993; Johnson 2002), the striped newt (*Notophthalmus perstriatus*) and the mole salamander (*Ambystoma talpoideum*). In years when GIWs fill in late fall and winter and dry down in early to late summer, larvae of these species transform into terrestrial juveniles (tail fins are absorbed, gills are lost, and lungs develop) and disperse into the uplands. In extremely wet years when GIWs hold water year round, larvae remain in the wetlands, becoming sexually mature and breeding the following fall/winter while retaining their aquatic body form (gills and high tail fins). This developmental adaptation allows salamanders to optimize breeding opportunities when conditions permit, while limiting the risks that are associated with migration to and from breeding habitat.

Amphibians that depend on GIWs as their primary breeding habitat are among the most threatened vertebrates in the Southeast, largely because their requirements include both the GIWs and the surrounding longleaf pine habitats (Dodd and Cade 1998; Dodd and Smith 2003). An example is the dusky gopher frog (*Lithobates sevosus*), which is federally listed as endangered, primarily because of habitat loss but also because of fire suppression and canopy closure within and around many of its breeding sites (Thurgate and Pechmann 2007. The global population of this species had been reduced to four breeding wetlands in Mississippi that support three populations (U.S. FWS 2014). Population declines in two other federally listed species—the endangered reticulated flatwoods salamander

(*Ambystoma bishopi*) and the threatened frosted flatwoods salamander (*A. cingulatum*)—are attributed to fire suppression in breeding wetlands combined with conversion of longleaf pine and slash pine (*P. elliottii*) flatwoods to intensively managed pine plantations (Means et al. 1996; Gorman et al. 2013). Habitat loss is also the primary cause of population declines for the striped newt; this inhabitant of longleaf pine sandhills and fishless GIWs is a candidate for federal listing as threatened (U.S. FWS 2011).

Reptiles

GIWs also provide critical habitat for numerous reptile species, including turtles (Buhlmann and Gibbons 2001; Steen et al. 2012; Semlitsch and Bodie 2003), snakes (Roe et al. 2004; Steen, Stevenson, et al. 2013), and alligators (Subalusky et al. 2009). Sharitz and Gibbons (1982) reported the occurrence of 6 turtle species, 9 lizard species, 19 snake species, and alligators in Carolina bays. Water snakes (*Nerodia* spp.), cottonmouths (*Agkistrodon piscivorus*), and red-bellied mudsnakes (*Farancia abacura*) have also been documented in GIWs (Roe et al. 2004; Eskew et al. 2009; Steen, Stevenson, et al. 2013), likely because the wetlands provide habitat for high densities of preferred amphibian prey.

Other Species

GIWs are also important habitat for birds and mammals in the longleaf pine ecosystem. Foraging and nesting have been documented in Carolina bays by wood ducks (*Aix sponsa*), wood storks (*Mycteria americana*), wading birds, and several neotropical migratory birds (Sharitz and Gibbons 1982; Mamo and Bolen 1999; Kennamer and Hepp 2000; Bryan 2005; Kilgo and Bryan 2005). Czapka and Kilgo (2011) found that richness of winter bird species was higher in pine forests surrounding Carolina bays than in pine forests that lack embedded GIWs, suggesting the importance of GIWs to the avifauna of the longleaf pine ecosystem.

The range of hydrologic conditions, soil conditions, and vegetative cover in GIWs also provides habitat for numerous mammal species (Lidicker et al. 1992). Thirteen small mammal species were documented using Carolina bays (Sharitz and Gibbons 1982); and a southeastern endemic, the round-tailed muskrat (*Neofiber alleni*), has been reported in other GIWs (Schooley and Branch 2006, 2009). A higher level of bat activity was reported over Carolina bays than over drained wetlands or forests (Menzel et al. 2005), and specifically, the presence of the southeastern myotis (*Myotis austroriparius*) has been strongly linked to the proximity of Carolina bays and bottomland hardwood communities (Ford et al. 2006).

HABITAT CONNECTIVITY

Wetland to Upland Connectivity

With the exception of a few fully aquatic salamanders, such as *Amphiuma* spp. and *Siren* spp., which may live exclusively within a single GIW and persist during dry periods by aestivating below ground, most amphibians in GIWs are truly semiaquatic. They breed and undergo larval development in wetlands; as juveniles and nonbreeding adults, they inhabit uplands (Snider and Bowler 1992). These species require a permeable ecotone through which they can migrate to and from adjacent upland habitats (Figure 9.5). Some species are particularly dependent on frequently burned longleaf pine forests to provide the microclimate, ground cover, belowground soil structure, and prey diversity that they need during their terrestrial life stage. The gopher frog (*Lithobates capito*) is a classic example of a longleaf pine specialist that relies on both systems. It breeds in fishless GIWs and inhabits longleaf pine forests as an adult, residing in gopher tortoise (*Gopherus polyphemus*) burrows or other belowground shelters (Blihovde 2006). Turtles also require wetland to upland connectivity. Buhlmann and Gibbons (2001) documented movements into and out of a GIW by eight turtle species. Five of these species use adjacent terrestrial habitats for shelter and/or nesting: the chicken turtle (*Deirochelys reticularia*), eastern mud turtle (*Kinosternon subrubrum*),

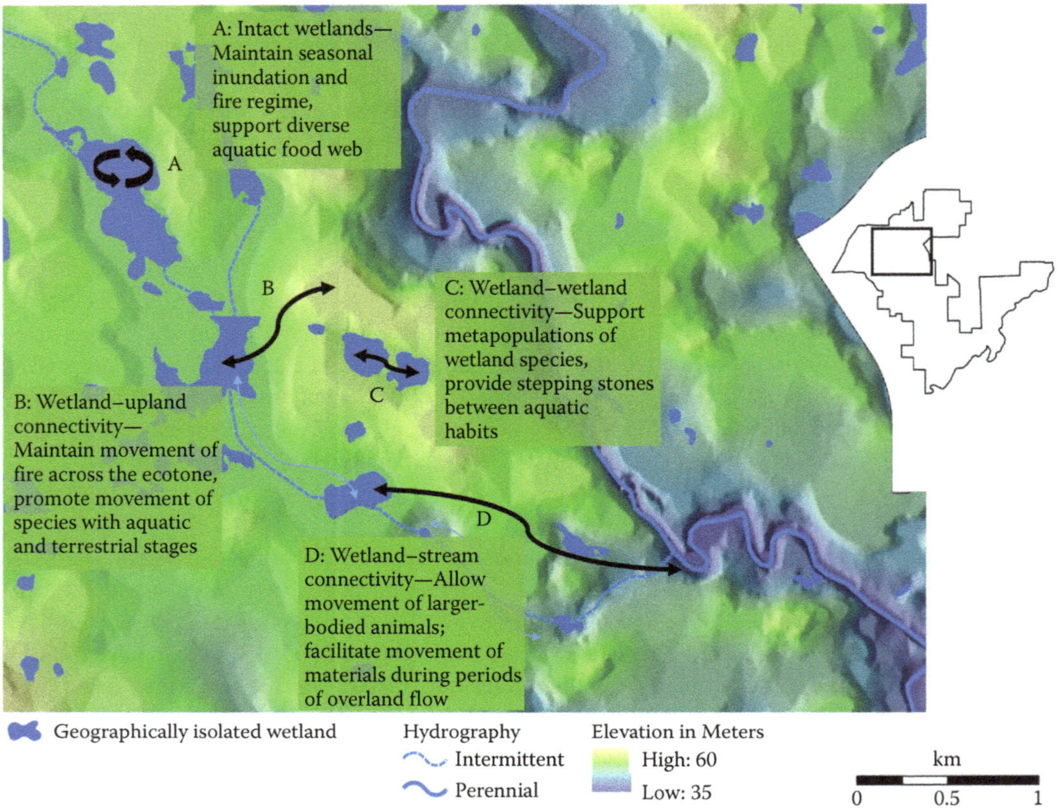

FIGURE 9.5 Topographic map showing a complex of geographically isolated wetlands and intermittent and perennial streams embedded within an intact longleaf pine landscape at the Joseph W. Jones Ecological Research Center in southwestern Georgia: (A) Intact wetlands have a temporally dynamic hydroperiod, canopy and emergent vegetation maintained by occasional fire and a complex food web with high species diversity; (B) wetland connectivity to upland habitats via an intact ecotone maintains appropriate disturbance regimes within the wetland and promotes the movement of fire from the uplands into the wetlands and between-habitat movement of animals that depend on terrestrial and aquatic environments for various life-history stages or other habitat requirements; (C) connectivity within a matrix of wetlands supports metapopulations of wetland species and provide stepping stones between aquatic habitats; and (D) connectivity between wetlands and perennial streams allows movement of animals that require permanent sources of water and occasional hydrological connections, allowing for overland flow of materials and nonterrestrial organisms. Note that letters and arrows represent ecosystem processes maintained by connectivity among intact components of the landscape. (Modified from U.S. Geological Survey. *Bethany quadrangle, Georgia and Elmodel quadrangle, Georgia [maps]. 1:24,000. 7.5 Minute Series*, United States Department of the Interior, Reston, VA, USGS, 1974; wetland location data provided courtesy of the Joseph W. Jones Ecological Research Center.)

striped mud turtle (*K. baurii*), common musk turtle (*Sternotherus odoratus*), and common snapping turtle (*Chelydra serpentina*). The other two—pond slider (*Trachemys scripta*) and Florida cooter (*Pseudemys concinna floridana*)—migrate through adjacent upland habitats toward bodies of permanent water. Likewise, semiaquatic snakes frequent adjacent upland habitats for nesting, shelter, or migration between wetlands (Roe et al. 2004; Willson et al. 2006). Willson et al. (2006) describe annual migrations of water moccasins to and from wetlands and found evidence that other species, such as banded water snakes (*Nerodia fasciata*) and green water snakes (*N. floridana*), rely on recolonization of GIWs via migration following drought.

Amphibians, reptiles, and other animals that depend on or move through GIWs during a part of their lives form a mobile link between wetland and terrestrial ecosystems (Lundberg and

Moberg 2003; Earl and Semlitsch 2012). As they migrate, they also move associated energy and materials (or subsidies) across habitat boundaries (Polis et al. 1997; Schreiber and Rudolf 2008; Capps et al. 2015; Tiegs et al. 2015). During shifts in life-history stages, many species also have a distinct shift in body chemistry, the result of which is a redistribution of energy and nutrients across the landscape (Regester et al. 2008; Capps et al. 2015). Fluxes of energy and nutrients (such as nitrogen and phosphorus) that move across habitat boundaries can be variable across space and time (Polis et al. 1997), a phenomenon that strengthens connections among heterogeneous habitat patches (Loreau et al. 2003; Leroux and Loreau 2008; Sitters et al. 2015). However, the contributions of animals to spatially and temporally heterogeneous biogeochemical processes have only been acknowledged recently (P. McIntyre et al. 2008; Atkinson and Vaughn 2015; Capps et al. 2015). The abundance of GIWs within longleaf pine forests (Tiner 2003), coupled with the large amount of amphibian biomass that disperses from GIWs into longleaf pine forests (Dodd 1992; Greenberg and Tanner 2005; Gibbons et al. 2006), suggests that GIWs may play an important role in these biogeochemical transformations.

Connectivity among GIWs and between Ecosystems

GIWs embedded within longleaf pine forests are structurally, temporally, and spatially variable habitat patches whose importance to individual wildlife species occurs along a continuum of habitat connectivity (Figure 9.5). For some species, an entire population is supported by a single GIW, with only limited gene flow from other nearby GIWs. An intact mosaic of wetlands within an upland forested matrix can maintain a metapopulation of these species. If a single GIW loses its entire population—either gradually through attrition or by a catastrophic event—it can be recolonized by individuals from another nearby GIW (Levins 1969). This pattern of immigration and emigration among GIWs has been confirmed by Schooley and Branch (2006, 2009) for populations of round-tailed muskrat, and is presumed to hold true for many amphibian species (Smith and Green 2005).

Furthermore, metacommunity theory proposes that colonization rates increase with the number and proximity of neighboring patches (Hanski and Ranta 1983; Kiflawi et al. 2003); and several studies show that habitat quality, proximity to dispersal corridors, and characteristics of the dispersal matrix are also important predictors of GIW occupancy (DeMaynadier and Hunter 1995; Roznik and Johnson 2009; Schooley and Branch 2009). Loss of connecting dispersal corridors and declining area and quality of GIWs increases the extinction risk for these species through reduced immigration opportunities, increased inbreeding, and vulnerability to random environmental perturbations (Hanski 1999). For example, Cosentino et al. (2011) demonstrated that connectivity, wetland area, and the presence of fish influenced genetic diversity and divergence in populations of eastern tiger salamanders (*Ambystoma tigrinum*).

On the other end of the population genetics spectrum are species with particularly high mobility, for which movement among GIWs during their terrestrial life stage can support a single population within the wetland mosaic. These movements can be driven by a range of factors which may vary for different species, such as ontogenetic shifts in habitat use or resource demand for amphibians and alligators, spatial distribution of preferred prey for water snakes and wading birds, or hydrologic regimes for turtles (Buhlmann and Gibbons 2001; Gawlik 2002; Roe et al. 2004; Steen et al. 2007; Subalusky et al. 2009; McKee 2012). For these species, multiple GIWs within dispersal distance of one another are important for providing sufficient resources (food and habitat) to maintain a viable population, particularly during prolonged droughts. The loss of individual GIWs can increase the amount of energy needed for dispersal and the risk of mortality for dispersing individuals (Roe et al. 2004). Roe and Georges (2007) reported that movement distances for reptiles among GIWs are considerable (499–1518 m), emphasizing the conservation importance of protected travel corridors.

Regardless of specific habitat connectivity requirements, maintaining viable populations of semi-aquatic reptiles and amphibians within longleaf pine ecosystems requires the following landscape mosaic: a complex of GIWs that are distributed within the range of species dispersal distances, sufficient habitat quality for breeding, and a matrix of frequently burned longleaf pine uplands that allow dispersal (Semlitsch 2002; Semlitsch and Bodie 2003). Because the importance of these elements will vary among species, a diversity of habitat types and spatial configurations is required to support the diverse assemblages that inhabit longleaf pine ecosystems. Although not all GIWs will be primary habitat for every wetland-dependent species, they may be important as stepping stones to facilitate movements across the landscape (Amazega et al. 2002; Baum et al. 2004) and in providing complementary resources, particularly during periods of environmental change (Roe and Georges 2007). Furthermore, during extreme rain events, episodic surface water connections and microtopographic features can also act as linkages among GIWs, facilitating overland flow of materials and greater dispersal distance of animals, particularly for species that lack a terrestrial life-history stage (Snodgrass et al. 1999; Schooley and Branch 2009).

An intact landscape mosaic of GIWs can also provide stepping stones to streams and rivers, which can be particularly important for larger-bodied animals that, at some point in their life history, require the habitat or prey offered by larger and more permanent water bodies. An example is American alligators, which use GIWs as nesting and nursery sites but spend most of their adult lives in nearby riverine systems (Subalusky et al. 2009). Furthermore, alligators often frequent multiple GIWs as they move between the two systems. Because adults accrue most of their body mass in the riverine system but nest in the GIWs, they can provide a substantial resource subsidy to the wetland system. Moreover, adult alligators act as ecological engineers by building nest mounds and digging burrows, both of which provide additional habitat for a range of wetland species and can alter ecosystem function through changes in hydrology (Mazzotti and Brandt 1994). The importance of both GIWs and riverine systems to the life-history stages of alligators, coupled with the impact of alligators on both systems through resource transfer and habitat modification, suggests a high level of functional connectivity between the two ecosystems (Subalusky et al. 2009).

RESTORATION OF CONNECTIVITY

Large contiguous tracts of frequently burned longleaf pine forest with embedded GIWs occur primarily on public lands such as Eglin Air Force Base, Fort Stewart Military Reservation, Apalachicola National Forest, Conecuh National Forest, and Blackwater River State Forest. Aside from these large publicly owned tracts, some longleaf pine forests are maintained on large private lands managed for northern bobwhite (*Colinus virginianus*), or as smaller patches of habitat within a gradient of increasingly human-dominated landscapes. Many former longleaf pine forests have been converted to stands of slash and loblolly pine (*P. taeda*). Because even the largest forested tracts with embedded wetlands are fragmented by roads, power lines, or urbanization, restoration of habitat connectivity is clearly needed. Restoring effective habitat connectivity among small, isolated tracts of longleaf pine in urban settings is likely not possible, because urban land uses are impermeable to many wildlife species and these habitat patches tend to favor generalist native species and invasive species (Delis et al. 1996; Delis and Mushinsky 2005). However, successful restoration requires a process that considers the full spectrum of benefits that can be derived from habitats in a range of conditions and then sets priorities among ecosystem functions.

ALTERED LANDSCAPES

The primary factors that disrupt the connectivity between longleaf pine forests and GIWs include hardwood encroachment from fire exclusion, the creation of roads or firebreaks around wetlands, and the conversion of longleaf pine forests to intensively managed plantations, crop or grazing land, and urban land uses. As shown in Table 9.1, these land uses tend to decrease habitat connectivity among wetlands in addition to eliminating habitat in adjacent uplands. In contrast, ditches and other alterations to natural drainage patterns can bring undesirable connectivity among GIWs by creating corridors that allow colonization by predatory fish or invasive plants (Zedler and Kersher 2004; Hohausová et al. 2010).

TABLE 9.1
Major Categories of Disturbance for Geographically Isolated Wetlands in Longleaf Pine Ecosystems. Note that many wetlands experience more than one of these disturbance categories, as one type of disturbance often leads to other types

Disturbance	Effect on Ecosystem Function	Restoration Actions
Hydrological alterations resulting from ditches, dikes, roads, and other constructed landscape features	Shortened or lengthened hydroperiod; community-assemblage alterations, such as introduction of fish or encroachment of hardwoods; changes in optimal breeding conditions for some amphibians	Fill in ditches; use hardened low-water crossings and culverts to restore natural hydrological connections where possible
Fire suppression	Hardwood encroachment into and around wetlands; loss of emergent vegetation; decreased quality of breeding habitat for aquatic fauna	Mechanically remove mature hardwoods and restore the natural fire regime; if necessary, treat hardwood resprouts with herbicide
Increased isolation	Loss of connectivity that supports metapopulations of wetland species or loss of stepping stone habitat for dispersing animals	Divert barriers to dispersal (roads, fences, and other landscape features) so that they circumvent rather than transect wetland complexes; restore longleaf pine uplands that connect wetlands
Invasion of nonnative plants and animals	Loss of native plant species, which can alter wetland climate, habitat availability, and ecotone structure; declines in native amphibian species through competition and predation by fish and other groups	Mechanically remove nonnative plant species; reseed native plant species during the appropriate time in the hydroperiod; remove artificial hydrological connections to more permanent water bodies

Hardwood Encroachment

In longleaf pine forests, both fire exclusion and repeated winter burning allow fire-intolerant hardwoods to become established in and around embedded GIWs. Winter burns exacerbate hardwood encroachment because wetlands are frequently inundated, and fires are less intense in winter than they would be in spring and summer. Once established, fire-intolerant hardwood trees or shrubs shade out ground cover and further impede the movement of fire into the GIW even during dry periods (Martin and Kirkman 2009). Nearby roads and firebreaks also facilitate hardwood encroachment by limiting the spread of fire from fire-maintained upland areas. The hardwoods in upland-wetland ecotones may not be a physical barrier for movement of semiaquatic species, but small vertebrates crossing through a hardwood-dominated forest would be more vulnerable to predators, such as raccoons (*Procyon lotor*) that use hardwoods for den sites (Jones et al. 2004). Moreover, the presence of mature hardwoods can significantly decrease hydroperiods in GIWs, with implications for amphibian larval development (Skelly et al. 1999; Clayton and Hicks 2007). Perhaps most significant is the change in vegetative structure associated with a closed hardwood canopy in wetlands. Shading reduces the density of emergent herbaceous vegetation, thereby reducing cover for larvae and potentially increasing the risk of predation (Werner et al. 1983; Babbitt and Tanner 1997).

Once hardwoods become established as trees within a wetland, treating them with fire alone is rarely an effective method for reversing this trend. Often, vegetation succeeds toward an alternative stable state as a hardwood-dominated depression that is resistant to the entry of fire (Martin and Kirkman 2009). Mechanical removal of hardwoods followed by chemical treatment of resprouts can successfully restore the herbaceous vegetation within and around hardwood-encroached GIWs, particularly if a persistent soil seed bank is present (Martin and Kirkman 2009; Gorman et al. 2013).

Several recent experiments have evaluated the response of amphibians to hardwood removal. Gorman et al. (2013) used mechanical hardwood removal as a surrogate for fire to restore wetlands used by the frosted flatwoods salamander, a longleaf pine specialist that requires herbaceous-dominated GIWs for breeding (Gorman et al. 2009). Two years post-treatment, they found that although the mechanical treatments reduced canopy cover, herbaceous cover did not increase, nor did amphibian species richness. Their explanation was that the treatments were confounded by drought, and that positive herbaceous vegetation and amphibian responses would likely require several years. In another study of longleaf pine forests on the Francis Marion National Forest in South Carolina, Klaus and Noss (2016) found that both generalist and specialist amphibian species responded positively to hardwood removal, mulching, and prescribed burning in fire-suppressed GIWs. They concluded that the specialists are adapted to wetlands that are frequently burned, have low detrital input, and high primary productivity (Klaus and Noss 2016), whereas the generalists respond to high temperature and high productivity rather than a particular wetland type.

Intensive Forest Management

GIWs embedded in commercial pine plantations likely retain some of the biological functions of their counterparts in longleaf pine forests, particularly if they have not been bedded and planted in pine. Best management practices on these plantations often include recommendations for buffer zones around wetlands (Jones et al. 2010). Although generally designed to protect water quality rather than wildlife habitat, these suggested practices are likely beneficial for some wildlife (Jones et al. 2010). However, buffers around wetlands to protect water quality may not be sufficient for semiaquatic or migrating wildlife species such as amphibians and reptiles (Semlitsch and Bodie 2003). Amphibians often avoid clear-cuts, and some salamanders also avoid stands that have been subjected to selective harvesting (Semlitsch et al. 2009). Amphibians dispersing through recently clear-cut stands experience increased risk of mortality from elevated temperatures, water loss, soil compaction, and reduction of shelters (Harpole and Haas 1999; Semlitsch et al. 2009). For frogs and toads, retaining coarse woody debris after harvesting ameliorates the effects of water loss and improves juvenile survival rates (Rittenhouse et al. 2008).

The use of alternative harvesting methods such as individual-tree or group selection is critically important for accommodating the terrestrial habitat requirements of semiaquatic species and facilitating habitat connectivity across landscapes (Semlitsch et al. 2008). Planted pine stands and the wetlands embedded within them can be managed to retain the structural conditions that resemble fire-maintained longleaf pine landscapes (Kirkman, Mitchell, et al. 2007), thereby enhancing their suitability as habitat for GIW-dependent wildlife. The use of prescribed fire (in both uplands and wetlands) is key to enhancing biological diversity and functions of GIWs embedded in intensively managed pine plantations.

Agriculture

GIWs in agricultural landscapes present a significant challenge for efforts to restore habitat for longleaf pine-associated wildlife. In addition to the loss of forest cover, native wetland vegetation may have been removed or fire suppression may have allowed encroachment by hardwoods (Stuber et al. 2016). Further, inputs of nutrients from fertilizers and livestock can degrade water quality (Verhoeven et al. 2006) and cause reduced reproductive success in amphibians (Knutson et al. 2004). Brühl et al. (2013) suggested that terrestrial pesticide exposure in agricultural landscapes could have a larger impact on amphibian populations than previously recognized. Agricultural chemicals applied to adjacent or nearby fields can have severe (and sometimes lethal) effects (Howe et al. 1998; Boone and James 2003; Hayes et al. 2010). Relatively little information is available on the levels of agricultural chemicals in GIWs, but studies suggest that concentrations of herbicides, such as atrazine, are sometimes higher in GIWs than in large permanent bodies of water (Solomon et al. 1996; Howe et al. 1998), and that these chemicals can accumulate across trophic levels (Hall and Kolbe 1980). Atrazine exposure has been associated with changes in behavior, reproduction, and development in amphibians (Rohr and McCoy 2010) and with endocrine disruption in amphibians and alligators (Hayes et al. 2011).

Beyond these direct impacts in agriculturally dominated landscapes is the potential loss or severe reduction of habitat connectivity among GIWs and between GIWs and suitable upland habitat. Amphibians experience increased risk of mortality from desiccation or predation if they attempt to traverse agricultural fields (Rothermel and Semlitsch 2006; Rittenhouse et al. 2008). Nonetheless, a study of GIWs in agricultural areas of southeastern Minnesota (where natural wetlands are scarce) found that in areas where livestock access is limited, small constructed wetlands could support similar numbers of amphibian species as their natural counterparts (Knutson et al. 2004). Moreover, amphibian species richness on agricultural lands is enhanced in GIWs that are close to remnant patches of native upland vegetation, especially if the patches have some original woodland components (Knutson et al. 1999; Babbitt et al. 2006; Medley et al. 2015). At a large cattle ranch in southern Florida, three amphibian species that also occur in longleaf pine forests—barking treefrog (*Hyla gratiosa*), pine woods treefrog (*H. femoralis*), and oak toad (*Anaxyrus quercicus*)—only bred in wetlands ≤200 m from a hardwood hammock (Babbitt et al. 2006). Although this study did not explicitly address the effects of proximity to longleaf pine forests, it clearly suggested the need for research on the dispersal capabilities of upland-wetland associated amphibian species.

Federal Incentive Programs for Restoration

Financial and technical support to restore wetlands in agricultural settings is available through several programs administered by the U.S. Department of Agriculture (USDA). The Agricultural Conservation Easement Program (USDA 2016), which replaced the Wetlands Reserve Program in the Farm Bill, offers enrollments that feature a range of time commitments. Enrollment options include permanent conservation easements and a voluntary program called the Wetlands Reserve Enhancement Partnership, which was designed to leverage resources for wildlife habitat improvement as well as protecting, restoring, and enhancing wetlands. This relatively new program has yet to be evaluated; however, a recent review by De Steven and Gramling (2012) found that GIWs were the focus of 5%–20% of the projects funded in Mississippi, Georgia, and South Carolina from 2000 to 2008.

Another incentive program that could benefit embedded wetlands is the Longleaf Pine Initiative, within the USDA Conservation Reserve Program, which promotes planting longleaf pine and native warm-season grasses on 100,000 ha across the nine southeastern states. The USDA Natural Resources Conservation Service, in partnership with the U.S. Department of the Interior, also offers landowners financial incentives to manage for wildlife through the Working Lands for Wildlife Program.

The scope of these programs is ambitious; however, the projects supported tend to be small, short term, and lacking explicit consideration of GIWs. Their ability to restore key ecological processes of GIWs would likely be more effective if they favored projects that target land adjacent to existing protected longleaf pine forests or projects that are part of conservation planning to increase habitat connectivity. Moreover, these programs would be more useful if they offered opportunities to measure plant and wildlife responses to restoration as part of the effort to improve biodiversity and functions in these systems (De Steven and Lowrance 2011).

The Challenges of an Uncertain Future

When setting priorities for restoration of GIWs in the longleaf pine ecosystem, considering scenarios of future climatic conditions is as important as evaluating the implications of altered landscapes (Golladay et al. 2016). Climate change is likely to have a large effect on ecological processes in GIWs (Brooks 2009), particularly for those within the range of longleaf pine. Predictions for the Southeast include not only increases in temperature, but also changes in the timing and abundance of precipitation (IPCC 2014). Some species can alter their behavior as temperatures change; for example, freshwater turtles can begin nesting earlier in response to a warming climate (Janzen 1994). Other behavioral adaptations in turtles, such as placing nests in shaded locations or creating deeper nests, can ameliorate the effects of higher temperatures (Refsnider and Janzen 2012). However, models predict that some turtle populations will be extirpated with even moderate temperature increases, because such increases would result in the production of all female offspring (Janzen 1994; Schwanz and Janzen 2008). Studies suggest that higher temperatures could also exacerbate the spread of disease in amphibians (Pounds et al. 2006), although evidence of direct links between climate change and disease has not been established (Lips et al. 2008; Rohr et al. 2008).

The predicted changes in the timing and abundance of rainfall and increases in evapotranspiration resulting from higher temperatures would significantly affect the biological functions of GIWs by altering the long-term hydrologic regime. Although species that depend on GIWs are adapted to periodic dry downs, long-term changes in hydroperiod would likely affect predator-prey interactions and, ultimately, wetland productivity (Blaustein et al. 2010). Most amphibians and other semiaquatic animals with a biphasic life history rely on particular hydrologic conditions to complete their life cycles. Because complete reproductive failure can occur if wetlands dry before larvae complete metamorphosis, perpetual long-term patterns of drought may affect population persistence. Moreover, the size of larvae at metamorphosis is positively related to the length of the hydroperiod and to the probability of juvenile survival (Pechmann et al. 1989; Crespi and Warne 2013).

Extreme rain events can also negatively affect habitat by facilitating colonization by fish or causing saltwater intrusion in coastal GIWs (Walls et al. 2013). Thus, changes in the duration of hydroperiod can have multiple implications for amphibian survival. Further, changing weather patterns that restrict or alter the ability to use prescribed fire as a management tool (see Chapter 13) would directly impact the wetland vegetation, including both structural and compositional characteristics that are important for habitat.

Guidance for addressing climate-driven changes emphasizes the need to protect adequate and appropriate space, reduce other stressors, use an adaptive management approach, and test adaptation strategies (Hansen et al. 2010). Recommendations specifically targeting amphibians, but also generally relevant for GIWs as specialized habitat, include augmenting water levels, supplementing natural or artificial shelters and canopy cover over wetlands, and constructing wetlands to provide diverse habitats (Shoo et al. 2011). However, the prospect of augmenting water levels as a long-term solution to changes in hydroperiod of GIWs is daunting. Likewise, expecting constructed wetlands to serve the myriad functions of natural wetland complexes that are embedded in frequently burned longleaf pine forests is probably unrealistic. A more practical approach would be to prioritize protection and restoration management of remaining wetland complexes that represent a wide range of hydrologic conditions and are located in the least disturbed landscapes (Kirkman et al. 2012; Walls et al. 2013).

SUMMARY AND FUTURE RESEARCH NEEDS

Maintaining existing stands of mature, frequently burned longleaf pine on protected areas is critical to protecting embedded GIWs and the biodiversity and habitat connectivity within these unique systems. These stands offer the best examples of reference conditions to guide restoration of altered habitats (see Chapter 11). At sites where reference conditions persist, additional research is needed to determine the scale at which embedded GIWs support wildlife species (McKee 2012) and to quantify the ecosystem-level contribution of biomass and subsidies produced in GIWs (Capps et al. 2015). In fire-suppressed landscapes, new research is needed to determine whether restoration of vegetative structure and reintroduction of fire will have long-term impacts on population dynamics and species richness. Also needed are evaluations of the effectiveness of longleaf pine restoration incentive programs and programs that specifically target restoration of wildlife habitat in agriculturally impacted wetlands.

Anticipating the effects of climate change on GIWs is critical for habitat management of vulnerable species, even on protected and managed conservation lands (Hayes et al. 2010). In addition, species-specific information is needed on the response and plasticity of wetland-dependent wildlife to temperature and precipitation shifts. Collectively, this information will be useful in developing conservation and management approaches for species that depend on GIWs and in maintaining ecosystem function and services of these important and underappreciated habitats.

Finally, maintaining and restoring ecological functions of GIWs embedded in the longleaf pine ecosystem will likely require policy changes that can happen only if the public becomes more aware of the values and services that these systems provide to society (Zedler 2003; Calhoun et al. 2005; Zedler and Kercher 2005).

ACKNOWLEDGMENTS

The authors acknowledge the many research technicians and graduate students at the Joseph W. Jones Ecological Research Center, where a large amount of the cited research was conducted. These include Jennifer Howze, Anna Farmer, David Steen, Melanie Kaeser, Kim Coffey, Lisa Giencke, Katie Martin, Glenn Martin, Stribling Stuber, Kathy Aleric, Heather Summer, and many others. We thank Kevin McIntyre, Steven T. Brantley, Matt Cohen, Betsie Rothermel, and Steve Golladay for their helpful comments on this chapter.

Part III

Ecosystem Restoration: Linking Ecological Understanding and Management

10 Restoring and Managing the Overstory

An Ecological Forestry Approach

Steven B. Jack and R. Kevin McIntyre

CONTENTS

INTRODUCTION

Few forest types in the United States offer aesthetic and ecological values that can compare to longleaf pine (*Pinus palustris*). Restoration of longleaf pine ecosystems often requires decades of exacting treatments to reach a state of sustainable management. For landowners and land managers who are interested in restoring their forests or cropland to longleaf pine, this means that the importance of realistic objective setting and well-informed planning cannot be overemphasized. The range of possible management objectives for longleaf pine forests is broad, falling along a continuum from preservation with little or no active management to intensive forestry that maximizes particular "products." These products have traditionally been wood based, but they could also include specific wildlife species, water yield, or some other resource value (Smith 1986). Restoration objectives also fall along a continuum, from relatively minor changes in a particular characteristic or attribute to drastically altering the current structure and function of an ecosystem to a more desired condition (Clewell and Aronson 2013). With many tools available for designing treatments, a well-defined set of objectives will help to identify the most appropriate tools to use; often the same tools can be used to meet multiple objectives by applying them in different ways or at different times along the restoration trajectory.

The reigning objective or paradigm of U.S. forest management during the last half of the 20th century was timber production and sustained yield (Duerr and Duerr 1975), a legacy of its European roots. Although maximizing timber production is not the primary management goal for many forest owners—and particularly for what are now referred to as "family forest" (and formerly as "nonindustrial private") landowners (Butler 2008)—this model was often applied broadly without regard for goals other than timber production, especially in the Southeast and Pacific Northwest where the forest industry has always had a prominent presence. In the Southeast, forest management in the 20th century predominantly followed an industrial plantation model that focused on managing trees and tree growth (see Chapter 3), sometimes to the exclusion of other forest attributes (Boyd 2015; Carter et al. 2015) such as wildlife habitat and ecological diversity. This widely implemented approach resulted in significant fiber- and timber-production gains as well as reforestation of significant acreages (Carter et al. 2015). In general, the industrial model as practiced in the Southeast (1) simplifies the forest structure and minimizes heterogeneity, (2) attempts to minimize the occurrence and impact of natural disturbances to prevent the loss of trees, (3) aims for full site occupancy by the trees so that they utilize all available resources, (4) frequently uses fertilizers and other site amendments to boost productivity, and (5) implements even-aged approaches with fixed (and frequently short) rotation ages. These attributes proved successful in meeting timber production objectives for many decades (Fox et al. 2007). Primarily because of its inconsistent seedling and sapling growth patterns, longleaf pine did not fit well within this industrial management model (Schmidtling 1987; Landers et al. 1995).

In the late 20th century, the focus on production of timber and default application of an industrial model began to be questioned (Kohm and Franklin 1997). Many private landowners, conservation organizations, and government agencies have broader objectives, including aesthetics, wildlife, recreation, and ecosystem services such as water yield, water quality, and carbon storage. This trend toward multiple objectives has accelerated in recent years with changing cultural and societal expectations (and subsequent regulatory actions) of what forests can and should provide (see Chapter 3). For many in the private sector, revenue from the sale of timber is frequently only one component of an overall suite of goals for owning and managing forestland (Butler and Leatherberry 2004). Landowners who have or who wish to establish longleaf pine forests often seek alternative approaches to forest restoration and management that incorporate broader goals as well as timber income into an overall management strategy for a working forest.

Management plans for longleaf pine forests often include some aspects of ecosystem restoration. If longleaf pine is not present, it will need to be introduced. If longleaf pine is present but the forest structure is not in the desired condition to meet landowner objectives, treatments can be implemented as needed to recruit additional age classes or increase longleaf pine stocking, enhance the ground cover plant community, reduce the hardwood component, improve wildlife habitat, or reduce surface fuel accumulation. In this chapter, we review the concepts of ecological forestry (Mitchell et al. 2006; Mitchell, Hiers, et al. 2009; Franklin et al. 2007, 2017), an approach that uses natural disturbances and natural forest development to guide silvicultural manipulations. Our focus is on the Stoddard-Neel approach, using its conceptual framework as an ecological forestry case study for promoting sustainable management of longleaf pine forests. We then provide examples of management regimes for achieving restoration and maintenance objectives that are based on a range of initial conditions. These examples, although not prescriptive or all-encompassing, demonstrate how ecological forestry concepts—and specifically Stoddard-Neel principles of management—can be applied to meet multiple objectives in longleaf pine forest ecosystems.

ECOLOGICAL FORESTRY

Conceptual Overview

Changing expectations for the resources and amenities from forests gave rise to several proposals for new forest management approaches, both from within the ranks of the forestry profession (Kohm and Franklin 1997; Puettmann et al. 2009; O'Hara 2014) and from related disciplines (Hunter 1990; Lindenmayer and Franklin 2002). Although some of these approaches build on accepted (albeit infrequently practiced) silvicultural practices, others call for new terminology and techniques. Some seemingly "new" ideas have strong parallels to some of the approaches to forest management (e.g., Toumey and Korstian 1937) that preceded the industrial model of the 20th century and resulted in simplified, homogeneous systems. What they all have in common is that they are based primarily on a redefinition of forest management objectives, leading to restoration of heterogeneity and complexity in ecosystem structures, age classes, ground cover composition, and other characteristics (Kohm and Franklin 1997; Puettmann et al. 2009). The concepts for ecological forestry (Franklin et al. 2007, 2017) emerged from this progressive development of silvicultural approaches that incorporate ecological objectives, provide a balance of resources, and increase complexity.

Ecological Forestry and Longleaf Pine

Legacies

Ecological forestry uses an understanding of natural disturbance regimes—including patterns of mortality, longevity, regeneration, and age structure—as a guide to management decisions and silvicultural prescriptions. Treatments are applied to approximate the frequency and scale of the disturbances that are most common for the forest type being managed (Franklin et al. 2007; Palik et al. 2002). Rather than trying to simplify forests, ecological forestry embraces and incorporates complexity. In contrast to production-oriented approaches, management of multiple age classes is frequently favored, resulting in a range of structural characteristics (Franklin et al. 2007). Although no silvicultural manipulation can perfectly replicate the effects of a natural disturbance, practices that are patterned closely after natural disturbance and that leave similar postdisturbance legacies are more likely to sustain the composition, structure, and function of the desired ecosystem (Franklin 1993; Palik et al. 2002; Drever et al. 2006). Although ecological forestry relies on detailed knowledge of a particular forest system and even a particular tract of land, some concepts can be generalized and applied to all situations. Four general principles are common in ecological forestry-based approaches to management: (1) retaining biological legacies to ensure the continuity of forest structure, function, and composition before and after silvicultural treatments; (2) maintaining or restoring structural complexity, including spatial heterogeneity and biological richness; (3) applying silvicultural interventions (such as selection harvesting and prescribed fire) at ecologically appropriate intervals; and (4) implementing stand-level silvicultural treatments within the context of a landscape-scale plan (Franklin et al. 2007, 2017).

In longleaf pine forests, biological legacies that provide ecological continuity following natural disturbances or harvest include forest structural elements such as snags (standing dead trees), fallen logs, stump holes, and tip-up mounds. All of these features have well-documented structural and habitat values that add to forest complexity and biological richness. One feature that is not always immediately thought of as a legacy is the largely intact, multiaged canopy that remains after fine-scale disturbance events or harvesting operations that emulate disturbance. Maintaining a perpetual forest cover, which is a fundamental result of silvicultural approaches that apply individual-tree selection methods (O'Hara 2014), contrasts sharply with more common practices, such as clear-cutting, that remove all or most of the canopy trees and replant. Other important legacies include a significant component of older (for example, >150 years old) trees, as well as cohorts of advanced regeneration that were established at multiple times and locations (R. McIntyre et al. 2008). The old trees provide unique structural and functional attributes, particularly for wildlife habitat (Mitchell, Engstrom, et al. 2009), and the regeneration provides a multiaged structure and potential future canopy trees.

Structural Complexity

Historically, fire was the most common disturbance during the stand development process of longleaf pine grasslands. Fire is so important to the maintenance of longleaf pine forests (see Chapters 2 and 13) that some consider it less as a disturbance than as an inherent ecological process in this forest type (Fill et al. 2015). Prescribed fire has become the primary tool for incorporating this process into ecologically based forest management. Silvicultural decisions have consequences for prescribed fire in longleaf pine forests (Mitchell, Hiers, et al. 2009) in that canopy manipulations affect the distribution of fine fuels. Thus, changes in canopy structure during the stand development process, whether through natural disturbance or management actions (such as thinning), can affect the structural complexity and spatial heterogeneity of the forest, the ability to continue a regime of frequent fire (Mitchell et al. 2006), and the behavior and effects of fire (see Chapter 6).

Timing of Interventions

Ensuring adequate time between harvests or other silvicultural manipulations allows the recovery of ecological functions and promotes a diverse biological community. In longleaf pine ecosystems, adequate recovery periods are critical for maintaining the integrity of the ground cover plant community, which is an essential ecosystem component (see Chapter 5). Ground cover facilitates prescribed fire by supplying fine fuels, provides habitat and food resources for game and nongame wildlife, and harbors many rare and endemic plant species (Glitzenstein et al. 1995; Hainds et al. 1999). With adequate recovery periods, ground cover communities can resume their functional roles and rebuild resilience after timber harvesting. Recovery periods should be long enough to build sufficient stocking of the overstory to enable commercially viable harvesting operations and adequate residual stocking to supply fine fuels for postharvest prescribed fire.

Integration

Ecological forestry integrates silvicultural management actions into the larger landscape context—treatments at the stand-level are not considered in isolation, but as a part of a deliberate plan to maintain or restore spatial and temporal heterogeneity at a broader scale. In upland landscapes that are dominated by longleaf pine, applying this principle can result in early successional habitats, inclusion of multiple stand-development stages, and protection of features such as geographically isolated wetlands or natural drainages that provide both diversity and ecological functions (see Chapter 9).

When timber production is the primary objective for longleaf pine management, the even-aged approach to management—specifically the shelterwood regeneration system or one of its variants (Croker and Boyer 1975; Brockway, Outcalt, Guldin, et al. 2005), combined with the frequent application of prescribed fire—is arguably the most appropriate and easily implemented approach (Boyer 1993). However, an ecological forestry approach that fosters multiple age classes is often appropriate when objectives are holistic and focused on maintaining multiple ecological attributes (Franklin et al. 2007; O'Hara 2014). Although uneven-aged silvicultural techniques were not regarded as appropriate for longleaf pine for much of the 20th century, managing the species in multiaged forests has more recently been proven effective (see Chapter 4).

STODDARD-NEEL APPROACH

Conceptual Framework

Through much of the 20th century, longleaf pine was often revered for its intrinsic aesthetics and its ecological attributes, as well as for its important role in the history of the Southeast. It was also considered by many natural resource professionals to be part of a bygone era, a niche species and therefore not sustainable within the context of modern forestry (see Chapter 1). Throughout this period, however, a few forest and wildlife managers from the private sector and the U.S. Forest Service

(Croker 1987; Way 2011) recognized the inherent value of the longleaf pine ecosystem. Their understanding of the dynamics and relative complexity that characterized regeneration, fire, wildlife, and sustainability in longleaf pine forests resulted in new approaches for fostering and maintaining the broad ecological characteristics of the ecosystem (Croker 1968). Stoddard-Neel is one such system that focused on restoring and maintaining the natural longleaf pine ecosystem even as southeastern forest management predominantly concentrated on other pine species and short-rotation plantations (Fox et al. 2007; Carter et al. 2015). The approach, which was developed in the 1930s, was based on Herbert Stoddard's observations on private hunting land in northern Florida and southern Georgia (Stoddard 1969; Way 2011). Leon Neel joined Stoddard in 1950 and continued the work into the next century (Neel et al. 2010). As interest in ecological forestry has grown, adaptations of the approach have spread to other parts of the Southeast.

Stoddard-Neel is as much a holistic, ecological philosophy for forest and wildlife management as it is a specific set of prescriptions and techniques. Grounded in a strong land ethic (Leopold 1949), it seeks to preserve all aspects of complex forest systems while still deriving economic benefits. Its focus on maintaining and enhancing as many ecosystem characteristics, components, and functions as possible inevitably results in trade-offs, with no single characteristic or output maximized at the expense of others. Significant consideration is given to how management actions, and particularly harvesting, affect other resources. For example, even though timber is harvested to produce income, the focus is not on maximizing individual tree or stand-level growth to the detriment of maintaining a diverse ground cover plant community, a viable wildlife community, or other resource values.

A few simple concepts, which align well with the ecological forestry concepts described earlier, characterize Stoddard-Neel (Neel et al. 2010). The ecosystem is managed to maintain a perpetual, multiaged canopy, and harvests are not considered as rotations in that they do not terminate the forest as preparation for restarting the process of stand development. This long-term perspective, and the recognition of time as an ecological factor, requires patience because one or two silvicultural treatments are not typically sufficient to meet multiple ecological objectives. On sites that require restoration, many years will be needed to realize the ultimate objective of a functioning, multiaged forest—a structure that typically cannot be achieved in a single manager's career. Stoddard-Neel takes a conservative approach to tree removals. The target of any harvesting entry is to remove no more than the periodic growth (Jack, Neel, et al. 2006; R. McIntyre et al. 2008), similar to maintaining the core principal for an investment fund. This conservative mindset, again, derives from a long-term focus in which the forest is managed with no set rotation or end; when managing into an indefinite future, higher stocking and more future management options are preferable to removal of trees for immediate economic gain.

Prescribed fire is integral to the application of Stoddard-Neel in southeastern pine forests with grass-dominated ground cover. Burning is objective-driven, frequent, and variable in season. The native vegetation in these ecosystems evolved with the frequent occurrence of fire; thus, the use of prescribed fire as a modern surrogate is essential to their restoration and maintenance. In particular, prescribed fire helps control hardwood trees and woody shrubs that would dominate many sites if fire were excluded (Mitchell, Hiers, et al. 2009). Although chemical or mechanical treatments can complement holistic restoration prescriptions that include prescribed fire and can shorten the time needed to achieve desired forest structures, they are not an adequate alternative for fire because they do not restore the same ecological functions (Outcalt and Brockway 2010). Application of prescribed fire will, over time, help to delineate a true upland site that is suitable for longleaf pine restoration from an area that would naturally support a larger hardwood component in the canopy. Ground cover is an indicator of the developmental stage and the health of a particular forested area; during active restoration, the ground cover composition changes with repeated fire and can be monitored as an indicator of progress toward the desired endpoint (Glitzenstein et al. 1995; Kirkman, Coffey, et al. 2004; Beckage et al. 2011).

The guidance for canopy manipulations (harvesting) under Stoddard-Neel is largely derived from patterns of natural disturbance. All historical evidence suggests that native longleaf pine forests had old-growth components and trees of many ages and sizes. The evidence also suggests even-aged patches of regeneration (usually <1 acre) within the matrix of an all-aged tree canopy (Platt, Evans, and Rathbun 1988). These gaps could have resulted from a range of mortality sources. Coarse-scale disturbance events such as tropical cyclones occur infrequently and are generally limited to areas nearest to the Atlantic Ocean and Gulf of Mexico (Stanturf et al. 2007). Fine-scale disturbances, such as lightning strikes or small windthrow events, are common across the range of longleaf pine and typically result in the death of a single tree (Figure 10.1), or less frequently, three to five trees per mortality event; over time, these frequent fine-scale disturbances coalesce to form canopy gaps (Palik and Pederson 1996; Outcalt 2008; Pederson et al. 2008). The result is variable-sized gaps that are large enough to sustain the recruitment of new regeneration cohorts, and thus sustain the demographic and canopy recruitment processes of longleaf pine forests. As much as possible, Stoddard-Neel seeks to mimic the natural disturbance dynamics of longleaf pine forests through a form of individual-tree selection that always maintains sufficient canopy coverage to supply the fine fuels that are needed to sustain frequent application of prescribed fire (Jack, Neel, et al. 2006; R. McIntyre et al. 2008).

Finally, the establishment of new longleaf pine age classes is an ongoing process and not the point-in-time event that characterizes many other silvicultural regeneration systems (Smith 1986). Given frequent fires, periodic seed crops from the overstory, and a ground cover dominated by forbs and grasses, longleaf pine seedlings will become established in both large and small seeding events—infrequently over large areas during region-wide masting years and frequently through localized seed production within areas that are not more than a few acres in size. Once established, seedlings can be maintained for extended periods and then released by harvesting operations (see Chapter 4).

Those interested in practicing multiage silviculture with longleaf pine for multiple objectives often ask how Stoddard-Neel compares to other multiage approaches in pine-grassland systems of the Southeast. Although many differences have been reported (Moser et al. 2002; Guldin 2006; Moser 2006; O'Hara 2014), principal among them is the importance of tree growth and timber production in defining objectives. Most multiage approaches apply harvesting to mature stands with the goal of adhering to a target structure or volume that will sustain a regular flow of timber products over the long term (O'Hara 2014). In contrast, Stoddard-Neel does not have a structural

FIGURE 10.1 A canopy gap created by the death of an individual tree. (Photograph courtesy of Richard T. Bryant.)

target to guide the selection of trees to be harvested and, although a form of volume regulation is used (Jack, Neel, et al. 2006), the focus is on maintaining a broad range of ecosystem characteristics over time.

Another obvious difference is that most multiaged approaches place a high priority on the establishment of regeneration for the desired species; thus seedlings and sapling-sized trees make up a large percentage of total stocking (Guldin 2006; O'Hara 2014), resulting in the classic reverse-J-shaped distribution of size classes. Management under Stoddard-Neel is typically based on much longer time scales; regeneration establishment is intended to be a continuous process and, when coupled with the conservative removals from harvesting, generally results in many fewer trees in the smallest diameter classes (Jack, Neel, et al. 2006). Thus, the structure of forests managed using Stoddard-Neel is characterized by fewer small trees and a larger component of old and large trees (Moser et al. 2002).

Applying the Stoddard-Neel Approach

The practice of silviculture involves both the art and science of managing a forest to meet a particular set of objectives (Smith 1986; Nyland 1996). Within this general framework, Stoddard-Neel relies very heavily on the qualitative, art aspects of the silvicultural practice, with the expectation that the practitioner has a deep understanding of the underlying forest ecology (Jack, Neel, et al. 2006; R. McIntyre et al. 2008; Neel et al. 2010). Most multiage approaches to management in the United States are oriented toward the science side of the definition and have a quantitative structural target for the residual stand after a harvesting entry (Baker et al. 1996; Farrar 1996; Guldin and Baker 1998; O'Hara 2014; Brockway et al. 2014). Stoddard-Neel does not have a similar quantitative structural target for residual stands; as discussed earlier in this chapter, however, it does describe a desired range of stand structures that are based on those resulting from natural disturbance regimes.

When applying individual-tree selection to mimic natural disturbance patterns, consideration of several stand-level factors is critical. First is the encouragement and cultivation of structural heterogeneity (vertical and horizontal). Second is the enhancement, when possible, of canopy gaps that have already begun to develop rather than harvesting to create new gaps. Individual-tree selection methods can create canopy heterogeneity that resembles the heterogeneity caused by natural dynamics, effectively creating a range of gap sizes. Gaps that have already begun to develop often have advanced longleaf pine regeneration that is ready to begin emergence from the grass stage once trees are removed, thereby releasing the next age group. Gaps that do not have established advanced regeneration provide a range of microsites, with variable competitive and climatic conditions (see Chapter 4), for establishing adequate regeneration to sustain the multiaged stand structure into the future (Coates and Burton 1997; Palik et al. 2002; Mitchell et al. 2006). Gaps are generally kept small, however, so that needle cast is adequate to supply the fuels needed to carry fire through the gaps, preventing eventual domination by hardwoods (Mitchell et al. 2006).

In most aspects, the selection of individual trees for removal under Stoddard-Neel is much like selections under other multiage approaches. The objective of most approaches is to "cut the worst and leave the best" trees, implying that defective and less vigorous trees are preferentially removed from the forest, and the higher-quality and faster-growing trees are retained (Guldin 2006). Stoddard-Neel is a bit more complicated, however, because some poorly formed or defective trees—called "character" trees—that would be removed if timber quality was the primary consideration, are left in the stand to provide wildlife habitat, structural diversity, a component of old trees (>100 years), and fine fuels for prescribed fire. These additional considerations, as well as the extended temporal perspective, distinguish Stoddard-Neel from other multiage approaches. Although many of the trees selected for removal will be the same regardless of approach, forests managed under Stoddard-Neel will follow a different trajectory of stand development over decades of management and multiple harvesting entries because the focus is on complex structure, retention of character trees, and use of natural disturbances as a model for canopy manipulations. Consequently, these forests can be readily and uniquely identified as different (Crofton 2001) from other forests that have a similar composition.

Some specific tree characteristics can be used in selecting which trees to remove (Figure 10.2). Defects such as fire scars, fusiform rust galls, and forking of the main stem are all factors that can lead to a tree being marked for removal. Trees with fire scars, and especially large scars at the stem base, are problematic in managed forests that use frequent prescribed fire because the scars tend to reignite each time the area is burned and smoldering tends to persist long after the fire; thus, these trees are often selected for harvesting. Because some individual-tree crown characteristics—such as foliar density, foliage color, and crown size—are good indicators of tree vigor and growth potential (McConville et al. 1999), they are used to help select trees for removal. Spacing of residual trees is important, more for the even distribution of needle fuels than for the maximization of individual tree growth (Mitchell, Hiers, et al. 2009; Neel et al. 2010). As long as fuels are adequately distributed, irregular spacing of trees is in fact preferable to uniform spacing because it helps provide the desired spatial heterogeneity.

Active monitoring of harvesting operations is vital. Damage to residual trees during the harvesting is not tolerated, and operators are penalized (or logging operations stopped) if damage becomes excessive. Also, special consideration is given to unique or special plant or wildlife populations,

(a)

FIGURE 10.2 Individual characteristics that could lead to trees being selected for harvesting during an individual-tree selection harvest: (a) Burned-out scars near the tree base. *(Continued)*

(b)

(c)

FIGURE 10.2 (*Continued*) Individual characteristics that could lead to trees being selected for harvesting during an individual-tree selection harvest: (b) a fork in the main stem, especially low on the tree; and (c) low foliar density, indicating reduced vigor of the tree on the left side as compared to other individuals in the canopy. (Photographs [a] and [b] courtesy of Stephen Golladay; photograph [c] courtesy of Richard T. Bryant.)

as well as to longleaf pine regeneration, to protect them from damage. Because Stoddard-Neel is a conservative approach that removes less volume per acre, individual timber sales may need to cover more acres than traditional harvesting approaches to accrue economically operable volumes. Similarly, intensive merchandizing of cut trees is important because it helps to support the conservative marking of trees; maximizing the value of trees at the landing by carefully allocating cut stems to different product classes can increase the return to the landowner and offset lower harvesting volumes. Codifying protocols in a written, executed contract will avoid potential confusion about expectations or issues that could possibly arise between the landowner and the contractor.

Finally, Stoddard-Neel is more difficult to learn and implement than other multiage methods (Guldin 2006; Moser 2006; O'Hara 2014). This is because of its relatively greater reliance on the art aspects of silviculture, the absence of a specific quantitative target to guide marking for harvesting, and the higher level of experience and understanding of the ecosystem required for practitioners to become proficient (R. McIntyre et al. 2008)—factors that continue to be a barrier to wider implementation (O'Hara 2014). However, integration of its fundamental concepts into operational objectives can do much to forward the goal of maintaining the ecosystem and its many components. Within the context of the longleaf pine ecosystem, the use of conservative harvesting, frequent burning, and retention of residual trees in appropriate volumes and spatial patterns to provide adequate fine fuel distribution are critical for meeting the broad goals of Stoddard-Neel.

The principal goal of Stoddard-Neel is to maintain or restore a functioning forest ecosystem with a broad suite of ecological characteristics. Sometimes this means that an existing forest is managed using a "light touch" while factoring in regeneration, ground cover, and wildlife to perpetuate the forest indefinitely. It can also mean that off-site species such as loblolly (*P. taeda)* or slash (*P. elliottii*) pines are retained in the canopy of a former plantation for decades to provide fuel for the prescribed fire that ultimately will help to restore native ground cover; in this example, the more immediate objective of ensuring adequate fuel for prescribed fire is a tool that contributes to the longer-term goal of the eventual conversion of the plantation to a longleaf pine ecosystem with a multiaged structure. Stoddard-Neel is applicable for many situations and objectives (examples below), but not for all. If objectives are at odds with the major tenets of Stoddard-Neel, another management system should be considered.

CHOOSING A RESTORATION TRAJECTORY

Restoration Objectives and Desired Condition

Interest in restoration and management of longleaf pine ecosystems is motivated by a diversity of objectives. Land managers select priorities from a range of goals that includes economic benefits, wildlife, biodiversity, recreation, and aesthetics. Although some may prioritize one or two goals, more often managers are challenged with balancing multiple goals across a given property or landscape, especially within a conservation context (O'Hara 2014). Although general principles for ecosystem restoration have been articulated (Society for Ecological Restoration 2004), site-specific conditions ultimately have strong influences on approaches and outcomes (Stanturf et al. 2014). Land use history affects the biological legacies and developmental trajectories that are present on a given site (Foster et al. 2003); for longleaf pine, fire history adds another layer of complexity (Glitzenstein et al. 1995; Jacqmain et al. 1999; Goolsby et al. 2005). In practice, longleaf pine restoration can begin from many starting points, depending on site conditions and land use history. Rather than attempting to cover all possible treatments exhaustively, the following paragraphs provide general examples of approaches that can achieve restoration or maintenance objectives. These approaches focus on the incorporation of ecosystem integrity, wildlife habitat, and biodiversity values into a balanced system that manages for multiple resources. As described earlier in this chapter, Stoddard-Neel provides a well-known example of this balanced perspective to management in the longleaf pine ecosystem. Although other viable silvicultural approaches can also fulfill a land manager's timber management objectives (Guldin 2006; Brockway, Outcalt, Guldin, et al. 2005), none of them incorporates the same levels of fire management, wildlife habitat, and biodiversity, and all focus less on restoration and more on maintaining a target stand structure in semimature and mature forests. In longleaf pine forests, canopies are typically spatially heterogeneous at finer scales if the forests have a sustained history of prescribed fire and if they support multiple age classes. At the stand scale, continuous canopies persist through time and incorporate this finer-scale heterogeneity in a shifting gap-phase dynamic (Brockway and Outcalt 1998; McGuire et al. 2001; Gagnon et al. 2003). The model vegetative structure—characterized by a diverse herbaceous ground cover dominated by graminoids, few trees in the midstory layer, and relatively low stocking of canopy hardwoods—is maintained by frequent, low-intensity surface fires (Glitzenstein et al. 1995; Kirkman, Coffey, et al. 2004; Mitchell, Hiers, et al. 2009; Neel et al. 2010; Brockway and Outcalt 2015).

Rather than a static and tightly defined structural objective, these general characteristics are intended to provide a range of desired conditions that vary depending on edaphic site characteristics, land use history, developmental stage of the stand, and management objectives. Management actions and desired future conditions will, of necessity, evolve through time in an adaptive management model that responds to changing abiotic and ecological drivers. For example, the desired basic structural objectives could begin to be approximated in younger longleaf pine plantations after a first or second thinning. Parameters such as basal area, overstory canopy cover, and midstory canopy cover can be manipulated through standard forestry management practices early in the stand development process. However, if other characteristics—such as multiple age classes of trees or grass-dominated ground cover—are desired, typical silvicultural manipulations may need to be replaced by less conventional interventions or by an approach that simply allows seed masting events, frequent fire, and other natural processes to unfold over time.

Common Factors for Restoration

In large part, the condition of the site at the beginning of the restoration process determines the treatments needed to achieve the desired structural and functional attributes. The sequence in which the treatments are applied is also an important consideration—for both efficacy and cost—that can affect progress toward the desired state. Unlike some ecosystems in which restoration is essentially a discrete event that alters the trajectory of development, longleaf pine restoration is a process that unfolds over relatively long periods of time, with the timing and sequence of multiple treatments employed to alter the restoration trajectory in an adaptive manner. Other general considerations that apply for any given starting point include assessing the suitability of longleaf pine both for the restoration site and for meeting landowner objectives, the potential for using prescribed fire over the long term, and the likely success of ground cover community restoration.

Site Suitability and Matching Objectives

A primary consideration when contemplating the restoration of a longleaf pine ecosystem is whether the site is suitable from a biophysical standpoint, and whether longleaf pine is an appropriate socioeconomic choice. Although enthusiasm for longleaf pine restoration is laudable, it should be recognized that there are environmental limitations for the species. Longleaf pine can tolerate a relatively wide range of conditions (Boyer 1990), but there are situations for which longleaf pine is ill-suited, such as sites that have alkaline or hydric soils and frequently flooded sites. These factors should be considered in assessing the suitability of any specific site for longleaf pine restoration or management. Social and economic factors are also critical factors to be evaluated in the initial site assessment. Perhaps the most important question from a socioeconomic perspective is forecasting, to the degree possible, whether prescribed fire will be feasible over the long term. Urban development is projected to increase substantially in the Southeast (Wear 2013). In the wildland-urban interface, prescribed fire is often restricted (see Chapter 13) or, if possible, is exceedingly complex and expensive to implement (Wimberly et al. 2006). In addition, longleaf pine may not always meet a landowner's economic objectives or needs. With some exceptions, it generally underperforms other pine species in discounted capital budgeting analyses (see Chapter 3). Ecologically oriented management strategies, such as frequent prescribed fire and maintaining lower stocking rates of timber to provide optimum wildlife habitat, can incur substantial opportunity costs. For some owners a portion of their land, regardless of ecological suitability for longleaf pine, may need to be dedicated to other species of southern pine that perform better economically, thus allowing them to manage longleaf pine on the remainder of their land.

Past, Present, and Future of Prescribed Fire

Across the wide range of conditions and developmental states found on sites that are suitable for longleaf pine restoration, perhaps the most important general principle is the primacy of maintaining (or introducing) a regime of frequent prescribed fire (Mitchell et al. 2006). This consideration should be a driving factor in all decisions and actions along the restoration pathway for the ecosystem. The role of fire—and by extension, the role of fuels—is the most critical factor in longleaf pine management, affecting decisions about stocking rates, midstory management (mechanical/chemical treatments), ground cover restoration, wildlife management, and other strategies for restoration of the longleaf pine ecosystem (Mitchell, Hiers, et al. 2009). Continuity of fine fuels (see Chapter 6), primarily fine fuels from pine litter and grasses, is a significant determinant of fire behavior in longleaf pine ecosystems (Loudermilk et al. 2009; Mitchell, Hiers, et al. 2009). The most immediate source of fine fuels that a land manager can control is the pine component of the overstory that contributes needle litter to the fuel bed (O'Brien et al. 2008; Mitchell, Hiers, et al. 2009). Decisions about the planting density of seedlings, overstory-stocking rates, the spatial distribution of mature trees, and overstory species composition all influence fuel loading, fire behavior, and fire effects (Mitchell et al. 2006; Neel et al. 2010).

During stand establishment, a critical issue for longleaf pine planting density is the competition on the site, particularly from hardwoods (Pecot et al. 2007). Higher planting densities can result in more fine fuels available more quickly, thereby enhancing the ability to manage hardwood competition with prescribed fire. This can be particularly important for a cutover site in which hardwoods have become firmly established in the understory stratum (<3 feet). For existing or mature stands, ground cover condition directly affects decisions about overstory stocking and the spatial distribution of longleaf pine because the amount and distribution of fine fuels are largely determined by the additive effects of pine needles and ground cover, particularly grasses (Mitchell, Hiers, et al. 2009). For example, bunchgrass-dominated ground cover communities with low or moderate levels of woody plants can be managed with lower overstory stocking because the grasses also play an important role in the level of fine fuels (Jack et al. 2005). If the woody plant component in the understory is greater, the stand will require higher canopy stocking levels and careful attention to the spatial distribution of overstory trees so that fine fuel levels are adequate to control these hardwoods with prescribed fire (Kirkman, Coffey, et al. 2004; Jack, Mitchell, et al. 2006). The species composition of the overstory also influences fire behavior and fire effects. Longleaf pine needles have a higher resin content than other pine species (Rebertus et al. 1989; Fonda 2001), making them a superior fuel for burning. Although mature hardwoods are a natural component of many longleaf pine ecosystems and provide habitat for wildlife, their ability to facilitate or impede fire is species dependent (Kane et al. 2008; Hiers et al. 2014; Varner et al. 2015). These considerations could lead managers to preferentially retain longleaf pine in mixed pine stands and can influence the species composition, stocking level, and spatial distribution of those hardwoods that are also retained in the stand.

Another fire-related issue for the restoration of established stands, whether degraded longleaf pine stands or stands dominated by other species, is the reintroduction of fire into fire-excluded stands and the stepwise process of reducing fuel loads to levels that are safe for a sustained regime of frequent prescribed fire (Mitchell, Hiers, et al. 2009; Varner et al. 2009; O'Brien et al. 2010). Fire-excluded sites typically have built-up layers of duff, particularly near mature pine trees, and the fine roots of these large trees grow in this duff layer (O'Brien et al. 2010). Prescribed fire application during dry conditions can consume duff and kill fine roots, stressing and oftentimes killing overstory trees (Varner et al. 2007, 2009; O'Brien et al. 2010). Reintroduction of fire needs to be a judicious process, with initial applications restricted to cool weather conditions and high levels of fuel, duff, and soil moisture (Varner et al. 2005). Before a typical fire regime can be safely established, accumulated fuels must be incrementally removed over multiple burning events, leaving adequate time between events for root adaptation and recovery. Fire-excluded stands often also have a high component of hardwood species, such as semideciduous oaks (*Quercus* spp.); their leaves are low-value contributors to fine fuels and can actually serve to suppress fire, limiting the ability to conduct successful burns (Kane et al. 2008). On mesic sites these oak species often must be removed by a combination of mechanical and chemical treatments to ensure that an effective regime of prescribed fire can be initiated and maintained; xeric sites, with a different community of oak species, often do not need the same mechanical and chemical treatments to successfully reestablish a frequent fire regime (Hiers et al. 2014; see Chapter 11).

Ground Cover Status

The ground cover community is an essential component of the longleaf pine ecosystem, serving as the primary source of plant diversity (Walker and Stilletti 2006). The ground cover also serves several important functional roles, including the maintenance of frequent fire (Hendricks et al. 2002), the cycling of nutrients (Hendricks and Boring 1999; Hiers, Mitchell, et al. 2003; Cathey et al. 2010), and as habitat and food for wildlife (Stoddard 1931; Miller and Miller 1999; McGraw et al. 2004). Aside from canopy reestablishment and a frequent fire regime, restoring ground cover is the most important component of a successful longleaf pine restoration (see Chapter 11).

Starting Points for Restoration

The restoration process can begin anywhere along a continuum that ranges from bare ground to a mature forest. For the purpose of this chapter, the continuum has been segmented into four categories. The first category consists of sites that have no existing forest cover, either as a result of recent timber harvests that cleared all (or most) of the forest cover or sites that have a history of frequent row crop farming or grazing. Although they share the common characteristic of no forest canopy, these two starting points require different restoration pathways. The second category consists of relatively young pine plantations, whether longleaf pine or (more commonly) loblolly or slash pine. A third category consists of a wide range of mature forests (often of natural origin) that have only a small component of longleaf pine or no longleaf pine at all and are dominated by slash or loblolly pine or, more typically, pine-hardwood mixes. The last category consists of forests that can be classified as longleaf pine-dominant but that are in a degraded state, often requiring one or more treatments to bring them into the desired structural condition.

Regardless of the starting condition, the ultimate goal for any longleaf pine restoration is to achieve a structure that transitions the stand into a maintenance phase. Given the increasing interest in longleaf pine restoration, considerable effort has gone into clarifying and communicating the concept of maintenance phase and establishing metrics for assessing the general condition of individual longleaf pine sites (McIntyre 2012; Ware 2014; Nordman et al. 2016). The result is a set of parameters (expressed as quantitative ranges) that define the characteristics of the desired maintenance phase for the canopy stratum (basal area, canopy closure, age structure, and hardwood stocking); the midstory stratum (hardwood stocking and shrub cover); ground cover (herbaceous cover, native warm-season grass cover, longleaf regeneration); and invasive plant cover (see Chapter 1). These parameters offer land managers general objectives that can be applied to a range of stand ages, with the assumption that individual characteristics of any longleaf pine stand will move into (and out of) the desired parameter ranges through time, and therefore require ongoing management interventions and adjustments to bring them back into the desired state. For most parameters, the desired condition is largely achieved through frequent application of prescribed fire. Canopy stocking of both pine and hardwood species also has a large effect on most of the midstory and ground cover characteristics through controls on both above- and belowground competition, by providing fine fuels and, in the case of some hardwood species, potential fire suppression.

Just as the continuum of possible starting conditions is broad, so also are the ranges of potential restoration trajectories and treatment sequencing. In the next section, possible scenarios for restoration and management will be framed within the context of Stoddard-Neel.

STARTING THE TRAJECTORY: EXAMPLES FOR DIFFERENT STARTING CONDITIONS

For all four of the general starting conditions described in the previous section, the long-term objective is a multiaged forest ecosystem with a predominantly longleaf pine overstory, little or no midstory stratum, and a grass-dominated ground cover. Each example that follows is meant to highlight the factors to consider during treatment selection—not to be all-encompassing or prescriptive—and to illustrate the management regimes that can begin the trajectory toward the desired condition. The examples are based on the cumulative experience of managers who have been practicing a modified form of Stoddard-Neel that was developed at the Joseph W. Jones Ecological Research Center, as well as observations at many other locations within the range of longleaf pine. Site quality can affect the overall success of restoration efforts, how long the process will take, and how much competition from undesirable plants will be present.

NO FOREST COVER

Sites in this category are predominantly former agricultural fields (old fields), pastures, or recently harvested forests in which all (or nearly all) of the canopy was removed. The first consideration is to control competing vegetation that would otherwise impede or prevent the establishment and survival of longleaf pine seedlings. A mix of chemical and mechanical treatments may be used to help control competing plants. Ideally, application of control treatments precedes planting because treatment options become much more limited once the seedlings are in the ground.

In old fields, competing vegetation tends to be invasive agronomic and ruderal species (typically annuals). The specific mix of species depends on the length of the fallow period (time since the last cropping), the season in which the soil was last disturbed (affects seed availability and time for germination), and site-specific seed bank characteristics (Keever 1950; Nichols et al. 2015). If the field was harrowed in autumn, the plant community will be dominated by ragweed (*Ambrosia* spp.), partridge pea (*Chamaecrista fasciculata*), and other species that are beneficial to wildlife (Coble et al. 1981); conversely, summer tillage produces less desirable species (Yarrow and Yarrow 1999). The particular plants present are important because they have different competitive effects on longleaf pine seedlings and they offer different types of wildlife habitat (Stoddard 1931; Miller and Miller 1999). If the field is left fallow for more than a growing season, perennial plants such as blackberry (*Rubus* spp.) can become established; they are both difficult to control and strong competitors to planted seedlings. Harrowing, if included as a treatment, is conducted far enough in advance of planting longleaf pine to allow the soil to settle and thereby provide better contact with the seedling root system. When planted at the same time as the seedlings, cover crops such as winter wheat (*Triticum aestivum*) have been found to help control initial competitors without suppressing the pine seedlings, principally because they mature in early spring and their cover declines during the pine growing season (J. Atkinson, personal communication). Some old-field and pasture sites have a dense, restrictive soil layer called the "hardpan" or "plowpan" that develops from repeated tillage over time. Subsoiler plows can break up the hardpan, allowing root growth to penetrate this restrictive layer (Allen et al. 2005; Johnson and Gjerstad 2006).

Compared to old fields that have been recently subjected to repeated tillage, pastures are often less problematic because they do not have a residual seed bank of undesirable plants. However, these sites also present unique challenges, most commonly from rhizomatous, sod-forming species such as Bermudagrass (*Cynodon dactylon*) or bahiagrass (*Paspalum notatum*)—both of which are aggressively invasive, creating extremely competitive conditions for the establishment of longleaf pine seedlings. Competition from rhizomatous grasses in actual planting rows can be controlled by scalping (described in next paragraph) or by band spraying of herbicides. However, a holistic restoration of the longleaf ecosystem that includes ground cover restoration would require a complete eradication of rhizomatous grasses with broadcast applications of herbicide (see Chapter 11); because many pasture grasses are not completely controlled by one application, a follow-up assessment and additional herbicide treatments may be required for complete control (Minogue et al. 2012). Scalping is an effective mechanical treatment for improving survival and early growth of longleaf pine seedlings on old-field and pasture sites (Barnard et al. 1995). The treatment uses a tractor-drawn plow blade to remove soil and sod from the planting row in a band that is 20–30 inches wide and 3–4 inches deep, after which the seedlings are planted in the exposed bare soil. Moving this top layer of soil, including sod and the upper strata seed bank, from the immediate planting zone provides an area that is relatively free of competition and a window of time for seedling root systems to extend below the rooting zone of the grass.

Recently, cutover forests differ from agricultural sites in that the primary form of competition that could be detrimental to successful establishment of longleaf pine seedlings is woody plant competitors (primarily hardwood species). Whether woody competitors are root and stump sprouts or suppressed individuals that were released by harvesting, controlling them is a high priority. Herbicides, either applied on cut stumps or broadcast over released plants and sprouts from roots (Fox et al. 2007), are usually effective. Mechanical removal of debris and stumps is rarely used to meet restoration objectives because it is expensive and could cause damage to the site, a concern of more importance if any desirable residual ground cover is present.

Once control measures are in place for competing vegetation, longleaf pine seedlings can be planted. Planting densities can vary widely, depending on landowner objectives—tending toward higher densities (≥700 trees per acre) if the objective is timber products but lower densities (400–450 and infrequently as low as 250–300 trees per acre) if the objective is wildlife and biodiversity. Higher densities accelerate crown closure—crowns of adjacent trees touching or nearly touching, resulting in canopy coverage of almost all ground area (Smith 1986)—and earlier self-pruning of individual trees (Adams and Clason 2002). The financial benefits of high density planting include the ability to conduct commercial thinning operations at an earlier age, income from pine straw raking, higher wood quality, and better tree form. In contrast, a lower planting density provides a longer period before canopy closure, maintaining early successional vegetation between trees to increase wildlife habitat value and plant diversity (Marsh et al. 2012). However, unless the planting density is extremely low, these habitat advantages only persist for 2–3 additional years, and the delayed canopy closure often results in poorer-quality trees due to larger and more numerous branches because the trees will not self-prune as readily. Regardless of planting density, the canopy will eventually close and the ground cover will be shaded (Figure 10.3). Changes in habitat structure will, however, occur at different rates and points in time depending on planting density. The accelerated crown closure at higher planting densities can be advantageous because shading suppresses some undesirable plant species (especially annuals on former agricultural sites) and ultimately provides some degree of competition control for future ground cover restoration. An accelerated crown closure also provides substantial needle cast earlier in stand development; this aids implementation of a prescribed fire regime (Morris et al. 1992), further helping to control hardwoods and other competing plant species whose growth was only temporarily constrained by initial site preparation treatments.

Beyond the timber/wildlife dichotomy, the planting pattern at establishment is determined by broader restoration objectives, anticipated use of equipment in the planted area, and aesthetic considerations. The most strai ghtforward and efficient pattern is to plant trees in rows at a spacing that will meet the desired planting density. This is a pattern that provides travel lanes for movement of equipment for mechanical or chemical treatments, but not one that is aesthetically pleasing. Alternative planting patterns—such as random seedling placement, spiral planting patterns, or

FIGURE 10.3 Ground cover response to canopy development at a longleaf pine plantation in southwestern Georgia. Key dates: Wiregrass seed hand sown at the time the trees were planted, 1996; decline of grassy cover after canopy closure, 2006; first thinning in early summer, 2014, and some recovery of grassy ground cover, 2015. (Photographs for 1999–2012 courtesy of John Brooks; photograph for 2015 courtesy of Jessica McCorvey.)

undulating rows—are available for landowners who do not like the look of young plantations with "pines in lines." The history of these alternative patterns has been mixed, with many landowners reporting logistical challenges and increased planting costs.

When introducing ground cover (see Chapter 11) on nonforested restoration sites, timing is the most important consideration, and one that is greatly influenced by site-specific objectives and site quality (Hess 2014). In pastures and cutover forests where competition from undesired vegetation is substantial, reintroduction of the desired ground cover community is best delayed until the other vegetation is reduced or eliminated and the site has adequate fuels to sustain frequent prescribed fires. The timing for ground cover introduction in old-field sites can be more flexible, with desired ground cover species established at the same time the trees are planted or after a first thinning (when the more open canopy does not suppress ground cover development).

Once longleaf pine seedlings are established, with or without desired ground cover, they will at some point begin to grow taller, emerging from the grass stage, and will eventually form a closed canopy. When they have reached this stage of development, they can be managed according to the concepts outlined below for plantation forests.

Plantations

Pine plantation forests—composed of longleaf pine or other pine species established on cutover forests or former cropland—are a frequent starting condition for restoration. When the long-term objective is an ecosystem with a multiaged longleaf pine canopy and a grass-dominated ground cover, all plantations will progress along generally similar trajectories of overstory development and can be managed using a similar suite of treatments. The major differences are (1) whether new age groups of longleaf pine are planted or established naturally, and (2) the measures that will be needed to control regeneration of the nonlongleaf pine canopy species.

Longleaf Pine Plantations

Young longleaf pine plantations can be managed in a manner that is appropriate for any pine plantation with one significant difference: the use of prescribed fire when trees are young. They can be burned as early as a year after planting, provided that prescriptions for initial applications specify cooler conditions. Winter burns with low ambient air temperature, high soil moisture, and firing techniques that minimize intensity of fire behavior (such as grid ignition) are preferred. In springtime, care must also be taken to avoid burning once young trees begin to initiate seasonal growth and have exposed terminal buds (also known as "candling").

As the trees grow taller the plantation reaches the stage of canopy closure, provided that initial planting densities were not extremely low. The time required to reach this stage depends on the initial planting density and site quality. An initial commercial thinning is generally conducted a few years after canopy closure, typically at 15–20 years (again, dependent on initial planting density and site quality). As with other pine species, this first thinning typically involves the removal of every third or fourth row along with some low-vigor trees in the remaining rows. Generally, these harvesting rates remove 30%–50% of tree density and basal area. After this first thinning, desired ground cover plant species can be introduced if not already established (see Chapter 11). Direct seeding of ground cover plants can be facilitated following thinning because (1) the harvesting operation typically leaves areas of bare mineral soil that can serve as a receptive seed bed, (2) competition has been suppressed by years of canopy closure and low light conditions, and (3) thinning has removed enough canopy cover to provide sufficient light levels for sustainable development of ground cover plants.

As tree crowns expand after thinning, they will eventually produce viable seed, either in broad-scale masting events or when individual trees produce cones. The key treatment for capturing this natural regeneration in longleaf pine plantations is a frequent regime of prescribed fire to ensure adequate bare mineral soil for seedling germination. These seedlings can be sustained by adjusting the timing and intensity of subsequent prescribed fires, after which they can be released by future thinnings. Successful germination should be followed by a respite from prescribed fire until the seedlings become large enough to survive burning (typically at least 2 years after germination); as is true for planted seedlings, initial fire prescriptions need to specify low fire intensity and cool ambient temperatures. In this manner new age groups are recruited into the developing plantation as the stand moves toward a multiaged structure.

Stand development continues after the first thinning operation until tree growth and stocking levels are sufficient to warrant another thinning. If the objective is timber products and a set rotation, the thinning will target uniform spacing between trees to maximize growth of individual stems. Conversely, if the objective is a sustainable multiage, individual-tree selection is used for this and all future thinnings (in accordance with Stoddard-Neel) and the initial row structure is, for the most part, ignored beyond what is needed for the movement of logging equipment (for example, using old removal rows for skid trails). Using individual-tree selection can introduce some heterogeneity into the spatial arrangement of trees, and small canopy gaps can be created or enhanced either to provide sites to begin recruitment of longleaf pine regeneration or to release any longleaf pine regeneration that might already be present. With multiple thinnings and continued stand development, older trees from the initial planting will be dispersed throughout the stand in a heterogeneous pattern and different age classes will become established. Thus, over time, the pattern of rows from the initial planting will likely become indiscernible. The development of a natural looking, multiaged condition will depend on the intensity of removals at each harvesting entry, but is likely to occur by the third or fourth thinning, roughly 30–40 years after plantation establishment.

Plantations of Other Pine Species

For plantations that have been established with other pine species, two options are available for restoring longleaf pine to canopy dominance. First, the existing plantation can be removed (clear-cut), followed by site preparation and planting of longleaf pine seedlings. This approach resets the system and eliminates any ground cover development that might have occurred as well as the structural characteristics and pine needle fuels for prescribed fire that would have been provided by the existing canopy trees (Kirkman and Mitchell 2002; Kirkman, Mitchell, et al. 2007; Hess 2014). Further, the resulting stand, although composed of longleaf pine, will retain a characteristic plantation structure and appearance for decades. The second alternative uses a gradual conversion approach in which longleaf pine seedlings are introduced by underplanting them beneath the canopy of the existing species (Kirkman, Mitchell, et al. 2007; B. Knapp et al. 2013; Hess 2014)—in effect, using artificial regeneration to mimic natural regeneration. This gradual approach preserves the existing structural characteristics, wildlife habitats, fine fuels, and ground cover plant communities that were already in place within the existing plantation. The multiple strategies that are available for underplanting longleaf pine seedlings include planting seedlings in canopy gaps to mimic the spatial patterns of natural disturbance and regeneration found in mature longleaf stands; removing several rows of existing pines and underplanting these strips; planting in regularly shaped/spaced patches within the existing stand; and underplanting across the entire area in a uniform pattern. The selection of an appropriate strategy will depend on (1) the long-term objectives for the stand, (2) the time frame allowed for implementing the conversion, and (3) the logistics of creating, planting, and maintaining the areas to be underplanted. Examples of different approaches can be found in southwestern Georgia at Ichauway (gaps), in the Florida Panhandle at Tyndall Air Force Base (strips), and in coastal South Carolina on properties owned by The Nature Conservancy (uniform underplanting).

PLANTATION SPECIES CONVERSION CASE STUDIES

The upland sites in the historical longleaf pine range have largely been converted to other land uses, with millions of acres converted to plantations of slash or loblolly pines (considered off-site species) that were managed under an industrial production-based model (see Chapter 3). Many private and public owners of these plantation forests are now interested in longleaf pine restoration. A traditional and commonly used approach to converting slash or loblolly pine plantations to longleaf pine is to clear-cut the off-site species of pine, apply appropriate site preparation treatments, and plant with container-grown longleaf pine seedlings. An alternative approach is a gradual conversion that retains some portion of the existing canopy of off-site pines for an extended period of time and establishes longleaf pine under the existing canopy (Kirkman, Mitchell, et al. 2007; B. Knapp et al. 2013; Hess 2014).

Broadly speaking, preserving the existing canopy offers a temporary surrogate for the desired forest structure and function that would have been (and presumably will be) provided by a longleaf pine canopy, allowing time for slow development of the longleaf pine planted beneath that canopy. Although they are less desirable in the long term, the off-site pines contribute needle fall as fine fuels for prescribed fire, provide important habitat structure for wildlife, and suppress the release of hardwood stems that are present in the understory stratum. As these gradual conversions have become more popular in recent years, a range of spatial planting and temporal sequencing patterns has been developed both to satisfy ownership objectives and to address specific site conditions. Below we present four case studies that reflect different ways in which this gradual approach was implemented.

Southwestern Georgia: Underplanting Gaps in a Mature Plantation

To mimic, as closely as possible, the natural demographic and spatial patterns for longleaf pine regeneration in multiaged forests, a gap-based approach for establishing new age groups was implemented at the Joseph W. Jones Ecological Research Center at Ichauway. The initial attempt—likely the earliest documented implementation of a phased conversion—began in 1998 on a 250-acre, 60-year-old slash pine plantation (Kirkman and Mitchell 2002). The original plan was to underplant longleaf pine seedlings in small canopy gaps (0.25 acre), which were created by harvesting overstory trees on about 10% of the stand area every 10 years. The assumption was that, after 10 years of growth, the seedlings would be sufficiently tall to be visible and would therefore not be damaged by logging equipment during subsequent harvesting operations. This early implementation revealed several opportunities for improving the approach. First, although longleaf pine seedlings can survive and grow in relatively small openings under a canopy of longleaf pine (McGuire et al. 2001), canopy gaps in slash and loblolly pine plantations need to be larger because the crowns of these species transmit less light to the forest floor (Kirkman, Mitchell, et al. 2007). Second, use of appropriate chemicals to control competing woody plants in the ground cover in the planting gaps would have improved the survival of longleaf pine seedlings and minimized the release of existing shrubby hardwoods. Finally, the number of entries required when only 10% of the stand was harvested at each entry became operationally impractical over the long run. Unless gap locations are carefully planned at the onset of the restoration, modern logging equipment and techniques simply do not have room to maneuver as more and more small openings are planted. Figure 10.4 shows the condition in 2015 following two iterations of thinning and underplanting.

Southwestern Georgia: Underplanting Gaps in a Younger Plantation

With a plan that was informed by the lessons from the first phased-conversion, the Joseph W. Jones Ecological Research Center implemented additional conversions in two 25-year-old slash pine plantations—one 45 acres and the other 60 acres. In 2012, the plantations were thinned a second time. Before trees were marked for harvesting, existing gaps and gaps that were beginning to develop were identified; gaps were enlarged and new gaps were created so that all ranged from 0.3 to 0.5 acre and they cumulatively comprised about 25% of the stand. Consideration for the location of canopy gaps was part of the planning process to allow easy access for future harvesting. After thinning operations were complete, a broadcast application of herbicides was used to control nonpine woody vegetation, and longleaf pine was planted in the gaps in February 2013. First-year longleaf pine seedling survival was high, and the seedlings began vertical growth during the second growing season (Figure 10.5). This is a long-term applied study/demonstration with monitoring that will continue into the future.

Florida Panhandle: Underplanting in Strips

Tyndall Air Force Base, located on a flatwoods site near Panama City, Florida, implemented a phased conversion in two slash pine stands that were about 45 years old, one previously thinned by removing every third row and applying individual tree selection in remaining rows, and the other never thinned. Base managers chose to use a row-removal approach for phased

(a)

(b)

FIGURE 10.4 In Southwestern Georgia: (a) Longleaf pine underplanted in canopy gaps created by harvesting in a 60-year-old slash pine plantation, and (b) emerging from the grass stage. (Photograph [a] courtesy of Richard T. Bryant and [b] courtesy of Jessica McCorvey.)

conversion. In the previously thinned stand, every third pair of remaining rows was removed, taking about a third of the volume and leaving a four-row gap (Figure 10.6). In the stand that had never been thinned, the pattern was to remove four rows, leave two rows, remove one row, and leave two rows (Figure 10.6). For both thinning operations, a few low-vigor individual trees were selected for removal in the rows that remained. All linear openings (former rows) were roller chopped 3 years after the harvest and planted with longleaf pine seedlings (Figure 10.7). This row-removal approach allowed relatively simple communication of concepts to timßber markers and logging crews, greatly facilitating implementation. Although perhaps less natural looking in the initial stages, the stands are being managed with the long-term goal of a multiaged, mixed-composition, naturally regenerating pine flatwoods ecosystem and are expected to appear less geometric over time and with additional management.

South Carolina Low Country: Uniform Underplanting

In 2007, The Nature Conservancy purchased several tracts, from a forest products company, which were predominately occupied by 12- to 16-year-old loblolly pine plantations,

(a)

(b)

FIGURE 10.5 In Southwestern Georgia, longleaf pine (a) underplanted in canopy gaps created during the second thinning in a 25-year-old slash pine plantation, and (b) emerging from the grass stage after three growing seasons. Photos were taken 2 weeks after a prescribed fire so seedlings have some scorched needles. (Photographs courtesy of Jessica McCorvey.)

on sites ranging from wet flatwoods to xeric sandhills. Because the original plan for these industrial forest tracts was to transfer them to the Francis Marion National Forest in 3–4 years, The Nature Conservancy decided to pursue an accelerated conversion process to ensure that conversion would be largely completed before the transfer. Retaining components of the existing canopy structure was desired for both ecological and wildlife considerations. The silvicultural prescription included a heavy thinning of the overstory loblolly pine to about 25–30 ft^2/acre, herbicide treatment as needed to control competing vegetation, and uniform underplanting of longleaf pine seedlings at approximately 600 trees per acre. The residual density of canopy trees satisfied recovery plan guidelines for the minimum basal area required for red-cockaded woodpecker (*Picoides borealis*) foraging

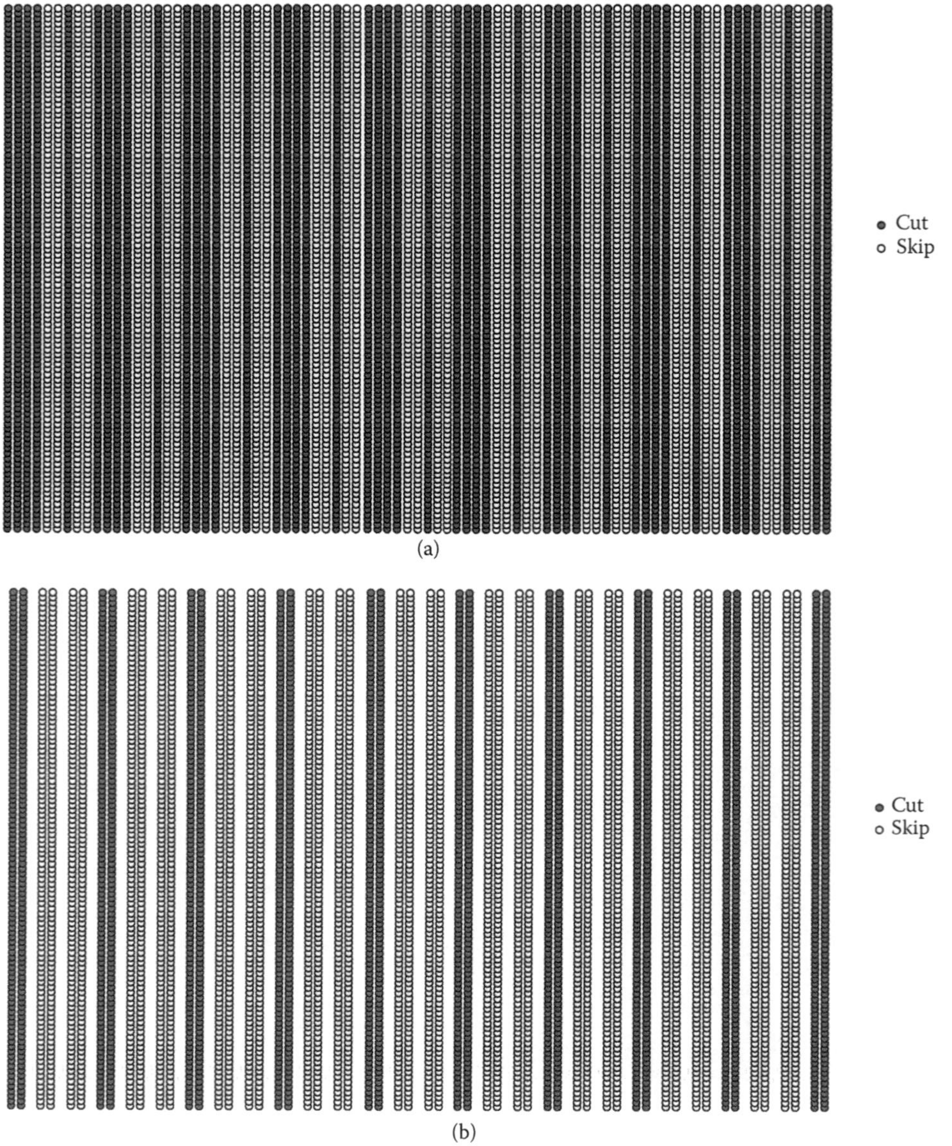

FIGURE 10.6 Longleaf pine restoration in a slash pine plantation in the Florida Panhandle. The thinning pattern (a) for a stand that had not previously been thinned, and (b) for a previously thinned stand. Note that thinning operations employed different combinations of removed rows and residual rows, depending on pre-treatment structure, with longleaf pine seedlings planted into the strips left by row removals.

habitat, and the planted longleaf pine seedlings will eventually contribute to planned increases in longleaf pine acreage on the Francis Marion National Forest. This accelerated approach has been particularly successful on drier sites (Figure 10.8). Restoration progress is slower on more fertile sites because of persistent competition from woody shrubs and overseeding from the residual loblolly pines. Perhaps the most important lesson learned from this effort is the critical need for strategically timed, frequent fire of adequate intensity to control the prolific seedlings that are established when loblolly pine is left in the canopy.

(a)

(b)

FIGURE 10.7 Longleaf pine restoration in a slash pine plantation in the Florida Panhandle. Longleaf pine seedlings (a) planted in linear openings created by row removals, and (b) emerging from the grass stage. (Photographs courtesy of Melanie Kaeser.)

FIGURE 10.8 Coastal South Carolina longleaf pine restoration in loblolly pine plantation; longleaf pine seedlings emerging from the grass stage two growing seasons after planting. Note that seedlings were uniformly planted under a low basal area canopy on a dry, sandhill site adjacent to the Francis Marion National Forest. (Photograph courtesy of Kevin McIntyre.)

Staffing levels and budgets strongly influence the choice of an underplanting strategy. For example, emulating the spatial patterns of natural gaps requires more time and labor to delineate where gaps will be created during harvesting and to supervise crews planting seedlings in the gaps. Underplanting in strips is likely preferable for some ownerships because it is relatively simple, less labor intensive, and easier to supervise. Regardless of underplanting design, some common considerations can improve the establishment, survival, and growth of planted seedlings. First, controlling competing vegetation can significantly increase seedling survival (Cain 1991; Knapp et al. 2014; Nelson and Bragg 2016); these treatments often include a preplanting herbicide application, either exclusively in the areas to be planted or throughout the stand. Second, frequent prescribed fire keeps competing vegetation in check and removes less fire-tolerant seedlings of the existing canopy species; this is particularly important when planting under a canopy of loblolly pine, a species known for fecundity (Hu 2011; Knapp et al. 2011; Kush 2016). Third, the targeted density of the residual canopy trees and the gap sizes will depend on the dominant species of the existing plantation because the canopies of different pine species allow differing levels of sunlight to fall on the forest floor (Pecot et al. 2007; B. Knapp et al. 2013; Hess 2014) with differing effects on longleaf pine seedling survival and growth (see Chapter 4); this means that a more open canopy is required when underplanting beneath a slash or loblolly pine canopy than when underplanting beneath a longleaf pine canopy. Finally, the logistics of future operations, especially harvesting, need to be considered when locating the areas to be regenerated. Patches of regeneration need protection from equipment for several years until the saplings are large enough to be easily seen and avoided; at the same time, equipment needs room to operate. Managing the number, size, orientation, and location of regeneration patches to accommodate entries by equipment will greatly enhance the ability to conduct future harvests.

Mixed Pine and Pine-Hardwood Forests

Forests in this category have (1) a mixed pine canopy, with or without a longleaf pine component; (2) a mixture of pines and various hardwood species in the canopy; or (3) a hardwood canopy with no pines present. Such stands often have a history of infrequent fire or are fire excluded. They often, although not always, have heavy fuel accumulations in the form of litter and duff, understories and midstories dominated by woody species, and few of the desired grasses and herbaceous species in the ground cover.

For fire-excluded forests, the first action is to reestablish a regime of frequent prescribed fire, a multistep process that requires careful implementation. If high fuel loads have accumulated, controlling the intensity of prescribed burns will be more difficult. In these situations, the first step is to gradually reduce fuel accumulations and begin to reduce woody plants in the understory and midstory. Depending on the levels of fuel present, several low-intensity fires could be required to reduce fuels to an acceptable level (Varner et al. 2009). If the accumulated fuel (such as hardwood leaves or large woody debris) does not burn well, other treatments (especially mechanical treatments) may be needed. Mowing, chopping, or masticating woody plant material in the midstory can transfer it to the soil surface where it will dry enough to burn, and subsequent fires can then help to control any posttreatment sprouting (Brockway et al. 2009). Felling or removing larger hardwoods can create canopy openings, increasing light availability at the soil surface, promoting grass and herbaceous vegetation, and decreasing the drying time for fine fuels so they burn more readily. With any of these mechanical treatments, supplemental herbicide treatments may be required to prevent resprouting. The necessity for these treatments will vary based on the growth characteristics of the particular species present and the efficacy of prescribed fires that follow the mechanical treatments (Addington et al. 2012; Hess 2014). As fire is introduced, the specific prescriptions and firing techniques used will depend on landowner objectives and site conditions (Waldrop and Goodrick 2012).

If the canopy has a mixture of pine species, their composition and density can be managed through planned harvesting and preferential retention of longleaf pine in the canopy. Regardless of species composition, the stocking of residual pines needs to be high enough and appropriately spaced across the site to supply sufficient fine fuels for prescribed fire (Mitchell, Hiers, et al. 2009). If longleaf pine is not dominant (or even present) in the canopy, retention of off-site pine species will be needed until longleaf pine stocking is high enough to supply adequate needle cast for fuels. The choice of species to retain depends on the current canopy composition and the site type. For example, slash pine is naturally found on wet sites in the Coastal Plain; loblolly pine and shortleaf pine (*P. echinata*) are more often found on finer-textured soils, typically in the upper Coastal Plain and the Piedmont. When necessary, longleaf pine seedlings can be introduced through planting once the fire regime is well established.

A canopy that has a mixture of pine and hardwood species requires a more aggressive restoration. Such stands are often in a degraded state, both structurally and functionally, and are usually fire excluded. If a stand has few canopy hardwoods but many small stems in the under- and midstories, a combination of chemical, mechanical, and fire treatments can be used to clear out these strata; undesirable canopy hardwoods can subsequently be reduced by felling or by herbicide treatment and leaving the stems as snags (only appropriate if the proportion of hardwood canopy trees is not excessive). Although timber markets vary geographically, the hardwood stems can often be sold in many areas, albeit at low prices. Absent a commercial hardwood market for chips or solid wood products, the stems can be chipped on-site and hauled for use as biofuels if that market is available, or as a last resort, piled and burned. Any pines in the canopy can be managed as described above for mixed pine forests, with an emphasis on favoring the retention of longleaf pine when present and evaluating all pine species for their relative contributions of fuel and structure.

HARDWOOD REMOVAL CASE STUDY IN SOUTHWESTERN GEORGIA

Degraded sites with partial stocking of residual longleaf pine offer excellent opportunities for restoration. Many of these sites were likely dominated historically by longleaf pine, with recent ingrowth of hardwood stems and other pine species caused by fire exclusion or fire that was applied with inadequate frequency, intensity, or both (Oswalt et al. 2012). About 1.24–1.82 million acres in the Southeast contain 20%–49% stocking of longleaf pine but are not currently classified as longleaf pine forests (Guldin et al. 2016). This level of stocking is considered high enough to provide fine fuels for prescribed fire and a seed source for natural regeneration in most cases, depending on the specific land use history and current site conditions. When compared to afforestation or reforestation scenarios, these degraded sites offer a quicker path to the desired forest structure. Restoration of degraded sites can also be more cost-effective because income derived from timber removals can partially offset the cost of restoration treatments. An illustration of this process comes from a restoration involving hardwood removals at the Joseph W. Jones Ecological Research Center at Ichauway in southwest Georgia, where most upland forest canopies range from 80 to 100 years old. Although longleaf pine is the dominant canopy species for most of the upland forests at Ichauway, some stands had significant hardwood stocking levels—a legacy of game management practices, particularly very low fire intensities, that were typical in the Southeast until the 1980s (Stoddard 1931). Demand and prices for low-grade hardwood products increased from 2001 to 2007, making restoration through mechanical removal of hardwood trees financially feasible.

In 2003, a commercial harvest of hardwood trees for pulp, cletewood (regional term for wood used to make pallets and boxes), and sawtimber products was conducted in a 188-acre stand that had roughly equal stocking of hardwoods and pines as well as on adjoining areas (461 acres in total), leaving residual longleaf pine. To reduce sprouting, hardwood stumps were sprayed with herbicides immediately after the harvest. After a full growing season, a broadcast herbicide was applied to control additional sprouting of hardwoods in areas where pine stocking was insufficient to provide adequate amounts of fine fuels to carry prescribed fire. Early in 2005, container-grown longleaf pine seedlings were planted into canopy gaps that were too large for adequate natural seeding from the mature longleaf pines. The stand has subsequently been managed using prescribed fire on a return interval of 1–2 years. Figure 10.9 shows the restoration development trajectory from one photo point in this stand.

Although income from harvesting did not fund all of the expenses for the restoration process, over half (53%) of the costs were offset (Table 10.1). Notably, not all of the 461 acres received all treatments—typically only 60%–65% of the acres were harvested in areas receiving the sequence of treatments described above. If only 65% of the 461 acres had the complete suite of treatments, including herbicides and planting, 82% of total restoration cost is offset by harvesting receipts. Although restoration treatments can be quite expensive, this case study shows that some (and sometimes most) of the restoration costs can be recouped by taking advantage of available revenue opportunities.

Contrary to many longleaf pine restoration approaches that seek to remove most hardwood stems, not all hardwood species are considered undesirable in the context of a healthy longleaf pine ecosystem (Hiers et al. 2014). Retaining some hardwood trees in the canopy offers benefits such as production of hard mast and habitat structure for wildlife, but not all hardwood species contribute these factors equally. Among the oak species, some such as southern red oak (*Q. falcata*), turkey oak (*Q. laevis*), and post oak (*Q. stellata*) are classified as pyric, and others

(a)

(b)

FIGURE 10.9 In Southwestern Georgia, the restoration chronology following hardwood removals from a mixed longleaf pine-hardwood canopy and underplanting of container-grown longleaf pine seedlings in the largest gaps: (a) Pretreatment conditions, (b) immediately after treatment. *(Continued)*

(c)

FIGURE 10.9 ***(Continued)*** In Southwestern Georgia, the restoration chronology following hardwood removals from a mixed longleaf pine-hardwood canopy and underplanting of container-grown longleaf pine seedlings in the largest gaps: (c) 8 years after treatment and 6 years after planting. (Photographs courtesy of Joseph W. Jones Ecological Research Center.)

TABLE 10.1
Income from Harvesting and Expenses for Restoration Treatments 2003–2005 in the Southeastern United States

Product Class	Harvested (Tons)	Stumpage (US.$/ton)	Revenue (US.$)
Pulpwood	7541	5.50	41,476
Cletewood[a]	189	23.00	4347
Logs (sawtimber)	1289	38.00	48,982
Total	9019	Various	94,805
Treatment	**Area Treated (acres)**	**Cost of Treatment (US.$/acre)**	**Outlay (US.$)**
Piling debris	461	150.00	69,150
Purchasing and applying herbicides	461	100.00	46,100
Purchasing longleaf pine seedlings	461	75.00	34,575
Planting seedlings	461	60.00	27,660
Total	461	385.00	177,485

Note that these data reflect stumpage prices and approximate expenses that were current at the time of implementation.
[a] Wood used to make pallets and boxes.

such as live oak (*Q. virginiana*) and water oak (*Q. nigra*) are considered nonpyric (Kane et al. 2008; Hiers et al. 2014). Compared to the nonpyric species, pyric species are better adapted to a frequent fire regime and their litter has a higher fuel value (Kane et al. 2008). On sites where an oak component is desired to meet ecological objectives such as wildlife habitat, the pyric group is preferentially retained. An exception is the nonpyric live oak, which is preferentially retained in regions such as the low country of South Carolina because of its aesthetic and cultural values. Certain hardwood species are particularly aggressive in achieving dominance; for instance, species like sweetgum (*Liquidambar styraciflua*) or hickories (*Carya* spp.) can be a persistent problem on more productive soils unless their density is controlled before they grow into the midstory.

In the end, no single compositional mix of pines and hardwoods is appropriate for all situations and objectives. Ownerships that place greater value on wildlife habitat might call for a higher proportion of hardwood stems than those that place higher value on other objectives.

Longleaf Pine-Dominant Forests

Of the four general starting points covered in this chapter, the longleaf pine-dominant state offers the most straightforward trajectory for sustainable management, primarily because the focus of restoration treatments is not on species conversion, but rather on choosing the management actions that will best move the forest toward the desired structure and function. Even though longleaf pine is the dominant canopy species, the stand often has functional or structural attributes that need improvement. Has the stand been fire suppressed? If so, begin fuel reduction treatments and increase the fire frequency. Is the stand multiaged? If not, apply prescribed burning to create acceptable seedbed conditions and foster established longleaf pine seedlings. Is the stocking too high or too low? If stocking is too high, simply employ careful harvesting using individual-tree selection. If too low, increase stocking over long time periods by recruiting natural regeneration to canopy positions or by planting longleaf pine seedlings if a faster outcome is desired. Does the ground cover plant community have the desired composition and density? If not, use frequent burning to improve its composition over time and supplemental planting or seeding to produce faster results. How these considerations are incorporated into the management regime across space and time is determined by the overall restoration objectives as well as the availability of staffing and financial resources to carry out the treatments.

Eventually, the forest will reach what some call the "maintenance phase" (Ware 2014; Nordman et al. 2016), after which the objective is to perpetuate its existing composition, structure, and function. Given the longevity and biological characteristics of longleaf pine and its associated ground cover plant species, this dynamic maintenance phase can be perpetuated far into the future. The single most important management treatment is the application of frequent prescribed fire (with specific objectives); in many situations, this may be the only treatment required.

Once the forest is classified as in a maintenance phase, most landowners employ some level of harvesting, whether for the purpose of producing income or for maintaining a desired structure. It is important to remember that the desired forest structure is not a single, quantitative point; rather, it more closely resembles a continuum of values, described by Kirkman et al. (2013) as a zone of possible structures in multifactorial space. The system is dynamic, not remaining at a fixed point indefinitely but shifting within the zone through time, with management actions helping to keep the structure within the desired range.

Depending on objectives, many silvicultural systems are available for consideration when managing multiaged longleaf pine forests. Methods such as Basal Area/Maximum Diameter/Diminution Quotient (BDq; Farrar 1996; Guldin 2006) and Proportional-B (Brockway et al. 2014) focus on producing multiaged stands with sustainable timber production, but their effects on plant diversity, wildlife habitat, and other ecological characteristics is not well-studied and they result in a different forest structure than the Stoddard-Neel (Moser et al. 2002). Thus, the applicability of these methods to restore the longleaf pine ecosystem is much less evident than that of the Stoddard-Neel with its focus on maintaining biological diversity, ecological functions, and heterogeneous structure.

SUMMARY

Longleaf pine ecosystem restoration occurs across a broad gradient of ecological and landscape conditions. Although many high-quality examples of longleaf pine ecosystems remain, none are free of human manipulations, whether they are the legacies of Native American burning, the logging of virgin forests, or the current realities of landscape fragmentation and silvicultural practices (see Chapters 1 and 3). The longleaf pine forests that remain are, of necessity, highly managed ecosystems that require frequent and ongoing management interventions. This is an accepted premise among those who know these forests well, but perhaps underappreciated by those less familiar with longleaf pine and certainly by most residents of the Southeast. For land managers and restoration practitioners, the premise of active management is fundamental to working with longleaf pine, whether dealing with a newly established plantation or a relatively pristine old-growth stand.

Regardless of their position along the restoration continuum, the most essential prerequisite for all longleaf pine restoration sites is the frequent application of prescribed fire. The commitment to a sustained, proactive prescribed fire program is critical. If wildlife or other ecological values are important management objectives, pursuing longleaf pine restoration without this commitment to prescribed fire is pointless. Inherent in the growing interest in longleaf pine restoration is the need for prescribed fire on increasing acreages. This requires increased professional capacity, higher funding levels, and an informed public consensus about air quality and smoke management issues (see Chapter 13). Compounding these challenges is a rapidly changing landscape and a growing wildland-urban interface—both of which tend to inhibit the use of prescribed fire, forcing strategic choices about where longleaf pine restoration can be sustained into the future.

A significant challenge for longleaf pine restoration is the slow pace with which the process unfolds. Longleaf pine restoration takes considerably more time than restorations in other ecosystems, such as a watershed in which streambank stabilization, repair of altered channels, or other discrete interventions can largely mitigate degradation and restore structure and function relatively quickly. When starting a longleaf pine restoration project from a reforestation or afforestation scenario, decades of persistent intervention and manipulation are needed to achieve the desired structure and function. Patience is a requisite; the desired results are seldom achieved in a manager's career or a landowner's lifetime. This high level of management intervention and prolonged time scale also requires a sustained commitment of financial resources.

The concepts of ecological forestry, in which forests are managed for a balance of both moderate economic returns and ecological values, have proven viable for addressing a broad range of goals that incorporate multiple resource values. These working forests require an ongoing management infrastructure to facilitate the frequent interventions that are part of the longleaf pine restoration process. Periodic income from timber harvesting can provide economic support for long-term restoration objectives and partially offset the sustained commitment required for a successful restoration.

A range of silvicultural tools is available for those who wish to practice less intensive approaches to forest management. Stoddard-Neel has a long record of success in sustaining ecological values for longleaf pine while producing moderate economic returns (McIntyre et al. 2010), arguably focusing more on ecological considerations than other restoration approaches. It is important to acknowledge that Stoddard-Neel, as well as other approaches that could be considered as ecological forestry, is not likely to meet the objectives of all landowners who are interested in growing longleaf pine. These approaches to forest management inherently involve higher opportunity costs than more intensive silvicultural systems (see Chapter 3). For many categories of private landowners, particularly larger-acreage ownerships, the juxtaposition of opportunity costs and fiduciary responsibility can become a barrier (McIntyre and McCall 2014). These economic challenges can be exacerbated by local and regional market characteristics, such as a lack of price differentiation for higher-quality products, mill restrictions on log size, and restricted access to mills resulting from closures and consolidations. However, many landowners, both private and public, have a broader suite of objectives and the latitude to balance economic considerations with other values. For these landowners, ecological forestry approaches such as Stoddard-Neel offer viable alternatives to more intensive systems of forest management and are underutilized relative to the land base on which they might potentially be applied.

Knowledge about longleaf pine restoration and appropriate methodologies has evolved considerably over the last two decades. Although adaptive management has proven beneficial for those interested in longleaf pine restoration, many questions remain; and the long time scales at which these forests operate require an iterative approach that can accommodate the inevitable challenges and refinements to guiding concepts and methodologies used for management and restoration of these ecosystems.

All of these factors reinforce the critical importance of clearly defining management and restoration goals before beginning management actions. Objective site assessment, including biophysical, social, and economic suitability for ownership goals, is an essential first step in the process. Longleaf pine restoration is not appropriate for every site or every ownership, even when all the biophysical characteristics are appropriate. Also of interest are questions about the future of longleaf pine in a rapidly changing climate (see Chapter 15) and whether altered management practices will be required. Although these uncertainties are challenging, the expectation is that management and restoration techniques and practices will continue to evolve and improve.

ACKNOWLEDGMENTS

We thank Mike Battaglia, Jerry Franklin, Stephen Pecot, and Clay Ware for their thoughtful and thorough reviews which greatly improved the final version. Jessica McCorvey assisted with preparation of the figures. We also are indebted to Jimmy Atkinson and especially Leon Neel for their mentoring and numerous discussions over many years about how to carefully and wisely apply ecological concepts to longleaf pine ecosystem management and restoration; the summary provided in this chapter would not have been possible without their considerable contributions.

11 Restoring and Managing a Diverse Ground Cover

L. Katherine Kirkman and Lisa M. Giencke

CONTENTS

INTRODUCTION

Reestablishing ecological processes in degraded or extirpated longleaf pine (*Pinus palustris*) communities inherently depends on the restoration and management of the native ground cover. The profound significance of the ground cover vegetation in this ecosystem is due in part to its extraordinary biodiversity, but also to the critical functional roles it serves as a source of pyrogenic fine fuels, a structural ladder for pine fuels (the primary source of energy), its strong regulation of nutrient cycling, and its importance in provisioning habitat for wildlife. Therefore, recovering or reassembling components of the native ground cover in longleaf pine restoration sites is a cornerstone for a developmental trajectory toward a diverse, sustainable, and fire-maintained longleaf pine forest.

Because longleaf pine sites occur across a wide range of soil types and respectively vary in community structure and species composition, a broad spectrum of approaches is necessary for successful ground cover restoration. Differences in soil moisture and fertility strongly influence ecosystem processes such as allocation of aboveground and belowground resources, competition, species recruitment, and fire regime (see Chapters 5 and 7). Understanding how land use history of a site will interact with ecosystem processes to govern species establishment and species interactions helps in deciding which techniques are appropriate for transitioning degraded ground cover to more desirable conditions.

In this chapter, we describe the general characteristics of the ground cover in longleaf pine communities, including patterns of endemism and rarity, life-history traits, responses to season and frequency of fire, and responses to canopy-cover variability. We describe the functional roles of specific groups of ground cover species as a means to prioritize species for reestablishment. We discuss the identification and use of conditions from reference sites, and we review current knowledge of soil seed banks and seed dispersal mechanisms relative to determining realistic restoration goals. We summarize site-based techniques for enhancing existing ground cover and reintroducing vegetation when necessary. And finally, we conclude with a discussion of the next steps that will be necessary for more effective ground cover restoration and the critical information needed for achieving success.

Characteristics of the Ground Cover

Numerous plant species within the diverse plant communities of the longleaf pine ecosystem are restricted to fire-maintained habitats. This endemism is partly attributable to the diversity of soils of the southeastern Coastal Plain and its history of frequent fire (Sorrie and Weakley 2006; Noss et al. 2015). Nearly 200 vascular plants that are strongly associated with longleaf pine communities are considered to be rare or in danger of extinction (Walker 1993). They represent the full spectrum of the longleaf-associated species pool—with composites (Asteraceae), grasses (Poaceae), and legumes (Fabaceae) prominent among the families that have the highest number of rare species. These species are predominantly perennial and are evenly distributed between upland and wetland habitats, with notable concentrations in shrub bogs and fire-maintained ecotones between uplands and wetlands.

In frequently burned upland longleaf pine stands, the ground cover is usually dominated by one or a few robust perennial grasses and a diverse mixture of forbs and low-statured grasses. Within its range, southeastern wiregrass (*Aristida stricta*, including *A. beyrichiana*) can strongly dominate the ground cover (as measured by percent cover and biomass), particularly on sites that have escaped human disturbances resulting from agriculture or intensive game management. Occurring in the absence of wiregrass, or occurring along with wiregrass, are other grass species such as little bluestem (*Schizachyrium scoparium*), broomsedge (*Andropogon virginicus*), slender little bluestem (*Schizachyrium tenerum*), pineywoods dropseed (*Sporobolus junceus*), yellow Indiangrass (*Sorghastrum nutans*), switchgrass (*Panicum virgatum*), and big bluestem (*Andropogon gerardii*). These "warm-season" grasses are particularly adapted to hot, dry conditions; they share a characteristic C_4 photosynthetic pathway that allows them to reach maximum growth in late summer (Doliner and Jolliffe 1979). Because many of these grasses have native ranges that extend north and west of the Coastal Plain (Figure 11.1), seeds sourced from outside the Southeast may have ecotypic adaptations to their local environment that can have negative impacts on restoration success (see the section Seed Provenance and Seed Transfer Zones). Other ground cover species are found infrequently at the local scale, positioned in the spaces between the dominant grass clumps (Kirkman et al. 2001; Clark et al. 2008).

Although numerous taxonomic families are represented in the ground cover of frequently burned longleaf pine communities, the most common are Asteraceae, Fabaceae, and Poaceae. Most species are herbaceous, with a sparse occurrence of small shrubs. Sprouts of hardwood species are usually present in subxeric to mesic sites; however, their growth is limited by fire and canopy tree competition, which restricts them from forming a midstory or canopy (as would happen in prolonged periods of fire exclusion or large tree gaps). Exceptions are the fire-tolerant oaks, such as southern red oak (*Quercus falcata*) or blackjack oak (*Q. marilandica*), that may occur sporadically throughout the midstory or canopy. Fire-tolerant oaks occurring in xeric sites (Hiers et al. 2007) include sand post oak (*Q. margaretta*), turkey oak (*Q. laevis*), and bluejack oak (*Q. incana*). In frequently burned flatwoods sites extending northward to southern South Carolina, low-growing saw palmetto (*Serenoa*

FIGURE 11.1 Range maps showing the distribution of common native warm-season grasses of the southeastern Coastal Plain: (a) Big bluestem, (b) broomsedge bluestem, (c) wiregrass, (d) switchgrass, (e) little bluestem, (f) slender little bluestem, and (g) yellow Indiangrass. (Modified from U.S. Department of Agriculture, Natural Resources Conservation Service (USDA NRCS). 2016. The PLANTS Database 2016. Online access: http://plants.usda.gov, National Plant Data Team, Greensboro, NC 27401–4901. Accessed January 2016.)

repens) and shrub species (primarily Ericaceae genera) are prevalent. Farther north and west, saw palmetto gives way to other shrub species, particularly gallberry (*Ilex glabra*) (Peet 2006).

Most of the herbaceous species are perennials that rapidly regrow following top-kill when burned (Garren 1943; and many others). Many of these resprouters are presumed to be long-lived species (Clewell 1989; Glitzenstein et al. 2012; Palmquist et al. 2014, 2015; Veldman et al. 2015). The perennating organs of herbaceous resprouters in this ecosystem vary morphologically, but they generally fall into two categories: either bud banks at or near the soil surface or rhizomes within the soil. Although most ground cover species have adaptations to withstand frequent fire (Beckage and Ellingwood 2008), they also may have other traits—such as chemical composition or spatial leaf arrangement—that actually promote the spread of fire (Nelson and Hiers 2008; Fill et al. 2016). This enhanced flammability may have evolved as a mechanism to reduce lethal levels of soil heating by facilitating the rapid burning of aboveground vegetation (Gagnon et al. 2010).

The agricultural history of a site is a primary factor impacting native ground cover, as the ability to recolonize after tillage varies significantly among the diverse suite of species (Hedman et al. 2000; Dale et al. 2002; Kirkman, Coffey, et al. 2004; Ostertag and Robertson 2007; Brudvig and Damschen 2011; Brudvig et al. 2013, 2014). For example, because wiregrass does not readily recolonize once it has been eliminated, its presence has been used as an indicator of sites that have not been intensively cultivated (Clewell 1989). Several other species have been repeatedly identified as indicators of minimally disturbed sites because their affinity to burned native communities lacking a history of severe soil disturbance is greater than that for disturbed sites; these include twinflower (*Dyschoriste oblongifolia*), scaleleaf aster (*Symphyotrichum adnatum*), Carolina wild petunia (*Ruellia caroliniensis*), dwarf huckleberry (*Gaylussacia dumosa*), goat's rue (*Tephrosia virginiana*), early blue violet (*Viola palmata*), white edge witchgrass (*Dichanthelium dichotomum* var. *tenue*), and sensitive brier (*Mimosa quadrivalvis*). Still others, such as certain varieties of broomsedge, occur in longleaf pine stands regardless of their disturbance history (Grelen 1962). Results from studies comparing the species affinities for disturbed sites versus relatively undisturbed sites have a high degree of concordance, but there are also notable inconsistencies among studies for some of the indicator species identified. In particular, narrowleaf silkgrass (*Pityopsis graminifolia*), variable witchgrass (*Dichanthelium commutatum*), and Carolina wild petunia (Hedman et al. 2000; Kirkman, Coffey, et al. 2004; Ostertag and Robertson 2007; Brudvig and Damschen 2011; Brudvig et al. 2013) have been documented as indicators of disturbed sites in some studies, but indicators of undisturbed conditions in other studies. Such discrepancies among studies in species affinities, as well differences in the degree to which recruitment of desired species occurs in a disturbed site, may be explained by disparities in the fire history and landscape characteristics of the sites that were being studied. Most remnant sites with a long history of fire exclusion would likely have much less reproduction and propagule availability than frequently burned tracts (Glitzenstein et al. 2001), perhaps with the exception of ultraxeric sites (Provencher, Herring, et al. 2001; Kirkman et al. 2013).

Many species associated with fire-maintained longleaf pine ecosystems can readily recolonize in areas disturbed by human activities (Kirkman, Coffey, et al. 2004), provided that propagules are present within dispersal distance of the site. Species that do not readily reestablish following soil-disturbing activities may be dispersal-limited (Mulligan et al. 2002; Kirkman, Coffey, et al. 2004; Brudvig and Damschen 2011; Brudvig et al. 2013; Veldman et al. 2014). Many of the ground cover species associated with longleaf pine communities are gravity dispersed (Figure 11.2) and consequently have a limited dispersal distance relative to species that are wind dispersed, consumed or carried by animals (including ants), or explosively dispersed. Further, restoration sites with a high degree of habitat connectivity would be more responsive to natural immigration processes than those occurring in fragmented landscapes (Brudvig and Damschen 2011). Experiments investigating corridors of clearcut vegetation linking patches of restored longleaf pine stands revealed that ground cover composition is influenced by the shape of the corridor, which regulates species assemblages based on modes of dispersal (Brudvig et al. 2009; Damschen et al. 2014); for more, see Chapter 5.

The persistence of a soil seed bank varies among soil types, as does its role in regeneration of ground cover vegetation following a disturbance. For example, in dry to mesic sites, most species appear to be transient or short-term residents in the soil seed bank (Iacona 2008; Kaeser and Kirkman 2012). A notable exception is a large group of species (primarily legumes) that have impermeable hard seed coats. Seed of some of these species can remain dormant in the soil for many years until exposed to air-temperature fluctuations, seed coat abrasion, or other

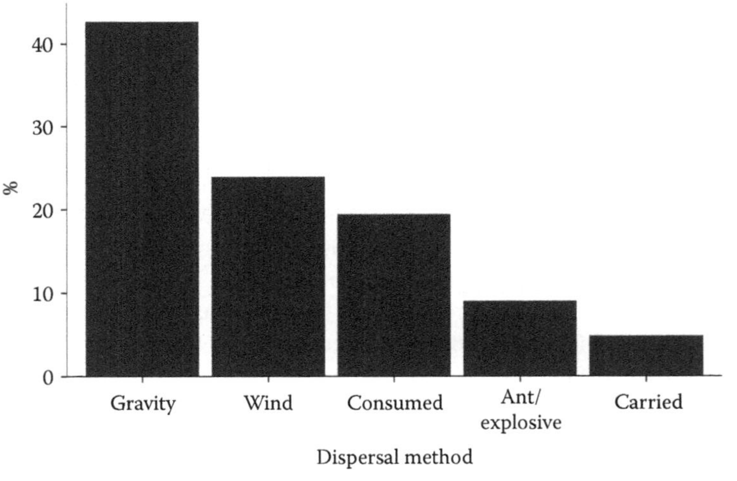

FIGURE 11.2 Dispersal methods of ground cover plants that are associated with longleaf pine ecosystems. (From K. Kirkman and L. Giencke, unpublished data.)

environmental stimuli that break dormancy (Baskin and Baskin 1998). In contrast, in degraded wet longleaf pine sites, seedling emergence of some obligate wetland species from soil samples indicates that they can persist in the seed bank (Cohen et al. 2008; Ruth et al. 2008; Andreu et al. 2009; Sharma 2012; J. Myers, personal communication). The duration of persistence and the ability of these seeds to contribute to natural recruitment processes (otherwise known as passive restoration) are unclear. Regardless, in wet savannas and flatwoods, the obligate matrix grasses do not have a seed bank (Cohen et al. 2008; Ruth et al. 2008; Andreu et al. 2009; Sharma 2012; J. Myers, personal communication), and many of these grasses are susceptible to soil disturbance (Glitzenstein et al. 2001). Consequently, these ground cover communities are not able to reassemble from the seed bank after a severe disturbance.

Response to the Season and Frequency of Fire

Season-of-fire (burning conducted during the summer growing season versus the dormant winter season) is controversial, especially within the context of ground cover responses and the efficacy of the fire in controlling hardwood encroachment (Drewa et al. 2002; Robertson and Hmielowski 2014; Fill et al. 2015). Although season-of-fire has no effect on ground cover species richness (Streng et al. 1993; Glitzenstein et al. 2008), numerous studies have linked it to species-specific changes in flowering responses (Biswell and Lemon 1943; Parrott 1967; Platt, Evans, and Davis 1988; Platt et al. 1991; Robbins and Myers 1992) and clonal growth (Hartnett 1987; Brewer and Platt 1994). For some species, increased flower production, delayed flowering, and more synchronous flowering (Platt, Evans, and Davis 1988) have been associated with growing season fires. Wiregrass is an iconic example of a species that depends on a combination of ambient temperature and season-of-fire for flower production; fertile seed is most readily produced in response to late spring or summer fires (Parrott 1967; Outcalt 1994; van Eerden 1998). Conversely, for multiple species of legumes and grasses, the response of peak flowering to season-of-fire varies considerably (Hiers et al. 2000; Shepherd et al. 2012).

Season-of-fire appears to have a cumulative effect on vegetation, with effects becoming detectable only after numerous prescribed fire events (Glitzenstein et al. 1995; Robertson and Hmielowski 2014). For example, studies of hardwood abundance after a single prescribed burn have found no differences in the number of resprouting stems as a result of timing of fire (Olson and Platt 1995; Cronan et al. 2015). However, in a study of shrubs across a gradient of upslope savannas to downslope seepage savannas, two successive dormant season fires resulted in heavier resprouting compared to the same regime of growing season fires (Drewa et al. 2002). This short-term response is consistent with other observations of reduced growth after an early growing season fire, presumably because of depleted root reserves at that time of the year (Robertson and Hmielowski 2014). In long-term experiments comparing season-of-fire effects on hardwoods, repeated growing season fires produced more top-kill and complete kill, less sprouting, and lower sprout density (Waldrop et al. 1992; Streng et al. 1993; Glitzenstein et al. 1995; L. Giencke, unpublished data). One of these experiments, a 14-year study of season-of-fire (eight dates during the year) and fire frequency (annual versus biennial) in longleaf pine sandhill sites (Glitzenstein et al. 1995) concluded that deciduous oaks were most vulnerable to early growing season fire and least vulnerable to dormant-season fire. Nonetheless, many of these studies confound the seasonality and intensity in experimental fire regimes (see Chapter 6), making conclusions difficult; further, definitions of growing season and dormant season vary so widely among studies that the term "growing season" has little clear meaning. While summer fire is not the major driver of understory dynamics as originally hypothesized (Platt, Evans, and Davis et al. 1988), variation in fire season appears to promote different species (Hiers et al. 2000). Many authors have concluded that frequent fire with variable fire seasonality is most likely to foster the diversity of grasses and forbs desired in restoring a degraded ground cover (Hermann et al. 1998; Hiers et al. 2000; R. McIntyre et al. 2008; Shepherd et al. 2012; Robertson and Hmielowski 2014).

Frequent burning over a long period is needed to create and maintain all pine-grassland communities (Waldrop et al. 1992; Glitzenstein et al. 2003, 2012). In most longleaf pine forests, ground cover dominated by grasses results from repeated annual or biennial fires, whereas less frequent fires result in more shrubs and woody sprouts (Walker and Peet 1983; Waldrop et al. 1992; Kirkman, Goebel, et al. 2004; Glitzenstein et al. 2012; Palmquist et al. 2015). Although periodic summer, periodic winter, and annual winter burning regimes result in a ground cover dominated by grasses and forbs, they also result in top-kill of hardwoods that subsequently resprout. In contrast, successive annual summer burning, which also promotes grassland communities beneath the pine canopy, eventually results in the reduction of small hardwoods and shrubs.

Response to Canopy Cover

Native ground cover species in longleaf pine ecosystems are generally adapted to the high light conditions that typify the multiage canopies of naturally regenerated mature longleaf pine stands (Lemon 1949; McGuire et al. 2001; Battaglia et al. 2002, 2003; Kirkman and Mitchell 2006; Platt et al. 2006). As is true with most warm-season grasses, wiregrass is more vigorous in an open canopy (Parrott 1967; Means 1997). This growth response was demonstrated in a 20-year-old longleaf pine plantation that was experimentally thinned to various degrees of canopy retention. Wiregrass seedling survival, which was 20% in closed-canopy conditions, increased substantially (along with growth and reproduction) in more open canopies (Mulligan et al. 2002). Despite decreased vigor, the surprisingly high level of survival under limited light of the closed canopy suggests adaptive attributes that allow wiregrass to persist for long periods when the absence of fire increases canopy cover of hardwoods.

Furthermore, in trenching experiments established to measure the effects of canopy removal, increased light was found to be more influential in regulating herbaceous ground cover biomass than increased availability of belowground resources (McGuire et al. 2001; Pecot et al. 2007). In contrast, the deep-rooted woody species respond mostly to the decrease in belowground competition from overstory pines; for more, see Chapters 4 and 7.

UNDERSTANDING FUNCTIONAL ROLES IN PROCESS-BASED RESTORATION

A process-based approach is key to establishing a successional framework that will result in the reassembly of a sustainable ecosystem. A few species and species groups have a disproportionate influence on the functional processes of the longleaf pine ecosystem. Thus, understanding and recognizing these influential roles can be helpful in identifying restoration priorities.

Role of Grasses in Flammability and Hardwood Exclusion

The availability of fine fuels to carry frequent, low-intensity prescribed fire is an obvious requisite condition for long-term maintenance of all longleaf pine sites, regardless of restoration objective. Fire and fine fuels are linked by a cyclical feedback mechanism: the application of frequent fire discourages hardwood dominance and promotes herbaceous growth, which in turn encourages subsequent fires (Beckage et al. 2011; Ellair and Platt 2013; Fill et al. 2015). In conjunction with pine needle litter, dominant pyrogenic grasses such as wiregrass, bluestems, broomsedges, and Indiangrasses serve as spatially continuous fuels that aid the spread of fire.

In general, grasses tend to be the most flammable of all ground cover species given their high surface-area-to-volume ratio and resulting flame contact (Simpson et al. 2016). With its high levels of lignin and fiber and its tendency to retain dead leaves for up to 2 years, wiregrass is particularly pyrogenic (Wells and Shunk 1931; Fill et al. 2016). About 85% of wiregrass leaves die each growing season (Parrott 1967). The long, fine leaves (up to 20 inches in length) arch upward from the clump, increasing the rate of fire spread by touching and igniting adjacent clumps. Importantly, erect clumps of live and dead wiregrass leaves tend to suspend the high energy-releasing pine needle litter (Outcalt et al. 1999; Fonda 2001; O'Brien, Loudermilk, Hiers, et al. 2016; O'Brien, Loudermilk, Hornsby, et al. 2016), synergistically increasing fuel loading (Platt et al. 1991; Mitchell, Hiers, et al. 2009).

Some data suggest that the tendency of wiregrass to form dense stands may directly retard woody plant establishment. On the Francis Marion National Forest in South Carolina, wiregrass plugs were planted closely spaced into rectangular plots within a longleaf pine seed orchard with a little-bluestem-dominated ground cover. More than 20 years later, midstory woody cover and woody stem density were lower in the wiregrass sections (Fill et al. 2017). Similar observations of canopy gaps through time (with and without densely planted wiregrass) also suggest that dense wiregrass restricts the encroachment of midstory and ground cover hardwoods (Kirkman, Jack, and Giencke, unpublished data), presumably because the wiregrass fuelbed allows a more continuous spread of fire or direct competition with hardwood seedlings preventing hardwood establishment (Figure 11.3).

Other common perennial grasses also readily carry fire (Abrahamson and Hartnett 1990; Gilbert et al. 2005; Wenk et al. 2013). Although the pyrogenic characteristics of wiregrass and other ground cover species have not been rigorously compared, anecdotal observations of land managers suggest that the flammability of the wiregrass-pine needle combination offers the maximum flexibility for prescribed fire under higher humidity and lower temperature conditions (M. Melvin, personal communication). Thus, reestablishing wiregrass within its native range would facilitate efforts to incorporate frequent fire into the management of restored longleaf pine sites. Nonetheless, promoting other pyrogenic grasses is also a high priority, particularly in areas that are outside the historical range of wiregrass.

FIGURE 11.3 An underplanting of wiregrass in a canopy gap (left side of photograph) located at a longleaf pine restoration site. The adjacent forest matrix (right side of photograph), absent wiregrass, is characterized by much denser clumps of hardwood sprouts. (Photograph courtesy of Joseph W. Jones Ecological Research Center.)

Providing Wildlife Habitat

Other ground cover species groups are important in providing habitat and food resources for wildlife in the longleaf pine ecosystem. A dense layer of herbaceous vegetation provides an important structural component (Engstrom 1993; Means 2006), especially for ground-nesting-birds—such as northern bobwhite (*Colinus virginianus*), Bachman's sparrow (*Peucaea aestivalis*), and wild turkey (*Meleagris gallopavo*)—that prefer continuous vegetation, both for nesting and as cover from predators (Stoddard 1931; Plentovich et al. 1998). Juvenile birds especially depend on the abundant insects available beneath ground cover vegetation as well as the plentiful seeds and fruits of many ground cover species (Stoddard 1931; Means 2006). Leaves of native legumes store large amounts of protein, rendering them a preferred forage for many herbivores (McGraw et al. 2004), including white-tailed deer (*Odocoileus virginianus*), gopher tortoises (*Gopherus polyphemus*), and pocket gophers (*Geomys pinetis*). Likewise, legume seeds are a nutritious dietary component of northern bobwhite, wild turkey, songbirds, and small mammals (Nestler 1949; Buckner and Landers 1979).

Nutrient Cycling

Legumes also play a key functional role in replacing nitrogen losses from frequent fires, because of their ability to fix atmospheric nitrogen (N_2) into ammonia (NH_3), a biologically useful form of nitrogen (Boring et al. 1991; Hendricks and Boring 1992, 1999; Hiers, Mitchell, et al. 2003; Lajeunesse et al. 2006; Cathey et al. 2010); for more, see Chapter 7. Given their abundance in mesic and xeric sites, legumes significantly increase nitrogen availability and serve an important role in maintaining long-term site productivity within nitrogen-limited longleaf pine ecosystems. Further, depleted nitrogen and carbon levels have been reported both in 14-year-old longleaf pine plantations and in naturally regenerated pine stands on former agricultural lands (Markewitz et al. 2002). Consequently, given the legacy of forests that have been degraded or converted to other uses in the Southeast, establishment of ground cover with an abundance of nitrogen-fixing legumes (Hainds et al. 1999) and a robust grass cover with associated high rates of belowground tissue turnover (Saterson and Vitousek 1984; McLauchlan et al. 2006) may be a key factor in reestablishing nutrient cycling processes, particularly in sandy soils of previously tilled restoration sites (see Chapter 7).

DEFINING OBJECTIVES OF GROUND COVER RESTORATION

The overarching goal of longleaf pine restoration is to develop a system that returns some degree of ecosystem functioning in a way that can be sustained with frequent prescribed fire. Restoration objectives for native ground cover in longleaf forests are often aimed at increasing or improving sustainable habitat for the diverse assemblage of plant and animal species associated with this fire-maintained ecosystem. Even if site-specific constraints eliminate restoration to historical conditions as an option, alternatives can be devised to enhance at least some desired functions (White and Walker 1997; Falk 2006; Hobbs 2007). Regardless of the initial conditions and specific restoration targets, frequent fire is the key process that regulates the close relationship of structure with function. Consequently, the first step in restoration of longleaf pine communities is to establish conditions conducive to frequent prescribed fire.

Consulting reference conditions (based on remnant sites, historical data, or both) can provide critical insights into the development of ground cover restoration targets and actions. For the most part, second-growth longleaf pine stands that have retained certain characteristics—including a diverse bunchgrass-dominated ground cover, minimally disturbed upland-wetland ecotones, and the presence of rare species—can be considered to have old-growth ground cover conditions and therefore can be used to determine the range of compositional and structural variability that is appropriate for the site (Kirkman and Mitchell 2006; Veldman et al. 2015). Using reference conditions to monitor restoration sites over time can provide a measure of change and a benchmark for restoration progress; for more, see Chapter 14.

Reference sites are temporally dynamic in response to fluctuating environmental conditions; therefore, comparing changes in reference sites to changes in a restoration site over time provides a more robust benchmark than comparing restoration change to a static reference concept (Hiers et al. 2012). The extensive regional loss of longleaf pine forests means that minimally altered reference sites may be scarce, particularly in areas of higher fertility where large tracts of forests have been converted to cropland (Williams 1989; Frost 2006; Kirkman and Mitchell 2006). In these situations, historical data coupled with observations from other parts of the Southeast can provide information on reference conditions that would help define attainable restoration targets, although such perspectives would not necessarily reflect current environmental conditions (Walker and Silletti 2006).

The resources and time required to implement a directed change in ground cover abundance or composition (or both) are governed by the initial site conditions resulting from cultivation, timber harvesting, or other former land uses, as well as history of prescribed fire (Walker and Silletti 2006). Severe soil disturbance, resulting from practices such as intensive cultivation, has the most important permanent effect, leaving no remnant species assemblages from which to reassemble the plant community. Further, legacies from agricultural or silvicultural activities (see Chapter 7) often include altered soil properties, including less organic matter, elevated phosphorus, and higher pH (Craft and Chiang 2002; Markewitz et al. 2002; Brudvig et al. 2013; Bizzari et al. 2015)—as well as competition from aggressive nonnative pasture grasses such as Bermudagrass (*Cynodon dactylon*) and bahiagrass (*Paspalum notatum*), or from other undesirable plant species that establish at the site. In contrast, sites that have been subjected to canopy removal or fire exclusion, but are otherwise minimally disturbed, likely will contain considerably more propagules of desirable species that can catalyze recovery processes. In fire-excluded sites, the restoration approach may be more passive (Palmer et al. 1997; Suding et al. 2004), focusing on restoring site structure to carry fire with the assumption that these conditions will eventually promote the repopulation of native ground cover vegetation. However, reintroduction of fire to fire-excluded sites may include substantial risks to the canopy due to smoldering duff; and such attempts must be implemented with judicious planning (Varner et al. 2009); for more, see Chapters 6 and 10. A more active approach for severely disturbed sites includes direct seeding or planting plugs of ground cover species (Walker and Silletti 2006; Fill et al. 2015).

Choosing the Right Tool for the Task

Midstory Reduction Options

Remnant natural longleaf pine stands often have a legacy of infrequent fire, resulting in dense midstory vegetation, and either fire-intolerant hardwoods, or in the case of pine flatwoods, palmetto-shrub thickets. Because a dense midstory alters fuel loads and fire behavior, and significantly diminishes light availability to ground cover, reduction of midstory vegetation is often a necessary step in restoration. Either alone or combined, treatments such as prescribed burning, mechanical reduction (roller chopping or mowing), and herbicide application have been used widely in young longleaf pine plantations to eliminate competition and enhance establishment and growth of desired seedlings. Only a few experiments have studied the effects of treatments, primarily herbicide application and midstory chopping, on nontarget species in natural fire-excluded longleaf pine stands where protection of native ground cover is a priority (Litt et al. 2001; Addington et al. 2012).

The selection of herbicides for midstory control, usually hexazinone, imazapyr, glyphosate, or triclopyr, depends on the species being targeted for removal (Hainds et al. 1999; Nelson and Cantrell 2002; Addington et al. 2012), as well as the species that are to be retained. Experiments examining the effects of herbicide-fire treatments on herbaceous vegetation (Table 11.1) reported no negative effects (Kush et al. 1999; Brockway and Outcalt 2000; Outcalt and Brockway 2010; Addington et al. 2012), but they were conducted primarily in plantation stands and only with a few of the herbicides listed above. Further, these findings must be interpreted with caution because most of the studies examined ground cover responses at large plot scales; their results may not have captured important

TABLE 11.1
Site-Specific Studies Showing the Effects of Midstory Reduction in Southeastern U.S. Longleaf Pine Stands

Site Type	Initial Stand Conditions	Objectives	Treatments	Outcomes
Sandhills and river dunes	Fire-excluded longleaf pine wiregrass	Response of ground cover to herbicide for hardwood reduction (Wilkins et al. 1993)	1-year study comparing 3 rates of herbicide treatment (hexazinone)	Oak understory and midstory decreased and wiregrass increased with increasing rates of herbicide
	Longleaf pine and wiregrass with oak encroachment	Season-of-fire and fire frequency effects on longleaf pine and oaks (Glitzenstein et al. 1995)	8-year study comparing fire season (8 dates) and annual or biennial fire frequency	Oaks rapidly decreased with frequent burning in early spring
	Naturally-regenerated 14-year-old longleaf pine stand (thinned)	Long-term understory and ground cover response to hardwood control (Kush et al. 1999)	23-year study using a factorial design to compare 3 hardwood control methods (herbicide [2,4-D amine], mechanical, and a control) and 4 burn treatments	No differences in hardwood midstory and understory biomass due to burning treatment; single herbicide treatment as effective as periodic mechanical treatments for hardwood tree control, but not shrubs; no difference in herbaceous biomass or diversity
	Fire-excluded planted longleaf pine stand with natural wiregrass	Vegetation responses to oak reduction (Brockway et al. 1998; Brockway and Outcalt 2000)	2- and 7-year studies comparing herbicide (hexazinone) rates and 2 application methods	None of the treatments affected wiregrass biomass; all treatments reduced oaks and initial species richness (followed by richness recovery)
	Fire-excluded natural longleaf pine stands	Response of ground cover to hardwood reduction (Provencher, Gordon, et al. 2001; Provencher, Herring, et al. 2001; Provencher, Litt, et al. 2001; Kirkman et al. 2013)	3- and 15-year studies comparing a single application of fire, herbicide (hexazinone) application, mechanical treatment, and a control—all treatments followed by frequent fire	None of the treatments differed in ground cover richness or composition; reference conditions changed over time
Dry to wet-mesic	Fire-excluded (10–16 years) mixed pine stand	Fuel load reduction (Outcalt and Brockway 2010)	8-year study comparing fire, thinning, fire plus thinning, herbicide (triclopyr), and a control	Herbicide treatment had the quickest initial response; thinning alone did not reduce fuel loading over time

(Continued)

TABLE 11.1 (*Continued*)
Site-Specific Studies Showing the Effects of Midstory Reduction in Southeastern U.S. Longleaf Pine Stands

Site Type	Initial Stand Conditions	Objectives	Treatments	Outcomes
	Clearcut former loblolly pine plantation	Vegetation response to herbicide treatment (Addington et al. 2012)	6-year study comparing 2 herbicide treatments (imazapyr/glyphosate; hexazinone) and a control	Herbicides increased longleaf pine seedling growth and reduced hardwood stem density
	Shrub-encroached pine savanna gradient (upland to seepage)	Vegetation response to herbicide plus fire versus fire only (Platt et al. 2015)	1-year study comparing herbicide (imazapyr and tricolpyr, triclopyr only), herbicide plus single fire, and single-fire only	Herbicide plus single fire resulted in slightly more shrub reduction and less flowering of C_4 grasses than single-fire only
Flatwoods	Loblolly pine with shrubby midstory and longleaf pine with saw palmetto	Effects of fire frequency on vegetation composition in South Carolina and Florida (Glitzenstein et al. 2003)	44-year study comparing multiple fire-return intervals (1, 2, 3, and 4 years) and an unburned control	The shorter fire-return intervals (1 and 2 years) resulted in a shift from woody to herbaceous species dominance in the South Carolina plots; reduced dominance of saw palmetto in Florida required a 1-year fire-return interval
	Clearcut second-growth longleaf pine with saw palmetto	Effects of saw palmetto reduction on ground cover and longleaf pine seedlings (Walker and Cohen 2009)	3-year study using a factorial combination of 2 mechanical treatments with 3 site preparation treatments and a control, enhanced by more intensive chopping plus herbicide (imazapyr/ triclopyr) application and bedding treatment	All treatments reduced initial percent ground cover; none of the treatments affected species richness at large scales or longleaf survival; herbicide application and bedding resulted in increased height growth in pines
	Longleaf pine and saw palmetto	Seasonal effects of fire and mechanical treatment on saw palmetto reduction (Willcox and Giuliano 2010)	2-year study comparing chopping, fire, and chopping plus fire	Growing season chopping was necessary to reduce saw palmetto density; fire alone had no effect
	Clearcut slash pine plantation with shrubs and wiregrass	Herbicide effects on shrub layer, ground cover, and longleaf pine growth (Freeman and Jose 2009; Jose et al. 2010)	5-year study comparing 4 low-rate herbicide treatments (imazapyr, sulfometuron methyl, hexazinone, hexazinone/ sulfometuron methyl) and a control	Longleaf pine survival was highest in control plots; only short-term effects of herbicide on shrub-cover reduction; no effects on ground cover richness or composition; increased herbaceous cover including wiregrass with herbicide

effects at smaller scales or for individual species. Factors that have been shown to determine the vulnerability of nontarget plants include the age of the plant, seasonal timing, and the rate of herbicide application (Kaeser and Kirkman 2010).

Different site conditions and restoration objectives require different midstory reduction techniques (Glitzenstein et al. 1995). Given that initial treatment responses are often short-lived and that changes in community structure usually occur only after multiple applications of fire, the most accurate comparisons of hardwood reduction treatments are those that measure changes in vegetation structure over time (decades). Furthermore, outcomes of reduction treatments also depend on the initial degree of midstory encroachment, the residual ground cover, and the intensity and frequency of subsequent prescribed fire.

For fire-excluded xeric sites where a longleaf pine canopy and native ground cover are still present, the long-term benefits of treating hardwoods with a single application of herbicide followed by prescribed fire, compared to prescribed fire alone, has been questioned (Kush et al. 1999; Provencher, Gordon, et al. 2001; Kirkman et al. 2013; Outcalt and Brockway 2010). Hiers et al. (2007) postulated that on such open and extremely infertile sites, herbaceous ground cover species are more susceptible to the buildup of litter from infrequent use of fire than to shading by a pyrophytic midstory of turkey oak, bluejack oak, or sand post oak (see Chapters 5 and 6). Their findings suggest that frequent fire alone (rather than mechanical or chemical reduction of the midstory) was sufficient to restore and maintain a diverse ground cover, unless evergreen oak encroachment was present. Furthermore, on xeric sites, the presence of a deciduous oak midstory can facilitate longleaf pine seedling establishment (Loudermilk et al. 2016); for more, see Chapter 4.

On sandhill sites where oak reduction is desired, frequent early spring burns can rapidly reduce oak density; the optimal fire-return interval in extremely infertile sites may actually be 2 years rather than annually because of low fuel accumulation rates (Glitzenstein et al. 1995). One advantage of herbicide treatment is that the immediate increase in fine fuel from herbaceous vegetation provides a wider range of weather conditions for effectively conducting prescribed fire (Menges and Gordon 2010; Outcalt and Brockway 2010; Addington et al. 2012). However, to counter the risk of damage to canopy trees the initial reintroduction of fire in sites with elevated fuel loads from herbicide treatments necessitates that prescribed fire be conducted within a conservative window of weather conditions (K. Hiers, personal communication).

In mesic sites, control of shrub midstory encroachment with fire alone is more problematic because of the rapid growth of hardwoods in more fertile conditions; in such situations, supplementary mechanical or chemical intervention may be needed (Table 11.1). Platt et al. (2015) speculated that in a wet-mesic savanna, shrub reduction might be best achieved if an early growing season burn is followed by herbicide application soon after postfire regrowth, presumably because of reduced belowground carbohydrate storage of the shrubs. This technique has not been tested.

In flatwoods, considerable research has focused on shrub midstory control as preplanting site preparation treatments for pine silviculture because these communities are often dominated by shrubs, such as saw palmetto, gallberry, swamp titi (*Cyrilla racemiflora*), Coastal Plain staggerbush (*Lyonia fruticosa*), and fetterbush (*L. lucida*). Several site preparation techniques can help reduce midstory woody species in these poorly drained sites. Imazapyr is one of the herbicides that can reduce shrub cover and increase longleaf seedling growth with minimal damage to grasses and forbs, but with possible increased mortality to longleaf pine seedlings (Jose et al. 2010). Bedding in flatwoods has been used to improve drainage, often resulting in increased longleaf pine seedling growth (Walker and Cohen 2009), but bedding complicates achieving other restoration goals as the stand develops. Chopping and other mechanical treatments can help reduce midstory saw palmetto and woody shrubs (Table 11.1), especially when combined with frequent prescribed fire (Willcox and Giuliano 2010; Duever 2011; M. Trager, personal communication). Roller chopping, however, should be used cautiously, especially when wiregrass is present in the ground cover (Lewis and Hart 1972; Outcalt 1992). Damage to wiregrass by chopping is less severe when treatment is limited to periods of high soil moisture (Outcalt 1992). Even though mechanical treatments do not appear to affect large-scale species richness (Walker and Cohen 2009), such disturbances could have important negative impacts on species richness at smaller scales (Glitzenstein et al. 2012; Kirkman et al. 2016).

Conversion Challenges on Off-Site Pine Plantations

Vast acreages of the historical longleaf pine range are now occupied by plantations of loblolly pine (*P. taeda*) and slash pine (*P. elliottii*), both on public and private lands. Thus, much of the land targeted for restoration to longleaf pine consists of even-aged stands of these "off-site" species. Such species are referred to as "off-site" because they are natively adapted to habitats with less frequent fire than that of longleaf pine. Silvicultural management of planted pine usually involves ≤ 30-year cycle of clearcutting, planting, thinning, harvesting, and replanting (Bennett 1980). Given the intolerance of slash and loblolly pines to fire in their early stages (≤15 years), these stands are usually not burned or only burned after an initial stand thinning (Dixon et al. 1984). As described above, deliberate fire exclusion to permit seedling establishment increases the growth and dominance of shrubs and midstory hardwoods and reduces grasses and forbs. In the absence of fire and as the planted pine canopy closes after the first 10–15 years of growth, little if any native ground cover vegetation will be present at the time of thinning due to shading. If prescribed fire is applied after thinning, some species of native ground cover may recolonize the stand, provided propagules have persisted at the site or are close enough to be dispersed into the stand (Kirkman, Coffey, et al. 2004).

For several decades, the prevailing technique for converting commercially planted stands of slash or loblolly pine back to longleaf was to remove canopy trees completely and to plant longleaf pine seedlings (Jack, Mitchell, et al. 2006); for more, see Chapter 10. More recent evidence indicates that this practice may be counterproductive (Kirkman, Mitchell, et al. 2007), creating a successional trajectory that cannot achieve the frequent-fire restoration target. Specifically, eliminating the existing overstory canopy of the off-site species, and with it the source of pine litter that serves as fine fuel, promotes the release of hardwoods in the ground cover or midstory and creates barriers for prescribed fire (McGuire et al. 2001).

One of the first studies to demonstrate a multistep approach to achieving a variably aged longleaf pine forest with diverse ground cover was conducted in a mature (65-year-old) slash pine stand in an upland mesic site in southwestern Georgia (Kirkman and Mitchell 2002; Kirkman, Mitchell, et al. 2007). Initially, the slash pine was thinned to encourage successful establishment of planted longleaf pine seedlings in small canopy gaps, while retaining enough canopy trees to provide continuous fine fuels from pine litter in the stand as a whole. In addition, experimental treatments for hardwood reduction (mowing, herbicide, and no treatment) were applied. Because pine litter was sparse or not present in the canopy gaps, wiregrass was established via seeding to promote continuous fuel. During the first 16 years after harvesting and planting, the stand was burned 11 times, longleaf pine seedlings emerged from the grass stage, and a second thinning of slash pines was coupled with additional planting of longleaf pine seedlings. No differences between hardwood reduction treatments were observed after 3 years.

Similar multistage approaches for conversion of off-site pine plantations to longleaf pine have been tested on other soil types. Knapp et al. (2014) examined the effects of multiple levels of loblolly pine canopy retention for planting longleaf pine seedlings and found that in loamy (but not sandy) soils, clearcuts produced more midstory woody stems than other canopy treatments. Similar results were observed by Hu et al. (2016). Hess (2014) examined longleaf pine seedling establishment in response to canopy thinning in a flatwoods slash pine plantation where infrequent use of prescribed fire had resulted in encroachment by woody shrubs and saw palmetto. He found that the dominance of a woody/saw palmetto midstory and resulting higher intensity fires increased longleaf pine seedling mortality. The combination of fire behavior, midstory shrub dominance, and excessive needle cast from retained canopy trees resulting in longleaf pine seedling mortality has also been observed at other flatwoods restoration sites (Kirkman and Giencke, unpublished data).

These observations suggest that in flatwoods sites, efforts to reduce the dense shrubby midstory and increase the dominance of herbaceous vegetation will be necessary restoration approaches to lessen the intensity of prescribed fires and increase longleaf pine seedling survival. Control of the midstory will likely include a combination of herbicide application, roller chopping, and diligent use of frequent fire (as described in the previous section, "Midstory Reduction Options"). Restoring and maintaining dominance of herbaceous vegetation not only will provide the fine fuels for frequent low-intensity fires to allow establishment of longleaf pine seedlings in flatwoods, but also will help promote increased ground cover diversity through species recruitment, particularly if propagule sources of desirable species are located nearby.

Another approach for conversion of slash pine plantations in flatwoods focuses on the use of extremely frequent fire and canopy retention and is currently under investigation. Based on initial observations, preliminary recommendations are outlined below (J. Glitzenstein, J. McGuire, and J. Stowe, personal communication). The first step is to thin the slash pine and root rake to extract shrubs (while leaving herbaceous plants), and then burn annually as aggressively as possible. The removal of shrubs will stimulate a burst in herbaceous plant growth, providing fuel for annual prescribed fire. The next step is to plant wiregrass and other dominant native grasses before shrubs are able to regain dominance. Additional grasses will increase fine fuel and deter shrub encroachment. If additional shrub reduction is needed, a contact (nonsoil active) herbicide, such as glyphosate, can be applied directly to target plants so that the impact on ground cover diversity will be minimal.

GUIDE TO REINTRODUCING NATIVE GROUND COVER

Unlike well-developed techniques for reestablishing native ground cover in tallgrass prairies and other U.S. grasslands (Baer et al. 2005; McCain et al. 2010; Rowe 2010), the development of successful protocols has been much more recent for longleaf pine sites. However, within the context of longleaf pine restoration, many reintroductions of native ground cover species have been successful. Restoration prescriptions have been built through numerous iterative trials and collaborative exchanges among practitioners throughout the Southeast, as well as with restorationists experienced in other U.S. grassland ecosystems.

The two most common approaches for establishing ground cover species are planting nursery-grown seedlings (plugs) and direct seeding. Direct seeding is considerably more economical, but the use of plugs may be desirable for more rapid results and is likely essential for reintroduction of rare plants. Planting plugs is also the preferable technique for enhancing residual native ground cover because populations of additional species can be introduced with less disturbance to existing plants than would occur with soil preparation for direct seeding (Glitzenstein et al. 2001). Last, planted plugs of warm-season grasses are better able to tolerate competition from Bermudagrass and other established undesirable species (J. Glitzenstein, personal communication).

Seed Provenance and Seed Transfer Zones

Regardless of the reintroduction method selected (plugs or seed), finding local sources is important to success in establishing native species. But a question that often arises for restoration projects is: How local is local? It has been long understood that plants can develop genetically based adaptations to their environment. Turesson (1922) coined the term "ecotype" to describe populations of a species that have unique genotypic responses to different local conditions. Thus, the common—but not unanimous (Wilkinson 2001)—consensus has been that when moving plant materials from a source population to a restoration site, a shorter distance between the two locations makes for a higher probability of establishment success (as measured by survival, growth, and fecundity).

Ecotypic differentiation is the basis for the "home-site advantage" hypothesis, which predicts that species have become adapted over time to specific local conditions, both biotic (such as soil microorganisms or pollinators) and abiotic (such as climate and soil). This hypothesis is often tested for individual species in common garden or reciprocal transplant studies. These studies, which have provided evidence for and against local adaptation, nevertheless share two important findings: (1) geographic distance is not always equal to ecological or genetic distance, and (2) the distance plant material can be moved without jeopardizing restoration outcomes is highly species-specific. Therefore, the question "How local is local?" will never have a "one-size-fits-all" answer (McKay et al. 2005).

Because of time and financial constraints, testing every candidate species in a common garden (plants from multiple sources grown in the same site) or reciprocal transplant study (plants from multiple sources introduced into home environment and distant sites) is impractical, particularly for such a wide ranging and diverse ecosystem as longleaf pine. In the absence of species-specific data, seed transfer zones or plant adaptation regions (Vogel et al. 2005) have been developed to guide the selection of sources that would most likely produce successful outcomes for a given locality (Figure 11.4). The need for seed transfer zones was originally recognized decades ago in Western U.S. restoration sites—trees grown from nonlocal seed were often found to have lower growth and survival rates than trees of locally sourced seed (Randall and Berrang 2002; Johnson et al. 2004). Only recently has the need for empirically defined seed transfer zones been formally recognized for ground cover restoration (Knapp and Rice 1994) and specifically, on longleaf pine sites (Walker and Hernandez 2010).

The following example clearly illustrates problems that may arise from using nonnative seed sources in longleaf pine restoration. A native species, partridge pea (*Chamaecrista fasciculata*) was recommended to be planted with native warm-season grasses under longleaf pine restoration guidelines from the Conservation Reserve Program, the vehicle through which the U.S. Department of Agriculture–Farm Service Agency offers incentives for establishing native ground cover along with longleaf pine seedlings planted in former croplands (Conservation Practice 36). Few local commercial seed sources were available for most of the recommended ground cover species, including partridge peas; thus, an Arkansas native, the "Lark" selection of this species was chosen as a substitute in seed mixes. When it was planted in Alabama and Georgia, "Lark" partridge pea grew taller and more densely than partridge pea from sources native to those states; it quickly overtopped and shaded out the longleaf pine seedlings, resulting in substantial pine mortality. In addition, "Lark" partridge pea has been associated with the development of *Rhizoctonia* blight, a fungal disease that also causes mortality of longleaf pine seedlings (GFC 2010). The substantial longleaf pine-establishment failure resulting from the use of this nonlocal ground cover seed came at considerable economic loss. The "Lark" partridge pea selection was replaced on planting lists by shorter-statured selections or by local ecotypes (when available) and has since been dropped for use in CP36 plantings (SC NRCS 2015).

Despite recognition by restoration practitioners of the probable advantages of using a local seed source, native seed companies have expressed hesitation about producing seed of multiple provenances for individual species, citing the additional time and cost involved for collection and production compared to the final price (Smith et al. 2007). In the trade-off between the size of the seed transfer zone (market area) and economic feasibility, small localized seed transfer zones result in more costly local seed, whereas larger zones offer lower costs, albeit at the risk of introducing maladapted plants into restoration sites. Therefore, to meet the needs of both restoration practitioners and commercial seed companies (whose involvement is vital for accomplishing regional-scale species reintroduction projects), seed transfer zones need to be small enough to result in successful restoration outcomes, but large enough to be profitable for seed producers.

After testing seed transfer zones of various sizes in the Great Basin floristic province, Kramer et al. (2015) found that the finest-scale zones—level IV ecoregions (Omernik 1987)—best represented the degree of genetic variation within source populations. However, using these criteria resulted in a fourfold increase in the number of zones compared to the broadest-scale zones used in their study (level III ecoregions). Instead of the finer-scale zones, they recommended using the level III ecoregions in combination with the provisional seed transfer zones of Bower et al. (2014), which delineate appropriate zones based on minimum winter temperature and aridity; the resulting seed transfer zones are intermediate in size and more realistic for seed production and marketing.

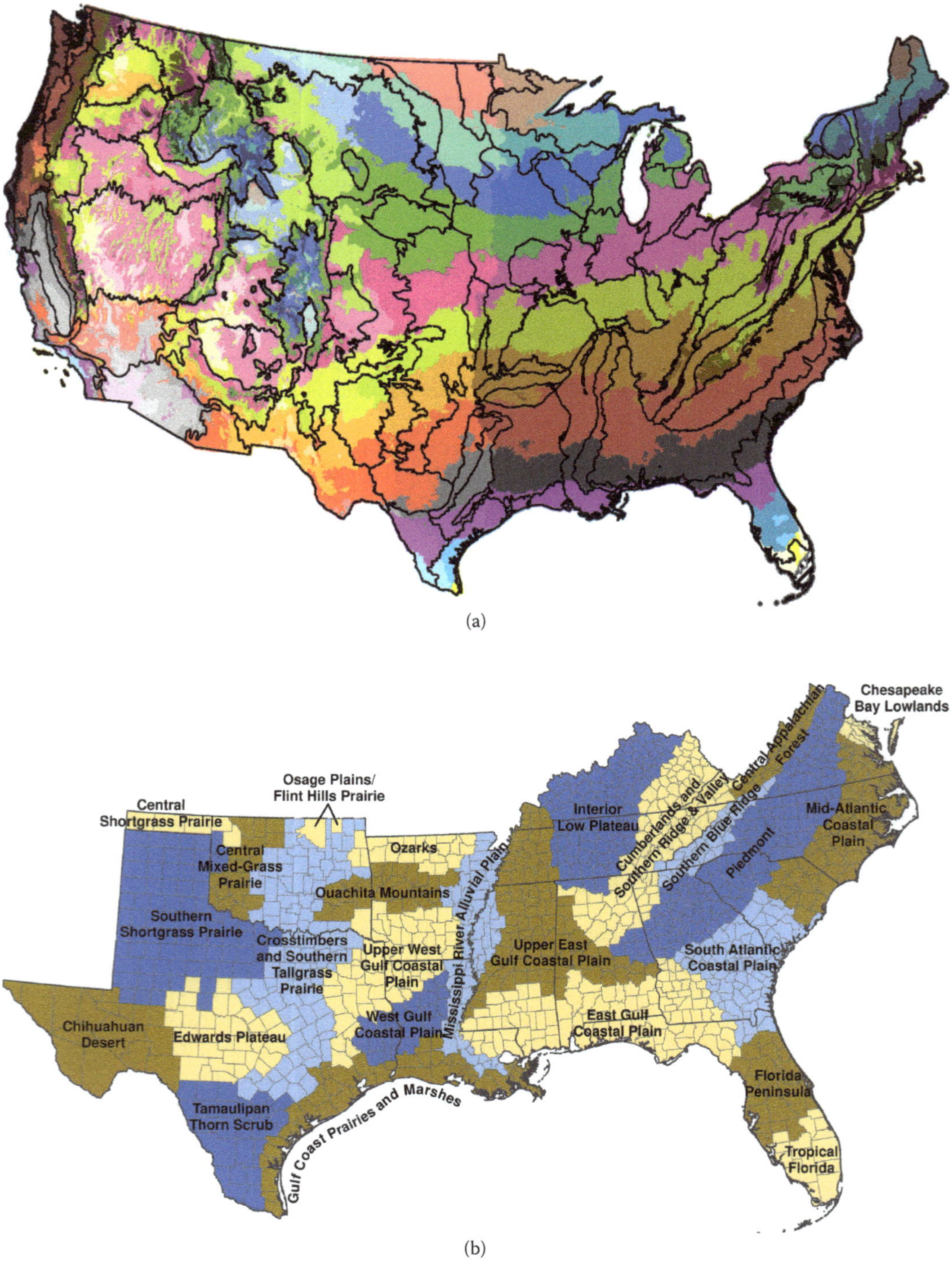

FIGURE 11.4 Generalized provisional seed zones for native plants: (a) In the contiguous United States, with seed zones represented by color polygons and ecoregions delineated by black lines. (Reprinted from Bower, A. D. et al., *Ecological Applications*, 24, 913–919, 2014; © 2014 Ecological Society of America. With permission.) (b) In the southeastern United States, with color polygons added to differentiate zones. Southeastern map. (Modified from unpublished USDA Forest Service map by Walker, J. and G. Hernandez. (2010) in a poster, Ecoregions and endemism to define gene conservation guidelines for longleaf pine ground-layer restoration, presented at the Longleaf Alliance Meeting, Columbia, SC.)

Studies evaluating level III ecoregions as seed transfer zones have reported mixed results. Miller et al. (2011) found that these ecoregions would be appropriate as seed transfer zones for four of the five species they tested in a western Oregon common garden setting, whereas Erickson et al. (2004) found that the ecoregion boundaries were ineffective at delineating optimal transfer zones for a grass species in the Blue Mountains of Washington and Oregon.

The provisional seed transfer zones from Bower et al. (2014) place most of the longleaf pine range into two seed transfer zones (of the 64 zones for the contiguous United States), covering latitudinal bands from east-central Texas through coastal North and South Carolina (Figure 11.4). Another provisional seed transfer zone map (Figure 11.4) proposed for the southeastern United States (Walker and Hernandez 2010) is based on The Nature Conservancy's map of 67 ecoregions within the contiguous United States (TNC 2009). Using these provisional zones, most of the longleaf pine range falls within four seed transfer zones: (1) East Gulf Coastal Plain (EGCP), extending from the Mississippi River to south-central Georgia; (2) South Atlantic Coastal Plain (SACP), from south-central Georgia to east-central South Carolina; (3) Mid-Atlantic Coastal Plain (MACP), from east-central South Carolina to the northern extent of longleaf and wiregrass in North Carolina; and (4) Florida Peninsula (FP).

A recent reciprocal transplant study was designed to test the validity of the southeastern provisional seed transfer zones for a small suite of longleaf pine ground cover species. The study examined six species across five geographic regions of the southeastern Coastal Plain: western, central, and eastern portions of the EGCP; the SACP; and the FP; it also included seed from Kentucky, when possible, to provide a comparison with more distant sources. Results showed earlier flowering of plants from Kentucky relative to those from southeastern seed sources, with much less variability occurring among the southeastern sources (Figure 11.5). For some species, survival and growth were positively associated with collection site and corresponding garden location, whereas others showed no differences attributable to seed source or garden site. Genetic analyses suggest a close relationship for some species between plants from the SACP and eastern EGCP regions compared to more distant seed sources, even within the Southeast (Giencke et al. in review).

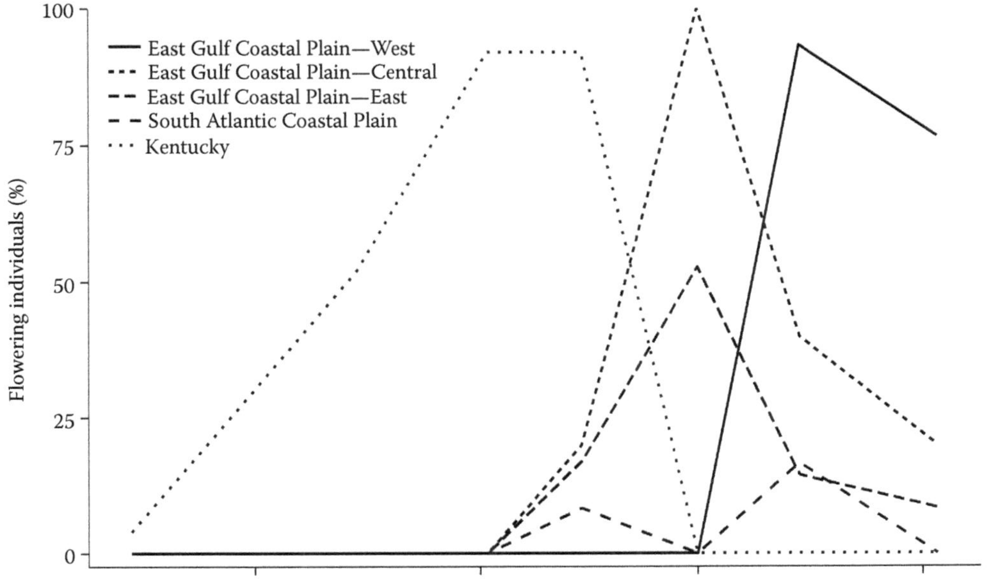

FIGURE 11.5 Comparison of flowering phenology of yellow Indiangrass from five source regions. Note that the peak flowering of the Kentucky population occurred 4–6 weeks earlier than the southeastern sources. (L. Giencke and K. Kirkman, unpublished data.)

Seed-Collecting Guidelines

Several unintended genetic consequences can occur when nonlocal seed is introduced into a restoration site (Table 11.2). These include founder effect, genetic drift, inbreeding depression, genetic swamping, and outbreeding depression (Montalvo 1994; Anttila et al. 1998; Lesica and Allendorf 1999; Hufford and Mazer 2003; Sgrò et al. 2011).

Table 11.3 summarizes seed-collection protocols most likely to minimize the genetic risks inherent to restoration projects, while capturing adequate genetic variation within a population (McKay et al. 2005; Basey et al. 2015). First, if species-specific genetic information is unavailable, seed should be collected or sourced from areas that are near the restoration site and have similar environmental conditions (such as soil moisture and texture). Having a local collection site that closely matches the restoration site will increase the chances that the seeds from that site will be genetically adapted for the restoration site.

Knowledge of species life-history traits, such as the degree to which a species cross-pollinates rather than self-pollinates can be useful in speculating about appropriate seed transfer distances. Optimal seed transfer zones of self-pollinated species are likely much smaller than those of out-cross-pollinating species because their gene flow distance is reduced (Hufford and Mazer 2003). Pollination mechanisms also govern the recommended number of individuals from which to collect seed. Crossa and Vencovksy (2011) suggested collecting seed from ≥25 unrelated individuals per population for cross-pollinating species, and ≥50 individuals when the species is self-pollinating or when the breeding system is unknown. Collecting from widely spaced individuals minimizes the chance of relatedness within the seed collection (Vekemans and Hardy 2004).

A seed collector may consciously or unconsciously choose to collect only in "good" years, from "good" sites, or from the healthiest looking individuals in a population—likely based on the assumption that these individuals produce seeds that will be the most vigorous at the restoration site. In reality, however, acting on these biases actually increases the likelihood that the range of

TABLE 11.2
Summary of the Genetic Risks That Can Jeopardize Efforts to Restore Forest Species Using Seeds from Nonlocal Sources

Risk	Definition	Cause
Founder effect	Reduction in the genetic diversity of a population that descended from a small colonizing population	Seed was collected from too few individuals in the source population
Genetic drift	Random changes in the frequency distribution of alleles	Small population size results in large changes in allelic frequency distribution and subsequent loss of genetic variation
Inbreeding depression	Reduction in fitness caused by the crossing of related individuals	Related individuals are crossed, which is more likely to occur when seed is collected from a very small number of individuals in the source population
Genetic swamping	Replacement of local genotypes with nonlocal genotypes	Hybridization between remnant and restoration populations that results in a loss of local genotypes
Outbreeding depression	Reduction in fitness caused by the crossing of genetically distant individuals	Hybridization between remnant and restoration populations that results in a dilution of local genotypes

TABLE 11.3
Summary of Guidelines for Collected Seed for Introduction into Restoration Sites

Recommendation	Rationale
Collect locally	Minimizes genetic swamping
Match the biotic and abiotic conditions of the donor and recipient sites	Based on the principle that the "ecological distance" can be as far or farther than the geographic distance
Know breeding system of focal species	Effect on gene flow: seed transfer zones of cross-pollinating species usually larger than those of self-pollinators (Hufford and Mazer 2003)
Minimize collection bias	Genetic variability often limited by collecting only from the most vigorous individuals, only once during fruiting period, or only during "good years" (Havens 2015)
Collect seed from ≥25–50 individuals in each source population	Ensures that 95%–99% of the genetic diversity of the source population will be captured
Collect from multiple populations	Increases the ability of the restoration population to adapt to single disturbances or new conditions, including climate change, at the restoration site

fitness characteristics will be narrow. Likewise, seed collection from early-, mid-, and late-fruiting individuals can help contribute to increased genetic diversity in the restored population, potentially building more resilience against future changes in climatic regimes or environmental conditions.

The actions and cautions described above focus on the health of the introduced population. However, maintaining the viability of source populations is equally important. Thus, as a precaution, the native seed collection program, Seeds for Success, administered by the U.S. Bureau of Land Management, recommends that the amount of seed collected be ≤20% of the seed produced at the site (BLM 2015).

Direct Seeding of Native Ground Cover

The interplanting of native ground cover with longleaf pine seedlings has become a widespread practice on privately owned land that has been removed from agricultural production. Often, these practices are tied to government-sponsored conservation incentive programs with wildlife management objectives, or goals to establish habitats that will support pollinator species (USDA NRCS 2008, 2016). Because of these federal- and state-funded programs, thousands of acres of privately owned properties are now planted in longleaf pine, particularly in Georgia (see Chapter 3).

Native warm-season grasses have been the preferred choice for these restoration sites for several reasons: (1) they were the original dominant ground cover species in native longleaf pine communities, (2) they provide critical fine fuels for carrying fire, and (3) they offer important cover for ground-nesting bird species. Other high-priority groups recommended for reintroduction in this situation are legumes (described above) and forbs that provide important resources for insect pollinators such as milkweeds (*Asclepias* spp.) and many species of composites.

Many of the guidelines for establishing native warm-season grasses and forbs on former agricultural fields were based on unreplicated trials conducted by land managers, property owners, and restorationists rather than quantitative tests in controlled experiments. Within this context and short-term perspective, most ground cover restoration has occurred simultaneously with the planting of longleaf pine seedlings.

Previously cultivated sites, where most ground cover reintroduction efforts to date have taken place, mostly occur on fertile upland sandy or loamy soils with intermediate soil water-holding capacity. Even decades after agricultural abandonment, these soils have been found to retain elevated phosphorus and pH (see Chapter 7), conditions that favor aggressive establishment of nonnative invasives over native species. This means that site preparation is needed to control existing competitors and undesirable species in the soil seed bank before seeding native grasses and forbs. In fields invaded by Bermudagrass or bahiagrass, adequate site preparation includes an aggressive removal treatment using glyphosate or imazapyr. Bahiagrass requires one application in spring (late February to mid-April) and another in late summer or early autumn. Bermudagrass requires an additional round of herbicide applications the following year: this nonnative grass is both extremely difficult to eliminate and extremely invasive—even small residual amounts can rapidly recolonize and regain dominance.

For successful establishment, native warm-season grass seed should be planted February through May in the Southeast (optimal times vary within the region). Planting in early spring allows seedlings to establish root systems so they are better able to withstand hot, dry summer days and effectively compete with aggressive ruderal species. Before planting, a firm seedbed should be prepared using conventional tilling methods. Proper seeding depth is critical for establishment of grass seeds; seeds planted too deep are unable to germinate. The correct depth depends largely on the soil type: seeds should be planted 1/4 to 1/2 inch deep on firmer seedbeds (loamy soils), <1/4 inch in sandy soils. It is important to check seeding depth frequently during the seeding process to ensure that seeds are not being planted too deeply.

Another successful method of direct seeding is to sow uncleaned seed using a standard hay blower. Bulk seed (ground cover material including stems and seeds of all plants collected from a donor site) is usually blown onto restoration sites that have been clearcut and then disked and packed using a cultipacker or bulldozer. Operational seeding rates for this method range widely among projects (such as Disney Wilderness Preserve, Fort Stewart Hunter-Army Airfield, and Apalachicola Bluffs and Ravines Preserve), from about 25 to 50 pounds per acre depending on the seed content and viability (Cox et al. 2004; Brockway, Outcalt, Tomczak et al. 2005). In addition to the target grass species, the bulk seed mixture contains seeds collected at the same time from numerous other ground cover plants, presumably resulting in a more diverse mix of species in the restoration site. One disadvantage of this method is that it requires storage and transport of very large quantities of bulk seed.

The recommended amount of pure live seed per acre (seeding rate) can vary, depending on planting objectives. The first step in calculating the amount of seed required to achieve the desired seeding rate is to evaluate the total viability of seed, usually via germination tests or tetrazolium (TZ) tests conducted by commercial laboratories and some state agencies. Viable seeds sometimes fail to germinate in germination tests; for these seeds, the TZ test serves as a more accurate measure (AOSA 2002), using color to sort between viable seed embryos (stained) and nonviable seed embryos (unstained). To ensure that enough seed is being sown to meet objectives, actual seeding rates should be adjusted based on the percentage of pure live seed, an industry standard that reflects seed viability and purity.

Finally, the no-till drill on the seed planter needs to be calibrated to deliver the required amount of seed per acre. No-till drills work best with cleaned seed (de-bearded seed and other plant debris removed). These drills use coulters and openers to create a furrow in the soil. Seeds are then sown into the furrows and covered with soil by press wheels, ensuring proper seed-to-soil contact (Figure 11.6). Various drill widths (3, 5, 8, 10, 12, 15, 20, and 30 feet) are available to meet a variety of planting needs and seeding depths. For situations in which uncleaned bulk seed collections are used, adding a specialized seed box that accommodates light, fluffy seed will ensure that the seed mix is properly mixed and sown at a precise rate. Use of this equipment requires a smooth planting bed, devoid of woody debris.

Another type of seed planter, a packer/seeder, is specifically designed to restore grasslands and accommodate fluffy uncleaned seed; however, the outcome of the calibration will be much less precise than a seed drill, given the narrow range of settings for delivery and the variation in composition of the bulk material. The packer/seeder has disks that scarify the soil and an

(a)

(b)

FIGURE 11.6 Establishment of native warm-season grasses: (a) Sowing with a no-till drill and simultaneous application of glyphosate to reduce competition from undesirable species that emerged following canopy thinning of planted longleaf pines, and (b) sowing uncleaned bulk seed with a tractor-pulled packer/seeder in a recently abandoned agricultural field. (Photographs courtesy of Joseph W. Jones Ecological Research Center.)

agitator in the hopper that forces seed down through pipes and drops it onto the soil surface. Deflated rubber tire casings then roll over the planted area to ensure good seed-to-soil contact (Figure 11.6). A packer/seeder is advantageous when considerable woody debris is present at the site.

Stands planted with native warm-season grasses typically take at least 2 years to become well established, depending on precipitation and soil moisture. After seeding and establishment, the application of selective herbicides (that will not harm the planted species) can reduce competition from ruderal plants (Kaeser and Kirkman 2010); mowing to a height of 8–10 inches can further reduce competition and prevent undesirable species from flowering and producing seed. During the second year after ground cover establishment, a spring burn can be applied to control competition from fire-intolerant plants, provided the site has enough plant material to carry a fire. With continued use of prescribed fire, native warm-season grasses will flourish; burning also serves to prune lower limbs of the young longleaf pine trees, which is important for tree growth form.

Wiregrass and other reintroduced ground cover species appear to thrive for many years under the open canopy of a young pine plantation, but vigor has been observed to decline as the canopy closes. Although the ability of individual species to tolerate a period of canopy closure is unknown, species such as wiregrass have been shown to survive low light conditions for several years and recover once canopy trees are removed and the site is burned (Mulligan and Kirkman 2002).

The debate about the appropriate planting density of longleaf pine seedlings has been considerable; with some arguments stressing the benefits for wildlife habitat of species that require an open canopy and grass-dominated vegetation, and others stressing optimal density for quality timber production (Longleaf Partnership Council 2013). Prescriptions for wildlife habitat usually aim for 400–500 seedlings per acre, which potentially extends the period of early stages of stand development (open canopy) and increases wildlife-habitat value. For stands that are planted for eventual timber production, initial establishment of a higher density (500–900 seedlings per acre) is recommended to achieve a more desirable tree growth form. Regardless of density, row spacing should provide long-term access for maintenance equipment such as tractors for mowing or seeding ground cover and harvesters for eventual canopy thinning (see Chapter 10).

Operationally, the thinning of planted pine stands often entails removing every third row of trees at about age 15–20. Given that most planted longleaf pine stands in the Southeast have not yet reached 20 years, few trials have been attempted to reestablish ground cover from seed after that first thinning. Experiments are underway at the Joseph W. Jones Ecological Research Center in southwestern Georgia to compare the long-term success of ground cover that was established in stands of different ages (Figure 11.7). Preliminary evidence suggests that establishment of native ground cover in a planted longleaf pine stand after the first canopy thinning has merit (S. Jack, K. Kirkman, L. Giencke, unpublished data); for more, see sidebar. Additional components of ground cover biodiversity may eventually find their way into the stand; but this usually happens over very long periods of time, even if the stand is on a remnant site or adjacent to an appropriate corridor for seed dispersal. If desired, additional species may be introduced depending on specific restoration objectives that may emphasize wildlife, aesthetics, pollinators, or rare species.

STAGES OF PINE PLANTATION DEVELOPMENT AND NATIVE GROUND COVER REINTRODUCTION

Comparison of the restoration of native ground cover planted at the same time as longleaf pine seedlings or after a thinning following canopy closure is being investigated. Study sites include plantation stands >15 years old and newly planted stands of longleaf pine seedlings. In the older plantations, the canopy was thinned by removing every third row of trees. The downed woody debris from the thinning operation was removed with a grapple rake and a mixture of native grass and forb seed (wiregrass, little bluestem, yellow and lopsided Indiangrass (*Sorghastrum secundum*), switchgrass, pineywoods dropseed, and goat's rue) was sown using a no-till drill. Shading and heavy needle cast of the previously closed canopy had killed many of the invasive plant species; but as a precaution, all sites were sprayed with imazapyr mixed with metasulfuron methyl before thinning to remove residual bahiagrass or Bermudagrass. A single treatment of glyphosate and ammonium salt of imazapic was also applied at the time of seeding to remove any species that had emerged after canopy harvesting had disturbed the soil. Seed of native warm-season grasses that were planted in the take-out rows produced a successful stand of grass the first growing season (Figure 11.7c). Frequent prescribed fires will be applied and the sites will be monitored periodically to examine ground cover composition, natural longleaf pine regeneration, and wildlife use in both ages of stands.

SUMMARY AND PATHWAYS FORWARD

The multistep introduction of trees and herbaceous ground cover described in this chapter establishes a basic structural platform of vegetation for the eventual development of a multiage longleaf pine stand that is sustainable with frequent fire. Although both initial site conditions and landowner objectives will dictate the degree of management intervention necessary and the rate of achieving the desired outcomes, ecological restoration is inherently a long-term process.

Given the important feedback interactions between ground cover vegetation and fire behavior, maintaining sufficient herbaceous ground cover to serve as fine fuels is the most basic prerequisite for achieving any long-term longleaf pine ecosystem restoration objectives. In recent decades with a greater understanding of longleaf pine ecology and through considerable trial and error, a more mature set of restoration approaches and operational techniques for ground cover restoration and management has emerged.

The effort (expense and long time frame) involved in reassembling diverse ground cover from scratch underscores the recommendation that protecting and enhancing remaining stands with native ground cover should be a high conservation priority, particularly for sites that are contiguous with or linked to larger tracts of fire-maintained longleaf pine stands.

Scales of Ground Cover Restoration

Repeated fire events are the key to developing successional pathways for the vegetative structure and ecological processes that determine reassembly success of longleaf pine ecosystems. Long-term monitoring of manipulative experiments across the longleaf pine range will be helpful in understanding the effects of restoration scale, the implications of surrounding land use, and the value of corridors that link restored patches of remnant longleaf pine with reintroduced ground cover (Figure 11.7). Establishing realistic goals at a landscape scale requires an understanding of how quickly plant species assemblages develop over time with frequent fire and whether restored sites will offer suitable habitat for wildlife. Insights into these processes can also inform efforts to set priorities for restoration sites and to estimate time frames for achieving long-term objectives.

Operational Issues

Innovation and documentation of success as well as failure are important for understanding the range of options available for managers under the many combinations of site history, fire regime, overstory retention, and basic site properties. Even though there remains considerable uncertainty regarding restoration approaches and many possible pathways, examples of successful efforts are emerging. Because so few planted longleaf pine stands are more than 20 years old, there are

FIGURE 11.7 Establishment of ground cover: (a) In an old field after one growing season, with equal prominence of yellow Indiangrass and undesirable plants that had become established at the site; (b) in an old field after eight growing seasons during which the planted wiregrass formed a dense stand; and (c) in a recently thinned longleaf pine plantation after one growing season, with planted lopsided Indiangrass in the foreground. (Photographs courtesy of Joseph W. Jones Ecological Research Center.)

currently few examples for comparison; obtaining useful information will require long-term observations and monitoring. Pine straw raking in the Southeast is a common, lucrative practice, but no assessments have been conducted on the benefits and disadvantages of this practice as a pretreatment for ground cover reestablishment. Existing agricultural drills or seeders are often too fragile to operate over the rough terrain and logging debris in recently thinned plantations; thus, seeding equipment will need to be modified or replaced with more robust models that can better handle the post-harvest site conditions.

Species Selection and Seed Acquisition

Successful regionally extensive reintroduction of ground cover, whether on public or private lands, depends on the commercial availability of seed from southeastern sources at affordable prices. Networking opportunities are emerging for comprehensive public-private partnerships to identify seed needs and ensure the reliable availability of genetically appropriate seed. To succeed, this effort will require a better understanding of the genetic variability and local adaptations of species, particularly those that are widely distributed across southeastern landscapes. It will also require the incorporation of output from climate prediction models into seed transfer zone recommendations.

Twelve federal agencies and several hundred private and nonprofit organizations have joined to establish the Plant Conservation Alliance, which published national goals to ensure the availability of genetically appropriate seed for restoration projects (PCA 2015). Southeastern natural resource organizations need to be active participants in the implementation of this effort and serve as advocates for inclusion of longleaf pine restoration initiatives.

Research/Management Partnerships

Finally, resolving specific challenges surrounding ground cover restoration of wet-mesic or hydric flatwoods is a high conservation priority, particularly given the vast acreages of shrub-dominated flatwood forests that occur on public lands targeted for longleaf restoration. Research has shown a strong relationship that connects midstory fuels management, ground cover structure, and regeneration of longleaf pine. Ultimately, advancing restoration goals in flatwoods sites will require multiphase approaches for increasing herbaceous ground cover and reducing midstory to ensure the successful survival and establishment of longleaf pine with lower intensity fires. Rigorously designed, replicated long-term studies are needed to better understand the complexities of reversing fire-suppressed conditions in flatwoods sites, and to incorporate ground cover diversity improvements into vegetation trajectories. Such long-term efforts will require collaborations between researchers and land managers as well as the commitment of resources across the Southeast for both the implementation of frequent prescribed fire and the research that supports that effort.

ACKNOWLEDGMENTS

We thank Robert Mitchell (deceased), Steve Jack, Mark Melvin, and Jimmy Atkinson at the Joseph W. Jones Ecological Research Center for numerous discussions, iterative trials, and collaborative research projects in the development of potential native ground cover restoration strategies. We acknowledge funding from the National Fish and Wildlife Foundation and the Dobbs Foundation in the development of a native seed production garden and outreach project and the attentive work of numerous field assistants. We also recognize vital research and outreach partnerships with natural resource staff of the St. Marks National Wildlife Refuge, the Georgia Department of Natural Resources, The Longleaf Alliance, the U.S. Forest Service, and the USDA Natural Resources Conservation Service that have contributed to our current understanding of effective restoration approaches. We also thank many private landowners for their willingness to participate in establishment trials of native warm-season grasses. Special recognition of the dedication and progress made by Cody Laird and John Cox of Lolly Creek Ecological Land Management and Restoration at Oakridge Farms and John Seymour of Roundstone Native Seed Company; we thank them for their generosity in sharing information. We are indebted to Kevin Robertson, Jeff Glitzenstein, and Kevin Hiers who provided thoughtful reviews of earlier drafts.

12 Management and Restoration for Wildlife

Lora L. Smith, James A. Cox, L. Mike Conner, Robert A. McCleery, and Elizabeth M. Schlimm

CONTENTS

INTRODUCTION

Longleaf pine (*Pinus palustris*) forests are widely recognized for their wildlife diversity (Engstrom 1993; Guyer and Bailey 1993; Means 2006). This diversity (Table 12.1) is tied to forest structure, habitat heterogeneity across multiple scales, and the physical conditions found in the southeastern U.S. Coastal Plain. At small scales, many amphibians and reptiles rely on ground-level structures (tip-ups, stumps, fallen trees, and burrows) and vegetation that offer refuge from fire, predators, and extreme temperatures; these structures also enhance the diversity of small mammal and invertebrate prey (Means 2006). At larger scales, long-lived pines and midstory hardwoods provide nest sites for songbirds, arboreal habitat for snakes and mammals, and important yet variable food resources. The open structure of these ecosystems also allows many grassland species to persist in a forested setting. Embedded isolated wetlands contribute additional heterogeneity at local and landscape scales and provide essential habitats for many reptiles, amphibians, and invertebrates (see Chapter 9). At an even larger scale, soils and landscape position create broad gradients, ranging from wet pine flatwoods to xeric sandhills that further shape distinctive and diverse plant and animal populations (Engstrom 1993; Guyer and Bailey 1993; Means 2006). This complex array of factors, coupled with dramatic changes in land use and land cover in the Southeast over recent decades, has made management and restoration of wildlife in longleaf pine ecosystems both extremely important and particularly challenging.

In this chapter we review the key factors to consider when restoring wildlife diversity in the longleaf pine ecosystem: restoration targets, starting conditions, spatial scale and context, frequent fire, the importance of structural legacies, and the role of wildlife in engineering those structures. We also discuss trade-offs that are inherent in the common strategies used in managing for rare and imperiled species throughout the range of longleaf pine. As an alternative to managing for individual species, we present a case study that demonstrates how a suite of wildlife species might be used to determine wildlife restoration actions. And last, we describe the key research and conservation needs for wildlife management and restoration, and the programs that focus on restoring the wildlife diversity of longleaf pine ecosystems.

CONSIDERATIONS FOR RESTORATION AND MANAGEMENT OF WILDLIFE

Restoration Targets

High-quality habitat for longleaf pine wildlife specialists (Table 12.1) consists of frequently burned, pine-dominated stands with an open canopy, a smaller component of hardwood trees and midstory shrubs, and diverse and abundant herbaceous ground cover (Engstrom 1993; Guyer and Bailey 1993; Sutter et al. 2001; Means 2006). Benchmarks for historical reference conditions are based on information provided by early explorers and land surveyors (Bartram 1791; Harper 1913), the few relatively intact longleaf pine forests that remain on the landscape (Noel et al. 1998; Sutter et al. 2001), and the experience of land managers. Although these are arguably the best resources from which to derive desired future conditions, they also have some limitations. Old-growth stands have provided unique information on both the long lifespan of longleaf pines and the wildlife that such forests support (Platt, Evans, and Davis 1988; Engstrom and Sanders 1997), but these stands are small, shaped by local forces, and do not reflect the full spectrum of historical conditions and communities that characterize the range of the species. Mature stands of longleaf pine with undisturbed native ground cover are exceptionally rare.

More broadly, pollen data also suggest that pine-dominated landscapes have waxed and waned at different times over the past 120,000 years (Watts 1980). Some extant fauna were likely influenced by these changes as well as by the altered landscapes surrounding the remnant tracts that remain (Watts et al. 1992). Information about these dynamic, historical systems is incomplete, suggesting that some restoration targets will be difficult to define, and even more difficult to achieve. In the absence of detailed information on historical benchmark conditions, restoring habitat for endemic wildlife requires realistic targets for achieving open canopy pine grasslands that can be maintained with prescribed fire.

Starting Condition

The area once covered by longleaf pine has been transformed by other land uses over the last century, with habitat loss, fragmentation, and degradation resulting in the decline of wildlife populations (Van Lear et al. 2005). This is especially true for species that require large areas, such as the red wolf (*Canis rufus*), mountain lion (*Felis concolor*), and bison (*Bison bison*); and for endemic species that are adapted to specific conditions, such as the red-cockaded woodpecker (*Picoides borealis*), gopher tortoise (*Gopherus polyphemus*), and Bachman's sparrow (*Peucaea aestivalis*).

As shown in Figure 12.1, forests are still the dominant land cover in the former range of longleaf pine (23%); however, most open woodlands and savannas have been converted to intensively managed plantations of loblolly pine (*P. taeda*) and slash pine (*P. elliottii*). From 1950 to 2000, pine plantations in the Southeast expanded from roughly 700,000 ha to almost 13 million ha (Fox et al. 2007). This

TABLE 12.1
Vertebrates Whose Principal Habitat Is Longleaf Pine Woodlands and Savannas

Vertebrate Group	Scientific Name	Common Name	Habitat[a]	Status
Salamanders	*Ambystoma bishopi*	Reticulated flatwoods salamander	F, SA, GIW	FE
	Ambystoma cingulatum [b]	Frosted flatwoods salamander	F, SA, GIW	FT
	Ambystoma mabeei	Mabee's salamander	SA, GIW	N
	Ambystoma tigrinum	Tiger salamander	F, SA, GIW	N
	Notophthalmus perstriatus [b]	Striped newt	F, SA, GIW	CT
	Eurycea quadridigitata	Dwarf salamander	A	N
Frogs	*Anaxyrus quercicus*	Oak toad	F, SA, GIW	N
	Hyla femoralis	Pine woods treefrog	AR, SA, GIW	N
	Hyla gratiosa	Barking treefrog	AR, SA, GIW	N
	Hyla squirella	Squirrel treefrog	AR, SA, GIW	N
	Pseudacris brimleyi	Brimley's chorus frog	SA	N
	Pseudacris nigrita	Southern chorus frog	F, SA, GIW	N
	Pseudacris ocularis	Little grass frog	F, SA, GIW	N
	Pseudacris ornata	Ornate chorus frog	F, SA, GIW	N
	Lithobates areolatus	Crawfish frog	F, SA, GIW	N
	Lithobates capito [b]	Gopher frog	F, SA, GIW	P
	Scaphiopus holbrookii	Eastern spadefoot	F, SA, GIW	N
Snakes	*Cemophora coccinea*	Scarlet snake	F	N
	Crotalus adamanteus	Eastern diamondback rattlesnake	F	P
	Drymarchon couperi [b]	Eastern indigo snake	F	FT
	Heterodon simus	Southern hog-nosed snake	F	P
	Lampropeltis extenuata	Short-tailed snake	F	N
	Micrurus fulvius	Eastern coral snake	F	N
	Pituophis melanoleucus mugitus	Florida pine snake	F	P
	Rhadinaea flavilata	Pine woods snake	F	N
	Tantilla relicta	Florida crowned snake	F	N
Lizards	*Ophisaurus mimicus*	Mimic glass lizard	F	N
	Plestiodon egregious[b]	Mole skink	F	N
	Rhineura floridana	Florida worm lizard	F	N
Turtles	*Gopherus polyphemus* [b]	Gopher tortoise	F	CT
Birds	*Colinus virginianus*	Northern bobwhite	G	N
	Peucaea aestivalis	Bachman's sparrow	G	N
	Picoides borealis [b]	Red-cockaded woodpecker	C	FE
	Sitta carolinensis	White-breasted nuthatch	C	N
	Sitta pusilla [b]	Brown-headed nuthatch	C	N
Mammals	*Sciurus niger* [b]	Sherman's fox squirrel	AR	N
	Geomys pinetis	Southeastern pocket gopher	F	N
	Podomys floridanus	Florida mouse	F	N

Source: Guyer, C., and M. A. Bailey, *Proceedings of the Tall Timbers Fire Ecology Conference*, no. 18, Tall Timbers Research Station, Tallahassee, 1993; Means, D. B., *The Longleaf Pine Ecosystem: Ecology, Silviculture, and Restoration*, Springer, New York, 2006.

FE = federally listed as endangered, FT = federally listed as threatened, N = not a species of concern, P = petitioned for federal listing CT = candidate for federal listing as threatened, P = petitioned for federal listing.

[a] AR = arboreal, F = burrow-dwelling, A = aquatic, SA = semi-aquatic, GIW = geographically isolated wetland principal breeding habitat, C = cavity nester, G = ground nester.

[b] This species is subject to reintroduction or translocation (Costa and DeLotelle 2006).

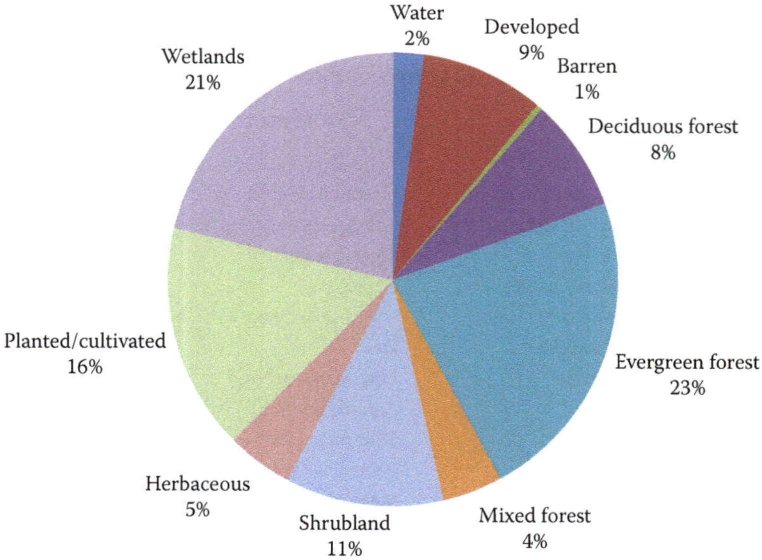

FIGURE 12.1 Major land use/land cover types within the former range of the longleaf pine ecosystem. (Data from Homer, C. G. et al., *Photogrammetric Engineering and Remote Sensing,* 81, 345-354, 2015.)

trend is expected to continue for several decades (Wear and Greis 2002a), and although young plantations benefit some wildlife, they do not support the rich diversity of wildlife that historically inhabited longleaf pine savannas (Brockerhoff et al. 2008). Specifically, the closed canopy that develops in a plantation reduces or eliminates the ground cover and midstory vegetation—a condition that is critical for longleaf pine-adapted wildlife (Trani 2002). Nonetheless, pine plantations provide opportunities for restoring components of the longleaf pine ecosystem (Kirkman, Mitchell, et al. 2007), and measures such as thinning and applying prescribed fire can quickly restore an open canopy structure within 20- to 30+ year-old plantations (Kirkman, Mitchell, et al. 2007; Jofré et al. 2016).

In comparison with pine plantations, fewer acres (16%) are currently devoted to production agriculture in the former range of longleaf pine (Figure 12.1). Recent government surveys (OCE 2015) show that the major crops include corn (*Zea mays*), cotton (*Gossypium hirsutum*), peanuts (*Arachis hypogaea*), and blueberries (*Vaccinium* spp.). Traditional farming techniques helped to support some of the wildlife that inhabited this pine-grassland ecosystem for many decades, but recent shifts toward larger-scale, more intensive agricultural practices have made these areas less favorable for wildlife (Warner et al. 2012). The acreage of cropland is predicted to decline over the next half century (Wear and Greis 2002a; Wear 2011), possibly providing opportunities for restoring additional acreages back to longleaf pine or making croplands more beneficial for wildlife.

Only 9% of the longleaf pine historical range has been converted to urban and residential landuses. However, pollution, roads, and the demands that urbanization places on natural resources have disproportionate negative effects on many longleaf pine forests (McCleery et al. 2015). Human population growth is expected to increase by >100% over the next 50 years, thus establishing urbanization as a growing concern for future decades (Terando et al. 2014). Restoration of many recently developed areas to suitable wildlife habitat will be unlikely, and in other areas it will be virtually impossible. However, the open canopy and abundant ground cover found in some developed areas mimic the structure of a longleaf pine savanna (McCleery et al. 2012). Species such as fox squirrel (*Sciurus niger*), brown-headed nuthatch (*Sitta pusilla*), southeastern pocket gopher (*Geomys pinetis*), and other longleaf pine specialists can often persist in some developed settings such as golf courses, utility corridors, and low-density residential areas (Jodice and Humphrey 1992; Means 2006).

In addition to losses from outright conversion, many tracts are succeeding to homogenized, closed-canopy hardwood-dominated forest that are unsuitable for longleaf pine specialists (Engstrom 1993).

The increase in woody biomass through shrub encroachment is a global phenomenon in savanna ecosystems (Van Auken 2000; Roques et al. 2001; Cabral et al. 2003) that can occur rapidly with the moderate temperatures and higher rainfall of the Coastal Plain. One reason for woody encroachment is the absence of fire in this fire-adapted ecosystem (Frost 1993; Van Lear et al. 2005). Growth of human populations and changes in land use in the early-to-mid 1900s have exacerbated fire exclusion and led to additional shrub encroachment (Frost 1993; Gilliam and Platt 1999) as well as challenges and constraints on the use of prescribed fire in longleaf pine forests (see Chapters 1, 13, and 17).

Patch Configuration

Restoration of longleaf communities ideally should take place at a scale that includes the full range of ecological processes (such as fire, grazing, and wind falls) that shaped this ecosystem and its heterogeneity. However, given the reduced extent and increased fragmentation that now exists, full restoration of large-scale animal movements and historically favorable disturbances is highly unlikely outside a few locations (such as the Apalachicola National Forest, Eglin Air Force Base, Blackwater River State Forest, Conecuh National Forest, Ocala National Forest, and Francis Marion National Forest). Still, strategic selection of restoration sites can improve habitat for wildlife. When restoration sites are limited to small patches (<100 ha), the placement and size of sites can be important both for species such as Bachman's sparrow that have small home range requirements (Taillie et al. 2015) as well as for species such as Sherman's fox squirrel (*Sciurus niger* var. *shermani*) that have larger home range requirements (Meehan and Jodice 2010). Restoration of small patches is likely to have greater benefits for wildlife when the patches are close to or adjoining other patches of longleaf pine forest and can function as part of a larger aggregation (Hansen and Defries 2007). The land uses of nearby tracts affects the function and management of restored patches. Intensive urbanization and agricultural production reduces both the functional size of a patch and its use by species occupying higher trophic levels (Hansen and Defries 2007). Alternatively, restored patches surrounded by less-intensive land use (such as wetlands or pine plantations) can provide unique habitat for longleaf pine specialists, nearby refugia for wildlife to use during disturbance events, and a buffer from human disturbances (Hansen and Defries 2007). Furthermore, sustaining viable populations of some wildlife requires that forest fragments occur in permeable landscapes that enable individuals to move among neighboring patches. In the longleaf pine range, wildlife disperse readily through open canopy forests (Roznik and Johnson 2009; Kesler and Walters 2012; Trainor, Walters, Morris, et al. 2013), but can be constrained by roads (Miller et al. 2012), fields (Jones 2013), and hardwood-dominated forests (Kesler and Walters 2012). More information is needed to understand how different land uses in adjacent tracts affect the movement of wildlife that inhabit restored longleaf pine patches.

Spatial Scale

The optimal patch size and configuration of restoration areas depends on the targets of the restoration. Taxa that have large space-use requirements need large patches and connectivity at much larger scales. Carnivores, including both mammals and large snakes, establish large home ranges and transverse long distances (200 to >500 m daily) (Conner 2000; Chamberlain et al. 2003; Franz 2005; Hyslop 2007; Howze and Smith 2015). For some species, large areas are needed to account for spatial and temporal variations in important resources. For example, Sherman's fox squirrels move to areas that have live oaks (*Quercus virginiana*) when mast production fails in longleaf pine settings (Kantola and Humphrey 1990). Some male Bachman's sparrows simultaneously hold distinct territories 1–2 km apart, a behavior that could be an adaptation to ephemeral ground cover conditions (Brown 2012; Cerame et al. 2014). For species such as the Florida pine snake (*Pituophis melanoleucus* var. *mugitus*), avoidance of roads reflects the need for landscape contiguity (Miller et al. 2012). Small, semi-terrestrial amphibians such as gopher frogs (*Lithobates capito*), flatwoods salamanders (*Ambystoma cingulatum* and *A. bishopi*), and striped newts (*Notophthalmus perstriatus*) also require large, interconnected patches of suitable habitat (Palis 1997; Johnson 2005; Roznik and Johnson 2009). As confirmed by Semlitsch (1998), Gamble et al. (2007), Pittman et al. (2008), and Hillman et al. (2014), these species have relatively low dispersal capabilities, they have a strong fidelity to breeding sites and terrestrial refugia, and they often function as meta-populations (see Chapter 9). Unfortunately, the appropriate patch sizes and configurations needed to support populations of these and many other longleaf pine specialists are largely unknown.

Fire

The wildlife diversity supported by the longleaf pine ecosystem is intimately linked to the periodic fires that maintain suitable vegetation structure, particularly at ground level (Landers 1987). The conditions that are suitable for some species disappear within 2 years after a burn (Jones et al. 2013), and the absence of periodic fire leads to hardwood encroachment and loss of herbaceous ground cover (Gilliam and Platt 1999), resulting in detrimental effects on most native wildlife (Engstrom et al. 1984; Brennan et al. 1998). Fire is essential for wildlife, but its direct effects on individual species vary and are thought primarily to include displacement and occasional mortality (Engstrom 2010). Restoration efforts can capitalize on this variation by manipulating fire-return intervals, seasonality, spatial scale, and fire intensity within the ecological context of restored sites. That frequent fire-return intervals maintain the vegetative conditions required by many longleaf pine endemics points to obvious long-term benefits. However, in some settings within the context of restoration, these long-term benefits can also entail short-term costs.

For most wildlife, short-term responses to fire (ranging from the actual fire event to as long as 3 months afterward) generally are not as well documented as long-term responses; however, studies suggest that short-term effects can contribute to wildlife diversity, especially among small mammals. Cotton rats (*Sigmodon hispidus*), cotton mice (*Peromyscus gossypinus*), and oldfield mice (*P. polionotus*) are common inhabitants of longleaf pine forests (Morris, Hostetler, Conner, et al. 2011; Morris, Hostetler, Oli, et al. 2011). Presence of cotton rats can reduce densities of the two smaller rodents (Turner and Grant 1987), but the species exhibit different short-term responses to fire that help to maintain overall small mammal diversity. All three species decline after a fire event, but fire-facilitated predation causes a larger reduction for rat populations than for mice (Morris, Hostetler, Conner, et al. 2011; Morris, Hostetler, Oli, et al. 2011).

Although the fire-return intervals that are beneficial for individual species can vary widely in longleaf pine forests, rare and declining wildlife have more narrowly defined requirements. Most recommendations suggest an interval <3 years, which is consistent with historical averages and also deters the build-up of woody shrubs that would likely have been culled by less frequent but more intensive fires (Means 2006; Cox and Widener 2008; Stambaugh et al. 2011; Glitzenstein et al. 2012). As further evidence of the importance of burn frequency, detailed assessments suggest that the buildup of grass and leaf litter that occurs in the 2–3 years between fires can produce conditions that are harmful for many ground-nesting species (Cox and Widener 2008; Jones et al. 2013). The most appropriate season for burning longleaf pine forests is debatable. Some ecologists and land managers prefer growing season (March through August) fires because they follow the historical, lightning-initiated fire regime and often produce desired vegetation responses (Platt, Evans, and Rathbun 1988; Huffman 2006; Stambaugh et al. 2011). Others focus on a broader range of seasons or the transition between seasons to maintain desired conditions for vegetation (Hiers et al. 2000; Cox and Widener 2008).

Data on the effects of season of burn on wildlife populations are mixed. Growing-season fires can be harmful to ground nesting birds (Rosene 1984). However, Little, Streich, et al. (2014) concluded that low-intensity growing-season fires have minimal impacts on wild turkey (*Meleagris gallopavo*) nesting. In contrast, dormant season (November to February) burns can eliminate nests of birds that breed early in the year (Cox and Slater 2007) and can remove understory vegetation used by rare wintering birds (McNair 1998). Although little is known about the direct effects of season of burn on other vertebrates, Humphries and Sisson (2012) described concerns about the effects of prescribed-burn timing on amphibians that migrate to and from isolated wetlands during the breeding seasons (fall, winter, and early spring). Because migrating amphibians typically move during rain events and at night, the risk of direct mortality from a prescribed fire is likely small regardless of season. Importantly, growing-season burns are most likely to burn through isolated wetlands, thereby maintaining the wiregrass and other herbaceous ground cover within and around these landforms (Brewer and Platt 1994; Outcalt 1994) that provide cover and foraging sites for amphibians (Gorman et al. 2009; K. Jones et al. 2012).

Although seldom considered in discussions of fire effects, prescribed burning has the potential to reduce predation of avian nests. Jones et al. (2004) found that raccoons, a common nest predator of ground-nesting birds (Staller et al. 2005; Melville et al. 2014), were 62% less likely to occupy recently burned longleaf pine stands than stands that were burned more than a year earlier. They hypothesized that soft mast is absent in recently burned stands, causing raccoons to forage elsewhere.

Prescribed fires can also indirectly affect wildlife by facilitating predation, a phenomenon that is influenced by season of burn. Small mammals seem particularly prone to predation after fire, presumably because they are forced to forage in areas where vegetative cover is absent (Komarek 1967; Morris, Hostetler, Conner, et al. 2011). Burning during the growing season can allow more rapid re-colonization after fire because vegetation regenerates more quickly than it would after dormant season burning, but confounding effects (such as burn unit size, burn unit shape, and weather conditions) make this difficult to assess without direct experimentation. Although prescribed burning increases the vulnerability of gopher tortoise nests to predation (Dziadzio, Smith, et al. 2016), it is essential for maintaining suitable burrow and nest sites (Aresco and Guyer 1999; Dziadzio, Smith, et al. 2016). Moreover, nests in the small openings that were created and maintained by fire experienced lower predation than those at burrows, regardless of time since burn (Dziadzio, Smith, et al. 2016). These examples highlight the trade-offs that are inherent in efforts to manage for an ecosystem rather than an individual species.

The extent of fire can also influence wildlife populations by altering rates of recolonization. An example is at the Joseph W. Jones Ecological Research Center at Ichauway, where relatively small (generally <50 ha) prescribed fires severely reduce small mammal populations by facilitating predation (Morris, Hostetler, Conner, et al. 2011). However, their numbers recover rapidly with green-up because source populations are available near the burned area if the fire is small. In contrast, the upland longleaf pine forests on Eglin Air Force Base have fewer and less diverse small-mammal populations than other longleaf pine forests; these low numbers could be the result of the low productivity of sandhill soils, a phenomenon that a large-area burn would exacerbate by hindering subsequent immigration (Conner et al. 2011).

Structural Legacies and Faunal Engineers

Burrow-Dwelling Wildlife

Nearly two-thirds of the vertebrates associated with longleaf pine forests either excavate burrows for themselves, use existing burrows, or inhabit stump holes and belowground root systems (Table 12.1). These structures protect wildlife from temperature extremes, fire, and predators—thereby contributing to the high diversity in longleaf pine ecosystems. Species such as the gopher tortoise (Figure 12.2) and southeastern pocket gopher, which excavate extensive burrows, are considered "ecosystem engineers" because their activities enhance habitat heterogeneity (Jones et al. 1994) in addition to aerating soil, mixing soil nutrients, and exposing bare mineral soil for potential establishment of plant seedlings (Simkin et al. 2004). Gopher tortoise burrows provide shelter for >360 other species (Jackson and Milstrey 1989), including the tortoise commensal noctuid moth (*Idia gopheri*), onthophagus tortoise commensal scarab beetle (*Onthophagus polyphemi*), Florida mouse (*Podomys floridanus*), and two rare snakes species, the eastern indigo snake (*Drymarchon couperi*) and eastern diamondback rattlesnake (*Crotalus adamanteus*). Southeastern pocket gopher burrows can be up to 150 m long and provide shelter for numerous frogs, lizards, small mammals, and endemic invertebrates

FIGURE 12.2 A gopher tortoise within its burrow. (Photograph courtesy of Ron Grunwald.)

(Mount 1963; Funderburg and Lee 1968; Skelley and Kovarik 2001; Blihovde 2006). Although not restricted to longleaf pine, the oldfield mouse also creates burrows used by other wildlife species, as does the nine-banded armadillo (*Dasypus novemcinctus*), a recent arrival to the Southeast (Taulman and Robbins 2014).

Numerous species also use stump holes as refugia; Means (1985, 2005) observed eastern diamondback rattlesnakes, eastern kingsnakes (*Lampropeltis getula*), garter snakes (*Thamnophis sirtalis*), black racers (*Coluber constrictor*), a gray ratsnake (*Pantherophis spiloides*), and coachwhips (*C. flagellum*) at longleaf pine stump holes in northern Florida. In addition to their value as shelter, stump holes can indirectly benefit snake populations by providing habitat for small mammals and other common prey (Timmerman and Martin 2003; Means 2005; Steen, Steen, et al. 2010). The seasoned stump from a longleaf pine can arguably provide the best subterranean cover for snakes because its resistance to heartwood decay allows it to persist for many years. Fires burn away the soft sapwood along its roots and work into the extensive, subterranean root system creating complex belowground refugia (Heyward 1933; Wahlenberg 1946). Although slash and loblolly pines are less fire resistant and lack the extensive tap roots of longleaf pine, large stumps of these pines also provide habitat for snakes (Means 2006). In upland forests, stump occurrence is limited by past and current removal practices. In the early 20th century, longleaf pine stumps were harvested to extract oleoresin (Wahlenberg 1946). On some sites such as Goethe State Forest in Florida, commercial extraction of stumps for oleoresin is still occurring (Eisenbies et al. 2009; Walmsley and Godbold 2010); on others, they are removed to reduce damage to equipment (J. Cox, personal communication).

Standing dead wood and coarse woody debris are also credited with enhancing wildlife diversity in longleaf pine forests (Kauffeld 1957; Loeb 1996; Means 2006). Means (2006) found that coarse woody debris offers shelter to migrating amphibians; and hatchling gopher tortoises regularly

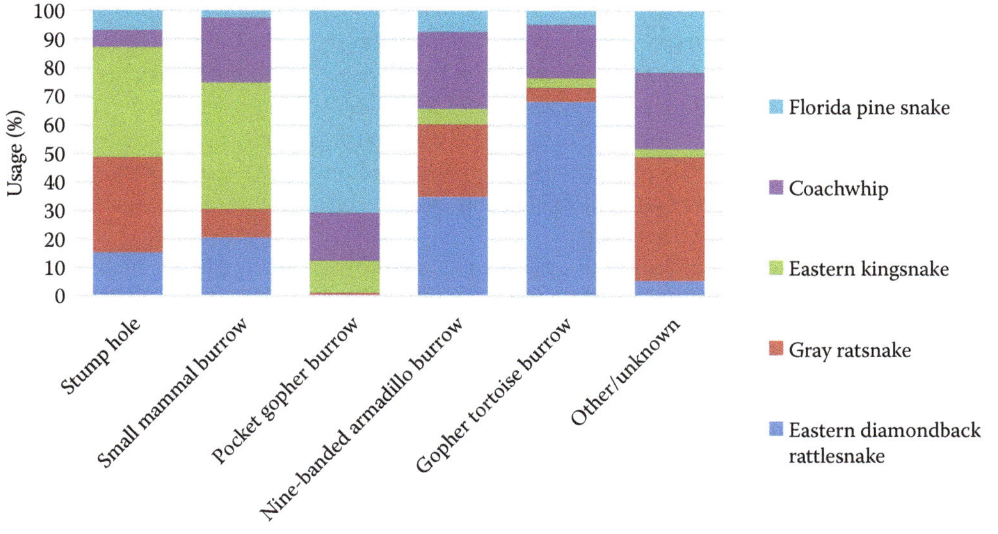

FIGURE 12.3 Proportion of belowground refuges used by five species of large snakes inhabiting a longleaf pine forest in southwestern Georgia: eastern diamondback rattlesnake, gray ratsnake, eastern kingsnake, coachwhip, and Florida pine snake. (From L. Smith, unpublished data.)

construct initial burrows under small, fallen limbs and other woody debris (T. Radzio, personal communication). The retention of large standing dead wood provides roosting habitat and maternity sites for bats (Miles et al. 2006) and roosting and nesting structures for about 20% of the birds that inhabit longleaf pine forests (Engstrom 1993), including the wood duck (*Aix sponsa*), American kestrel (*Falco sparverius*), and brown-headed nuthatch (*Sitta pusilla*). Snags also provide nest sites and refugia for southeastern flying squirrel (*Glaucomys volans*) and pileated woodpecker (*Hylatomus pileatus*), which can help offset competition for the cavities excavated by the red-cockaded woodpecker in living trees (Saenz et al. 1998; U.S. FWS 2003).

The importance of structural diversity provided by animal burrows, stump holes, and woody debris within longleaf pine forests is demonstrated by the refuge selections of large snakes in southwestern Georgia (Figure 12.3). All species spent more than a third of their time in belowground refugia (Hoss et al. 2010; Steen, Linehan, et al. 2010; Miller et al. 2012; Howze and Smith 2015). Florida pine snakes predominantly used southeastern pocket gopher burrows, whereas kingsnakes and coachwhips relied on burrows dug by oldfield mice and other small mammals (Steen, Linehan, et al. 2010; Miller et al. 2012). Gray ratsnakes used exposed root systems of hardwood trees, but also frequently used stump holes (L. Smith, unpublished data). In southwestern Georgia, eastern diamondback rattlesnakes most often used burrows dug by gopher tortoises and nine-banded armadillos (Hoss et al. 2010), but in a study in north Florida, they used stump holes (Means 2005).

Grazing

Savanna and grassland ecosystems have co-evolved with large grazers that move in herds across many landscapes (Stebbins 1981), undoubtedly including early longleaf pine ecosystems. With the extinction of large grazers, which likely followed the arrival of humans (Burney and Flannery 2005; Faurby and Svenning 2015), vegetation of the Coastal Plain likely shifted to a more wooded system with shrinking coverage by grasses (Webb 1974; Van Lear et al. 2005; Gill et al. 2009; Bakker et al. 2015). Evidence from archeological sites shows that bison were in the longleaf pine range in the Pleistocene (Webb 1974). Apparently widespread in the 1500s and 1600s, they persisted into the 1700s (Rostlund 1960) but were gradually replaced by domestic cattle (*Bos taurus*) and feral hogs (*Sus scrofa*) that European settlers brought to the Southeast (Grelen 1978).

The specific relationship between early grazers and the longleaf pine community cannot be known without costly and controversial reintroductions. However, we do know that grazing reduces woody encroachment and thereby plays an important role in other savanna ecosystems (Bond et al. 2003), increasing both grass productivity (Pandey and Singh 1992) and ground cover plant diversity (Olff and Ritchie 1998). Furthermore, fire and grazing work synergistically elsewhere to create more stable savannas that are dominated by grasses and trees (van Langevelde et al. 2003). Accordingly, grazing could be both an important disturbance and an underutilized tool in the maintenance and restoration of the longleaf pine ecosystem.

Few studies have examined the use of grazing as a disturbance for longleaf pine restoration or maintenance; the studies that exist suggest that moderate grazing by domestic cattle has no influence on plant diversity (Brockway and Lewis 1997; Albin 2014), possibly prevents sand pine (*P. clausa*) invasions (McCay 2000), and increases the net primary production of grasses in recently deforested longleaf pine ecosystems (Gaines et al. 1954). However, unlike the movements of large herds of migratory grazers, domestic cattle are confined to limited areas for extended periods, which can lead to overgrazing and reduced pine seedling survival (McCay 2000), soil compaction (Linnartz et al. 1966), and increased invasive plants (Williams 1952). Still a larger threat to the longleaf pine ecosystem is the proliferation of invasive feral hogs (Siemann et al. 2009), which can decimate pine seedlings, hinder pine regeneration efforts, reduce ground cover, and alter soil nutrients (Lipscomb 1989; Siemann et al. 2009).

RESTORATION AND MANAGEMENT FOR RARE, THREATENED, AND ENDANGERED SPECIES

Single Species Restoration Targets

The endangered red-cockaded woodpecker, often the focus of longleaf pine restoration efforts (Breckheimer et al. 2014), forages over large areas (70–140 ha) in mature timber stands and excavates cavities that are subsequently used by many other species (Blanc and Walters 2008). Frequent fire affects habitat quality, both for red-cockaded woodpeckers and for scores of other rare species (Jackson 1994; James et al. 2001). Dependence on live, mature trees for cavity building in extensive tracts of open canopy pine forests makes the red-cockaded woodpecker an important umbrella species as well as a potential indicator of well managed forests (Blanc and Walters 2008; Breckheimer et al. 2014).

Hardwood midstory removal is a procedure that is sometimes used to restore habitat conditions for the red-cockaded woodpecker (Conner et al. 1996). Hardwoods generally have been culled from sites where years of previous fire exclusion resulted in their dominance of the midstory. Treatments generally have been applied to small areas that contained cavity trees (generally <20 ha), or more infrequently to the area immediately surrounding an individual cavity tree (Conner et al. 1996). In 1998, for example, midstory oaks were removed within 25 m of active cavity trees on the Withlacoochee State Forest in central Florida (Figure 12.4). Treatments were applied to about 35 occupied sites, and the 0.5–1.0 ha openings were considered a stop-gap measure until open conditions could be established more broadly using prescribed fire (V. Morris, personal communication).

In 1994, Eglin Air Force Base began an extensive attempt to restore suitable woodpecker habitat by removing hardwoods across a large area (Provencher, Herring, et al. 2001). Previous work on the site suggested a link between hardwood densities and reduced red-cockaded woodpecker productivity (Hardesty et al. 1997). Hardwoods were removed from about 425 ha that contained clusters of cavity trees (Petrick and Hagedorn 2004). Additional hardwoods and sand pines were removed over 11 030 ha in an area where fire exclusion had been in effect for >20 years (Petrick and Hagedorn 2004). Hardwoods were also removed from 30 research plots (each measuring 81 ha) in conjunction with detailed monitoring to assess the effects of different removal methods (fire, chainsaw, and

(a)

(b)

FIGURE 12.4 Red-cockaded woodpecker habitat at the Citrus tract of the Withlacoochee State Forest near Brooksville, Florida: (a) Before hardwood reduction, and (b) after hardwood reduction. (Photographs courtesy of Vince Morris.)

herbicide) on soils, vegetation, and wildlife. Fire management was implemented concurrent with hardwood removal treatments.

Results included an expansion of active woodpecker territories into the treated areas and habitat improvements for Bachman's sparrow and other bird species (Provencher et al. 2002; Petrick and Hagedorn 2004). Follow-up assessments conducted 15 years later found that prescribed fire was likely as effective in restoring wildlife communities as the mechanical hardwood treatments had been (Steen, Conner, et al. 2013; Steen, Smith, et al. 2013). However, the mechanical treatments initially elicited a more rapid response for the red-cockaded woodpecker, which could be an important consideration when managing small populations of this imperiled species.

In light of the extensive loss of longleaf pine habitats, restoration generally is focused on creating conditions that favor desirable (endemic and specialist) species. Promoting habitat suitable for one species can lead to concomitant declines for other species that have different habitat and management needs. For example, hardwood removal can discourage the use of habitats by migratory songbirds (Wood et al. 2004), an issue that is possibly mitigated by the ability of most migratory songbirds to thrive in a range of forest types (Hunter et al. 1994; Tucker et al. 2003). Hardwoods provide important food resources for Sherman's fox squirrels and other rare species that primarily occur in longleaf pine forests (Perkins et al. 2008), where midstories contain many mast-producing fire-adapted hardwoods.

The effects of hardwood removal on these and other nontarget species are not well documented, but there is increasing support for a more ecosystem-based approach to hardwood removal treatments (Hiers et al. 2014). For example, a large-scale hardwood removal now underway on the Ocala National Forest in central Florida strives to improve conditions for red-cockaded woodpeckers and other longleaf pine inhabitants by focusing narrowly on oaks that have encroached into fire-excluded upland areas from adjacent scrub habitat (C. Sekerak, personal communication). The treatments are tailored to achieve the densities of hardwoods that were documented by early timber sales, historical aerial imagery, and maps of oak forests dating back to the early 1900s. Hardwoods retained in treatments include all non-oak native tree species as well as bluejack (*Q. incana*), sand post (*Q. margaretta*), and running oak (*Q. pumila*). Removal of scrub oaks (*Q.* spp.) is limited to those that occur singly and are <20 cm d.b.h. They will resprout quickly from rootstock, thereby contributing mast as a shrub.

Whether a single species, such as the red-cockaded woodpecker, serves as an adequate umbrella species is another question to consider when this species is the major focus of restoration efforts. Breckheimer et al. (2014) tested this assumption by comparing the dispersal habitat used by red-cockaded woodpeckers with that of two rare species, the St. Francis' satyr butterfly (*Neonympha mitchellii* var. *francisci*) and the Carolina gopher frog (*Lithobates capito* var. *capito*). Dispersal habitat that was modeled using movement data for each species revealed a substantial overlap in the spatial distributions of important areas despite different life history strategies. They concluded that umbrella species can help conserve or restore connectivity simultaneously for multiple threatened species.

However, the red-cockaded woodpecker does not necessarily provide appropriate synergistic conservation benefits for the brown-headed nuthatch, a longleaf pine endemic and species of conservation concern in portions of its range. Nuthatches, which are uncommon in the mature sandhill forests that are occupied by red-cockaded woodpeckers in the Ocala National Forest, are much more common nearby in flatwood forests dominated by younger slash pines (Cox et al. 2012). Distinctive habitat requirements for the two species can also be seen in the fact that nuthatches disappeared from sites in Missouri and southern Florida decades before the disappearance of red-cockaded woodpeckers (Cox and Slater 2007; Slater et al. 2013). In addition, Kale et al. (1992) found no nuthatches in some large Florida conservation areas that supported >30 woodpecker territories. The red-cockaded woodpecker provides umbrella services for this species elsewhere within the longleaf pine range, but as noted by Simberloff (1998), the use of "umbrella species" has limitations and is a blunt tool at best.

Ecosystem Approaches to Restoration

Ecosystem approaches to restoration regularly emphasize ecological processes, large spatial scales, key structural components, and the use of indicator wildlife species to assess progress and success (Simberloff 1998). Key ecosystem structures and processes that characterize longleaf pine ecosystems include the presence of many old trees and herbaceous ground cover (Platt, Evans, and Davis 1988), frequent low-intensity fire (Brockway, Outcalt, Tomczak, et al. 2005), and infrequent catastrophic events such as tropical storms and higher intensity fires (Myers and Van Lear 1998; Provencher, Litt, et al. 2001). These factors likely produced a range of tree densities and ground cover conditions in historical longleaf pine sites, and the large spatial scale typically used in ecosystem management makes provisions for the many types of natural communities that might be embedded within the longleaf pine ecosystem.

Ecosystem restoration efforts for the wildlife associated with longleaf pine forests generally focus on thinning over-stocked timber stands to allow sunlight to reach the ground, establishing frequent fire regimes, and promoting lush ground cover conditions that favor grasses and forbs rather than palmetto or hardwood species. Interestingly, the large-scale hardwood-removal treatment regime at Eglin Air Force Base was initially considered to be an ecosystem approach (Petrick and Hagedorn 2004) even though it subsequently drew criticism for its removal of fire-adapted oaks (Hiers et al. 2014). At such an infertile xeric site, the accumulation of litter was driving the change in ground cover rather than the presence or absence of oaks; thus, frequent fire restored ground cover (see Chapters 6 and 11).

Whether the ecosystem approach will serve restoration needs in today's fragmented landscape is uncertain. An evaluation performed for Eglin Air Force Base concluded that the spatial pattern of remaining longleaf pine forests could compromise the ability of reintroduced fire regimes to restore the landscape (Gordon et al. 1997). Movement of prescribed fires across adjacent land forms (for example, sandhills with embedded isolated wetlands or longleaf pine next to bottomland hardwood forests) is a critical process for many rare species, yet often must be augmented by mechanical or herbicide treatments. Moreover, fire is not being applied across the range of conditions that took place historically. Prescribed fires are banned during periods of drought even though these were the times, historically, when fires were likely most extensive. Because the period of fire exclusion has been lengthy, restoration will be best accomplished in conjunction with mechanical or herbicide treatments as opposed to prescribed fire alone.

Use of indicator wildlife species or species assemblages is a common practice for monitoring the effectiveness of longleaf pine restoration (Litt et al. 2001; Provencher et al. 2002; Steen et al. 2013a, b). However, selecting appropriate indicators is a complicated process that frequently leads to missteps if not approached carefully (Simberloff 1998; Niemi and McDonald 2004). The red-cockaded woodpecker and Bachman's sparrow have been recommended as indicator species for other bird species of conservation concern (Hannah 2015), but more thorough assessments are needed as suggested by the exhaustive work that went into the identification of indicator species in other ecosystems (Niemi et al. 1997). Artificial cavities can be used to increase red-cockaded woodpecker populations, but similar benefits might not be seen for other cavity-nesting birds in the area (Blanc and Walters 2008), and neither the red-cockaded woodpecker nor the Bachman's sparrow is likely to be an effective indicator of habitat condition for other vertebrates (Litt and Gordon 2003). Another difficulty arises when indicator species are absent from a restoration site, a condition that is likely to be common when they are poor colonizers.

Selecting a suite of species as indicators of habitat condition can provide information on a wider array of habitat features than use of a single indicator species (Steen, Conner, et al. 2013; Steen, Smith, et al. 2013). Different organisms often reflect specific habitat characteristics based on their physiology, natural history, and spatial requirements. Those that require large spaces and habitat connectivity within the landscape can provide insights into habitat condition attributes at a large scale. Those with smaller spatial requirements but with complex life histories (such as semi-terrestrial amphibians) can shed light on smaller-scale connectivity. Organisms that can respond quickly to management or restoration, such as birds that have high dispersal ability and lizards or small mammals that have high reproductive rates, can provide opportunities for rapid assessment and adaptive management. However, empirical data on habitat use can provide structural metrics on appropriate conditions for wildlife that are more useful for managers than data that are limited to the presence or absence of species.

WHAT WILDLIFE CAN TELL US ABOUT HABITAT CONDITION IN A LONGLEAF PINE SAVANNA

The presence of indicator wildlife species is often used to evaluate the success of management or restoration of longleaf pine forests. However, simply documenting species presence does little to inform managers about the range of habitat conditions (pine basal area, canopy cover, and percentage of herbaceous cover) that are needed to sustain wildlife populations through time. Knowledge of this range of conditions would be useful in assessing whether a site is within a "desired condition" for wildlife. The general recommendations for desired conditions in longleaf pine forests are based on a few individual studies and expert opinion; however, spatially explicit data are rarely used to examine the range of conditions that favor wildlife in historical longleaf pine sites (McIntyre 2012).

As part of a research initiative funded by the Gulf Coastal Plain and Ozarks Landscape Conservation Cooperative (GCPO LCC 2013), we used spatially explicit habitat data for longleaf pine indicator species from the 11 700 ha Joseph W. Jones Ecological Research Center to quantify relationships between indicator species and habitat-condition metrics. The indicator species were Bachman's sparrow, brown-headed nuthatch, northern bobwhite, southeastern pocket gopher, eastern diamondback rattlesnake, Florida pine snake, gopher tortoise, Sherman's fox squirrel, and pine warbler (*Setophaga pinus*). Habitat-condition metrics included pine basal area (m^2/ha), large pine basal area (≥35.5 cm d.b.h.), percent herbaceous cover, canopy cover, and shrub cover.

Indicator species locations and long-term habitat monitoring data were used to develop individual species models and a composite species-habitat model using a Mahalanobis distance approach (Clark et al. 1993). We compared averages and standard deviations of condition metrics from our models to desired condition metrics from the GCPO LCC's draft integrated science agenda (GCPO LCC 2013). Figure 12.5 shows results for one of the species (Bachman's sparrow) and for the composite of all nine species.

Output from the models revealed that some of the desired condition metric ranges were likely too restrictive, suggesting a need to reevaluate habitat metrics for indicator species (McIntyre 2012). Moreover, additional empirical data are needed on habitat condition for wildlife across a gradient of longleaf pine landscapes. To address the considerable overlap of habitat conditions for the nine indicator species, a subset of these species (such as the songbirds, which can be sampled efficiently using point counts) could be used as a surrogate for the entire suite of species.

This approach of using species-habitat models and habitat metrics to estimate the range of conditions that are appropriate for wildlife species can provide defensible, measurable targets for land managers.

Introducing Populations

Fragmentation of the longleaf pine ecosystem means that certain parcels will likely never be used by some wildlife even though habitat restoration is effective. Wildlife populations in small habitat patches that can support only 30–50 individuals will likely succumb to stochastic events and a reduction in genetic diversity over many generations. Some, such as the brown-headed nuthatch, are longleaf pine specialists whose genetic structure may already be affected by habitat fragmentation (K. Han, unpublished data). Although endemic species such as Bachman's sparrow and red-cockaded woodpecker seem able to disperse among habitat fragments (Stangel et al. 1992; Cerame et al. 2014), most other species of conservation concern will likely need intervention, such as translocation, to overcome the distances that separate their populations.

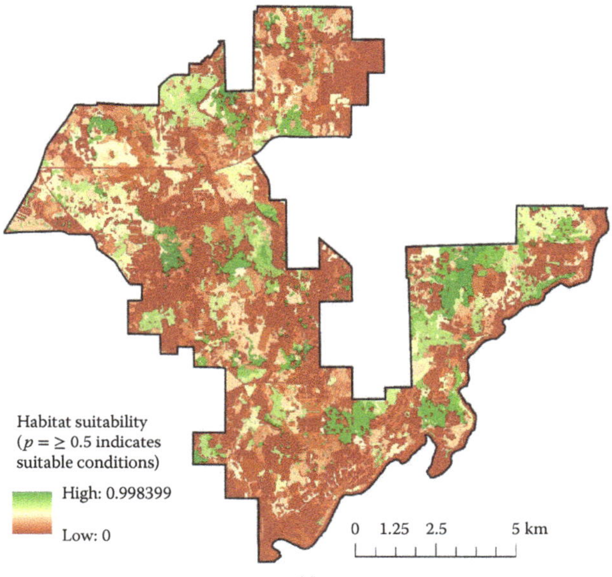

(a)

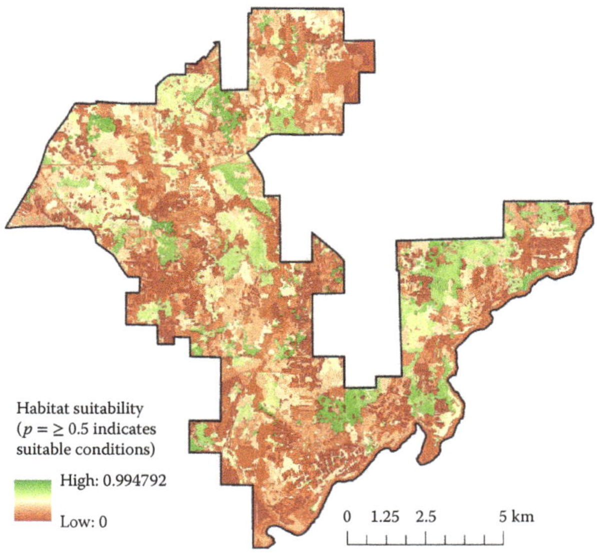

(b)

FIGURE 12.5 Mahalanobis distance model output for habitat suitability for longleaf pine indicator species at the Joseph W. Jones Ecological Research Center in southwestern Georgia: (a) Predicted habitat suitability for Bachman's sparrow; (b) predicted habitat suitability for a composite of longleaf pine indicator species—Bachman's sparrow, brown-headed nuthatch, northern bobwhite, pine warbler, southeastern pocket gopher, eastern diamondback rattlesnake, Florida pine snake, gopher tortoise, and Sherman's fox squirrel.

(Continued)

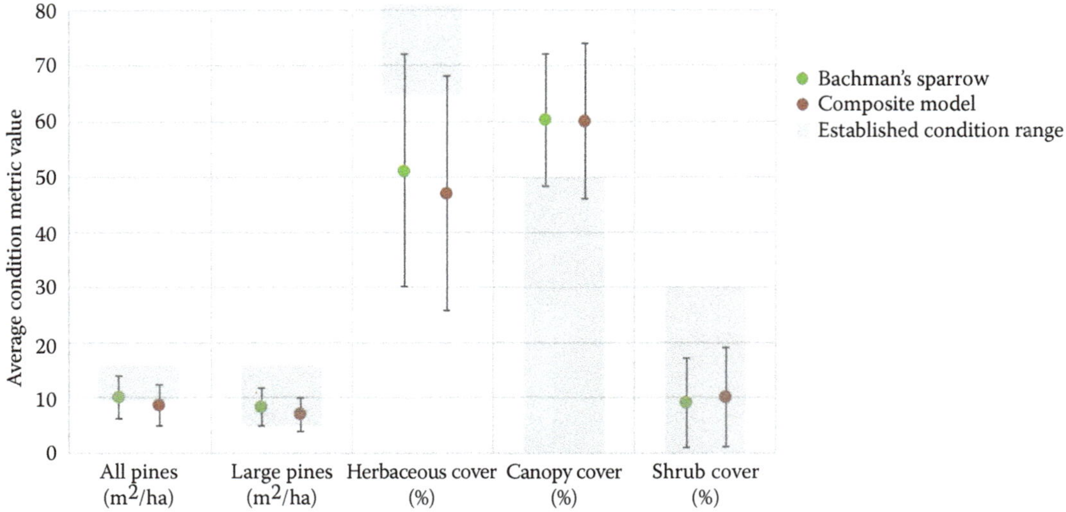

(c)

FIGURE 12.5 (*Continued*) Mahalanobis distance model output for habitat suitability for longleaf pine indicator species at the Joseph W. Jones Ecological Research Center in southwestern Georgia: (c) Average values and standard deviation bars for condition metrics—basal area (m^2/ha) for all pines, basal area (m^2/ha) for large pines (≥35.5 cm d.b.h.), percentage herbaceous cover, percentage canopy cover, and percentage shrub cover—from predicted suitable habitat ($P \geq 0.5$) as compared to condition metric ranges established in the GCPO LCC Integrated Science Agenda. (Data from Gulf Coastal Plains and Ozarks Landscape Conservation Cooperative (GCPO LCC), 2013, Integrated Science Agenda, Version 4. http://gcpolcc.org/assessment; accessed December 2016.)

Although translocation is in a state of infancy for many organisms, it has been proposed or performed for at least nine species (Table 12.1) that historically inhabited the longleaf pine ecosystem (Bendel and Therres 1994; Costa and DeLotelle 2006; Terhune et al. 2006; Dawson et al. 2009; Slater et al. 2013). For species such as red-cockaded woodpecker and Sherman's fox squirrel, reintroduction has been attempted at multiple sites, with high chances of successful establishment when conditions are suitable (Bendel and Therres 1994; Costa and DeLotelle 2006, Dawson et al. 2009). Establishing an important cavity excavator such as the red-cockaded woodpecker can improve habitat for other cavity-nesting species (Blanc and Walters 2008), and it can also help bolster neighboring woodpecker populations (Cox and Engstrom 2001). Translocation of gopher tortoises is now common but success is less certain; >70,000 individual gopher tortoises have been translocated in Florida to mitigate hundreds of residential and commercial development projects (Sullivan et al. 2015). Because this saves individuals without preventing a net loss of habitat, defining the success of such translocation activities has proven difficult (Bauder et al. 2014). Further, moving gopher tortoises into existing populations has raised concerns about stress on existing populations and the spread of diseases (Berish et al. 2000; McGuire et al. 2014). However, the recent emphasis on establishing viable populations across the range to reverse population declines (U.S. FWS et al. 2013) will likely continue the focus on strategic translocation efforts for this species.

Efforts to reintroduce the eastern indigo snake, a federally threatened species that historically inhabited much of the longleaf pine range, began in 2010 at the Conecuh National Forest in Alabama. Offspring from populations in eastern Georgia were used for the reintroduction (Stiles 2013). Initial monitoring indicated that individuals selected open canopy longleaf pine stands and used gopher tortoise burrows for refuge during the spring, summer, and autumn months; this suggests that the use of prescribed burning to create and maintain open-canopied habitats and gopher tortoise populations will be critical for the successful reestablishment of eastern indigo snake populations. Reproduction of released individuals was confirmed (Stiles 2013), but the long-term viability of the population is unknown because this large, highly mobile species is difficult to monitor (Hyslop 2007). Recent genetic analyses have raised questions about the propriety of releasing individuals from eastern Georgia to Alabama (Krysko et al. 2016); however, source populations in the western part of the range are extremely limited and recovery of the species is unlikely without the reintroductions.

A model for reintroducing rare amphibian populations is underway in the Apalachicola National Forest, where striped newts appear to have disappeared from several isolated wetlands that are embedded in longleaf pine forests. A 5-year reintroduction plan developed by the Coastal Plains Institute (R. Means, personal communication) includes an initial assessment to confirm the absence of populations, an assessment of diseases that could have decimated the population initially, and establishment of a captive breeding population using genetically similar populations. Since drought is thought to have exacerbated declines, liners were placed in the wetlands selected as release sites to retain water for appropriate periods. Monitoring is a large component of the plan.

In 2013, larvae produced by the captive striped newt population (n = 58) were released into a single pond, and bucket trap arrays later captured three efts (juveniles) exiting the pond. Another 433 larvae were released at four sites in 2014, and the proportion of efts captured leaving the ponds was much higher. The first adult returning to a pond was captured in 2015 (Means et al. 2015).

Translocation efforts conducted for the northern bobwhite (*Colinus virginianus*) have also been successful throughout the former range of longleaf pine. Some translocation distances were short (<5 km) and were designed to augment established populations (Terhune et al. 2006). Other attempts spanned several states and subjected translocated populations to starkly new environmental conditions; for example, individuals were moved from northern Florida to pinelands in New Jersey and Maryland (Chanda et al. 2009). The survival and productivity of populations that were subjected to such long-distance translocations will need to be assessed over many years.

The general recommendations and insights emerging from the introduction efforts involving longleaf pine ecosystem inhabitants are to (1) move organisms the shortest appropriate distance to reduce stress, transport costs, genetic variability, and potential for disease transmission; (2) improve the likelihood of translocation success by maximizing the number of individuals that are released; (3) concentrate on habitat quality and long-term commitment to management and monitoring on recipient sites; and (4) where possible, approach translocation at a region-wide scale and across agency boundaries (Saenz et al. 2002).

Although debate continues about the relative benefits of species-based approaches versus ecosystem approaches, aspects of both are likely needed. For example, the effort to develop high-quality foraging habitat for the red-cockaded woodpecker has largely focused on forest structure as measured by the basal area and size distribution of trees. More recently, broader ecosystem functions that involve the frequent use of prescribed fire and its effects on available calcium are also included in discussions about habitat quality for the species, and red-cockaded woodpecker populations in areas with a higher percentage of wiregrass in the ground cover were more stable than areas with saw palmetto (*Serenoa repens*) and woody shrubs (James et al. 2001). Forest structure variables are still considered to be extremely important, but not to the exclusion of the productivity that is characteristic of ground cover vegetation (James et al. 2001).

THE FUTURE

INCENTIVE PROGRAMS FOR MANAGEMENT OF RARE LONGLEAF PINE SPECIES

Private lands make up about 60% of the terrestrial area found in the United States (Eno et al. 2006). These properties provide habitat for most (80%) of the federally listed species found in the United States, 35% of which are wholly dependent on private lands (Langpap 2006). Habitat on private lands is clearly important, and programs that strive to conserve them have grown tremendously since the late 1990s. Rather than enforcing regulatory conditions, these voluntary programs either provide assurances or incentives to help promote habitat conservation (see Chapter 3). Because they often help to conserve habitat at reduced costs, they are generally much better received by landowners than regulatory enforcement actions (Langpap 2006).

The Safe Harbor Program is the largest public-private program that provides assurances to landowners within the former range of longleaf pine. Safe Harbor Agreements are voluntary agreements designed both to reduce the regulatory responsibilities that are inherent in endangered species protection efforts and encourage private property owners to manage lands in a way that restores, enhances, or maintains habitat for rare plants and animals. Within the longleaf pine ecosystem, potentially beneficial management activities include prescribed burning, developing longer timber rotations, maintaining older (residual) trees, restoring ground cover, and installing artificial cavities in the case of the red-cockaded woodpecker. More than 150 000 ha have been enrolled, and the program has led to restoration of red-cockaded woodpecker populations on many private lands while also enhancing connectivity among populations on public lands (Trainor, Walters, Urban, et al. 2013).

Another such program uses Candidate Conservation Agreements to provide assurances to landowners. Candidate species are those that might warrant listing as endangered or threatened, but formal listing has been delayed by higher priority issues or agency funding constraints. Candidate status represents a holding pattern. However, because it also serves as a notification to landowners, it can be used to encourage habitat management and restoration through voluntary agreements. Much like the Safe Harbor Program, this program can lead to fewer regulatory responsibilities if the species is ultimately listed and can avert some of the negative consequences that have resulted from species listings, such as the practice of preemptively cutting longleaf pines trees when the red-cockaded woodpecker was listed (Zhang and Mehmood 2002).

Conservation easements are a government incentive that can help conserve and restore habitat for rare species. Easements are voluntary, legally binding agreements between a landowner and a qualified land trust or government agency. The agreements allow landowners to retain ownership of a property while protecting ecological resources by permanently limiting some land uses such as mining, constructing new buildings, or engaging in unsustainable timber harvesting. The limits imposed on a property typically reduce its overall resale value, and that devaluation can be claimed for tax benefits.

Although private land initiatives are valuable and have many positive benefits, variation in landowner sentiments can complicate the use of these tools when trying to restore strategic longleaf pine areas. For example, only a quarter of the landowners surveyed in Texas were willing to participate in land conservation measures when the value of incentives offered was small (such as 50% cost share with no incentive payments to offset opportunity costs of alternative land uses (Sorice et al. 2011). Most landowners either required much larger incentives (44%) or were not interested at any level of compensation (31%). The inability to establish restoration sites strategically on private lands can yield a configuration of restored habitat patches that are too small or too disconnected to support species such as the red-cockaded woodpecker (Cox and Engstrom 2001) and gopher tortoise (Hermann et al. 2002).

CONCLUSIONS

Restoration is key to retaining wildlife diversity of the longleaf pine ecosystem. Clearly, prescribed fire and management of encroaching hardwoods are important tools for restoring wildlife habitat, as is promoting structural diversity through ecological engineers and by retention of standing dead wood, woody debris, and stumps. However, when faced with restoration from the ground up, as is the case with America's Longleaf Restoration Initiative (ALRI 2009) and other important incentive programs, these components of the ecosystem may not be present and therefore require considerable effort for reestablishment as well as research to develop effective and efficient methods for restoring wildlife and to determine the temporal scale at which wildlife will respond.

Finally, the value of restoring and establishing old-growth stands needs to be emphasized in discussions about the longleaf pine ecosystem. To our knowledge, there is no comprehensive effort to establish a series of representative old-growth sites on the many public lands that currently support longleaf pine. The information emerging from old-growth stands that have been studied extensively suggests they harbor unique characteristics and resources for wildlife that are not found on many second-growth forests (Platt, Evans, and Davis 1988). For example, data provided in Cox et al. (2001) showed that remnant old-growth longleaf pine stands in southwest Georgia typically provided 18–45 natural cavities for each red-cockaded woodpecker territory. This shows that the cavity limitations that typically exist in second-growth forests (Walters et al. 1992) may not have been nearly as common in old-growth settings. Restoration of a mosaic of old-growth sites distributed across multiple physiographic areas will take decades or centuries to complete, but future generations will certainly applaud such foresight.

ACKNOWLEDGMENTS

We thank Brandon Rutledge, Rob Sutter, D. Bruce Means, and Carrie Sekarak for their thoughtful comments on an earlier draft of this chapter. Jennifer Howze and Jean Brock created the figure documenting land use change across the range of longleaf pine.

Part IV

Restoration and Practical Issues

13 Air Quality and Human Health Challenges to Prescribed Fire

Mark A. Melvin and R. Kevin McIntyre

CONTENTS

INTRODUCTION

Fire plays a central role in all aspects of longleaf pine (*Pinus palustris*) ecology, management, and restoration (Wahlenberg 1946). Because the flora and fauna that inhabit longleaf pine forests have evolved with frequent low-intensity surface fires, fire is an "ecological imperative" for the survival of the ecosystem (Wade et al. 2006). Today, prescribed fire is the modern surrogate for the major ecological process that historically structured and maintained longleaf pine forests and is essential for preserving the conservation values of these ecosystems (Figure 13.1). Prescribed fire is also a cost-effective forest management tool for promoting forest health and maintaining acceptable fuel loads. Because many forestlands of the Southeast accumulate significant fuel loads on relatively short time scales (McNab et al. 1978), they are likely to burn at some point, regardless of whether they are managed primarily for ecological or economic objectives. Using prescribed fire can lessen the risk and severity of wildfire, as well as reduce negative public health impacts from smoke when wildfires do occur. Although the scientific knowledge needed to understand the role that fire plays in the structure and function of longleaf pine ecosystems has been available for decades (Heyward 1939), a land manager's ability to continue to use prescribed fire for management and restoration has never been so tenuous.

This chapter examines a range of social, political, and economic challenges to the use of prescribed fire as a management tool. Although many of these challenges are common throughout the United States, they are of particular relevance to the Southeast, given the importance of fire to its many ecosystems that are of conservation interest. Although smoke management and air quality issues present the most significant challenges to implementing prescribed fire throughout the longleaf pine range, they often are accompanied by other compounding and interacting issues, including rapidly increasing urban sprawl, climate change with more frequent and severe drought, lack of trained prescribed fire practitioners, tightening budgets for fire programs in both the public and private sectors, liability concerns, limited numbers of available burn days, and lack of public acceptance (Kobziar et al. 2015).

BACKGROUND

In the southeastern United States, land managers are faced with many, often complex challenges to the implementation of prescribed fire. These challenges are diverse, and many share the common characteristic

FIGURE 13.1 Smoke from a prescribed burn conducted in a longleaf pine forest for ecological benefit. (Photograph courtesy of Richard T. Bryant.)

of being social or political to some degree. Compared to other regions of the United States, the Southeast has maintained a "fire culture" (a tradition of using prescribed fire as a land management tool). As a consequence, tolerance of prescribed fire use by the general public is relatively high (Jacobson et al. 2001; Winter et al. 2004; Fowler and Konopik 2007). Many Southeastern states have passed laws and regulations that support prescribed fire (Brenner and Wade 2003), essentially conveying to landowners a "right to burn" or "social license" to implement prescribed fire as a land management tool. However, political support is vulnerable to erosion due to many factors, but there is perhaps no single challenge that can regulate prescribed fire use in the future more than the impacts on air quality from smoke and emissions. If smoke from prescribed fire is not properly managed, the potential and scale of negative impacts on public health and safety could be significant. Fire practitioners must rely on modern tactics and tools to minimize these impacts, because the responsibility for any smoke produced from prescribed fires—and any subsequent impacts from emissions—ultimately falls to the fire manager, the landowner, or both.

Overview of Prescribed Fire

The longleaf pine ecosystem of the Southeast is among the most fire-dependent found in temperate climates. For millennia, these forests evolved with frequent, low-intensity surface fires from lightning and from anthropogenic ignitions. At the time of the first European contact, longleaf pine forests were estimated to cover approximately 92 million acres (see Chapter 1). By 1930 most of the longleaf pine forests of the Southeast had been harvested, and the extent of longleaf pine continued to shrink to <4% of the original range by the mid-1990s (Frost 2006). Because the fragmented patches that did remain were largely left to regenerate by succession, subsequent species composition and function changed drastically, with the scale and frequency of the ecosystem's ancient fire regime forever altered.

Lightning-ignited fires are no longer sufficient to maintain the remaining fragmented longleaf pine forests of the Southeast. Instead, southeastern land managers now rely on the strategic use of prescribed fire for maintaining, restoring, and establishing longleaf pine. As a consequence, the Southeast leads the nation in the use of prescribed fire: for example, of the 11.7 million U.S. acres that were treated with prescribed fire in 2014, 8.1 million acres (76%) were in the 13 southeastern states (Melvin 2015). Even though the overwhelming majority of prescribed fire use occurs in the Southeast, it is not enough to meet the region's ecological needs (Figure 13.2). Efforts are needed that go well beyond maintaining the current scale of prescribed fire.

Set in one of the most rapidly growing regions in the United States with increasingly complex patterns of land use (Wear and Greis 2012), the Southeast faces serious future challenges for the application of prescribed fire in longleaf pine ecosystems. Although these constraints cover a range of issues—including policy, legal considerations, and social acceptance of fire—many relate back to implications of smoke on public health and safety (Winter et al. 2004; Melvin 2012; Kobziar et al. 2015). In a recent survey, smoke management was a top impediment limiting the use of prescribed fire by southeastern state forestry agencies (Figure 13.3).

SMOKE, EMISSIONS, AND HUMAN HEALTH AND SAFETY

It is important that fire practitioners understand what key constituents are found in smoke from prescribed fire (Robertson et al. 2014) as well as their possible health impacts. Smoke from prescribed fires can have both acute and chronic effects on human health. Forest fuels are composed of both living and dead vegetation. When this fuel burns (Table 13.1), about 95% of its emissions are composed of carbon dioxide and water vapor, with the remainder made up of a diverse and complex mixture that includes carbon monoxide, particulate matter (PM), and volatile organic compounds that may be ozone precursors (Urbanski 2014). PM is made up of airborne solid particles and liquid droplets

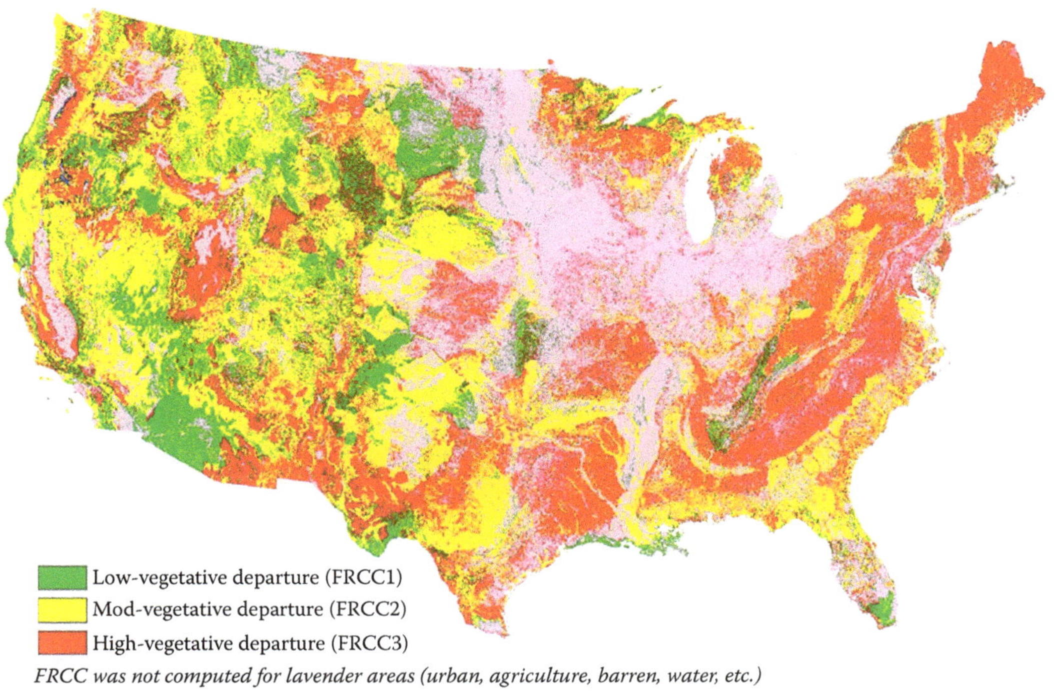

FIGURE 13.2 U.S. map of fire regime condition classes (FRCC), which compare vegetation structure and composition in 2001 to an estimate from the pre-European settlement period. Returning to the earlier vegetative state would require prescribed fire or other intervention, more for areas with higher vegetative departure classes and less for areas with lower classes. (From LANDFIRE. 2008. LANDFIRE national fire regime condition class (FRCC). https://www.conservationgateway.org/PublishingImages/LF_1%200_FRCC_map.jpg. Accessed February 2017.)

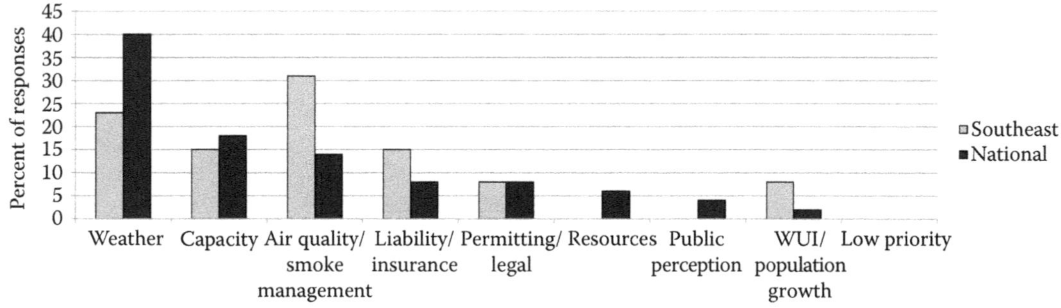

FIGURE 13.3 Comparison between Southeast regional and nationwide responses from state forestry agencies on the factors that limit the use of prescribed fire; note that WUI is the wildland-urban interface. (From Melvin, M, National prescribed fire use survey report. Technical Report 02-15. Coalition of Prescribed Fire Councils Inc., Newton, GA 2015.)

TABLE 13.1

Fire Average Emission Factors for Common Pollutants Resulting from Prescribed Fires and Wildfires in Different Fuel Types for Three Conifer Forests, Western Shrublands, Grasslands, and Boreal Forests

	Prescribed Fire Pounds/Ton (Estimated Uncertainties)					Wildfire Pounds/Ton (Estimated Uncertainties)	
Pollutant	**SE Conifers**	**SW Conifers**	**NW Conifers**	**Shrublands**	**Grasslands**	**NW Conifers**	**Boreal Forests**
Carbon dioxide[a]	3406 (342)	3306 (68)	3196 (78)	3348 (76)	3410 (88)	3200 (38)	3282 (214)
Carbon monoxide[b]	152 (30)	174 (36)	210 (26)	148 (36)	122 (42)	270 (22)	190 (72)
Methane[a]	4.6 (2.2)	6.3 (1.8)	9.7 (2.7)	7.4 (2.7)	3.9 (2.1)	14.6 (1.2)	6.8 (2.9)
Organic compounds other than methane	32.1 (21.8)	37.3 (34.7)	54.0 (31.1)	35 (26.9)	33.5 (23.2)	67.7 (34.7)	46.3 (26.3)
Particulate matter < 2.5 μm[b]	25.2 (8.0)	28.8 (10.0)	35.1 (10.3)	14.1 (1.6)	17.0 (10.2)	46.4 (20.8)	43 (9.6)
Nitrogen oxides	3.4 (1.9)	3.8 (2.1)	4.1 (0.1)	4.4 (1.6)	4.4 (1.6)	4 (2)	2 (0.2)
Ammonia	0.3 (0.3)	1 (1.4)	3.1 (0.8)	3 (2.9)	3 (2.9)	3 (1.5)	1.6 (0.8)
Nitrous oxide	0.3 (0.4)	0.3 (0.4)	0.3 (0.4)	0.5 (0.4)	- (-)	0.3 (0.4)	0.8 (-)
Sulfur dioxide[b]	2.1 (0.8)	2.1 (0.8)	2.1 (0.8)	1.4 (0.3)	1.4 (0.3)	2.1 (0.8)	2.1 (0.8)

Source: Urbanski, S. P., *Forest Ecology and Management, 317*, 51–60, 2014.

SE = southeast; SW = southwest; NW = northwest.

[a] Greenhouse gas.

[b] Criteria pollutant.

that have a wide range of sizes; some can be seen with the naked eye but others are microscopic (Figure 13.4). They are generally classified based on size: $PM_{2.5}$ are particulates < 2.5 μm; and PM_{10} are particulates < 10 μm, including all particles classified as $PM_{2.5}$. These small particles absorb and refract light, which can reduce visibility. The smallest of them can enter into the human body through the respiratory system, causing potential health concerns. Given that up to 70% of smoke particles emitted during prescribed fires are $PM_{2.5}$ (Ward and Hardy 1991), PM is typically the pollutant of concern for fire managers and air quality regulators because of its potential impacts on human health and safety. Medical research has linked exposure to PM to premature death in people with heart or lung disease (CAA 1990). Those with preexisting conditions such as irregular heartbeat, aggravated asthma, or decreased lung function are susceptible to irritation in the airways, which can lead to difficulty breathing (CAA 1990). Children and older adults are most affected, but healthy individuals can also experience temporary symptoms from elevated levels of PM (CAA 1990).

Particles that are emitted directly from known sources (such as fields, smokestacks, and wildland fires) are called primary particles, whereas particles that change or react in the atmosphere (such as sulfur dioxide and nitrogen oxides emitted predominately from power plants and automobiles) are known as secondary particles. As aerosols, they are often $PM_{2.5}$ and can serve as ozone precursors. Sunlight stimulates a reaction between secondary particles and hydrocarbons such as turpenes to form ozone, especially in summer months due to natural (forest) and fossil fuel emissions (Urbanski et al. 2008). At ground level (lower atmosphere), exposure to ozone is problematic for human health because it poses serious threats to individuals with compromised respiratory systems (U.S. EPA APTI 2016). Further, increased ground-level ozone has been linked to vegetation stress in trees used for landscaping and agricultural crops, as well as reduced forest yields (Avnery et al. 2011; Tai et al. 2014).

In addition to climate and weather, factors that can influence particulate emissions include fuel quantity and composition as well as the amount of woody debris from dead trees and branches. In general, emissions from fires vary greatly based on the type and amount (tons per acre) of fuel consumed as well as the ratio of smoldering to flaming combustion: fuels consumed in the smoldering phase are capable of producing several times the amount of pollutants as compared to fuels consumed during the flaming phase (Ottmar et al. 2008). Therefore, heavy fuel loads (> 10 tons per acre) and fuels consumed primarily during the smoldering phase emit significantly higher levels of PM.

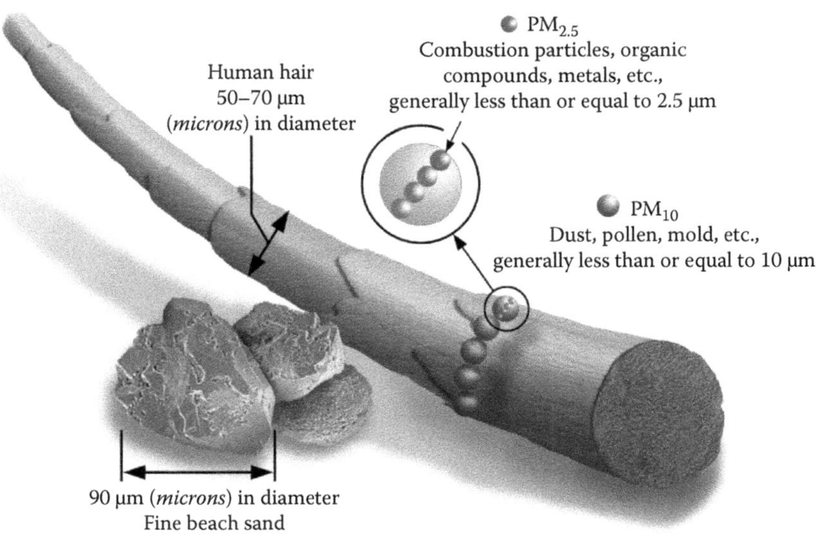

FIGURE 13.4 Two size classes of particulate matter (PM) compared to a strand of human hair and a grain of sand. (Modified from U.S. Environmental Protection Agency (U.S. EPA). 2016b. Particulate matter (PM) basics. https://www.epa.gov/pmpollution/particulate-matter-pm-basics. Accessed February 2017.)

Public safety is another major concern associated with prescribed fire. PM in smoke from prescribed fire can reduce visibility, creating hazards for motorists. Although the majority of prescribed fires conducted in the Southeast are incident free, they sometimes produce smoke that contributes to traffic-related accidents (Mobley 1989). In fact, of all the issues surrounding the use of prescribed fire, smoke on roads has the greatest potential to result in a fatality or serious injury to the public. The worst and most dangerous driving conditions occur when smoke and fog combine to create "superfog" (Achtemeier 2003). Thus, when burning near highways and other smoke-sensitive areas, smoke management concerns will govern the planning process.

Predicting emissions and their potential impacts is a complex task, given that the processes involved depend on numerous interacting variables. Burning forest fuels, never a complete combustion process, is influenced by a dynamic set of variables, including fuel moisture, fuel loading, and numerous parameters of weather that govern fuel consumption during the different phases of combustion (Ottmar 2014). This makes every fire unique and ensures that the complex interaction of these variables cannot be duplicated on a given acre or on any given day.

AIR QUALITY REGULATION AND POLICY

The Environmental Protection Agency (EPA) is the federal agency responsible for the administration of the Clean Air Act, which defines EPA's responsibility for protecting and improving air quality and authorizes the agency to regulate air emissions from stationary and mobile sources (CAA 1970). Along with its 1977 and 1990 amendments, the act also identifies standards and legal requirements that must be met by EPA and other federal agencies, states, and private industries. These regulations, which primarily focus on acute and chronic air quality problems linked to urban and industrial emissions, have been effective in improving urban air quality. Although EPA does not directly regulate prescribed fire, the latest amendments contain several sections that go beyond the traditional urban sources of emissions with both direct and indirect potential impacts on the use of fire.

The 1990 amendment required EPA to set the National Ambient Air Quality Standards (NAAQS) for pollutants to protect or improve human health and the environment (CAA 1990) using two types of standards, primary and secondary (Table 13.2). Primary standards are designed to protect human health, whereas secondary standards are directed toward protection of human welfare, including harm to animals, crops, vegetation, and visibility (CAA 1990).

Although EPA is the regulatory agency that sets the NAAQS, the states are largely responsible for monitoring air quality and enforcing compliance with standards. States establish and maintain a monitoring network to measure levels of "criteria pollutants." Criteria pollutants are regulated pollutants whose allowable levels are determined using science-based criteria for human health or environmental quality. Monitoring sites are required in Metropolitan Statistical Areas (MSAs), defined as municipalities with populations > 65,000 people. Any MSA that fails to meet the NAAQS based on the most current 3 years of monitoring data is classified by EPA to be in "nonattainment" status; once this classification is issued, controls and measures must be established to achieve attainment.

Each state is also required to develop a State Implementation Plan (SIP) to demonstrate that basic air quality management programs are in place for implementing new or revised standards and to identify the emission control techniques that will be employed to attain or maintain primary and secondary standards (CAA 1990).

Section 169 of the Clean Air Act requires states to address adverse visibility impacts in "Class 1 Areas," defined as National Parks and wilderness areas > 5000 acres. The agency's Regional Haze Rule issued in 1999 requires states to establish goals for returning visibility at Class 1 areas to natural background conditions. The rule includes provisions that require states to consider establishing a Smoke Management Program (SMP) to reduce visibility impacts from wildland and prescribed fires. To date, none of the southeastern states have determined that an SMP is needed to address adverse visibility impacts. However, the primary pollutants contributing to regional haze are sulfur dioxide, nitrogen oxides, and PM. But similar to NAAQS compliance, as emissions of the pollutants

TABLE 13.2
2016 National Ambient Air Quality Standards by NAAQS Criteria Pollutant and Regulated Levels

Pollutant	Primary/ Secondary[a]	Averaging Time	Level	Form
Carbon monoxide	Primary	8 hours	9 ppm	Not to be exceeded more than once per year
		1 hour	35 ppm	
Lead	Both	3 months[b]	0.15 μg/m^3	Not to be exceeded
Nitrogen dioxide	Primary	1 hour	100 ppb	98th percentile of 1-hour daily maximum concentrations, averaged over 3 years
	Both	1 year	53 ppb	Annual mean
Ozone	Both	8 hours	0.070 ppm	Annual 4th highest daily maximum 8-hour concentration, averaged over 3 years
Particulate matter < 2.5 μm ($PM_{2.5}$)	Primary	1 year	12.0 μg/m^3	Annual mean, averaged over 3 years
	Secondary	1 year	15.0 μg/m^3	Annual mean, averaged over 3 years
	Both	24 hours	35.0 μg/m^3	98th percentile, averaged over 3 years
Particulate matter < 10 μm (PM_{10})	Both	24 hours	150.0 μg/m^3	Not to be exceeded more than once per year on average over a 3-year period
Sulfur dioxide	Primary	1 hour	75 ppb	99th percentile of 1-hour daily maximum concentrations, averaged over 3 years
	Secondary	3 hours	0.5 ppm	Not to be exceeded more than once per year

Source: U.S. Environmental Protection Agency (U.S. EPA). 2016a. NAAQS Table. https://www.epa.gov/criteria-air-pollutants/naaqs-table. Accessed December 2016.

[a] Primary standards are established to protect human health; secondary standards are established to protect human welfare, which can encompass animals, crops, vegetation, and visibility.

[b] Rolling 3-month period.

are reduced from industrial sources through control strategies, other source categories (such as wildland and prescribed fire) will likely have increasing relative contributions to visibility impacts. This could be another impediment to the use of prescribed fire over the next 20–30 years.

States use SMPs to manage smoke from prescribed fires. SMPs specifically describe basic guidelines, procedures, and requirements for managing and minimizing prescribed fire smoke impacts (U.S. EPA 1998). They are usually developed with input from state forestry agencies and stakeholders from the fire community as well as the state air quality agencies, but the responsibility for program administration usually rests solely with the state air quality agency. If smoke from prescribed fires is directly impacting attainment of the NAAQS, a state may incorporate its SMP into its SIP. To date, none of the southeastern states in the longleaf pine range has done so because prescribed fire has not yet been identified as causing or contributing to violations of the NAAQS.

Once developed and approved, or "certified" by a state, the SMP is submitted to the appropriate regional EPA office and may be used by the state if a prescribed fire event violates air quality standards and is determined to be an "exceptional event." In 2007, EPA finalized the Exceptional Events Rule by defining an exceptional event as an event that affects air quality, is not reasonably controllable or preventable, and is either naturally occurring or arises from human activity that is unlikely to recur at a given location (U.S. EPA 2007). The rule was primarily implemented to address air quality impacts from events such as volcanic activity, chemical spills, earthquakes, and wildfires. Although a prescribed fire can qualify under the rule, the burden of proof is difficult for a state. Among other qualifiers, the state must demonstrate that an SMP was in place or that the burn was conducted using accepted basic smoke management practices. Recognizing the difficulty of making exceptional event demonstrations for prescribed fires, EPA proposed revisions to the Exceptional Events Rule in November 2015 that were finalized in 2016 (U.S. EPA 2016c). Under the new revisions, EPA clarifies that prescribed fire is a human-caused event eligible for treatment as an exceptional event, streamlines a path to show how air quality agencies can satisfy rule criteria for prescribed fire related events, and articulates the ecosystem benefits from prescribed fire and its role in reducing the risk of catastrophic wildfire:

> EPA has continued to express an understanding of the importance of prescribed fire, noting that it can be used to mimic the natural process necessary to manage and maintain existing fire-adapted ecosystems and/or return an area to its historical ecosystem (or another natural ecosystem if the historical ecosystem is no longer attainable) while reducing the risk to public safety and the risk of uncontrolled emissions from catastrophic wildfires. (U.S. EPA 2015b)

SMOKE MANAGEMENT TOOLS

The decision matrix that a fire manager must follow to plan a prescribed burn has become more complex in recent decades. Landowners of just two generations ago could burn on their property anytime they chose without the need to acquire governmental permission; it was the sole decision of the landowner. Historically, many fires were often set late in the day or even at dusk and allowed to burn into the night, with little thought given to smoke. This practice allowed for better fire control because relative humidity and fuel moisture increase in the overnight hours. Thus, fires typically extinguished themselves before dawn and resulted in low-intensity, often patchy burns. Today's smoke management concerns, within the context of a more populated southeastern landscape, prevent most landowners from employing traditional nighttime burns on their land.

Urban sprawl, with the expansive development of residential areas closer to (and often bordering) woodlands, has important implications for fire management. Where these areas meet and merge is called the Wildland Urban Interface (WUI). Lands considered as WUI are further divided into two categories, intermix and interface. Intermix lands are defined as areas with > 16 houses per square mile that are intermingled with wildland vegetation. Interface lands are defined as areas with > 16 houses per square mile in close proximity to large blocks of wildland vegetation (Martinuzzi et al. 2015). The spatial patterns of housing development are expected to continue decentralizing, with forecasts of an additional conversion of 30–43 million acres of land to urban use from 1997 to 2060 (Wear 2013); for more, see Chapter 3. Contributing factors include continued local population growth, immigration to the Southeast from other regions, economic growth, tax policies, land use policies, and impacts of land use planning (Macie and Hermansen 2002). In 2010, the WUI was home to about a third of the U.S. population, and of all Forest Service regions, the Southeast had the highest percentage of its land base classified as WUI (Martinuzzi et al. 2015).

The potential for smoke impacts has made prescribed burning in a WUI area much more challenging than in extensive undeveloped natural areas (Wade and Mobley 2007). Higher population and housing densities in WUIs increase a landowner's exposure to risk and liability from unwanted or unintended smoke impacts.

WUI expansion has also led to increased road density (Terando et al. 2014). The overall improvement of highway transportation infrastructure, along with an increase both in traffic and in highway speeds, have made using prescribed fire more difficult throughout the Southeast. Recognizing the problem of managing smoke on roadways, many southeastern states began to develop and implement burn approval processes in the late 1980s as an effort to decrease the number of smoke-related traffic accidents. With the exception of Louisiana, all states within the historical range of longleaf pine require that landowners obtain an authorization from their state forestry agency before burning (Melvin 2012). By establishing permitting systems and restricting burning to optimal weather conditions, state agencies have been successful in reducing the number of smoke-related traffic accidents and improving overall motorist safety. By incorporating smoke modeling into the authorization process, the state permitting systems have evolved from simple transportation-safety tools into the most commonly used comprehensive programs for managing emissions from prescribed fire and ensuring compliance with the NAAQS.

The degree of planning and implementation difficulty of smoke management is defined by scale (acreage), fuel loading, geography, weather, and the details of where and when a prescribed fire will take place. Collectively, these factors define the complexity of the individual burn and the tools necessary for implementation. A prescribed burn in light fuels located in a rural setting, such as burning agricultural crop residue, may require minimal tools and planning. Larger prescribed burns near WUIs, or in sites with heavier fuel accumulations, may require sophisticated models to evaluate potential impacts. Because each burn is unique in this way, prescribed fire managers must use the appropriate planning and screening tools to identify smoke-sensitive areas so they can take the necessary measures to minimize potential smoke impacts. Therefore, proper smoke screening is both a necessary and critical process.

One universal tool that helps guide a fire practitioner through the planning process is a set of Basic Smoke Management Practices that were developed by the Forest Service and Natural Resources Conservation Service to help fire managers evaluate burn complexity and identify the tools needed to minimize smoke impacts (O'Neill et al. 2011). These guidelines, which can be scaled up or scaled down depending on complexity of the site conditions, were built on six principles important to any prescribed burn: (1) evaluating smoke dispersion conditions, (2) monitoring effects on air quality, (3) record-keeping, (4) communication and public notification, (5) consideration of emission reduction techniques, and (6) sharing the airshed—coordination, planning, and communication among agencies and with the public. By following these established guidelines, fire managers can minimize smoke impacts and reduce the risks associated with using prescribed fire (O'Neill et al. 2011).

Two other technical guides for managing smoke were developed by Wade and Mobley (2007) and Waldrop and Goodrick (2012). Although similar to smoke management practices described above, these guides were developed specifically for managing smoke from prescribed fires in the Southeast, based on its unique geography and climate; they are considered to be the standard protocol for planning or conducting prescribed fire in the region.

FUTURE CHALLENGES

Results from the last 50 years of research have produced an undisputable body of evidence that supports the use of fire as an irreplaceable resource management tool (see Chapters 1 and 2). A primary challenge that the fire management community faces is a combination of misguided past policies and increasing difficulties with social acceptance. For most Americans today, fire in a forest has destructive and fearful associations, and is viewed as something unwanted and unnecessary (McCaffrey 2006). Furthermore, public policy is guided by the paradigm that fires can, and should, be extinguished by human intervention. Of course neither viewpoint is entirely true, and together they create a negative feedback cycle in which public sentiment toward fire is mostly negative, and further limitations are regularly placed on the use of prescribed fire (JWJERC 2015).

One such constraint is the disproportionate assignment of public resources to fire suppression, primarily in the West, to the detriment of funding for prescribed fire in the Southeast, particularly on public lands. For example, in 1995 wildfire costs accounted for 16% of the Forest Service budget, rising to 52% in 2015 and expected to reach 67% by 2025 if current trends continue (USDA FS 2015). The most expensive wildfires usually occur in the WUI; as noted earlier in this chapter, WUI acreage is projected to increase dramatically in the Southeast. Wildfire suppression costs reduce available funding for all other areas of the Forest Service organization, including prescribed fire treatments.

New research will be needed that distinguishes how carefully implemented prescribed fire might mitigate human health impacts from uncontrollable wildfires occurring during adverse climatic conditions or during the summer months that comprise the ozone season. Although it is generally accepted that emissions from a prescribed fire release less carbon into the atmosphere than a wildfire, only a few quantitative studies have compared how these differences might affect human health (Stanturf and Goodrick 2013).

The impacts that climate change could have on southeastern forest fuels will require future study as well (Mitchell et al. 2014). According to a recent climate assessment report on southeastern forest impacts (Vose et al. 2012), the summer and autumn potential for drought, with consequent increases in fire activity, is expected to rise from low to moderate levels in the eastern portion of the region, and from moderate to high levels in the western portion. Because the number of days in a year with acceptable weather parameters for permitting prescribed fire is limited (Liu, Prestemon, et al. 2013), the predicted changes in climate could lead to fewer opportunities for prescribed fire because of conditions that are at times too dry and at times too wet for prescribed fire (Mitchell et al. 2014). This scenario would likely lead to further reductions placed on the application of prescribed fire and result in higher fuel loadings, thereby exacerbating the increased potential for wildfire. Weather and fuels are the two most important factors that influence fire behavior and smoke emissions; changes in climate can directly affect both.

Funding for prescribed fire programs is another growing challenge that natural resource managers face, particularly when working in fire-dependent ecosystems such as longleaf pine. In addition to the growing pressure that the diversion of resources to wildfire suppression places on federal budgets, prescribed fire budgets and staffing at state agencies are also under similar downward pressure (GFC 2015). From 2004 to 2010, fire staffing levels of state agencies in the Southeast decreased by 24% (Stanturf and Goodrick 2013). At the same time that funding is decreasing, costs for prescribed fire continue to grow. Current estimates for contract burning in southeastern forests are about \$25–\$30 per acre (Mississippi State University Extension Service 2016). Using the lower figure, simply maintaining the 4.3 million acres of existing longleaf pine on a 3-year burn rotation would cost about \$36 million per year. The range-wide conservation plan for longleaf pine sets a goal of reaching 8 million acres, an increase of 3.7 million acres (ALRI 2009). This would increase burning costs by another \$31 million per year, for a total annual cost of \$67 million. Although some economies of scale can be gained by burning larger tracts, particularly on public lands, the overall costs of meeting prescribed fire needs for longleaf pine will continue to increase.

Capacity for prescribed fire is a related and emerging challenge, both for maintaining current levels of fire use and for increasing implementation of prescribed fire to meet resource needs. As outlined above, federal and state natural resource agencies face funding shortfalls for fire programs, and "fire borrowing" (reappropriating funds to wildfire suppression) during severe wildfire seasons in the West compounds the problem, constraining financial resources needed for prescribed fire. These funding challenges are directly linked to capacity shortfalls both for maintaining adequate numbers of existing trained personnel and for providing training and recruitment opportunities to meet future needs. The situation is exacerbated by the fact that prescribed fire has never been as technically complex as it is today. Fire managers, who in the past could focus mostly on site-level considerations, are increasingly challenged at broader scales by smoke management in the face of tightening air quality regulations and a rapidly growing WUI. Already expected to have a fairly sophisticated understanding of subjects related to the direct implementation of fire—such as fuel loading, fire behavior, basic on-the-ground fire weather, and safety—they will face emerging technical challenges into the future that will likely require advanced training in meteorology, computer modeling, emissions inventory, and public safety well beyond the burn unit boundaries. More than ever, modern fire professionals require a unique combination of education, training, and experience, an "education fire triangle" (Kobziar et al. 2009).

Training opportunities for prescribed fire practitioners within federal agencies are well developed and administered through the National Wildfire Coordinating Group. Long considered the standard for wildland firefighting, this program fosters a training path for prescribed fire but its primary focus is to prepare professionals for wildfire suppression. Further, these training opportunities are nearly impossible for private landowners and others outside the federal system to access, and often do not integrate with other fire training venues (Kobziar et al. 2009). Many state forestry agencies offer days- to week-long prescribed fire training courses that teach the basics of prescribed fire in a classroom setting. However, moderately rigorous courses that fall between the basic introductory courses offered by states and the comprehensive wildfire-oriented federal curriculum are limited. More opportunities specifically targeted at the application of prescribed fire in the Southeast are needed to adequately train more natural resource professionals.

Increasing prescribed fire capacity in the private sector presents a different set of challenges. Owners of larger tracts often have existing land management staff or the financial resources to hire contractors for prescribed fire. For the group of landowners who have fewer acres or resources, the demographics of its members present additional obstacles. For example, about 76% of the forested acreage in the Southeast is owned by individuals over the age of 55, and 46% is owned by those over 65 (Butler 2016). For many in this group, the physical challenges of conducting prescribed fire will likely be a future barrier.

Liability issues surrounding prescribed fire have long been a concern, especially for private landowners. In a 1999 survey, land managers from Florida reported that liability was a factor in the four most common reasons for not burning (Brenner and Wade 2003), and most current surveys still indicate liability as a leading impediment (Melvin 2012, 2015; Kobziar et al. 2015). Although landowner liability has many forms, including the ever-present risk of fire escape and resulting property damage, the greatest exposure to liability is from unintended smoke impacts, specifically smoke intrusion on roadways that creates transportation safety hazards. As air quality restrictions tighten, human health impacts may emerge as the leading liability concern. If this occurs, whether fire or land managers will be held liable is untested in court. For example, will fire managers be held liable for medical expenses incurred from smoke exposure? If this ever happens, it would represent a new category of "liability" imposed on prescribed fire.

Despite the numerous obstacles over the last 20 years—and particularly in the last decade—new private and public organizations and partnerships have been created in the Southeast to better address the complex issues surrounding fire management (Melvin 2015). Additionally, numerous regional groups are working to improve existing critical habitats and expand conservation efforts using prescribed fire. New and improved state laws and regulations support more prescribed fire, sometimes providing liability protection. Additionally, in recent regulatory actions, including the 2015 ozone standard revision (U.S. EPA 2015a) and the 2016 Exceptional Events Rule revision (U.S. EPA 2016c), EPA has acknowledged the ecological benefits of prescribed fire and the important role prescribed fire can serve in reducing the risk for catastrophic wildfires. Collectively, these factors may positively influence the level of support for prescribed fire in the future, particularly when applied for the conservation of longleaf pine.

Although the U.S. public is unlikely to reach the level of understanding needed to accommodate the use of prescribed fire at the scale that is needed, broader support for its use—where allowable—is possible. To accomplish future goals, integration of policy and science around the use of prescribed fire will be crucial and require collaboration as well as public support. The one certainty of the 21st century is change itself, which will likely occur faster than in any other period of history.

Even though the land use and policy patterns of the past laid a fragile foundation for the use of prescribed fire nationally, the Southeast seems to be well positioned to address future challenges.

In reality, however, the Southeast (and even the United States as a whole) is only a small part of a larger fire deficit problem. Fire is a global issue and if emissions from biomass burning are to be addressed at that scale, the international community will need to work together for a favorable outcome. What the future holds for the role of prescribed fire in maintaining and restoring the longleaf pine and other frequent fire ecosystems will largely depend on what society accepts and policy allows.

ACKNOWLEDGMENTS

We thank Rick Gillam, Morgan Varner, and Lisa Giencke for providing comments, edits, and suggestions on an earlier draft of this chapter. Jessica McCorvey formatted all figures and tables.

14 Monitoring and Adaptive Management

Robert D. Sutter and Brandon T. Rutledge

CONTENTS

INTRODUCTION

Restoration of the longleaf pine (*Pinus palustris*) ecosystem has become a regional conservation priority over the last two decades, with restoration taking place on public and private lands throughout the southeastern United States. In 2014, restoration efforts were underway on an estimated 1.55 million acres, including application of prescribed fire on 1.22 million acres and new plantings on 153,000 acres (Longleaf Partnership Council 2014). This recent increase in investments and focus can be attributed to the development of a regional conservation plan (ALRI 2009), which in turn led to the formation of the Longleaf Partnership Council, a 33-member group of federal agencies, state agencies, nongovernmental conservation organizations, and private organizations that guides and coordinates restoration efforts throughout the Southeast (see Chapter 1).

An essential component of longleaf pine restoration, monitoring assesses outcomes and determines if restoration actions are achieving the desired ecological condition, which can include metrics such as a minimum survival rate of seedlings, a specific forest structure, species richness of ground cover, or abundance of specific species of concern—such as the federally listed red-cockaded woodpecker (*Picoides borealis*) and the northern bobwhite (*Colinus virginianus*). Monitoring also provides information that improves the implementation of restoration actions, including the use of fire, herbicides, and other forestry practices. Monitoring is useful in assessing expenses and selecting the most cost-efficient restoration methods. And with effective data management, monitoring provides an institutional memory of the restoration project, making the information robust to changes in the status of the individuals and organizations involved. Ultimately, the availability and use of monitoring data allow land managers to make more informed restoration decisions.

Monitoring the outcomes of longleaf pine restoration has received increased attention. Eglin Air Force Base began an extensive, base-wide adaptive management program in 2001 (see Case Study 14.1). The Joseph W. Jones Ecological Research Center at Ichauway uses data collected from several monitoring activities to inform and document restoration decisions (see Case Study 14.2). State agencies and private landowners have begun monitoring to assess the survivorship of longleaf pine plantings and the effects of fire (see Case Study 14.3). Monitoring populations of red-cockaded woodpecker and gopher tortoise (*Gopherus polyphemus*)—a federally listed species west of the Mobile and Tombigbee Rivers and a candidate species farther east—often includes measures of the management and restoration of surrounding longleaf pine (U.S. FWS 2003; FFWCC 2012). And numerous monitoring efforts are underway to assess the management and restoration of longleaf pine by federal and state agencies, private landowners, and nongovernmental organizations such as The Nature Conservancy.

Even with increased attention, monitoring of restoration projects has not been widely implemented (Williams 2011; Westgate et al. 2013). Funding for monitoring is often insufficient or nonexistent, and when available, may be used instead to augment resources budgeted for management and restoration—many managers opt to manage and intuitively measure change rather than quantitatively assessing outcomes. Some managers are confident that they know the trajectories of restoration and therefore do not view monitoring as valuable. Additionally, monitoring is often viewed as too complex and time-consuming to implement. Even when monitoring data are available, a manager may not receive a meaningful evaluation within an appropriate time frame to incorporate the data into a restoration project. Last, site conditions such as fuel loads, and external forces such as timber prices and wildfires, are more immediate management concerns and addressing these issues may take priority over monitoring efforts.

The goal of this chapter is to offer techniques for improving and expanding the implementation of monitoring and adaptive management in the restoration of the longleaf pine ecosystem. First, we define monitoring within the context of adaptive management and describe how to determine the appropriate level of monitoring. Next, we present a generalized framework for implementation of monitoring and adaptive management to improve restoration decision making. We provide guidance on the development of a monitoring design and outline the key components of effective restoration plans and monitoring protocols. Using restoration and management case studies we illustrate different goals, methods, and lessons learned; and finally, we describe the importance of monitoring restoration within the context of a changing and uncertain future.

DEFINING MONITORING AND ADAPTIVE MANAGEMENT

Monitoring is the process of acquiring and evaluating information, accumulated through repeated measurements over time using comparable data from multiple samples, to improve restoration decision making (Elzinga et al. 2001; Nichols and Williams 2006; Lindenmayer and Likens 2010). This objective-based approach is most effective in situations where modifying restoration or reassessing the achievability of goals is possible (Elzinga et al. 2001; Nichols and Williams 2006).

Monitoring can provide important information for the restoration of longleaf pine, primarily in evaluating restoration actions to assess whether they are meeting or moving toward the desired ecological condition, either for the ecosystem as a whole, populations of interest, or specific attributes. Additional values include assessing the efficiency and costs of restoration investments and tracking the impacts of threats on ecosystem conditions and restoration activities. The information derived from monitoring reduces the uncertainty of restoration actions and improves overall understanding of longleaf pine restoration. Thus, monitoring is a key component within the adaptive management cycle (Figure 14.1).

Monitoring over many years, or even decades, is necessary to assess the survival and growth of longleaf pine seedlings and trees, changes in ground cover biodiversity, alterations in forest structure, dynamics of populations, and ecosystem responses to climate change. Long-term monitoring requires a systematic and comprehensive approach to collecting and managing data, ensuring data quality, and archiving data (Fancy and Bennetts 2012; Rüegg et al. 2014; Sutter et al. 2015). To be successful, the documentation of the methodology must withstand changes in personnel involved in the project (Michener et al. 1997; Lindenmayer and Likens 2010; Vines et al. 2014; Sutter et al. 2015).

The concept of adaptive management is defined differently by various scientists and practitioners (Williams et al. 2009; Williams 2011; Runge and Knutson 2012; Williams and Boomer 2012). We define adaptive management as a structured and sequential learning process that increases knowledge and reduces uncertainty, iteratively leading to more effective restoration, conservation, and programmatic decisions (Clark County 2016). As an approach for action, adaptive management allows the implementation of restoration actions within the context of uncertain future conditions. As a learning process, adaptive management uses available knowledge and the results of management to modify restoration actions. As a tool for achieving results, adaptive management leads to more effective, efficient, and enduring restoration outcomes.

This interpretation of adaptive management is broader than many definitions in the scientific literature, but this does not mean that an adaptive management approach is appropriate for every restoration situation (Runge and Knutson 2012; Williams and Brown 2012). An adaptive management approach to restoration is most valuable for more formalized, quantitative assessments in which (1) measurable objectives can be established, (2) multiple restoration options are possible and a mechanism is needed for assessing outcomes, (3) consequential decisions are necessary for the future of a species or system, (4) opportunities exist for shared learning, and (5) a monitoring design can be established to assess and institutionalize outcomes (Runge and Knutson 2012).

Adaptive management can be active or passive. Active adaptive management is the simultaneous implementation of two or more restoration or management options, structured by an experimental design, to test competing hypotheses about predicted outcomes (Larson, Belote, Williamson, et al. 2013) in an approach that includes replication, controls, and randomization. For example, adaptive management might entail testing different seasons of prescribed burning and recruitment of new individuals of a rare plant species into a population. Active adaptive management concentrates on learning to understand the causal factors for restoration responses and on selecting the most appropriate restoration option. This approach, which requires extensive planning and coordination, is the explicit focus of many published adaptive management papers (Keith et al. 2011; Larson, Belote, Wiliamson, et al. 2013).

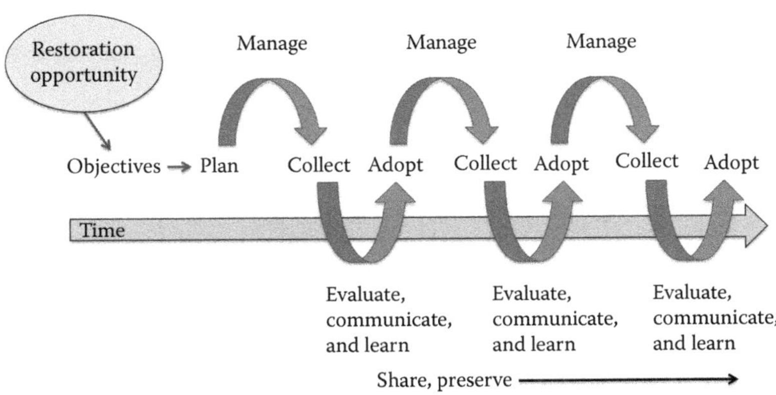

FIGURE 14.1 Adaptive management cycle for a forest restoration project. (Modified from Montambault, J.R. et al., *Conservation Biology*, 29, 1279–1289, 2015.)

Practitioners of passive adaptive management use the best available science and restoration data, select and implement a single restoration option, assess results, and modify restoration actions as needed (Williams and Boomer 2012). If a selected restoration option fails to meet restoration objectives, others are applied sequentially over time; examples include changing the season of burn or selecting different fuel conditions to increase fire intensity. Passive adaptive management has less scientific rigor than active adaptive management. Because learning is not the primary focus, knowledge acquisition is slower than experimentally testing multiple restoration actions in active adaptive management; however, relaying information to managers is usually quicker than publishing results of active adaptive management studies in peer-reviewed outlets. Passive adaptive management may not be appropriate for determining the best restoration regime within the context of rapidly changing conditions. Nevertheless, it is widely used, often as the default when managers are challenged by limitations in geography (unable to test multiple restoration options within a management area) or resource availability (unable to fund a more complex experimental design).

Sharing and communicating restoration results, also part of the adaptive management cycle, improves the work of other landowners and managers and informs decision makers and selected audiences (Figure 14.1). The many opportunities to share and communicate restoration results range from publishing in regional or national restoration journals and presenting at conferences and meetings, to education and outreach efforts including websites and newsletters, local field days, pamphlets, and social media.

One of the challenges associated with adaptive management is the communication of monitoring information to the land managers involved in on-the-ground restoration. This is often the result of insufficient outreach, the reluctance of land managers to change long-standing procedures, or lack of resources needed to incorporate recommendations obtained from monitoring. The use of monitoring data in restoration would likely increase if efforts were made to address monitoring questions that are pertinent to managers, include land managers in the planning of monitoring activities, embed monitoring professionals with land management staffs, provide a liaison between monitoring personnel and management personnel, offer information in timely and understandable formats, provide training opportunities, and establish an organizational structure that facilitates the exchange and use of information from monitoring to management.

Spatial and temporal scales are additional considerations for monitoring and adaptive management. Restoration actions that occur across large heterogeneous areas and over extended periods of time often require more robust monitoring and employ several different intensities of adaptive management and assessment. The application of monitoring and adaptive management over shorter time periods and at a finer scale could provide suitable models for broad-scale restoration application. The longleaf pine is a long-lived species and the ecosystem is characterized by complex biological processes and structures. For these reasons, the effects of longleaf pine restoration are often not fully realized in the short term (see Chapters 10 and 11). Monitoring and adaptive management should be designed to address this complexity.

The adaptive management cycle is not complete without preserving information so that it is available to the project and to future project designers (Figure 14.1). In the short term, preserving data is essential for analyses that compare multiple years. In the longer term, the information from a project can be used to repeat sampling or data analysis. Information that needs to be preserved includes certified data, metadata, reports, and analysis products (Sutter et al. 2015). All need to be stored in lossless and nonproprietary file formats using an appropriate storage media, with appropriate record retention schedules and archive locations (Sutter et al. 2015).

CHOOSING THE APPROPRIATE LEVEL OF MONITORING

Most restoration projects are multidimensional, with multiple objectives that change over space and time. For example, the monitoring of seedling establishment may evolve into increasing the abundance and fuel consistency of the ground cover, the restoration of an open woodland or grassland may include monitoring ground cover and the presence of specific rare species, and the monitoring of a rare or invasive species may be limited to one section of the project area. Because multiple objectives are not met by a single monitoring design or level of monitoring, monitoring efforts will vary across a project area and may change over time.

A land manager uses many sources of information, or lines of evidence, when assessing the level of monitoring needed for a restoration project (Diefenderfer et al. 2016). These lines of evidence include summaries from the scientific literature, information from other restoration projects, predictive ecological models, results from specific research projects, personal experience, and communication with other experienced managers. Often this assessment process is intuitive and somewhat unsystematic, rather than structured and closely documented.

These lines of evidence help managers select the restoration actions that have the highest level of certainty, with minimal risk. If a restoration action has been tested and its outcome is known with very little uncertainty, then the level of monitoring can be minimal (Figure 14.2). One efficient and inexpensive approach is to assess changes in forest structure, ground cover, and site conditions using a series of photographic images (Figure 14.3); the collection of photo point data is one of the least labor-intensive methods of monitoring a restoration project. Other methods include assessing forest structure and species composition by collecting data from one or several index plots (see the section that follows, "Developing a Monitoring Design"), or conducting qualitative assessments along transects of forest structure, species composition, seedling survivorship, and fuel loads. More detailed monitoring is critical when the outcomes of a restoration action are largely uncertain, especially when geographic and ecological variations cause project sites to respond differently, when a restoration action can be refined, or when environmental conditions (such as climate) or social conditions (such as population growth) are rapidly changing and will influence future restoration success. In these situations, an approach that involves more detailed monitoring and adaptive management is recommended (Figure 14.2). The level of monitoring is also influenced by external factors such as the occurrence of state- or federally listed species, reporting requirements of funding sources, and any legal constraints on the project (Sutter et al. 2015); all of these factors would increase the required level of monitoring.

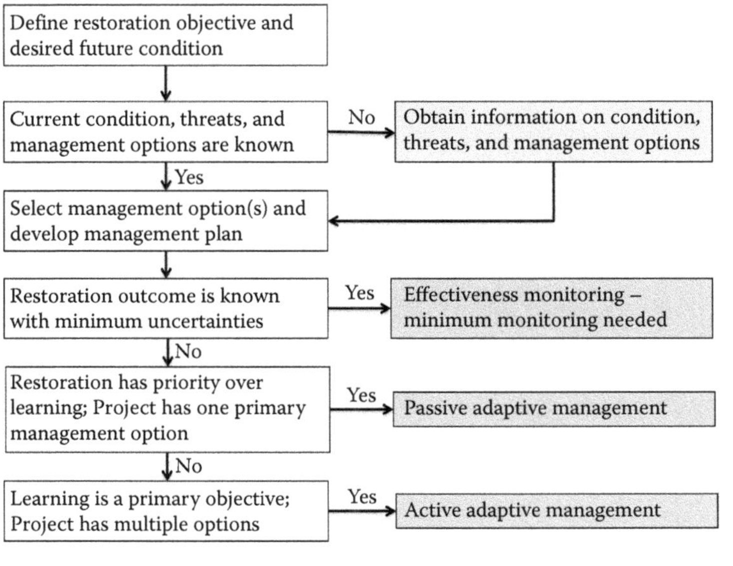

FIGURE 14.2 Decision tree for selecting an appropriate level of monitoring for a restoration project. (Modified from McDonald-Madden, E. et al., *Trends in Ecology & Evolution*, 25, 547–550, 2010.)

(a)

(b)

FIGURE 14.3 A series of photographs from a long-term monitoring plot located at the Joseph W. Jones Ecological Research Center in southwestern Georgia: (a) A fallow agricultural field in 2007, a year before longleaf pine and native ground cover restoration; (b) the same site in 2011. *(Continued)*

(c)

FIGURE 14.3 (*Continued*) A series of photographs from a long-term monitoring plot located at the Joseph W. Jones Ecological Research Center in southwestern Georgia: (c) The same site in 2015, converted to a longleaf pine plantation with native ground cover. (Photographs courtesy of Joseph W. Jones Ecological Research Center.)

The information gathered from monitoring provides managers with the information they need to learn and adapt restoration practices. In the Florida Division of Recreation and Parks, monitoring longleaf pine seedling survival in upland and flatwood restoration sites on the Crystal River Preserve State Park showed that the method used for planting resulted in a high survival rate and that drought was the primary driver of mortality (K. Morin, personal communication). Postfire survival of seedlings within 2–3 years after planting was lower in flatwood sites that were heavily encroached by saw palmetto (*Serenoa repens*). These findings suggest that mechanical treatments of palmettos and the application of fire are needed in flatwood sites to reduce fuels and midstory competition prior to future plantings. Another project, which studied the reintroduction of historical fire regimes at the Fort Bragg Army Base in Eastern North Carolina, showed that oaks (*Quercus* spp.), persimmons (*Diospyros virginiana*), and other masting trees were greatly reduced within burn blocks, but they appeared in disproportionately large numbers near firebreaks (Lashley et al. 2014). This suggests that the presence of firebreaks creates a zone of lower fire intensity that facilitates the linear establishment of fire-intolerant species and that modifications of the fire prescription may be warranted. And finally, at the Fort Benning Army Base, on the Georgia-Alabama border, monitoring data showed that higher prescribed fire intensities were needed on fine-textured soils to maintain an open midstory in longleaf pine sites (Addington, Knapp, et al. 2015).

IMPROVING RESTORATION DECISION MAKING

Implementing restoration monitoring within an adaptive management context requires thoughtful preparation. The following is a framework for developing such a program, outlining the primary steps for collecting the data that will improve restoration decisions.

Define Restoration Goals and Objectives

Defining the restoration goals and objectives for a site is critical to successful monitoring (Elzinga et al. 2001). A restoration goal is a general description of the ultimate desired ecological condition of the site. Expected outcomes are used to set the standards for measuring performance, for interpreting the monitoring data to reduce uncertainty about management decisions, and eventually for improving the restoration outcome. The expected outcome of management can be economic (sustainable yield), ecological (forest structure or presence of specific species), or a combination. The necessary time frame for assessment can span years to decades depending on the specific goal; likewise the frequency of data collection will vary widely among projects.

Shorter-term objectives are set as benchmarks toward the goal; each has components that are expressed so that they are descriptive and can be easily measured. An adequate objective usually meets the following criteria: specific, measureable, achievable, realistic, and time delimited (thus the mnemonic, SMART). Objectives can be thresholds, a specific quantitative condition, or a more general trajectory such as increasing the breeding males in a songbird population, increasing native ground cover, or decreasing the cover of an invasive plant species.

An example of a longleaf pine restoration goal might include the establishment of a longleaf pine stand on a former pine plantation or other previously timbered site. Another might be to establish at least 600 saplings per acre within a given time frame or to replace a planted pine stand of off-site species—species natively adapted to habitats that have different conditions than the historically resident species—by establishing 25% of the stand in planted longleaf pine saplings over a gradual period. If the goal is to create an open forest floor, the objective might be to reduce midstory hardwoods to less than 20% and increase fuel continuity across the site (through new plantings or manipulations of existing vegetation). An appropriate goal for wildlife restoration would be to increase or maintain a satisfactory population of a desirable species; for northern bobwhite, the objective might be 0.25 birds per acre within 5 years.

Select Restoration Options

Restoration options are the management actions available to attain the desired future condition. For longleaf pine restoration, the most common option is the application of prescribed fire, with numerous methods available depending on the restoration objective. Other restoration options include mechanical control or use of herbicides for hardwood reduction, planting longleaf pine seedlings and establishing ground cover, and wildlife habitat provision or translocation of animals.

The selection of restoration options requires an understanding of the current condition of the site. Options would be different for a former cropland or pasture site, a cutover site previously in native forest, an unburned mature stand, or a site that has been managed with fire. Restoration of agricultural sites requires an assessment of soil pH and fertility, aggressive nonnative species, diseases and grubs, and other environmental conditions (The Longleaf Alliance 2016). Cutover sites usually need some form of hardwood control; midstory vegetation and deep duff layers are concerns in unburned stands.

The selection of restoration options is also influenced by past and current threats and the current human context. The common threats for a longleaf pine restoration project include high fuel loads and duff depth, altered midstory structure, altered overstory, off-site planted pine, altered ground cover, and invasive species. Threats will influence restoration options or require the implementation of specific actions such as invasive species control. The human context of the restoration site—such as adjacent development and roads that limit fire prescriptions, especially in relation to smoke—will also have an impact on restoration options (see Chapter 13).

Select Indicators for Monitoring

Indicators are the variables that are measured to assess whether restoration is meeting or moving toward the objectives and goals. Thus, the indicators need to align with the goal and objectives. Indicators can measure the condition of the dominant species (for example, density or age structure of longleaf pine), the reduction of a threat (such as the decrease in a duff layer or decline of an invasive species), or the results of a restoration treatment (such as area burned, average burn frequency, or flame length). A comprehensive assessment of restoration outcomes usually requires multiple indicators.

The following are criteria to help select indicators. Indicators should be:

- *Relevant*—Effective indicators are directly related to the conservation or restoration goals and objectives.
- *Easily understandable*—Effective indicators are easy to understand by a wide range of stakeholders, including the general public.
- *Easily communicated*—Effective indicators are easy to communicate, including both raw data and the resulting information.
- *Reliable*—Effective indicators provide information that is consistently accurate and trustworthy.
- *Ecologically realistic*—Effective indicators are readily restored or promoted.

The number and type of indictors should also be informed by efficiency, which is defined as the ability to make repeatable measurements easily, quickly, and at a reasonable cost. The more efficient a monitoring project, the more likely it will be continued over time.

Develop a Restoration Plan

Restoration plans identify the actions needed to reach the desired future condition. They explicitly link management actions to the restoration goals and objectives, establish overall restoration direction, and compile the primary information on restoration in one document or file with links to others for easy access. They serve as a guide for activities and clarify priorities and management responsibilities. Additionally, restoration plans set the stage for measuring effectiveness and implementing adaptive management. Overall, restoration plans identify and track the implementation of restoration actions.

A comprehensive restoration plan has many components (Table 14.1). In general, the plan outlines the restoration goal, provides background on the current condition of the site, identifies the factors that will influence restoration, and delineates restoration objectives and the actions that will support them. Sections that are especially valuable include a detailed description of the physical and biological condition of the site, any permit requirements for burning, safety information (such as the locations and contact information of the nearest fire station and medical facility), and estimates of costs and funding sources (Clark County 2015). Ideally, the restoration plan should allow someone who is unfamiliar with a project to understand the restoration activities. Developing a restoration plan compels the landowner or agency official to specify the ecological condition that is desired and how that condition will be attained.

Restoration plans should be updated whenever a significant new development modifies or adds to the list of restoration actions, including translocation of animals, use of a new restoration action, fire and storm damage, or other significant changes in ecological conditions (Clark County 2015). Reviewing, updating, and revising the restoration plan should be the responsibility of a single person in an organization.

TABLE 14.1
Checklist for Developing a Comprehensive Restoration Plan

Section	Subsections	Qualifiers (If Needed)
Introduction	Restoration plan purpose and scope	
	Property ownership	
	Guiding documents and regulations	
	Roles, responsibilities, and funding sources	
	Plan approval	
	Protocol for update and revisions	
Background	Location	Boundaries
		Access
	Physical resources	Topography
		Soils and geology
		Hydrology
		Climate and weather
	Biological resources	Natural communities and community condition
		Species and current condition: rare and significant plant and animal species
	Cultural resources	Native American
		Historical
	Land use	Historical
		Existing and allowable uses
		Adjacent land use
		Future adjacent land use
	Permit requirements	
	Public services and safety	Fire and medical
		Law enforcement
		Utilities
		Safety
Factors that influence restoration	Timber resource extraction	
	Fire, past use, and ability to use in the future	
	Regulated (such as federally or state-listed) species	
	Nonnative and invasive species	
	Recreation	
	Climate change	
	Transportation infrastructure current and future	
	Adjacent and regional development, current and future	
	Utility corridors current and future	
	Others	
Restoration goals, objectives, and actions	Summary of goals and objectives	
	Discussion of objectives and restoration actions	
	Priority restoration areas	
	Funding sources and budget	
Restoration action implementation plan	Restoration action	
	Location	
	Time frame	
	Permits and authorization	
	Priority	
	Cost	
References	None	

Select a Monitoring and Adaptive Management Approach

The selection of the monitoring and adaptive management approach should be considered an investment choice. Figure 14.2 presents a decision tree for selecting the approach that is best suited for restoration projects.

The primary criterion for selecting the type of management assessment is the degree of certainty that is associated with the restoration outcomes. If the outcomes are relatively well understood, with little uncertainty, then a rapid assessment of restoration results is appropriate. If the outcomes of restoration are largely uncertain, then an adaptive management approach is strongly recommended. If adaptive management is the best option, the next step is to select between a passive or active adaptive management approach, based on whether a primary objective is to learn from the effort, what resources are available, and whether the project involves a single or multiple restoration options.

DEVELOPING A MONITORING DESIGN

This section provides a brief overview of the primary decisions needed to develop a design for monitoring. The intent is to introduce specific decision points so that the reader will understand their importance and is not intended as a replacement for seeking the assistance of a knowledgeable professional.

The section uses the following specialized terms:

- *Sample units*—The units from which data are collected, expressed as individuals such as longleaf pine saplings or red-cockaded woodpeckers or as spatial units such as quadrats, transects, or plots
- *Sample*—The total number of sample units from which data are collected
- *Statistical population*—The total of all possible sample units, such as all longleaf pine saplings or the entire area being restored

Statistical Inference

A key concept in sampling is statistical inference, collecting data from a subset of an area or population and extending that information to derive an estimate of the whole area or assemblage (e.g., total number of longleaf pine trees, quail density, or cover of a rare plant species within a defined area). The assumption is that the information collected from the sample is an accurate reflection of the statistical population.

Random Sampling

Because random sampling is a very powerful and efficient way of generating a set of characteristics, it is an acceptable substitute for measuring all components of the assemblage or area—a process that would be time-consuming, expensive, and never complete.

To ensure that statistical inferences can be made from a sample, the sample units must be randomly allocated. This means that every location within the statistical population has an equal opportunity to be selected as part of the sample. Random allocation of sample units eliminates bias, the conscious or unconscious decisions that managers make when selecting where to sample. Biases can occur when a sample unit is in an area that the manager wants to sample (abundant regeneration of longleaf pine or large number of individuals of a rare species) as well as in areas that the manager prefers to avoid (dense thickets or likely presence of venomous snakes).

Many methods are available for randomly allocating sample units in a population, from simple random sampling and systematic sampling with a random start to multitiered and cluster sampling (Sutherland 1996; Elzinga et al. 2001). In addition, a well-interspersed arrangement of sample units—so that they are distributed widely in the area being sampled—is advantageous; achieving this sample dispersion without introducing bias usually requires the advice of a knowledgeable professional.

Nonrandom Sampling

Although not appropriate for statistically extending data from a sample to describe a characteristic in the whole population, nonrandom sampling (also called index sampling or representative sampling) can provide important insights into the trend and response of a population, community, species or area. Data from nonrandom sample units can be useful in estimating the condition of the population. The closeness of the estimate to the actual condition depends on the representativeness of the sample units and the biological intuition of the manager.

Precision and Accuracy

Precision is a characteristic of sampling that reflects how close in value repeated measurements of the same attribute are to one another under the same or very similar conditions (Stapanian et al. 2016). The closer repeated measurements are to each other, the easier it is to detect change and the less likely the data have been compromised by human carelessness or instrument failures. Accuracy describes the closeness of any measure to its true value, such as the exact number of gopher tortoises or rare plant species on a site, but the true value of most populations is usually unknown. The focus of sampling is reaching an acceptable level of approximation.

In developing an effective monitoring design, reducing measurement errors and increasing the sample size can increase data precision. A monitoring design that specifies larger square or rectangular plots or longer belt transects can reduce the effects of high heterogeneity in a population. By designing sample units that are larger or longer sided, managers can capture more variability within a single sample unit than across sample units; this means that fewer sampling units would be needed to meet the desired level of precision. Finding a balance between reducing the variability of the data and increasing the number of sample units (and their cost) is a core challenge in developing a monitoring design.

Repeatability

Repeatability is a key component of monitoring, one that is often overlooked. Repeatability is important at several stages in the monitoring process, including the ability to locate key elements of the site (such as sample units or specific populations), locate and identify species, and repeat monitoring methods and measurements (Sutter et al. 2015; Stapanian et al. 2016). Lack of repeatability is one of the most common problems attributed to monitoring projects that fail to detect changes over time (McEachern and Sutter 2010).

The following steps can help to ensure the repeatability of a monitoring protocol:

- Training and certifying field staff
- Extensive documentation to ensure that the monitoring plan maps and describes—in as much detail as possible—the site, the location of the sample units (with GPS coordinates), and the monitoring methods; acceptable images include hand-drawn maps, aerial photographs, or satellite images
- Validation by a third party so that descriptions, maps, and documentation are clear enough to be replicated
- Storage of all documentation materials in a safe and accessible location

Data Quality

The level of effort required for successful data management depends on many factors. For longleaf pine, the long-term horizon of most restoration projects requires significant planning to ensure that data are available and of high enough quality to be useful for many decades. Ensuring the quality of data starts before data collection and includes field crew training, establishing a well-designed file management system, developing accurate data collecting systems, and reviewing data as they are accumulated (Sutter et al. 2015; Stapanian et al. 2016).

Data quality is ensured by thorough assessments at regular intervals, using a straightforward and transparent process that includes the following three essential steps (Sutter et al. 2015):

1. *Data verification*—Evaluates newly acquired data for completeness, correctness, and conformance with acquisition specifications—a process that includes assessing data once they have been entered into a database program, ensuring that all required information is present, and removing duplicate records; verification is best done as soon as the data are collected so that the collection can be repeated if necessary.
2. *Data validation*—Evaluates the quality of the data, checking that data are within a reasonable range and have structural integrity and logical consistency.
3. *Data certification*—Represents a benchmark indicating that data are in a finalized state and can be used for analysis and shared with others.

Monitoring Protocol

The monitoring protocol outlines all the components of the monitoring design, as well as other factors that affect the implementation of restoration monitoring and data management (Table 14.2). The first section (introduction) either refers to or repeats the information included in the restoration plan (described in the earlier section, "Develop a Restoration Plan"). The core part of the protocol provides details of the monitoring design, the field methods used to collect data, and the procedures in place for managing and accessing data. Details on field methods allow efficient planning for data collection and budget development.

A key concept that underlies the protocol is proactive decision making: that all the steps of monitoring, analysis, data management, and communication of results are considered before any data collection takes place.

TABLE 14.2
Checklist for Monitoring Protocol Content

Section	Subsection	Additional Information (If Needed)
Background and objectives (linked to restoration plan)	Background on species, habitat, or ecological system	
	Study area	
	Restoration action(s)	
	Monitoring goals, objectives, and assumptions	
	Sampling approach	References cited
	Protocol for update and revisions	
Variable selection	How variables were selected	
	List of variables, each with hypothesis, sampling method, and sampling timing and frequency	
Monitoring design (if appropriate)	Definition of area or population being sampled	Map
	Sample unit size, shape, and permanency	
	Number and allocation of sample units	
	Timing and frequency of sampling	Time table

(Continued)

TABLE 14.2 (*Continued*)
Checklist for Monitoring Protocol Content

Section	Subsection	Additional Information (If Needed)
Field methods	Staffing qualifications, training, teams, roles, and responsibilities	
	Field forms	Organized as data should be entered
	Equipment needed	
	Schedule	
	Permitting and authorizations	
	Safety	Emergency phone numbers, protocol
Data management and quality	Amount of detail required on data management and data quality—a graded approach	
	Data acquisition	Data capture system
		Calibration and quality control for instruments and equipment
	Data management	Database design
		File names and version control
		Backup
		Work space and file management
		Metadata
	Data quality	During data collection
		After data collection: verification, validation, and certification
	Data analysis	
	Data interpretation and evaluation	
Other	Archiving	
	Budget and funding sources	
References	None	

RESTORATION OF THE LONGLEAF PINE ECOSYSTEM IN AN UNCERTAIN FUTURE

The restoration of a longleaf pine forest, whether for timber or biodiversity (or both), is inherently a long-term endeavor. Time is a primary variable when growing trees, managing forest structure, and restoring species richness. A restoration project that takes many years or decades develops within the context of a changing ecological environment (Golladay et al. 2016). A dynamic environment results from a suite of interacting disturbances such as the introduction of nonnative species, disruption of predator-prey relationships, altered fire and hydrologic regimes, increasing landscape fragmentation from roads and development, deposition of atmospheric nitrogen, and increases in soil salinity (Hallett et al. 2013). Predicted climate change in the Southeast, including increasing temperatures and more variability in rainfall (Pachauri et al. 2014), would exacerbate these threats and create new stresses, thereby magnifying the uncertainty involved in restoring a longleaf pine stand or ecosystem.

The uncertainty of future conditions increases the challenges of restoration. Historical or current reference conditions may be unrealistic goals for the future, and instead add confusion to the selection of appropriate standards with which to evaluate success (Suding 2011). The concept of a dynamic reference model (Hiers et al. 2012; Kirkman et al. 2013) has been proposed to address future uncertainty, which involves measuring changes in restoration sites simultaneously with changes in reference or near-reference sites to assess the ecological dynamism of the ecosystem. Monitoring in these situations focuses on ecological processes (especially fire), forest structure, and community composition rather than individual species.

Considering the frequent fire regime and the wide edaphic and climatic gradients of these ecosystems, longleaf pine communities are likely to be relatively resilient to climate change (Diop et al. 2009; Mitchell et al. 2014); for more detail, see Chapter 15. Even so, anthropogenic threats will alter fuel loads, fire behavior, fire frequency, the distribution of species, and the occurrence and aggressiveness of nonnative species (Golladay et al. 2016). These are important factors that will determine the restoration potential of sites and influence restoration options. Restoration planning will need to anticipate ecological responses to change and include flexible and iterative goals, objectives, and actions to ensure the resilience of longleaf pine restoration sites.

Within this context, long-term monitoring and adaptive management is even more critical (Lindenmayer and Likens 2010; Golladay et al. 2016). At the site level, monitoring helps determine whether desired outcomes are achieved and documents changes in fuel loads, fire effects, species composition, abundance of native species, and encroachment by nonnative species. Monitoring reference sites and consulting reference conditions, especially those that have an old-growth component, can provide a measure of how longleaf pine historically responded to change (Hiers et al. 2012). Monitoring can assess the resilience of a restoration site by measuring fuel loads, identifying fire units and fuel breaks, and assessing connectivity and geophysical diversity (Anderson et al. 2014).

Monitoring must also include measures of social and economic conditions—such as new road corridors, increased road traffic, and increased rural and urban development—that could affect restoration activities in the future (Golladay et al. 2016). At a regional scale, a network of monitored restoration sites could assess population shifts and changing fire regimes and provide insights into potential desired ecological condition across environmental gradients.

Case Studies for Monitoring and Adaptive Management for Longleaf Pine Restoration

CASE STUDY 14.1 Adaptive Management at Eglin Air Force Base

Eglin Air Force Base is located in the Florida Panhandle, 45 miles east of Pensacola. The 464,014-acre installation is the largest forested military reservation in the United States (Wiens et al. 2009), supporting both the largest acreage of old-growth longleaf pine and the largest public ownership of a longleaf pine sandhill ecosystem (Hiers, Laine, et al. 2003). The fourth-largest population of the federally endangered red-cockaded woodpecker also occurs on the base (Wiens et al. 2009).

Based on the overarching management goal at the base—to conserve significant natural resources while supporting the military mission—the ecological management goals for longleaf pine are (1) to maintain high-quality sites and restore low-quality sites, ultimately establishing a functioning landscape of longleaf pine and associated communities; and (2) to maintain and increase the abundance of rare and endangered species. Although some longleaf pine stands are in reference condition, many have experienced decades of fire suppression and encroachment by Choctawhatchee sand pine (*Pinus clausa* var. *immuginata*) (Sutter et al. 2001). Frequent low-intensity fire has been reintroduced across the base using prescribed burns to maintain and restore ecological communities (Hiers, Laine, et al. 2003).

To address the restoration of such a large and diverse landscape, the base has developed a range of products, or management decision tools. Detailed desired future conditions (DFCs) for targeted species and ecological systems were compiled first (Sutter et al. 2001). DFCs are spatially explicit and time-delimited ecological goals that are established to guide conservation, land management, and restoration efforts (Wiens et al. 2009). Best available science and the expertise of 27 scientists and practitioners were used to develop the DFCs. A three-tiered ranking system was used to describe the condition of each targeted species and ecological system. A fourth tier was added to represent the most degraded sandhill stands.

Next, a spatially explicit Ecological Condition Model was developed to identify and track the condition of the longleaf pine sandhill habitat on the base. Using the information compiled in the DFC, eight criteria were identified as indicators of site condition. These included (1) canopy density, (2) deciduous cover, (3) time since last burn and fire frequency since 1972, (4) longleaf pine cover, (5) patch size, (6) red-cockaded woodpecker habitat, (7) road density, and (8) sand pine cover (Wiens et al. 2009). Data for the baseline map were derived from satellite imagery and historical and recent field data, with the resulting map showing the different tier conditions across the base. The map is updated annually using current imagery and field data.

The Ecological Condition Model was further modified to identify spatially specific management objectives and prioritize the application of prescribed fire across the base. The model annually prioritizes the management units for prescribed fire using and expanding on the same criteria and adding scoring values and weights (Hiers, Laine, et al. 2003).

More than 200 randomly placed, permanent, 1.6-acre monitoring plots were established to validate remote sensing, inform long-term management priorities, and examine the results of management actions on ecosystem condition and biodiversity. Data are regularly collected on tree density, species richness, density and cover of grasses, forb cover, legume density, multiple measures of fuels (longleaf pine litter cover, grass litter cover, and litter depth), and other metrics. Data are analyzed through the ordination of community data using nonmetric multidimensional scaling and using the Mahalanobis distance metric to compare the condition of restoration plots with reference plots. Spatial data analyses are used to adaptively select stands for fire management.

The results from analyses of the monitoring data have altered the concept of appropriate reference conditions that had been developed for longleaf pine sandhills on the base (B. Williams, personal communication). The data analyses showed that fire frequency is the primary predictor of ecological condition: the more frequent the fire, the closer a site is to the desired structure and the greater the abundance of desirable indicator species. The data also showed that season-of-burn does not have a strong effect on the abundance of desirable indicator species or ecological condition. Results showed that turkey oak (*Quercus laevis*), along with some other deciduous oaks, is a positive indicator of longleaf pine condition on these sandhill sites and that attempts to reduce oak density with higher fireline intensities often increase longleaf pine mortality (Loudermilk et al. 2016).

The data, coupled with further experimentation, also showed that in sites that have had a period of fire exclusion, longleaf pine mortality following prescribed fire is caused primarily by consumption of accumulated duff, rather than needle scorch. Thus, development of strategies for careful reduction of duff became a major management target.

Last, the results showed that the presence of sand pine is the primary indicator for a degraded longleaf pine sandhill ecosystem and that fire alone is insufficient in reducing sand pine at a landscape scale. Such information was used to redirect management efforts.

CASE STUDY 14.2 Ecosystem Restoration and Quail Management

The Joseph W. Jones Ecological Research Center at Ichauway is located in the Dougherty Plain physiographic province of southwestern Georgia. Ichauway is the 30,000-acre land base for the research center. It was assembled in the early 1900s as a quail-hunting plantation for Robert W. Woodruff, the longtime leader of Coca-Cola. Ichauway is located in the center of the historical range of the longleaf pine ecosystem. A large portion of the property still retains natural stands of longleaf pine-dominated forest (about 18,000 acres). Much of this forest contains highly diverse native ground cover, boasting more than 1100 documented vascular plant species (Drew et al. 1998; Kirkman et al. 2001). The remainder of the property consists of slash pine (*P. elliottii*), loblolly pine (*P. taeda*), mixed pine-hardwoods, riparian hardwood forests, geographically isolated wetlands, agricultural fields, and shrub-scrub uplands (Goebel et al. 2001). The 13-mile eastern boundary of Ichauway is formed by the Flint River, and about 15 miles of the Ichawaynochaway Creek flows through the center of the property. About 50 endangered, threatened, or special-concern species are found on-site, many of which are endemic to the longleaf pine ecosystem.

An ecological goal for Ichauway is to maintain, restore, and perpetuate natural communities associated with the longleaf pine ecosystem. This overarching management goal falls within the mission of the Jones Center: "To understand, to demonstrate, and to promote excellence in natural resource management and conservation on the landscape of the southeastern Coastal Plain of the United States."

The purpose of natural resources management at Ichauway is to facilitate research and provide an example of positive land stewardship. Actions include both the promotion and restoration of natural communities. Prescribed fire is an important management tool employed on-site. Fires are conducted with a 2–3 year return interval and in both the dormant and growing seasons; a greater emphasis is placed on frequency than seasonality. Throughout the 20-year history of the Jones Center, about one-third of all prescribed fires have been implemented during the growing season based on the burn objectives of a particular burn block (see Chapter 17).

Forest resources are managed using a modification of the Stoddard-Neel approach (see Chapter 10) to maintain a multiaged longleaf pine forest in perpetuity and to restore degraded sites. Timber is harvested using the conservative individual-tree selection method. Longleaf pine is replanted into areas converted from either agriculture or hardwood-dominated upland. The philosophy toward wildlife management is holistic rather than focused on a single species, and management actions are intended to provide benefits to the system as a whole. Objectives are formulated to manage the Ichauway property in a manner providing quality habitat for the range of longleaf pine-associated species.

Prescribed Fire Monitoring

The annual goal is to burn half of the burnable acreage (12,000–14,000 acres), producing a 2-year fire-return interval for most of the Ichauway property. Prescribed fire is the most influential and important management and restoration tool. As such, a rigorous monitoring program is employed to document the use of prescribed fire and the effects of individual burns, and to inform future fire management decisions.

Aspects of prescribed fire monitoring precede the actual implementation of fire on the ground. Meteorological information is gathered from many sources including an on-site weather station. Weather information includes daily measures of wind direction and speed, transport wind direction and speed, relative humidity, Keetch-Byram Drought Index values, fine fuel moisture, air temperature, precipitation, number of days since rain, and smoke dispersion. Specific weather parameters are written into burn plans and weather information is recorded on the day of the burn.

Each burn within a given unit has its own set of primary objectives, which can include: fuel reduction, hardwood control, propagation of fire-dependent species, research, aesthetics, education demonstration, seedbed or planting preparation, wildlife habitat management, or native seed production. The specific burn objectives for a unit vary depending on its current condition. Meteorological information and postburn assessments recorded from previous burns assist in determining the success of the previous burn and objectives for the current burn. Previous burn data, fire intensity, weather conditions, and fuel consumption all provide insights that are valuable for planning the desired outcome and parameters of future fires.

Further monitoring is conducted following a prescribed fire to assess fire effects and document fire history. Within 3–4 weeks after the fire, each unit burned is mapped using GPS equipment and entered into a GIS database, and crown scorch is mapped and quantified. Additionally, an evaluation is conducted to determine the amount of fine fuels consumed, top-kill of woody shrubs and hardwoods, and vegetative fuel consumption. This information is entered and maintained in a database along with information required to obtain burn permits.

The adoption of a 2-year burn regime, coupled with extensive monitoring, has resulted in several pertinent observations about the overall management of the Ichauway property. Biennial prescribed fires not only facilitate the development of adequate longleaf pine regeneration but also control the woody component. Burned areas are suitable to capture seed in the months after the fire; to ensure survival, the resulting seedlings are allowed at least one fire-free growing season. In the 2-year fire-return interval, woody stems are susceptible to being top-killed and consequently do not require high-intensity fires that would be expected with a longer burn rotation. Further, fire frequency and burn conditions become more flexible with a regime of shorter fire-return intervals. If a unit is not burned during the prescribed year (most likely because of weather considerations), a history of shorter intervals promotes conditions that still allow for a relatively low-intensity fire the following year. Further, managers are not pressured to burn under less than optimal fuel conditions because the lower fuel loads of a frequent fire regime provide a larger window for burning with regard to weather conditions and season. Additionally, a 2-year burning interval is advantageous because the lower fuel load limits crown scorch, which can negatively impact tree growth.

Quail Monitoring

The northern bobwhite (*Colinus virginianus*) plays an important historical, economic, cultural, and ecological role in the longleaf pine-wiregrass ecosystem of the Southeast (Burger et al. 1999; Fleckenstein 2013, 2014). If not for interest in this game species, much of the privately owned remnant tracts of longleaf pine might not have escaped land use conversion. Additionally, interest in the restoration and management of northern bobwhite has increased both regionally and nationally. Federal- and state-level conservation programs have been implemented to promote early successional habitats beneficial to quail and other species that depend on these fire-maintained habitats (Burger et al. 2006).

The Ichauway property was initially assembled and historically managed for quail hunting, with prescribed fire applied annually and other management actions implemented to promote quail populations. A portion of the property is still managed to support an adequate wild quail population for hunting—approximately 0.5–1.0 quail per acre. Quail management also falls under the broader context of the overall management of the property, where appropriate management objectives for the longleaf pine ecosystem as a whole, combined with very conservative hunter-access levels and bag limits, have maintained viable quail populations across the property. The quail management program also includes outreach about the wise and sustainable use of natural resources in the longleaf pine ecosystem.

Although promotion of wild quail populations is not the primary management objective of the Jones Center, monitoring quail populations is an integral aspect of its management program. Covey call counts are conducted annually during periods of peak calling (mid-October through early November) to estimate population density. A covey count consists of an observer monitoring and recording the number of quail coveys heard from an established fixed-radius point count plot. These points are located at 0.3-mile intervals in a grid covering the entire property. The location of each covey at a given point is recorded on a map, and the frequency of calling is monitored (Wellendorf et al. 2004). This information is used to estimate quail density for the property and as an index to track quail populations through time and across the site. Additionally, biological data (age and sex) are collected from all individuals bagged as part of the hunting program. Primary feathers of juvenile birds are also examined to determine the date that they hatched (Petrides and Nestler 1943).

Jones Center land managers use quail monitoring data to guide management actions such as timing of prescribed fire, habitat improvement (such as roller chopping, winter disking, or herbicide applications), supplemental feeding, and control of predators. Changes in quail population trends influence future decisions about managing specific areas on the property. Population decreases might lead managers to decrease hunting pressure within an area and increase management actions to promote quail populations the following year. If a large portion (>30%) of the harvest in a given year is adult quail, the explanation could be a poor hatching year or an above-average carryover of adults from the previous year. Coupled with population trends, this information can be used to prioritize the location and focus of quail management activities. Determining the ages of juveniles helps determine peaks in the nesting season and can lead to altered management actions during optimal nesting periods.

Monitoring has illustrated that quail can be managed on-site under the broader context of longleaf pine ecosystem management at a level that is sustainable for hunting, education, and demonstration. Although higher densities are sustainable on other lands managed primarily for quail, doing so would have consequences for other resource values. Management at Ichauway also maintains a large volume of high-quality timber and native diverse plant communities, provides habitat for other sensitive species, and promotes the ecosystem as a whole. To achieve these values, quail management is focused in former agricultural areas partially converted to longleaf pine restoration, using a ratio of 70% forest to 30% farmland. Fields are irregularly shaped and maintained at 3–5 acres each. Two treatments used within this matrix are winter disking to promote beneficial early successional habitat, and mechanical and chemical manipulation of habitats. In addition to the application of frequent prescribed fire across the landscape, these management techniques have led to the production of viable quail populations.

CASE STUDY 14.3 Longleaf Pine and Gopher Tortoise Habitat Restoration Project

The Long County Mitigation Site is a project that the Georgia Department of Transportation established in partnership with the Georgia Department of Natural Resources, Georgia Forestry Commission, and The Longleaf Alliance to restore longleaf pine on sandhills and enhance a population of gopher tortoises (Free and Smith 2012). The project is part of a multistate sandhill restoration project led by the Georgia Department of Natural Resources that targets restoration and monitoring on >38,000 acres of sandhills in Alabama, Florida, Georgia, South Carolina, Mississippi, and Louisiana (Elliott 2015). The Long County site is a matrix of sandhill habitat and riverine sloughs bordered by the Altamaha River in southeastern Georgia. Previous intensive management practices had extensively altered the upland vegetation through removal of native species and establishment of sand pine plantations. The management goals for the site are to reestablish longleaf pine-turkey oak communities, restore prescribed fire, and enhance the local population of gopher tortoises.

The gopher tortoise is a keystone species found within xeric habitats primarily associated with longleaf pine forests throughout the southeastern United States (Eisenberg 1983). Its burrows provide refuge for over 300 species of invertebrates and several rare vertebrate species, including obligate gopher tortoise commensals (Jackson and Milstrey 1989; Moler 1992). West of the Mobile and Tombigbee Rivers gopher tortoise is federally listed as threatened; in the eastern portion of the longleaf pine range, it is a candidate for threatened status (U.S. FWS 2011). The primary threats to gopher tortoises throughout their range are habitat loss, degradation, and fragmentation (Auffenberg and Franz 1982; Birkhead and Tuberville 2008), hence, restoration and management of habitat for this species is of considerable conservation interest.

When the eastern population of the gopher tortoise was petitioned for federal listing in 2006, the Department of Defense, whose land holdings support among the largest remaining populations of tortoises, spearheaded development of a Candidate Conservation Agreement (CCA). A CCA is a nonbinding voluntary agreement among stakeholders to address threats to the species and prevent further decline and possibly to preclude the need to list the species; for more details, see Chapter 12. One of the primary outcomes of the CCA was the development and implementation of standard survey and monitoring methods for gopher tortoise populations using line-transect distance sampling (Buckland et al. 2001) with burrow scoping to determine occupancy (Smith, Linehan, et al. 2009; Smith, Stober, et al. 2009). The Long County Mitigation Site is one of the first sites where this monitoring protocol was implemented in Georgia.

Prior to restoration, the gopher tortoise density on the Long County Mitigation Site was approximately 1.68 tortoises per acre, measured using the line-transect distance sampling method, and many tortoises were located along roadsides due to the closed canopy conditions in the sand pine plantations (Figure 14.4). Management actions began in 2013 with the clearing of about 700 acres of sand pine and planting of longleaf pine seedlings (Figure 14.4). Vegetation changes are being monitored using a standardized methodology for longleaf pine sandhills developed by the Georgia Department of Natural Resources. The methodology uses point intercepts along randomly placed transects to measure cover of canopy, midstory, shrubs, and ground cover species and litter. Breeding birds are also being monitored through time.

After the 2013 sand pine clearing operation, increases were reported for cover and species richness of the herbaceous layer, an important food resource for the gopher tortoise. The most dramatic result was observed in the distribution of gopher tortoise burrows. Within a year after treatment, the number of burrows doubled in the areas that had been cleared of sand pine, representing a marked shift from the previous distribution that was restricted to roadsides. This information indicates that the restoration efforts resulted in increased habitat for gopher tortoises. Line-transect distance sampling of gopher tortoises to determine changes in density and population size is the next step in this ongoing monitoring program.

(a)

(b)

FIGURE 14.4 Restoration of Long County Mitigation Site in Georgia, a multistate sandhills restoration/monitoring effort to benefit gopher tortoises: (a) Sand pine stand prior to restoration, (b) site conditions following clear-cutting of pine plantation. *(Continued)*

(c)

FIGURE 14.4 (*Continued*) Restoration of Long County Mitigation Site in Georgia, a multistate sandhills restoration/monitoring effort to benefit gopher tortoises: (c) Planted longleaf pine seedlings. (Photographs courtesy of Lisa Westberry, Georgia Department of Transportation.)

ACKNOWLEDGMENTS

The authors thank the many individuals who shared their insights and data, including Rob Addington, George Matusick, Cheryl Millett, Brian Pelc, Susan Cohen, Lisa Kruse, Matt Elliott, Keith Morin, Katrina Noland, Brett Williams, and Carrie Sekerak. The authors are especially thankful to the four reviewers of the manuscript, L. Mike Conner, Lora L. Smith, J. Kevin Hiers, and Jacob Thompson, whose comments greatly improved the chapter.

Part V

Longleaf Pine Ecosystem Restoration Perspectives

15 Planning for an Uncertain Future

Restoration to Mitigate Water Scarcity and Sustain Carbon Sequestration

Steven T. Brantley, James M. Vose, David N. Wear, and Larry Band

CONTENTS

INTRODUCTION

The desired future conditions of longleaf pine (*Pinus palustris*) can be described by ecosystem structural characteristics as well as by the provision of ecosystem services. Although the desired structural characteristics of restored longleaf pine ecosystems have been described at length, these characteristics deserve a brief review here because ecosystem structure directly contributes to the provision of key ecosystem services and helps differentiate these forests from other land uses in the southeastern United States. Briefly, upland longleaf pine stands managed with frequent fire are characterized by low basal area, an open canopy, a sparse midstory, and a diverse uninterrupted herbaceous layer (Walker and Peet 1983; Platt 1999; Kirkman et al. 2001; McIntyre 2012). Over the long term, emphasizing the single-tree selection method of canopy harvesting will produce an uneven-aged stand structure that adds to complexity and maintains both ecosystem services and long-term economic value (Mitchell et al. 2006). Achieving this characteristic stand structure often serves as a first indicator that restoration goals are being met (Rasser 2003; McIntyre 2012).

Many of the known longleaf pine ecosystem services—such as biodiversity, wildlife habitat, and occupancy by endangered species—depend on the characteristic ecosystem structure. Although the achievement of desired conditions can often be assessed based on only stand structure or species composition (or both), the full value of longleaf pine restoration for the restoration of other ecosystem services is largely unknown.

In this chapter, we review characteristics of longleaf pine forests in the context of protecting and enhancing water resources in the Southeast and sequestering carbon (C) for climate change mitigation. Although our primary focus is on protecting water resources, we also review C sequestration issues and describe the trade-offs between managing lands to mitigate water scarcity while simultaneously promoting long-term C sequestration. We suggest that the societal benefits of protecting water and C resources by restoring longleaf pine are often unrecognized and undervalued, but have the potential to complement goals already established for current restoration programs. Thus, a more complete understanding of how longleaf pine restoration affects fluxes and pools of both water and C could lead to increased incentives for longleaf pine restoration projects and enhanced opportunities for larger-scale restoration efforts.

CONTEXT FOR FOREST RESTORATION: CONTEMPORARY ISSUES OF WATER AND CARBON BUDGETS

Forests and Water

Forest Effects on Water Yield

Because forests are a critical source of clean and abundant water, forest management is an essential tool for management of water resources in the Southeast (Jackson et al. 2004; Lockaby et al. 2013; Caldwell et al. 2014; Marion et al. 2014). Although water quality is also a major concern in the Southeast, the primary focus of this chapter will be on managing quantity, specifically water yield. Here, we define water yield as the difference between incoming precipitation and water exiting the system as evapotranspiration (ET). Water yield is most often quantified at the watershed level, but the water budget concept can also be applied to the forest stand. Water yield contributes to multiple pools of water including streamflow, groundwater recharge, and replenishment of soil water storage.

Water Scarcity in the Southeast

In recent decades, a combination of population growth, expanding water use for agriculture, and increasing climate variability have stressed water supplies in the Southeast (Lockaby et al. 2013; Sun et al. 2013; Caldwell et al. 2016). Although inputs from precipitation are relatively high (> 1000 mm/year) in the Southeast compared to many arid regions, increasing demand has challenged long-standing policies and existing political structures in a region not historically accustomed to water scarcity (Ruhl 2005). Increasing water demands by the municipal and agricultural sectors have had a particularly large impact on streamflow and aquatic ecosystem services (Sun et al. 2008).

Much of the current focus on water management in the Southeast is on municipal water use. However, a relatively large percentage of these withdrawals are returned into surface waters, resulting in lower consumptive use compared to agricultural withdrawals (Richter 2014). Because agricultural water use is highly consumptive, it is more likely to have direct effects on stream and river discharge. This issue has been particularly apparent in the Coastal Plain (Figure 15.1), where many watersheds once dominated by longleaf pine have experienced exponential growth of center-pivot irrigation since the late 1960s (Pierce et al. 1984; Couch et al. 1996; Golladay et al. 2007). As demonstrated in Figure 15.2, row crop cultivation in the Southeast relies heavily on irrigation, and the proliferation of center-pivot irrigation systems has had a demonstrable effect on flow regimes (Couch et al. 1996; Golladay et al. 2007; Rugel et al. 2012; Golladay and Hicks 2013). The effect of agricultural water use in the Flint River and adjacent Chattahoochee River watersheds on downstream flow has been a major controversy for water managers and policy makers in Alabama, Florida, and Georgia since the early 1990s (Ruhl 2005), including a recent U.S. Supreme Court case (*Florida v. Georgia*) filed in 2013 that examined water apportionment in the region.

Ecological Consequences of Declining Streamflow

Changes in flow regimes of streams and rivers not only have social and economic impacts on human population centers and agriculture, but also impact aquatic ecosystems. In the Coastal Plain, Golladay and Hicks (2015) and McCormick and Baron (2015) demonstrated that aquatic ecosystems were at greatest risk from the historic low summer flows that dewatered critical habitat during droughts and reduced habitat quality in the pockets of water that remained. Such departures from

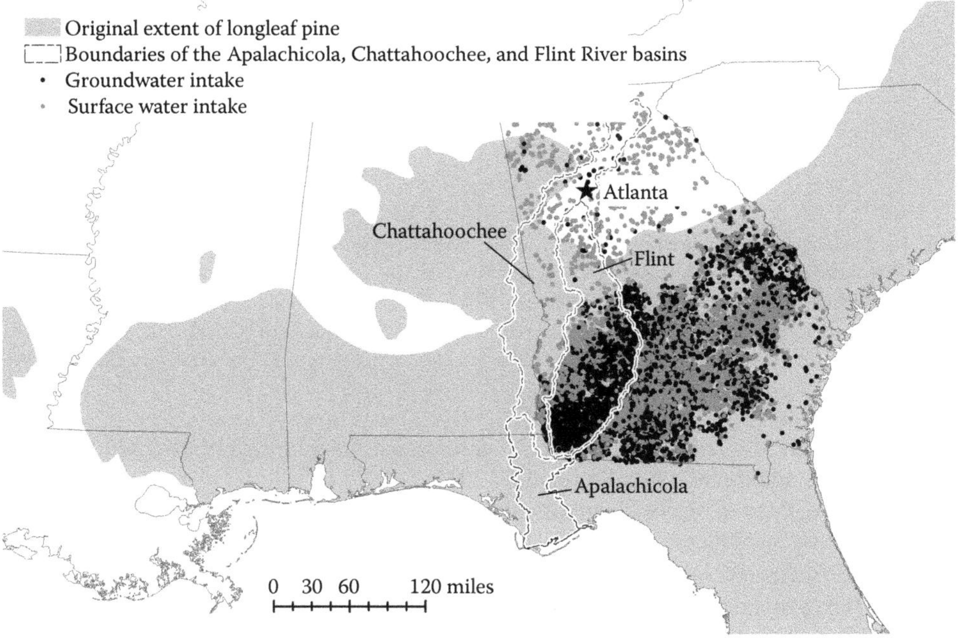

FIGURE 15.1 The overlap of the Apalachicola-Chattahoochee-Flint River basin with the historical range of longleaf pine in Georgia showing permitted groundwater and surface water intakes. (Map modified from Ware, S. et al., *Biodiveristy of the Southeastern United States: Lowland Terrestrial Communities*, John Wiley & Sons, New York, 1993. Source of data: Georgia Department of Natural Resources, Environmental Protection Division.)

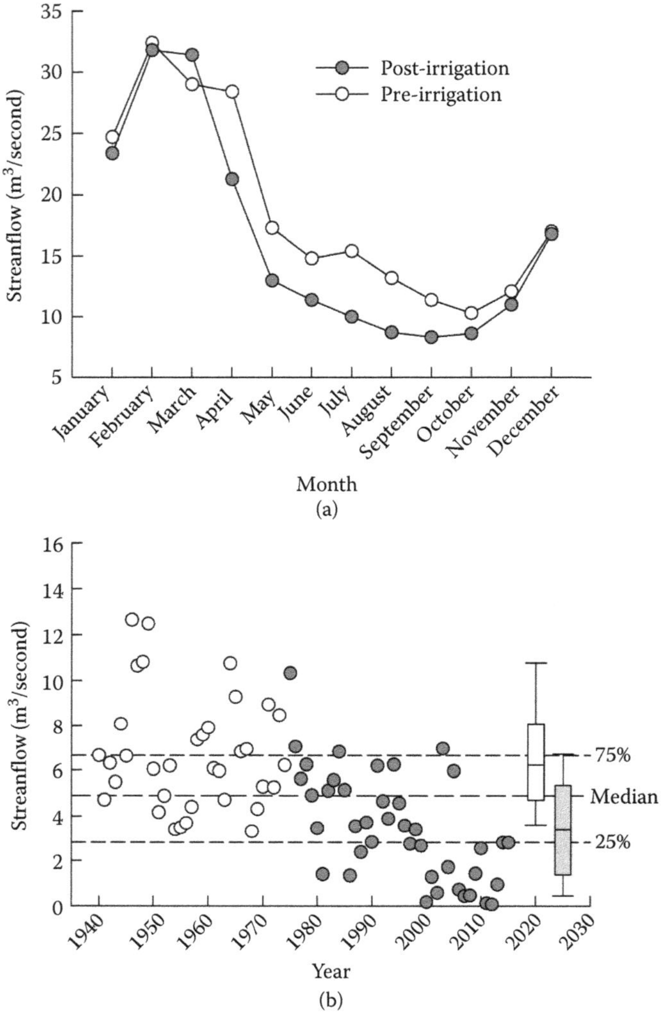

FIGURE 15.2 Observed changes in streamflow in the Ichawaynochaway Creek at Milford, GA (USGS 02353500) after the proliferation of center-pivot irrigation in the Southeast during the 1970s: (a) Differences in monthly median flow between pre-irrigation (1940–1974) and post-irrigation records (1975–2004), and (b) changes in 1-day minimum flows compared to long-term median. (Modified and updated from Golladay, S. W. et al., *Proceedings of the 2007 Georgia Water Conference*, University of Georgia, Athens, Georgia, 2007.)

historical minimums can harm fish, mussels, crustaceans, macroinvertebrates, and other aquatic life (van den Avyle and Evans 1990; Smith et al. 2015).

Altered flow regimes in major rivers also have negative consequences for estuaries and other downstream ecosystems (Fitzhugh and Richter 2004). Much of the current controversy surrounding flows in the Apalachicola-Chattahoochee-Flint River basin centers on damage to the health of Apalachicola Bay, a diverse and productive downstream ecosystem that supports an important oyster industry (Livingston 1991). Reduced freshwater input during the 2007–2008 drought resulted in significant increases in salinity followed by increases in oyster mortality (Petes et al. 2012). If issues of water scarcity in such systems are not substantively addressed in the near future, continued decreases in freshwater inputs—combined with sea level rise and increasingly frequent and severe droughts predicted under climate change—would have substantial deleterious effects on stream, river, and estuarine habitat and water quality—impacting both ecological and social systems (Sun et al. 2013).

Link between Water Scarcity and Land Management

Agricultural withdrawals have been the primary cause of water scarcity in recent decades, but some evidence suggests that the water balance in southern Georgia was changing even before the rapid expansion of center-pivot irrigation. Harper (1956) reported a lowering of the groundwater table in southern Georgia and cited similar earlier reports in many other parts of the United States following European settlement. Harper (1956) attributed the change in the water table, the reduction in cover of swamps, and the resultant loss of biodiversity to increasing water withdrawals for growing towns and cities. However, he also hypothesized that increases in ET, which resulted when natural vegetation such as longleaf pine and wiregrass was replaced by row crops, also had a role.

Globally, ET represents the second largest flux of water in terrestrial systems (after precipitation), with plant transpiration accounting for the majority of this flux (Jasechko et al. 2013; Good et al. 2015). By its very nature, ET represents consumptive use. Consequently, changes in land cover that affect ET at the watershed scale have direct effects on water yield. ET is also generally considered a conservative process in that it shows little variation from 1 year to the next; thus, natural interannual variations in rainfall tend to be reflected in water yield (Oishi et al. 2010). The earlier observations by Harper (1956) suggest that increases in ET from changes in land use since the beginning of European settlement have had a major impact on regional water balance. More importantly in the context of current regional water scarcity issues, this close coupling between land cover and water yield also suggests a potential path for mitigating water scarcity by emphasizing restoration projects that focus on reducing ET.

Carbon Sequestration

Forests and Carbon

Forests also play a critical role in the global C cycle (Jackson et al. 2005; Bonan 2008; Canadell and Raupach 2008; Lockaby et al. 2013; Caldwell et al. 2014). The potential of forests to sequester C has been the subject of considerable attention, both from scientists and from policy makers, as a tool to mitigate climate change by reducing net emissions of carbon dioxide—by far the most common greenhouse gas. U.S. forests currently sequester 173 Tg C/year, which is about 10% of C emissions from the U.S. energy and transportation sectors (Wear and Coulston 2015). Land use changes resulting from reforestation and afforestation shifted about 44 Tg C/year into the forest sector from 2010 to 2014 (Wear and Coulston 2015). Eastern forests, which provide nearly 80% of the C sink, are projected to provide nearly 90% by 2030 (Wear and Coulston 2015). Forest C accumulation is largely a function of forest type and age distribution, with the condition of eastern forests reflecting an ongoing recovery of a landscape that was largely cut over in the early 20th century.

Because of the critical role that forests play in C sequestration nationally, climate change mitigation policy logically focuses on protecting the vast C stores in forests and expanding their potential role as a C sink. Current U.S. projections show a gradual slowing of C accumulation overall (Wear and Coulston 2015) and the possibility that C transfers into forest ecosystems could slow and then reverse, thereby defining some serious challenges to the current policy goal. Policy initiatives focused on retaining forest uses will require incentives that overcome market-driven shifts in rural land uses and practices that increase the average annual C storage over the course of the management regime. C taxes would likely affect these outcomes, but other forms of inducement can influence landowner behaviors (van Kooten et al. 1995; Lubowski et al. 2006).

Carbon Sequestration in the Southeast

The Southeast has been identified as a particularly strong C sink based on high rates of net ecosystem exchange (NEE), where NEE represents the net balance between C gained from photosynthesis and C lost from respiration (Ingram et al. 2013; Novick et al. 2015). A detailed analysis of recent C dynamics (Coulston et al. 2015) confirmed the influence of management and regrowth in southeastern forests, where C sequestration in unharvested and undisturbed stands (144 Tg C/year) was nearly 100% higher than in harvested stands (77 Tg C/year). They also found that natural disturbances dampen the rates of C accumulation but do not lead to net emissions of C and that southeastern forests in particular have "room to grow," although at decreasing rates of accumulation than the rates observed in recent decades.

Land use change continues to reshape forests and their C pools. While forest area has increased slightly over the last 20 years, the flows into and out of forest use have been much larger (Coulston et al. 2015). Land use changes, especially exchanges between agriculture and forests, occur in response to changes in the relative economic returns from these uses in specific places. Overall, the steady transfer of forests to urban use has been offset by net flows of agricultural land into forests.

Environmental Trade-Offs of Carbon Sequestration

Much of the current management focus in the Southeast emphasizes intensive management for maximal biomass production. This type of management has both economic benefits for landowners and environmental benefits for climate change mitigation through C sequestration. However, it does little to address regional water scarcity issues, and it may exacerbate regional water stress under some circumstances based on the linkages among NEE, ET, and water yield (Jackson et al. 2005). In the following sections, we summarize the current state of knowledge about water and C cycling in longleaf pine, and we suggest future directions for studying the potential role of longleaf pine ecosystems in combining water scarcity mitigation with C sequestration.

CAN LONGLEAF PINE RESTORATION HELP REDUCE WATER SCARCITY?

Managing Forests for Water Yield

The concept of managing forests to augment water supplies is not new (Douglass 1983); however, several recent severe droughts and growing populations in the Southeast have revived awareness of these related issues and opportunities (Ford et al. 2011; J. Jones et al. 2012; Sun et al. 2015; Vose et al. 2016). Because forest thinning and harvesting often lead to increased annual water yield (Bosch and Hewlett 1983; Brown et al. 2005; Sun et al. 2015), the effects of droughts in the Southeast could potentially be mitigated by maintaining lower density forests with inherently lower ET (McLaughlin et al. 2013). Thus, large-scale restoration of the drought tolerant, low basal area, frequently burned longleaf pine forest could benefit water resources by reducing ET, translating to increased stream runoff or groundwater recharge (or both). Furthermore, lower densities would also increase resilience to drought by reducing water stress on the trees that remain after stand reductions (Kush et al. 2004; McDowell and Allen 2015).

Increasing forest cover reduces annual water yield under some circumstances. However, compared to alternative land cover types such as row crop agriculture and urban development, forest cover is better able to maintain water quality, reduce runoff pollution, stabilize baseflow, and mitigate stormflow (Anderson et al. 1976; Jackson et al. 2004; Sun et al. 2005; Lockaby et al. 2013; Ford et al. 2011). Although these benefits are realized with most types of forest cover, a possible unique benefit of restoring longleaf pine may lie in the potential of this forest type to serve as a source of relatively high water yield. Compared to other forests in the Southeast, restored longleaf pine has low annual ET rates (Figure 15.3), particularly when stands are maintained at low densities similar to those found in natural, frequently burned stands (Ford et al. 2008; McLaughlin et al. 2013; Novick et al. 2015; Whelan et al. 2015). This characteristically low ET, when coupled with a historical range that largely overlaps areas facing the most immediate threats from water scarcity (Figures 15.1 and 15.2), suggests that large-scale restoration of the longleaf pine ecosystem could represent a promising strategy to mitigate water stress in some southeastern watersheds.

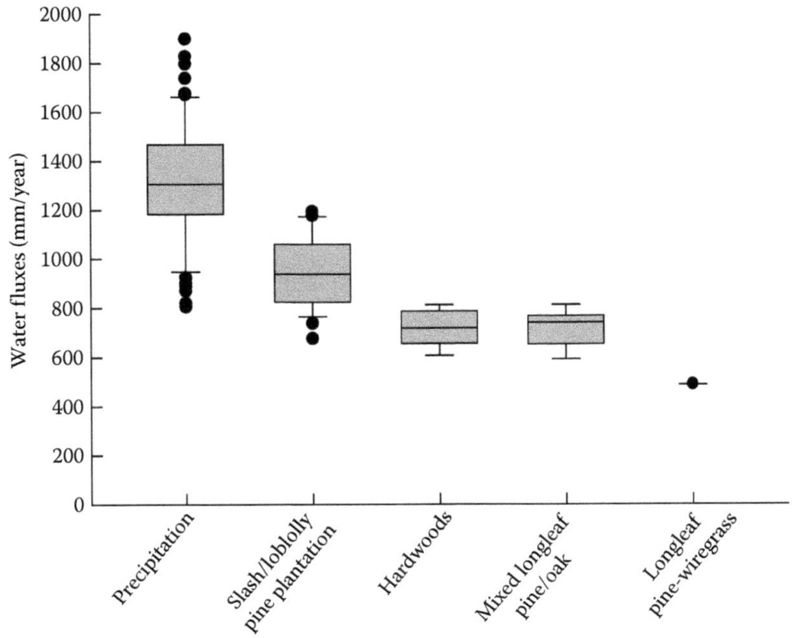

FIGURE 15.3 Annual precipitation distribution (1941–2010) at Dawson, GA (NOAA Station ID 092570) compared with annual evapotranspiration (ET) rates of dominant forest types in the Ichawaynochaway Creek watershed. (From S. Brantley, unpublished data. Precipitation data source: https://www.ncdc.noaa.gov/cdo-web/datatools/findstation. Evapotranspiration values are from a literature review of published ET values from the southeastern United States.)

Water Saving Characteristics of Longleaf Pine Ecosystems

Although specific data on ET partitioning in longleaf pine ecosystems have not been published, much of what we currently know about the structure and function of these ecosystems suggests several ways in which restoration and management with frequent fire would reduce ET. First, ET is likely to be lower in longleaf pine simply because its stand density is much lower than other southeastern forest types. Lower basal area translates into lower sapwood area and lower leaf area index values; these important structural characteristics interact to limit stand transpiration. Second, interception losses (the loss of incoming precipitation through direct evaporation during and after rainfall) are lower in the typically open-canopy longleaf pine with a midstory cover that is kept relatively sparse by frequent burning (McIntyre 2012); this results in further reductions in overall stand leaf area (Figure 15.4). This is a particularly important advantage for a fire-maintained longleaf pine forest, because midstory ET can be an important component of overall water use for other forest types (Johnson and Kovner 1956; Hamada et al. 2004; Powell et al. 2005; Brantley et al. 2013, 2015). Last, the reduced litter production of an open canopy and sparse midstory and the periodic consumption of forest floor biomass by frequent fires (Figure 15.5) may combine to reduce interception losses from the litter layer.

The effects of frequent prescribed fire on community composition also tend to favor lower ecosystem ET. Fire tends to select against aggressive, fast growing hardwood species such as water oak (*Quercus nigra*), laurel oak (*Q. laurifolia*), and black cherry (*Prunus serotina*). These species occur more typically in mesic or wet-mesic sites with richer soils, but they often encroach on upland longleaf pine stands when fire is absent or when cool weather or poor fuel conditions prevent fire from reaching adequate intensity (Jacqmain et al. 1999; McCay 2000; Varner and Kush 2004). These species are characterized by higher sapwood area (Figure 15.6) than southern red oak (*Q. falcata*), turkey oak (*Q. laevis*), and other typical ring-porous pyrophytic oak species that can only

FIGURE 15.4 Comparison between (a) a typical longleaf pine stand managed with a 2-year fire-return interval, and (b) a nearby site after 14 years of fire suppression. Note that fire suppression resulted in a rapid increase in mid-canopy leaf area, a shift in ground cover species composition, and an increase in forest floor biomass—all of which can contribute to higher rates of short-term carbon storage and stand evapotranspiration. (Photographs courtesy of Steven Brantley.)

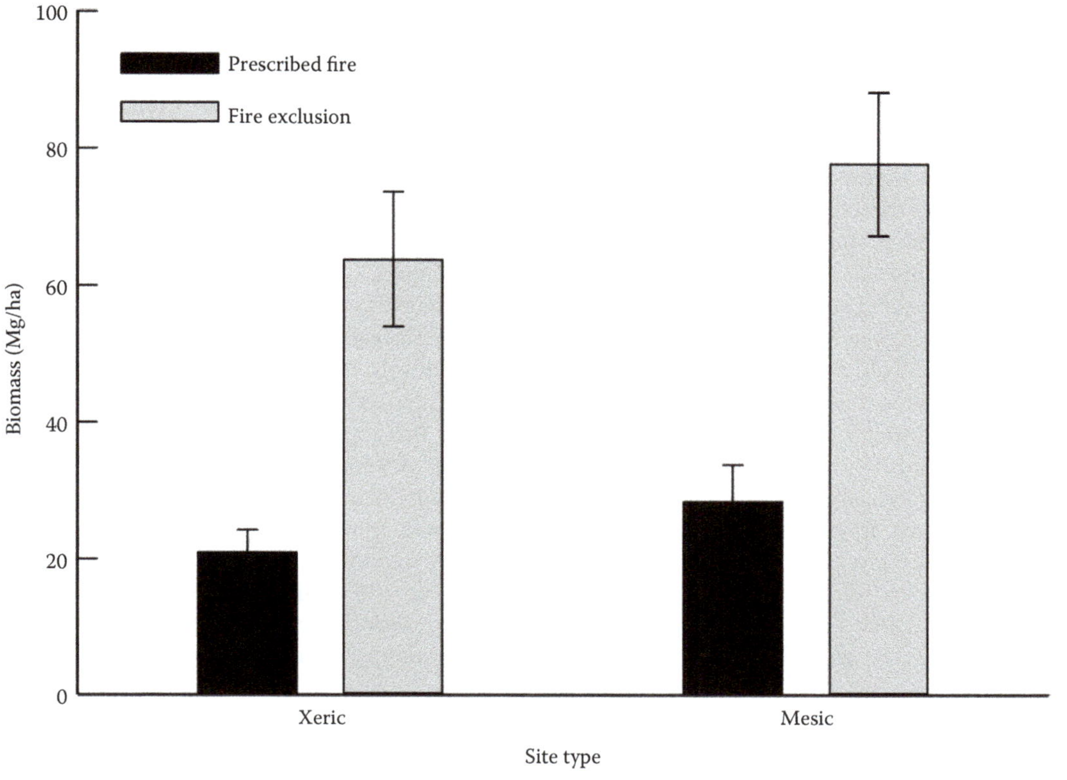

FIGURE 15.5 Forest floor biomass (litter) in regularly burned longleaf pine (2-year fire-return interval) and in stands subjected to 14 years of fire exclusion (From L. Boring, unpublished data.)

FIGURE 15.6 Differences in sapwood area, shown by red dye, between ring porous species (a), which tend to use relatively little water per unit of basal area and tend to be favored with frequent prescribed fire; and diffuse porous species (b), which tend to use substantially more water per unit of basal area. (Photographs courtesy of Ava Hoffman.)

transport water in their outer growth rings (Wullschleger et al. 1998). Shifts in species composition to more mesophytic species have been shown to result in relatively higher stand ET and concurrent decreases in water yield (Caldwell et al. 2016).

Finally, both longleaf pine and its frequent herbaceous codominant wiregrass (*Aristida stricta*) are conservative with respect to plant-level water use, although in different ways. Longleaf pine often demonstrates lower per-tree water use than other dominant, faster growing southeastern pine species under similar soil and climatic conditions (Martin 2000; Ford et al. 2008; Gonzalez-Benecke et al. 2011). In a comparison of longleaf pine and slash pine (*P. elliottii*), average daily transpiration for longleaf pine was 33% lower, mostly because of its relatively lower leaf area but also because of its higher stomatal sensitivity to soil moisture (Gonzalez-Benecke et al. 2011).

Water savings in longleaf pine ground cover are realized through higher water-use efficiency (the quantity of C fixed per unit of water consumed) and the relatively low leaf area of grasses and herbs. Many codominant herbaceous species in the longleaf pine ecosystem (Figure 15.7) tend to demonstrate higher water-use efficiency than the woody species that replace the herbaceous layer in the absence of fire (Ford et al. 2008; King et al. 2013; S. Brantley unpublished data). This is especially true of wiregrass and other warm season grasses (those with a C4 metabolism), which have photosynthesis pathways adapted to maximizing C fixation and minimizing water loss. Concurrently, when the herbaceous layer is grass-dominated, leaf area and productivity of the grasses are likely limited by their basal meristem anatomy, which severely limits exposure to sunlight (Knapp and Smith 2001; Knapp et al. 2008). These characteristics combine to reduce overall water use on the forest floor compared to sites dominated by the woody species that tend to encroach during prolonged fire intervals.

Stand Level Water Budgets in Longleaf Pine

All of the characteristics described above suggest that ET is lower in fire-maintained longleaf pine, but relatively few direct measurements of ET have been collected to test this hypothesis. To our knowledge, only three studies have reported annual ET in stands where longleaf pine was either a dominant or codominant canopy species. Using sap-flux measurements and scaled physiology data,

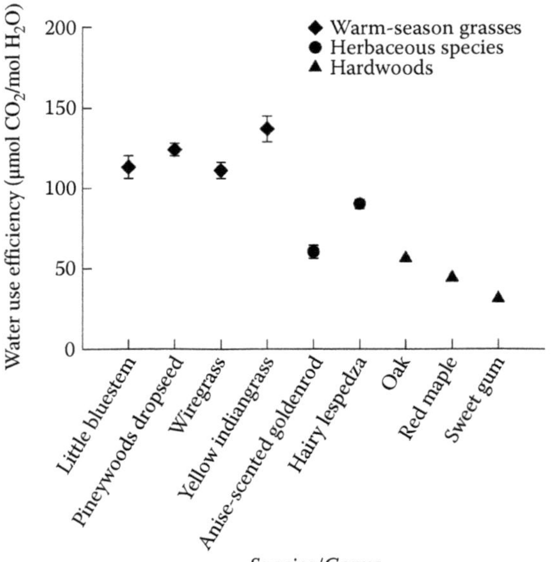

FIGURE 15.7 Water-use efficiency, defined as the number of carbon atoms fixed per unit of water used, for six herbaceous species that are commonly used in longleaf pine restoration programs (from S. Brantley, unpublished data) and three woody plants that are commonly recruited into the longleaf pine ecosystem under a regime of fire suppression. (From King, J. S. et al.,. *Bioscience,* 63, 102–117, 2013.)

Ford et al. (2008) reported that pure stands of longleaf pine and wiregrass use only 489 mm/year in mesic soils and even less in drier, sandhills soils. Powell et al. (2005) reported ET of 754 mm/year in a mature, low-density stand of mixed slash pine and longleaf pine in Florida. Whelan et al. (2015) reported a range of similar values (594–816 mm/year) for three southwestern Georgia sites in which longleaf pine was either dominant or codominant; this study provides the best range of potential spatial and temporal variation in ET for longleaf pine because multiple years of data were collected and the sites were located across a soil moisture gradient that is representative of longleaf pine habitat.

The forest ET values reported above and in Figure 15.3 were generally lower than most other land cover categories in the Southeast. However, the range of annual ET values reported was quite large, even among nearby sites (Powell et al. 2005; Ford et al. 2008; Whelan et al. 2015). Some of these inconsistencies can be attributed to methodological differences. For example, the studies that used stand-level energy balance methods reported higher ET than the estimates reported by Ford et al. (2008) for similar, nearby sites. Errors in scaling from tree and leaf-level measurements to the stand (Ford et al. 2008) could explain some of the inconsistencies between these results and the more integrated methods of Powell et al. (2005) and Whelan et al. (2015).

The three studies were different in several other important aspects. The longleaf pine stands studied by Powell et al. (2005) were also dominated by slash pine, which is known for higher per tree ET than longleaf (Gonzalez-Benecke et al. 2011). Additionally, ET values reported in Whelan et al. (2015) may not accurately represent ET for restored longleaf pine-wiregrass stands because the footprint (the area of forest sampled for ET) measured by the eddy covariance towers also contained patches of hardwood forest, forested wetlands, and agriculture.

Such inconsistencies indicate that analyses of ET data derived from eddy covariance need to be more spatially explicit for restored longleaf pine ecosystems. Eddy covariance and energy balance techniques collect large, nearly continuous data sets, which could be partitioned into discreet periods when ecosystem exchanges of water (and C) are primarily originating from segments of the footprint more strongly dominated by longleaf pine. Such an analysis would be supported by additional, semi-integrated measurements of the various ET components—such as sap flux, interception, and ground cover—in codominant species. These types of measurements would help account for transpiration by different species and provide estimates of soil evaporation, litter evaporation, and herbaceous transpiration as they relate to stand structure. The semi-integrated measurements of ET that partition water use among various components at the stand level could prove particularly valuable in helping watershed managers better understand the influence of fire (and thus fire suppression) on specific hydrologic fluxes and overall ET.

Despite their methodological differences, all of the studies described above agreed that longleaf pine has one of the lowest ET rates among various southeastern land cover types and that restored longleaf pine could be an ideal land cover for increasing water yield while maintaining the ability of forests to protect water quality.

LONGLEAF PINE POTENTIAL FOR CARBON SEQUESTRATION: BENEFITS, LIMITATIONS, AND UNKNOWNS

Current State of Knowledge

C sequestration by forests is generally expressed as net ecosystem productivity (NEP), the net accumulation of C in an ecosystem, and is often approximated using atmospheric measurements of NEE of CO_2 using eddy covariance. Assessments of C budgets in longleaf pine must also incorporate the effects of fire-induced C loss with NEE to quantify net ecosystem carbon balance (NECB). The relatively few studies that have quantified NECB in longleaf pine indicate that stands managed with frequent prescribed fire tend to be relatively small C sinks and can become short-term C sources under certain conditions (Whelan et al. 2013; Starr et al. 2014, 2016; Martin et al. 2015). For example, longleaf pine stands located in mesic soils in southwestern Georgia were small C sinks, whereas stands on drier (sandhills) soils were C neutral in the absence of fire (Whelan et al. 2013; Starr et al. 2015). However, the C neutral stands became small C sources when C losses from prescribed fire were factored into C budgets (Whelan et al. 2013).

Other forest types common in the Southeast—such as short-rotation pulpwood plantations and young (<100 years) fire-suppressed hardwoods—sequester substantially more C over the short term than longleaf pine (Clark et al. 2004; Bracho et al. 2013; Novick et al. 2015). Although short-term studies of NEE are useful for demonstrating the potential of ecosystems to sequester C, calculating annual NEE and NECB does not fully represent the long-term benefits of longleaf pine as a C sink. Studies of NEP and C accumulation in longleaf pine using forest survey methods have demonstrated substantial C gains in longleaf pine forests over time (Mitchell et al. 1999; Samuelson et al. 2014, 2017); and simulation modeling has shown long-term positive C accumulation in fire-maintained longleaf pine systems, although at a slower rate than other forest management strategies (Martin et al. 2015). Still, many questions remain regarding the value of longleaf pine restoration for sustainable C sequestration.

Uncertainties in Longleaf Pine Carbon Budgets

Climate Change, Ecosystem Resilience, and Reduction of Catastrophic Fire Risk

One of the greatest uncertainties in C science is how ecosystems will respond to changing climate conditions and how these changes will feedback to C sequestration. Although using frequent prescribed fire in longleaf pine can result in lower annual C accumulation in this forest type compared to others, prescribed fire may improve ecosystem resilience to changing climate and buffer against catastrophic events that would have large and long lasting effects on C budgets (Wiedinmyer and Hurteau 2010). Climate projections for the Southeast suggest that mean annual temperature will increase 4°C–8°C by 2100 (Carter et al. 2014). Projected precipitation changes are less certain. However, longer, more frequent, and more intense droughts are expected, and hotter and drier weather increases the likelihood of catastrophic wildfires (Carter et al. 2014). Frequent prescribed fire reduces fuel loading, thereby greatly reducing the risk of catastrophic, stand-replacing fires that would accompany hotter and drier future conditions. Preempting such wildfires through sustained fuel reduction would allow the resources normally directed toward fire suppression or postfire restoration to be directed toward proactive management programs with possible long-term cost savings, reduced risk to human life and property, reduced risks to human health from smoke emissions, protection of water quality, and long-term reduction in fire-related CO_2 emissions (Kush et al. 2004; Wiedinmyer and Hurteau 2010; Mitchell et al. 2014).

Knowing the balance between frequent (but small) C losses from prescribed fire and the potential large C losses from wildfire is paramount to future policy decisions. Although prescribed fire results in an immediate loss of biomass C as CO_2 through combustion, the risk of massive and long-term C loss from catastrophic stand-replacing fires is greatly reduced when longleaf pine stands are actively managed with prescribed fire. In fire-prone forest stands of western states, Wiedinmyer and Hurteau (2010) estimated that the targeted application of prescribed fire in dry forests reduced fire-related CO_2 emissions ≥ 18%–25% over the long term by mitigating the risks of catastrophic wildfires, suggesting that C releases during wildfires can more than offset the C sequestration benefits of fire suppression.

Black Carbon in Longleaf Pine

A second uncertainty surrounding the use of frequent fire in longleaf pine C budgets is the role of black C. Although a large proportion of fuel is converted to CO_2 through combustion, a substantial proportion of biomass C is transformed by incomplete combustion into a large and diverse set of molecules often collectively labeled black C (Goldberg 1985; Forbes et al. 2006; Czimczik and Masiello 2007; DeLuca and Aplet 2008). Black C has generally been characterized as a comparatively stable form of C that can represent a major C sink, making up 5%–40% of total soil organic C when it becomes an integrated part of the soil matrix (Goldberg 1985; Czimczik and Masiello 2007; DeLuca and Aplet 2008; Liang et al. 2008).

Although total soil C can represent a relatively large fraction of total ecosystem C in longleaf ecosystems (Samuelson et al. 2014), little is known about fluxes of soil black C in frequently burned longleaf pine sites and its contribution to soil C pools. Ike (2010) reported that rates of soil black C accumulation were higher in frequently burned sites, but she also concluded that land use history has a larger influence on the overall size of the soil C pool than fire (see Chapter 7). Gonzalez-Benecke et al. (2015) reported that black C made up < 5% of soil organic C in longleaf pine stands and suggested that new inputs of black C from prescribed burning are small. However, the suggestion that black C fluxes and pools can be accurately estimated from shallow (often < 60 cm) surface soil analysis is based on several assumptions that have not been adequately supported by field studies. For example, the assumption that soil mixing in longleaf pine forests is unimportant is questionable based on the large number of burrowing species—such as gopher tortoise (*Gopherus polyphemus*), pocket gophers (*Geomys pinetis*), and ants—that characterize wildlife diversity in these systems.

These studies indicate that the standing pool of black C in frequently burned longleaf pine forests is small. However, recent research suggests that some fractions of the soil black C pool are actually more soluble than previously thought and that a large proportion of black C can be rapidly mobilized and transported through soils to aquatic systems (Dittmar et al. 2012; Jaffe et al. 2013). Jaffe et al. (2013) reported that about 10% of dissolved organic C flowing into and out of rivers originated from soluble charcoal. How these fluxes factor into terrestrial C loss from prescribed fire is unknown, and many questions remain about the factors that affect the storage and movement of black C in ecosystems. Mobilization and transport of black C are particularly important in the sandy soils that characterize much of the longleaf pine range. These questions provide a blueprint for studying black C transport and transformation in throughfall, soils, wetlands, and streams in watersheds that are dominated by longleaf pine.

Carbon Accounting in Longleaf Pine

The third critical area of investigation needed is development of optimal assessment metrics to quantify the value of longleaf pine restoration in projects that might consider C offsets as a restoration goal. C accounting (measuring the value of C emissions mitigation programs, such as forest management projects, in terms of total C impact) differs markedly from basic C science. Whereas C science focuses on net stand-level C exchanges (NEP, NEE, and NECB), C accounting protocols can also account for C sequestered in wood products removed from the site and for a project's secondary effects such as increased harvesting in nearby forested stands. Perhaps more importantly, the net outcome (in terms of the value of C offsets) of projects using C accounting principles reflects how the project performs against a modeled baseline of C stocks that reflects current regional forest conditions.

In a hypothetical longleaf pine restoration project in southwestern Georgia, Remucal et al. (2013) applied a forest project protocol that had been developed by the Climate Action Reserve to address greenhouse gas emissions in California. Their results showed a net emission of CO_2 over a hypothetical 100-year project lifespan. Although they proposed many reasons for this outcome, the primary factor was that prior restoration activities at the study site had reduced stand C stocks far below the expected C stock baseline for the Southeast. These same restoration activities had been considered successful in terms of restoring ecosystem structure and function, but the resulting stand structure (with characteristic low basal area and an open canopy) deviated from the typical C stocks observed in other nearby forests. As discussed previously, high C stocks in southeastern forests are a reflection of past reforestation of agricultural land, promotion of heavily stocked and productive forests through intensive management, and fire suppression. This means that current baseline forests do not represent the pre-European landscape in which the longleaf pine savanna was the dominant land cover.

We therefore suggest that fire-maintained longleaf pine should generally be considered the baseline condition for much of the southeastern Coastal Plain and for parts of the Piedmont. Because of the historical extensive coverage by longleaf pine, all changes in land cover since European settlement—including those that have contributed to the region's role as a major C sink—should be viewed as a result of human disturbance. Because current C accounting practices fall short when applied to longleaf pine ecosystems, further consideration must be given to how baseline conditions are determined.

CHALLENGES IN MEASURING COUPLED CARBON AND WATER CYCLES

Trade-Offs between Carbon and Water

Effective stewardship of ecosystem services requires land managers to recognize the substantial trade-offs that exist among alternative future land management scenarios (Daily and Matson 2008; Lindenmayer, Hulvey, et al. 2012). This dilemma is particularly important for land managers who are trying to balance C sequestration and ET. Ideally, simple metrics based on easily measureable plant physiology parameters would be adequate to understand these trade-offs and make informed decisions. For example, many researchers and land managers have given consideration to optimizing trade-offs between NEE and ET by selecting plant species for C offset projects and biofuel production based on water-use efficiency (King et al. 2013).

However, comparing water-use efficiency (WUE) alone does not adequately address the impacts of high annual net ecosystem productivity and high ET on water yield (Vose et al. 2015). At both the tree and stand levels, plant sensitivity (or insensitivity) to drought, rooting depth, soil water partitioning among species, and the physical effects of plant canopies on interception fluxes all play a role in balancing NEE with ET. At the watershed scale, feedbacks among vegetation, climate, soils, and other aspects of ecosystem function can also affect water yield, but not in the same manner as would be expected if estimates were based on WUE alone (Ukkola et al. 2016). Although knowledge of WUE among dominant species is a valuable first step, total ecosystem ET and the variables that affect ET are more critical measurements because it is the total amount of water used that affects C-water trade-offs and water supply (Sun et al. 2011; Sun and Vose 2016). Thus, as with other ecosystem services, managing NEE/ET trade-offs requires a better understanding of the ecosystem processes and feedbacks that affect the balance between productivity and water use (Carpenter et al. 2009).

Challenges in Coupled Carbon and Water Modeling

Projecting long-term C and ET dynamics in longleaf pine ecosystems and alternate land cover types will require models that couple the C and water cycles and account for ecosystem responses to climate variability under different projected climatic regimes. Predicting short-term responses to moderate drought is generally straightforward, especially if forest structure remains unchanged. However, climatic dryness (potential ET divided by precipitation), terrain characteristics, land cover types, biomass, soils, and characteristics of dominant species all influence the potential impacts of droughts on forest ecosystems. This complexity poses challenges for predicting the impacts of drought on ecosystem processes.

One of the major limitations of physically based modeling approaches is that changes in vegetation structure (such as reduced leaf area or changing root distributions) and function (such as shifts to species with different mechanisms for regulating water use) that occur in response to severe drought are rarely explicitly incorporated into modeling frameworks (Powell et al. 2013; Luo et al. 2008; Tague et al. 2013). Quantifying these changes often requires direct empirical investigation. Lumped parameter ecosystem models that were designed to describe the effects of soil water on ecological processes (such as C cycling) often oversimplify soil water and nutrient movement.

However, modeling approaches that account for subsurface hydrologic connectivity suggest strong spatial controls on ecosystem processes (Hwang et al. 2009; Emanuel et al. 2010). More accurate model predictions of severe and longer-term drought impacts will require a coupling of hydrologic and ecosystem processes within a dynamic context that includes appropriate feedbacks (Law 2015). This is not a trivial expectation; it requires the linking of leaf-level physiology, whole-tree responses, root dynamics and soil water access, stand level responses, and physical hydrology (Tague et al. 2013) into complex models that are difficult to parameterize and calibrate over large spatial scales.

Net water and C cycling in forested landscapes also depends on the spatial variation and covariation of forest species, structure, terrain, and soil conditions. Because management activities that alter water balances in upland forests can influence water balance elsewhere, a modeling approach to coupled water and C cycling in longleaf pine-dominated areas needs to consider the net landscape-level cycling—particularly with respect to lateral redistribution of subsurface water—over and above the scale of individual stands. In longleaf pine-dominated areas, slopes are typically gentle and topoclimate is constant, but soil variations can be substantial (Mitchell et al. 2014). These local variations depend on the parent material of soils and groundwater depth, which can range from near surface to well below the rooting zone. Local groundwater circulation redistributes rainfall recharge from uplands to bottomlands, distributing water and nutrient subsidies to some areas and maintaining more mesic conditions—with the potential for higher ET and C sequestration—in other areas.

Landscape Models Coupling Carbon and Water

Several models have been designed to address these challenges for some combinations of C, water, and nutrients, but relatively few models simultaneously address C and water at scales that are large enough to be useful for complex, mixed land cover watersheds. One of the most useful models is the Regional Hydro-Ecological Simulation System (RHESSys) (Band et al. 1993; Tague and Band 2004). The RHESSys developers initially combined a stand level water and C cycling model, Biome-BGC (Running and Coughlan 1988; Running and Gower 1991) with a distributed hydrologic base using flow path routing methods—which were adapted from TOPMODEL (Beven and Kirkby 1979) or the Distributed Hydrology Soil Vegetation Model (Wigmosta et al. 1994)—and soil biogeochemistry adapted from CENTURY (Parton et al. 1987; Parton 1996). RHESSys couples water and C cycling by building a network of hydrologic flow paths among ecosystem patches to derive aboveground and belowground primary productivity and water use, as well as surface and subsurface flows. It produces a template that is useful for mapping the gradient of more xeric to mesic edaphic conditions, including long-term changes and patterns in soil organic C.

Structurally, RHESSys is set up as a nested hierarchy of ecosystem patches, hillslopes, and catchments. Within patches, multiple strata (such as species or plant functional groups) can be incorporated; water, C, and nitrogen cycling can be solved for each stratum; and interaction and competition for resources can be tracked. Bottomland or riparian conditions (higher soil water and soil organic C) do not need to be prescribed, but can evolve over varying time periods by the lateral redistribution of moisture and net effects on aboveground and belowground C cycling. RHESSys can incorporate input on shrub and herbaceous layers into canopy gap statistics to provide distributed radiation between canopy and forest floor (Song and Band 2004; Song et al. 2009), with mixed species composition influencing stomatal and aerodynamic conductance. Further, the different strata included within an ecosystem patch can incorporate different root depths, allowing the tracking of interaction and competition for soil water. Disturbances, in the form of fire or management, can be incorporated by altering drainage flow paths or by scheduling events that alter the state variables used to describe canopy and soil conditions, or by altering drainage flow paths. Forest floor litter layers are maintained by C and water-mass balance, providing both total biomass and water content (which can interface with fire models).

Soil condition data must be at the scale of hydrologic flow paths because edaphic gradients can serve as strong drivers of forest growth and exchanges of C, water, and nutrients. However, these scales are often below the resolution of soil mapping—which is typically limited to a specific scale of mappable soil bodies and incorporates multiple inclusions (different soil series and phases) within a polygon. The RHESSys framework uses a knowledge-based method of inferring higher resolution soil properties that reflect the resolution of the available terrain data. If adequate terrain data are available, the Soil Landscape Inference Model (SoLIM) (Zhu et al. 1996, 2010) can be used for inferring spatial variations of critical soil properties in low-relief landscapes; these data are now becoming widely available in the Southeast through state-level LiDAR (light detection and ranging) mapping. Mapping canopy conditions and inferring soil variation at these scales would be required before the landscape-level patterns that characterize and determine edaphic gradients can be incorporated.

Other Scaling Challenges

Importance of Accurate Land Cover Estimates

Landscape-level models, such as RHESSys, rely on accurate land cover data for reliable predictions of NEE or ET. One reason that the potential of longleaf pine to improve water yield may not have been more thoughtfully considered in the past is the general failure to differentiate longleaf pine from other pine-dominated systems at the landscape scale. Much of what we know about pine ET in the Southeast comes from studies of slash pine, loblolly pine (*P. taeda*), and eastern white pine (*P. strobus*). Although these pines generally have higher WUE than hardwoods, they also tend to have higher annual ET rates because of their high leaf-level transpiration rates, year-round transpiration, and high winter interception resulting from the evergreen leaf habit and high stand density (Swank and Miner 1968; Swank and Douglass 1974; Ford et al. 2011; McLaughlin et al. 2013; Novick et al. 2015). However, these generalizations about high leaf-level ET and stand density do not apply to restored longleaf pine, suggesting that an accurate differentiation between longleaf pine cover and other pine land cover is needed.

Using spectral profiles to differentiate among pines is considerably more difficult than distinguishing between pines and hardwoods (van Aardt 2000). Thus, when stand-level ET is scaled to the watershed using remotely sensed data, all pine cover is generally lumped into a single category and potential differences in ET between longleaf pine and other pine species cannot be quantified. In an attempt to separate longleaf pine from loblolly pine, Nieminen et al. (2014) reported slight differences in spectral signatures that were collected using images from the DigitalGlobe WorldView-2 satellite; however, these differences were influenced by season and by the substantial differences in subcanopy vegetation that characterize the stands dominated by the two species. Alternately, Martin et al. (2013) used aerial photography to distinguish planted pine from naturally regenerated forest, but they could not visually distinguish among the pine species. To minimize the time involved in visually assessing land cover for individual photographs, they relied on a subset of available photographs. Although their results are useful, the relatively small geographic area represented could introduce errors into land cover estimates at larger scales. If better methods for distinguishing longleaf pine from other pines could be developed, then more accurate estimates of longleaf pine land cover could be incorporated into watershed scale ET estimates.

Model Validation

Models that link C and water can be useful, but the ability to validate modeled results with reliable independent estimates would add considerable value and veracity to modeled results. However, validating models to determine how land cover change and land management actually affect streamflow in longleaf pine restoration areas is difficult. Unlike controlled watershed experiments—which rely on small, confined, and well-defined catchments—much of the southeastern Coastal Plain does not lend itself to the water balance measurements typically used to balance basin water budgets (calculating ET by subtracting water yield from precipitation). For example, in much of southwestern Georgia and most of northern Florida, significant interactions between groundwater and surface water complicate efforts to study interactions between ET and water yield. In much of the longleaf pine range, reducing ET by managing land cover might increase either stream runoff or groundwater recharge (or both), but our understanding of groundwater and surface water interactions in these systems is incomplete (Rugel et al. 2012). Addressing these challenges will require an interdisciplinary approach and inclusion of experts—in soil hydrology, surface water hydrology, and groundwater hydrology—who work in concert with forest ecologists and landscape modelers.

CAN WATER AND CARBON POLICY PLAY A ROLE IN RESTORING LONGLEAF PINE?

Numerous strategies have been proposed for managing water scarcity in the Southeast. The emerging hypothesis that longleaf pine restoration could contribute to increased water yield, coupled with a growing policy focus on water-scarcity mitigation strategies, suggests that longleaf pine restoration could play a new and important role in the development of "water-wise" conservation easements. Despite this potential opportunity, the extent of restoration could be limited by several factors, including a current policy emphasis on C sequestration, public reluctance to accept frequent fire, and the long-term commitment required for successful restoration (see Chapter 3). Ultimately, the greatest challenge to expanding restoration at scales large enough to have a measureable effect on regional water scarcity may simply be that alternate land uses can generate substantially more income. Effective policies to encourage restoration will be needed to overcome these income and price barriers.

Unlike much of the forested land in the Southern Appalachian Mountains and in western states, forests in the Coastal Plain and Piedmont are primarily controlled by private landowners (Conner and Hartsell 2002). Landowner choices reflect the valuations placed on various land uses and on perceptions of risk. In such a landscape, the variety of land uses often reflects differences in site productivity and comparative advantage as well as variations in the nonmonetary benefits that owners expect from their holdings. Land use in the Southeast has shifted substantially over several decades as crop agriculture has generally trended downward—and is now concentrated in a few areas (such as Florida, the Mississippi Delta) or elsewhere in the United States (Wear 2013). Agricultural commodity prices can largely explain the shift in land uses, with prices for major crops trending downward during the 1980s and 1990s, reaching a nadir and then trending upward over the past decade, and remaining strong in recent years. Evidence from various data sources—including the National Resources Inventory (USDA 2015) and U.S. Forest Service survey results—indicates a concomitant shifting of agricultural land area to forest uses and some increase in forest land area shifting to agricultural uses. These recent price dynamics do not represent a projection of future price paths but they do indicate that in rural areas—where land use changes may be dominated by demands for developed land—the area deemed suitable for forest management, and thus for longleaf pine restoration, is influenced by the condition of competing agricultural markets.

Competition among wood products also influences a landowner's forest management choices. Since the Great Recession (2007–2009), demand and prices for sawtimber have been substantially lower than historical averages. In comparison, pine pulpwood returns have remained strong and even increased somewhat, potentially explaining the substantial increase in loblolly pine plantations in the Southeast. This reflects a combination of cyclical factors and long-term trends in demand (Wear et al. 2016). However, it also highlights higher returns from traditional intensive pine management and much less variance in pulpwood returns, at least in the short term; both of these factors favor short-rotation pine management over either less intensive forest management or longleaf pine restoration (Wear et al. 2016).

Because agricultural commodity prices, timber commodity prices, and price variance all play into land use decisions, they need to be addressed in the design of policy incentives. Conservation-oriented tax incentives influence land use choices for owners of marginal lands or for owners who are more interested in conservation than maximizing income generation. But with few exceptions, current tax incentive programs are not designed to emphasize low-density forests and frequent applications of prescribed fire. Although they offer considerable support for longleaf pine restoration to improve wildlife habitat, these incentive programs would have more value if they included offsets to forest water use. Such programs, however, could conflict with current or future C policy that may or may not favor longleaf pine restoration. Ultimately, land use portfolios represent a combination of uses that collectively attempt to hedge against future risks. Longleaf pine, even where it does not define the highest return to management, could be a part of the portfolio given its comparatively low risk and high value profile. Adding additional financial benefits through tax incentives focused on water-based ecosystem services (such as higher streamflow) could further increase that value.

For water valuation programs to be successful, scientists and policy makers will need to balance C sequestration goals with water supply needs, especially in areas that are (or could be) affected by water scarcity. This shift in policy would require acknowledgement that some natural and restored ecosystems will likely have lower C sequestration—at least in the short term—than the rates observed over the past century. Such a shift would also suggest that rather than focusing on C sequestration or any single ecosystem service, policy makers may need to focus more on the overall suite of services, including ecosystem resilience. Of course, this type of "bundling" may be optimal for some locations—depending on local hydrology, soil characteristics, or land use history—but specialization may serve better in others, emphasizing the need for better scientific assessment in policy design.

SUMMARY

Numerous successes in improving ecosystem function and increasing the availability of ecosystem services in the Southeast can be attributed to the restoration of longleaf pine. The large majority of these successes have been measured in relation to wildlife conservation, with additional value realized in benefits to human health (Gleim et al. 2014). Even with these successes and continued efforts to restore longleaf pine, more information is needed to understand fire-managed longleaf pine restoration and appreciate its value in addressing regional environmental issues such as water scarcity and global environmental issues such as climate change mitigation. Our review suggests that pursuing such restorations would likely contribute to water scarcity mitigation while also contributing modestly and sustainably to long-term C sequestration. However, these benefits come at the expense of short-term C sequestration and would require refinement of policies that currently use short-term metrics (such as NEE) to evaluate C sequestration potential. Otherwise, these benefits are highly complementary to established restoration goals and thus represent added returns on investments in restoration projects.

We have identified several critical needs and next steps for improving the metrics used to quantify the effects of longleaf restoration on bundled ecosystem services. These include additional measurements of NEE and evapotranspiration across a wider geographic range, continued support for existing studies that collect data on long-term NEE and evapotranspiration, improvements in modeling, and better coordination and synthesis of research across the region. Addressing these needs and using subsequent findings to inform policy would provide one avenue for increasing restoration opportunities while moving toward water sustainability in the Southeast.

Finally, we suggest better mechanisms for communicating the potential benefits of longleaf restoration to land managers, policy makers, and other stakeholders—especially those concerned with mitigating water scarcity.

ACKNOWLEDGMENTS

We thank Steve Golladay, Dennis Hallema, Paul McCormick, and Chelcy Miniat for providing valuable insights during discussions about the links between terrestrial ecosystems and water yield. Lindsay Boring, Louise Loudermilk, Daniel McLaughlin, and Ge Sun provided thoughtful reviews of our preliminary drafts. Haley Ritger, Stribling Stuber, Scott Taylor, and numerous other technicians and students contributed to the data presented in this chapter.

16 Longleaf Pine Restoration in Context
Comparisons of Frequent Fire Forests

Seth W. Bigelow, Michael C. Stambaugh, Joseph J. O'Brien, Andrew J. Larson, and Michael A. Battaglia

CONTENTS

INTRODUCTION

To see a frequent-fire forest burn for the first time is to experience a remarkable feat of nature. Most people are accustomed to the slow change of forests with the seasons, not the instantaneous conversion of green and brown plant mass to smoke and char. Yet to visit such a forest a week after it burns is to see bright green shoots emerging, highlighted against a background of charcoal. Frequent-fire forests, or forests that regularly experience low-intensity/low-severity fires, although surprisingly common, challenge commonly held notions about what forests are and how they function. They are found in North and Central America including the Caribbean basin and U.S. landscapes such as the upper Midwest, the central hardwoods area, the Rocky Mountains, the Intermountain West, the eastern Cascades range in the Pacific Northwest, and the southeastern Coastal Plain. Despite their drastic differences in range, ecology, anthropogenic alterations, and conservation challenges, these forests share many similarities.

The purpose of this chapter is to place the structures and processes of the frequent-fire longleaf pine (*Pinus palustris*) ecosystem in the broad context of other forest ecosystems that historically experienced frequent fire. We also compare the restoration challenges of longleaf pine with those in other frequent-fire forests. First, we address the ecological commonalities among frequently burned forests and the kinds of degradation that threaten them. We provide vignettes of five other frequent-fire forests; examine how their ecology, restoration goals, and restoration approaches differ from longleaf pine; and evaluate whether restoration trends and ideas in other frequent-fire areas might be relevant for the longleaf pine range (and vice-versa).

COMMON FEATURES OF FREQUENT-FIRE FORESTS

At large spatial scales (such as 1000 km), climate is the most important influence on where frequent-fire forests occur (Hawbaker et al. 2013). Higher average annual temperatures are negatively correlated with fire-return interval (Guyette et al. 2012), but this relationship can be modified by soils (Murphy and Bowman 2012): for instance, droughty excessively well-drained soils can support frequent-fire forests and savannas in cooler areas such as the sandy glacial outwash plains of the upper Midwest. In mountainous areas, slope and aspect alter the interception of sunlight, thereby influencing forest type and fire-return interval over fine spatial scales (Cansler and McKenzie 2014). Because lightning is the sole source of ignitions that are not caused by humans, the incidence of lightning strikes is important; for example, the extremely high incidence of strikes in the Southeast contributes to the development and maintenance of its frequent-fire forest (Outcalt 2008). Annual precipitation has a U-shaped relationship with fire-return interval, in that forests with very short (≤2 years) fire-return intervals only occur in places that experience either very low (Southwest) or very high (Southeast) annual precipitation (Guyette et al. 2012). The timing of precipitation delivery and warm temperature occurrence is also important; despite high precipitation, frequent-fire forests can develop in western U.S. areas that have high mountains and Mediterranean climates because most of the precipitation takes the form of snow and is lost in the spring snowmelt (Stephenson 1990). Although such relationships are recognized at local to national scales, a global synthesis on what determines fire regimes in world ecosystems has not yet been developed (Bond and Keeley 2005).

Frequent-fire forests possess convergent structural characteristics regardless of the climatic and soil factors that originally shaped their development. In savannas and woodlands, one of the most notable characteristics is the open canopy condition, which permits abundant sunlight to pass through to the forest floor (Battaglia et al. 2003; Bigelow et al. 2011). Open canopies are the result of complex feedbacks among fire, climate, and vegetation that can contribute to limitations on fire behavior (Collins et al. 2009; Mitchell, Hiers, et al. 2009; Scholl and Taylor 2010; Parks et al. 2015). Surface fires help to maintain an open canopy by scorching and killing lower branches and branch tips; they allow overhead winds to ventilate the stand and dry fine fuels, dissipating heat and smoke that would otherwise build up (Albini and Baughmann 1979). Equally important, frequent surface fires kill many of the small trees that would otherwise eventually occupy available canopy space (Grace and Platt 1995a). In comparison to the tree species that characterize crown-fire regimes, the species of frequent-fire forests tend to have thick bark at the base of their trunks, the potential to become tall, and self-pruning ability (Pausas et al. 2004).

The role of fire as an agent of mortality for young trees in frequent-fire forests has striking consequences for species composition and forest structure. Young trees (saplings) are disproportionately susceptible to mortality and the variation among species in sapling bark thickness and other heat-insulating qualities is considerable, which means that frequent fire favors fire-resistant species (van Mantgem and Schwartz 2003; Hammond et al. 2015; Pausas 2015). Such species tend to produce pyrogenic litter that creates a positive feedback to the fire regime (Nowacki and Abrams 2008; Platt et al. 2016). In frequent-fire forests that have active fire regimes, the tree populations often have a broad irregular distribution of sizes with several peaks representing age classes that have established in the canopy gaps formed by disturbances (Arno et al. 1995; Bailey and Covington 2002; Moser et al. 2002; North et al. 2005).

The cultural traditions involved in human use of fire vary among the frequent-fire forests across the North American continent, but one common theme is the universal practice among indigenous peoples of applying fire as a tool for modifying the environment (Ryan et al. 2013). This practice has persisted and developed in virtually all areas of the continent (Pyne 1982). Fire histories of western landscapes show an abrupt decrease in fire frequency following the period of rapid settlement that characterized the latter half of the 19th century (Arno et al. 1995; North et al. 2005; Sherriff and Veblen 2007). The indigenous practice of woods-burning was adopted by European settlers in the Southeast to a far greater extent than in other regions. These burning practices have served as a model for reintroduction of fire from the mid-20th century onward. The preeminent 20th century California fire ecologist Biswell (1989) described how the course of his career was changed in 1940 by a day spent with a Georgia timber company employee who was assigned to burn pinelands; the experience convinced him that prescribed fires could be used beneficially in forest management. Today, prescribed burning is still done on a much larger area in the Southeast than in any other region (Melvin 2015).

FOREST DEGRADATION: DISTURBANCE OUTSIDE OF HISTORICAL RANGE OF VARIATION

Fire maintains the structure and function of frequent-fire forests (Bond and Keeley 2005), and fire exclusion constitutes a damaging disturbance that has been all too common over the past 200 years (Stephens and Ruth 2005; Fill et al. 2015). Fire exclusion can take the form of active suppression of naturally ignited wildland fires, or neglect of prescribed fire as a management tool. Fire regimes have also been altered by more intensive harvesting than was typical during the development of frequent-fire forests and by landscape fragmentation, which prevents the spread of fire; such fragmentation can occur from roads, fences, housing, and wholesale conversion to agriculture (Duncan and Schmalzer 2004).

One common consequence of fire exclusion in frequent-fire forests is the establishment and growth of fire-intolerant shrub and tree species (Parsons and DeBenedetti 1979; Arno et al. 1995). This midstory development has the effect of decreasing transmittance of light to the forest floor, thus competitively eliminating many ground cover species because most plants of frequent-fire forests are adapted to abundant sunlight. Effects include increases in aboveground live biomass, total leaf area, and canopy-cover of trees that are less drought-tolerant and use more water per unit of leaf surface area than the trees they are replacing (Nowacki and Abrams 2008). Similarly, regeneration of dominant fire-adapted tree species may sharply decrease, either because their establishment depends on bare mineral soil, or because they are poor competitors and are sensitive to above- or belowground competition. This compositional change leads to increased whole-ecosystem water use, and can also result in decreased ecosystem resilience and increased drought vulnerability (Niinemets and Valladares 2006; Ganey and Vojta 2011; Dobrowski et al. 2015; van Mantgem et al. 2016).

Parsons and deBenedetti (1979) observed that fire exclusion in frequent-fire forest leads to a buildup of dead fuels and an altered canopy structure that forms "ladders" for fire to climb into tree crowns (the fire exclusion/fuel buildup perspective). Some of the evidence for this fire exclusion/fuel buildup hypotheses is that the proportion of a landscape burning at high intensity depends on the amount of time since the last fire or the amount of time since the departure from typical fire-return intervals (Harris and Taylor 2015; Steel et al. 2015). Others caution that a distinction must be made between fire exclusion in frequent-fire forests and fire exclusion in fire-dependent forests that have longer and more irregular fire-return intervals; for the latter, fire exclusion may not necessarily lead to larger or more intense wildfires (Johnson et al. 2001; Noss et al. 2006).

Expansion of shade-tolerant native trees from fire-free areas into fire-dependent forests is both a cause and a consequence of fire-regime interruptions. Under normal frequent-fire regimes, these trees are restricted to moist, shaded microsites (such as riparian areas or narrow ecotones between vegetation types). In part, these invading trees alter the fire cycle by shedding litter that holds moisture well or otherwise decreases flammability (Stephens et al. 2004). The shape and size of their litter, principally dead leaves, is a key determinant of their pyrogenic properties (Kane et al. 2008). Some nonnative plants can also disrupt the fire cycle. They carry fire exceedingly well and burn more intensely than the ground cover plants that they replace, predisposing canopy trees to increased mortality and thereby destabilizing the basic scaffolding of the forest (Brooks et al. 2004).

Isolated relict shade-intolerant trees of the historical forest are interspersed with dense, clumped stands of shade-tolerant trees in some fire-excluded forests (Gilliam and Platt 1999; Taylor 2004). In others, fire exclusion may simply result in higher density of shade-intolerant trees (Laughlin et al. 2011). These shade-tolerant trees tend to be Douglas-fir (*Pseudotsuga menziesii*) and the true firs (*Abies* spp.) in western landscapes, and mesic hardwoods in central and eastern landscapes (Larson and Churchill 2012; Hanberry et al. 2014). Dense stands of shade-tolerant trees result in a light-deprived ground cover that impedes the regeneration of shade-intolerant canopy trees (Veblen and Lorenz 1991; Gilliam and Platt 1999; Stambaugh and Muzika 2007; Bigelow et al. 2011).

Excessive harvesting of trees or other plants in frequent-fire forests also constitutes a disturbance that may be outside the historical range of variation. Commonly, frequent-fire forests are dominated by very large and old trees (legacy trees), which exert control on ecosystem properties and are therefore classified as keystone structures (Lindenmayer, Laurance, et al. 2012). This control can include suppressing competing trees, providing specialized habitats for mammalian or avian wildlife, and sequestering large amounts of carbon (Lutz et al. 2012). Large trees generally increase ecosystem heterogeneity both vertically (canopy height variability) and horizontally (within-stand patchiness). Common vegetative responses to the removal of large trees can include the release of younger age classes either of the same species or of more shade-tolerant species, setting the forest on a different successional trajectory. Such responses to large-tree removal usually result in stands that have more homogeneous structure in both their horizontal and vertical dimensions (Churchill et al. 2013). Once excessive harvesting of large old trees has occurred, restoring the characteristic ecosystem structure and function is difficult, especially in low productivity sites.

Another major consequence of changes in the fire-vegetation cycle is the alteration of dead wood dynamics. Dead wood, both as standing snags and as coarse woody debris, is a major habitat element in frequent-fire ecosystems (Harmon et al. 1986). Even though altered frequent-fire ecosystems continue to produce dead wood, its quality and dimensions may preclude its use as specialized habitat by many wildlife species. Many woodpecker species will only use large-diameter snags for excavating their nests (Zarnowitz and Manuwal 1985). Large down wood is used for many purposes by wildlife, including as runways for small mammals and as subnivean runways and refuges in snowy areas by the American pine marten (*Martes americana*) and other mid-sized mammals (Haggstrom and Kelleyhouse 1996). Because down wood in frequent-fire forests changes constantly and is susceptible to consumption by surface fires, a frequent-fire forest that loses live large-diameter trees can require many decades to restore its large dead wood component (Knapp 2015).

FIVE FREQUENT-FIRE FORESTS

The examples described above show some of the many disruptions threatening the characteristic fire-vegetation cycles that maintain structure and diversity of frequent-fire forests. To provide multiple points of reference for understanding the longleaf pine ecosystem, we present vignettes of five other U.S. and Caribbean basin frequent-fire forests (Figure 16.1). They represent unique ecosystems, each with its own conditions and restoration approaches. The selection, which provides a broad geographic representation of the frequent-fire forest, is in no way intended to be comprehensive but rather to show how the interplay between ecosystems and human concerns determines restoration needs and priorities.

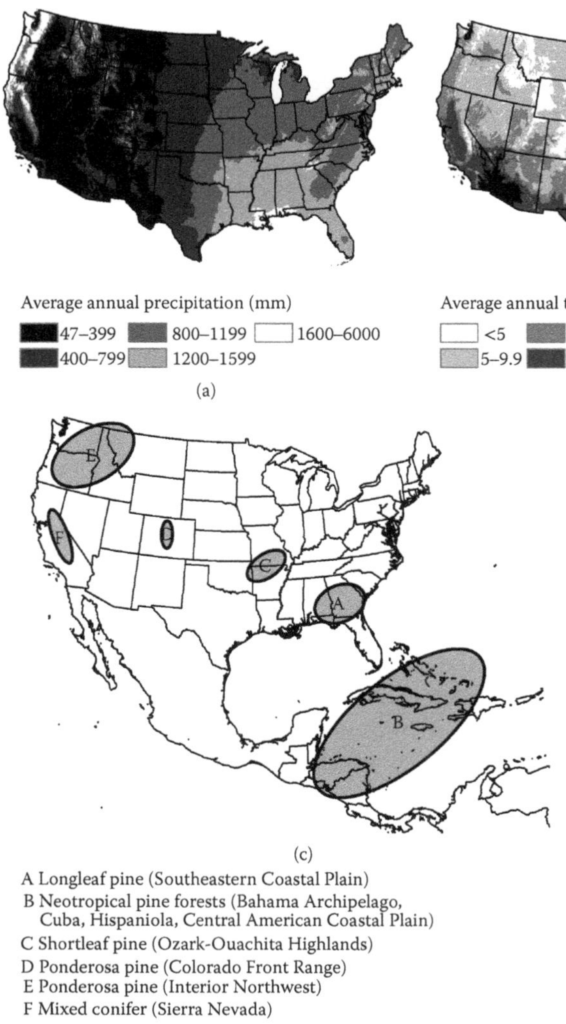

FIGURE 16.1 Continental U.S. climate gradients, 1981–2011, and location of the frequent-fire forests: (a) Average annual precipitation (b) average annual temperature, and (c) location of five frequent-fire forest areas in the Caribbean and North America. (Data from Prism Climate Group, Oregon State University, http://prism.oregonstate.edu, created 15 December 2015.)

Neotropical Pine Forests of the Caribbean Basin

The Caribbean Basin is home to several tropical pine ecosystems that depend on frequent fire to maintain structure and species composition, with fires typically recurring at 1–5-year intervals (Myers and Rodríguez-Trejo 2009). These forests occur in the Bahama Archipelago, Cuba, Hispaniola, and the Coastal Plain of Central America (Figure 16.2). The overstory is composed of one or two of several species—the Caribbean pine (*P. caribaea*), the West Indian pine (*P. occidentalis*), or the tropical pine (*P. tropicalis*)—all members of the southern pine group, subsection Australes; the midstory is sparse or absent; and the ground cover has a diverse assemblage of palms, shrubs, and herbs. In the Bahama Archipelago (Commonwealth of the Bahamas and Turks and Caicos Islands), the Bahamian variety of Caribbean pine (*P. caribaea* var. *bahamensis*) forms the canopy. On Hispaniola, the dominant pine is West Indian pine; in Central America, it is the Honduran variety of Caribbean pine (*P. caribaea* var. *hondurensis*). Cuban pine forests are dominated by the Caribbean variety of Caribbean pine, (*P. caribaea* var. *caribaea*), by tropical pine in sandy dry sites, or occasionally by the two species occurring together (de las Heras et al. 2005). The tropical pine forests are the ones most functionally similar to longleaf pine forests: they have a grass stage and their ground cover is dominated by graminoids.

Forest soils include acidic quartz sands in Cuba and the Central American Coastal Plain and volcanic clays in Hispaniola. Pines also occur on exposed limestone outcrops in the Bahama Archipelago, and portions of Hispaniola and Cuba. Limestone is an unusual substrate that has contributed to high levels of endemism in these forests (Cano Carmona and Cano Ortiz 2012).

Fire Regime

Fire-return intervals in the area are typically <10 years and can be as frequent as annually (Myers et al. 2006; Harley et al. 2013). Canopy-derived fuels are critical in maintaining fuelbed continuity in the many Caribbean pine forests growing on exposed limestone. For example, in pine rockland forests—such as those of the Bahamas and Hispaniola—fire frequency is largely determined by overstory fuel production because the exposed limestone interrupts the continuity of other fuels. Vegetation occupies pockets of soil separated by patches of bare rock. If mature trees are present, pine needles carpet the rock allowing fire to spread across the landscape by creating fuel continuity among the patches of vegetation. Overstory productivity is driven primarily by rainfall. In areas with Ultisols and other better developed soils, ground cover that carries fire more evenly and regularly can often develop (Kellman 1984). Such areas include Honduran coastal areas and the western highlands of Cuba (de las Heras et al. 2005).

Changes Resulting from Fire Exclusion, Grazing, and Logging

When fire is excluded, shrub-form broadleaved species are released and can replace pines as the dominant overstory in as little as 25 years. Litter shed by these broadleaved species contributes to changes in the ground cover moisture, which make broadleaf-dominated areas much less likely to burn. In areas where fire is customarily applied, overly frequent or intense fire can produce undesired outcomes. For example, fires lit annually at the height of the dry season in the Miskito savannas of Honduras result in intense and complete burns that inhibit pine regeneration, causing the conversion of extensive pine forests into grasslands. Elsewhere, lower-intensity annual fires can result in regeneration failure and ultimately, conversion to grassland (Myers et al. 2006).

In pine rockland ecosystems, the loss of the pine overstory results in fuel discontinuity, interrupting the fire regime and allowing the release of fire-suppressed broadleaved species. Nevertheless, these frequent-fire forests can often recover from overstory-removing disturbances such as hurricanes if regeneration persists and reestablishes fuel continuity (O'Brien et al. 2008). In areas where annual fire is applied in combination with logging, conversion of woodland to prairies can occur rapidly, as has happened in some areas of Honduras (Myers et al. 2006). Another form of ecosystem degradation is the nearly range-wide replacement of tropical pine with plantations of the Caribbean variety of Caribbean pine, which does not have a grass stage (de las Heras et al. 2005).

(a)

(b)

FIGURE 16.2 Frequent-fire pine forests of the Caribbean basin: (a) Ground cover fire in a Caribbean forest, Abaco National Park, Bahamas; (b) Caribbean pine, Rio Platano Biosphere Reserve, Honduras.

(Continued)

(c)

FIGURE 16.2 *(Continued)* Frequent-fire pine forests of the Caribbean basin: (c) West Indian pine, Sierra de Bahoruco, Dominican Republic. (Photographs courtesy of Joseph O'Brien.)

Restoration, Management, and Conservation

Few conservation measures are in place for tropical pine ecosystems. However, an even more critical issue for management of Caribbean basin pine forests is the maintenance of appropriate fire regimes. Wildfire is a major concern, and countries such as the Dominican Republic and Honduras have forbidden the intentional burning of forests. The exception is the Commonwealth of the Bahamas, which has recognized the value of fire as a management tool and has recently instituted legal changes to allow prescribed burning.

Sea level rise resulting from climate change is placing stress on these ecosystems and sometimes hindering recovery from disturbance. Inundation of islands by salt water following hurricane storm surges exacerbated by high sea levels can result in the complete loss of pines (Ross et al. 2009; Maschinski et al. 2011). In the Turks and Caicos, an infestation of an invasive scale insect resulted in the near complete loss of both the pine overstory and the regeneration in the ground cover (Malumphy et al. 2012). The sea level rise likely decreased the availability of fresh groundwater, which in turn would have increased the vulnerability of the pines to the invasive insect. In both situations, formerly pine-dominated areas are likely to transition to tropical broadleaved forests in the absence of intensive management intervention. Ultimately, the shift to dominance by broadleaved, fire-sensitive trees will lead to the loss of endemic fire-dependent plants and animals. The interaction of sea level rise and the ecology of fire-dependent ecosystems in low islands is a major conservation concern.

Shortleaf Pine Forests in the Ozark Highlands

Shortleaf pine (*P. echinata*) is the northernmost member of the southern pine group, extending beyond the Piedmont Plateau and into the central hardwoods forest. Its broad range from

Connecticut to West Texas is indicative of its ability to grow in a multitude of climates, given favorable site conditions. The Ozark-Ouachita Highlands area of Missouri, Arkansas, and Oklahoma represents the northwestern range margin (Figure 16.3). Its climate is transitional between continental and subtropical. The northern margin of the shortleaf pine range aligns

(a)

(b)

FIGURE 16.3 Shortleaf pine forests: (a) Ozark-Ouachita Highlands in Missouri, and (b) Boston Mountains in Arkansas. (Photographs courtesy of Michael Stambaugh.)

with the 43 cm winter precipitation isohyet; tree growth is limited by extreme winter temperatures (Fletcher and McDermott 1957; Stambaugh and Guyette 2004). Historically, shortleaf pine was a dominant forest type in this area, occurring in pure stands even near the outer margin of its range (within about 100 km).

Shortleaf pine is highly tolerant of stressful conditions, which provides an advantage over its oak (*Quercus* spp.) and hickory (*Carya* spp.) competitors on steep slopes and poor soils (Fletcher and McDermott 1957). Like eastern redcedar (*Juniperus virginiana*) it can occupy cliff edges and rock fissures. Southerly slope aspects with high solar radiation provide warm microsites where shortleaf pine thrives in northern parts of its range. It can occur naturally on gentle slopes, but usually only those with impoverished soils. Many shortleaf pine sites are sandy, or have a fragipan (a hardened, brittle soil horizon) that restricts growth of competing species (Graney and Ferguson 1972).

Shortleaf pine is an early successional shade-intolerant species, with little capacity to replace itself in closed-canopy forests, and only limited potential for recruitment through gap-phase succession in mixed oak-pine stands (Stambaugh and Muzika 2007). Historical sources—such as explorer notes, surveyor records, tree-rings, and photographs—consistently portray shortleaf pine as occurring in open, fire-maintained plant communities (Schoolcraft 1821; Guyette et al. 2006). Pine-bluestem (*Andropogon* spp.) savannas or pine-oak/oak-pine woodlands were common historical shortleaf pine communities, a sharp contrast to the closed-canopy shortleaf pine forests of today (Hanberry et al. 2014).

Fire Regime

Frequent surface fire regimes (1–15-year intervals) characterized historical shortleaf pine communities throughout the Ozark-Ouachita Highlands (Stambaugh et al. 2013). Historical fire events were sometimes extensive, particularly during droughts and times of increased human population size and economic activity (Guyette et al. 2002, 2006). Shortleaf pine fire tolerance is based in part on a thick bark and the ability to resprout when small, for example, < 20 cm d.b.h. (Garren 1943). Other indicators of fire tolerance and dependence include its self-pruning ability, the presence of axillary buds, and the inability of seedlings to establish on deep litter (Grano 1949). Frequent burning promotes shortleaf pine establishment by reducing litter, killing fire-sensitive hardwoods, and increasing transmission of light to the ground cover thereby minimizing damping-off by reducing moisture near the ground surface. Other poorly quantified effects of repeated frequent burning include decreases in coarse woody debris, changes in organic matter, and fuel type changes from leaf litter to forbs and grasses. Burning to restore shortleaf pine provides ancillary benefits of increased soil nutrient availability, wildlife forage, and plant diversity (Masters et al. 1993; Liechty et al. 2005).

The history of fire in shortleaf pine forests since the 17th century provides insights into the conditions necessary to promote the species. Research on historical fires in shortleaf pine stands indicates that long-term, very frequent fire (1–4-year intervals for decades) can be detrimental to the survival of shortleaf pine seedlings (Dey and Hartman 2005; Stambaugh and Muzika 2007). The current understanding is that shortleaf pine requires intermediate and variable fire frequency to allow for regeneration and recruitment while controlling hardwood competition.

Changes Resulting from Fire Exclusion, Grazing, and Logging

Challenges with sustaining shortleaf pine often arise from deep shade conditions, inadequate seed sources, and hardwood competition. Much of the present hardwood-dominated forest in the Ozark-Ouachita Highlands originated as shortleaf or mixed shortleaf-hardwood forests and underwent succession to hardwoods as a result of fire exclusion.

When hardwoods are clearcut on sites formerly dominated by shortleaf pine, their sprouts and small residual stems nearly always outcompete planted shortleaf pine even when the pines were established in advance of harvesting. This hardwood-shortleaf pine competitive asymmetry is the primary reason for the decline of shortleaf pine in the absence of fire. Furthermore, current prescribed fires may be less effective at killing or top-killing competing plants than fires of historical regimes because they typically occur under different weather conditions and burn through altered fuel types and thus are less intense and cover smaller areas.

Restoration, Management, and Conservation

Fire suppression policies of the 20th century have caused widespread failure in shortleaf pine regeneration and recruitment. Site preparation techniques used for restoration from bare-ground include the ineffective (and possibly counterproductive) practice of mechanically ripping subsoil layers and burning to enhance growth and survival (Gwayze et al. 2007). Restoration techniques for fire-suppressed, hardwood-choked shortleaf pine forests include silvicultural thinning to 13 m^2/ha, removing midstory hardwoods but retaining some oak and hickory clumps, and prescribed burning every 3–4 years (Hedrick et al. 2007). Shortleaf pine regeneration and growth is difficult to promote on sites that are dominated by hardwoods.

Perhaps the greatest challenge for shortleaf pine management is to increase public acceptance of prescribed fire and other means of controlling hardwood competition (Hedrick et al. 2007). Its history of successful regeneration on highly disturbed sites indicates that shortleaf pine can tolerate extreme disturbances and suggests that areas being restored from hardwoods to shortleaf pine may require more frequent or more severe disturbances than prescribed fire alone. This is particularly true during the initial restoration phases of forests that have high hardwood density. In such situations, higher severity effects can be achieved by felling, girdling, or applying herbicides. Hardwood competition arising from the 20th century propensity for fire suppression may be unprecedented, demanding flexible and innovative management strategies.

Ponderosa Pine and Mixed-Conifer Forests of the Interior Northwest

Dry ponderosa pine (*P. ponderosa*) and mixed-conifer forests make up the major frequent-fire conifer ecosystems of the Interior Northwest and northern Rocky Mountains (Figure 16.4), defined as the area east of the Cascade crest and west of the Continental Divide within the Columbia River watershed (Weaver 1943; Habeck and Mutch 1973). On the hottest and driest forest sites, ponderosa pine forms pure or nearly pure forests and savannas. On more mesic sites, which is the dominant condition for the area, ponderosa pine forms mixed-species stands with Douglas-fir, grand fir (*Abies grandis*), and western larch (*Larix occidentalis*). Douglas-fir and grand fir are both shade-tolerant short-needled species that become dominant in the absence of fire. Grand fir prevails on moister, milder sites compared to Douglas-fir, which is more tolerant of lower temperatures and larger moisture deficits. The thick bark of mature Douglas-fir makes the species moderately fire tolerant (Arno et al. 1995; Clyatt et al. 2016). Western larch is a highly shade-intolerant early-seral species that is exceptionally tolerant of fire (Harrington 2012; Hopkins et al. 2014). It forms extensive even-aged or multiaged stands originating from stand replacement or mixed-severity wildfires (Marcoux et al. 2015), in addition to being a component of mixed-species, frequent fire-maintained stands on drier sites.

The characteristic structure of Interior Northwest frequent-fire forests is a fine-grained mosaic of individual trees, tree clumps, and open areas (Larson and Churchill 2012). Tree densities are typically low, ranging from about 10 to 300 per hectare; higher stand densities tend to occur on moister sites, north-facing aspects, and valley bottoms (Hopkins et al. 2014; Clyatt et al. 2016). This structure, which is mediated by topography, emerges from a cycle of patchy tree mortality and regeneration that is driven by native bark beetles and frequent fires (Weaver 1943; Larson and Churchill 2012). Forests of north-facing aspects and canyon bottoms burn less frequently but at higher severity than forests of adjacent south-facing aspects (Figure 16.4). Large, widely distributed old trees of fire-tolerant species dominate the canopies of open dry sites and dense multistory patches (Lutz et al. 2012; Hagmann et al. 2014; Hessburg et al. 2015).

Fire Regime

Tree-ring based estimates for historical point fire-return intervals were as low as 1 year and as high as >50 years in the dry forests of the Interior Northwest (Weaver 1959; Heyerdahl et al. 2001; Wright and Agee 2004), typically ranging from 5 to 20 years for ponderosa pine forests and 10 to 50 years for mixed-conifer forests (Arno 1980; Agee 1993; Arno et al. 1995). Virtually all of these tree-ring based fire histories show a cessation of frequent fire from about 1880 to 1905, although a few sites with continuing or restored frequent-fire regimes exist in large wilderness areas (Larson, Belote, Cansler, et al. 2013; Clyatt et al. 2016).

Most frequent-fire forests in the Interior Northwest occur in heterogeneous, topographically complex landscapes (Figure 16.4) that also contain mixed- and high-severity fire regime forests (Hessburg et al. 2015). Historical reconstructions show that the fire regime is strongly controlled by climate; years with widespread fires had warm springs and warm-dry summers, and years with fewer or smaller fires had cool springs and cool-wet summers (Wright and Agee 2004; Heyerdahl et al. 2008). The synchronizing effect of climate on fire regimes persists in the modern record, with warm-dry springs and summers being strongly associated with burning on larger acreages (Morgan et al. 2008; Littell et al. 2009), larger areas burned at high severity, and greater spatial aggregation of high-severity areas within fires (Cansler and McKenzie 2014).

(a)

FIGURE 16.4 Ponderosa and mixed-conifer forests of the Interior Northwest: (a) Ponderosa pine and mixed-conifer forest on steep topography 15 years after the 1994 Butte Creek fire, North Cascades National Park Complex, Washington. *(Continued)*

(b)

(c)

FIGURE 16.4 ***(Continued)*** Ponderosa and mixed-conifer forests of the Interior Northwest: (b) Old-growth ponderosa pine forest with pinegrass ground cover on gentle topography, Dugout Research Natural Area, Oregon; (c) ponderosa pine and Douglas-fir woodlands and forests, Imnaha River Canyon, Oregon. *(Continued)*

(d)

FIGURE 16.4 ***(Continued)*** Ponderosa and mixed-conifer forests of the Interior Northwest: (d) Fire-excluded ponderosa pine/mixed-conifer forest—with an overstory dominated by ponderosa pine with occasional Douglas-fir and ground cover dominated by grand fir and Douglas-fir seedlings, Minam River Valley, Oregon. (Photographs, in order of appearance, courtesy of Alina Cansler, Derek Churchill, Andrew Larson, and Andrew Larson.)

Changes Resulting from Fire Exclusion, Grazing, and Logging

Successional changes to forest structure and composition caused by disruption of the historical frequent-fire regime were already apparent by the 1930s (Weaver 1943). In the decades-long absence of fire, heavy fuel loads accumulated while forests became denser and more dominated by shade-tolerant trees (Lunan and Habeck 1973; Arno et al. 1995). These changes were exacerbated by selective harvesting of the largest and most fire-resistant mature pines (Hessburg et al. 2005; Naficy et al. 2010; Merschel et al. 2014). The combined effects of past management includes increased surface fuel loads (Agee and Lolley 2006), increased late-successional multistory stand structures (Figure 16.4), and fewer large trees across the landscape (Hessburg et al. 2015).

Restoration, Management, and Conservation

Five principles for restoration apply broadly to frequent-fire pine and mixed-conifer forests of western mountain landscapes. First, planning and management needs to be conducted at appropriate scales to restore multilevel landscape patterns and processes. Restoration of these frequent-fire forests cannot be achieved through a stand-based approach; the complex topography and landscape heterogeneity of arid western mountainous forests require planning at larger scales (Hessburg et al. 2015). Second, topography is the best guide for restoration of successional and habitat patchworks. Patch sizes and stand structures need to be tailored to ridge, valley, and aspect topographies because these topographic settings give rise to contrasting forest communities and fire regimes (Heyerdahl et al. 2001). Third, spatial patterns of trees need to reflect the expected fine-scale heterogeneity that is appropriate for the natural disturbance regimes and biophysical setting. Historical reconstructions of fine-scale tree spatial patterns are useful for achieving this effect (Larson and Churchill 2012; Churchill et al. 2013; Clyatt et al. 2016). Fourth, successful reintroduction of frequent fire requires close coordination of fire and vegetation management specialists for the management of fuel amount and configuration so as to avoid undesired fire behavior. Methods for reducing risk of crown fire include increasing canopy base height and reducing both crown bulk density and surface fuels (Agee and Skinner 2005). Finally, large old trees—the structural backbone of dry, frequent-fire forests—need to be retained and recruited. Often the most fire-resistant individuals in the population (Wyant et al. 1986; Regelbrugge and Conard 1993), they provide ecosystem services such as habitat for vertebrates and long-term carbon sequestration.

Ponderosa Pine-Dominated Forests of the Colorado Front Range

The Front Range is the easternmost range of the southern Rocky Mountains, rising from the Great Plains, in a series of ridges and valleys that become the foothills of the Rocky Mountains and eventually reach 4300 m at the Continental Divide. This highly dissected landscape provides a range of elevations and aspects that influence the distribution of forest types.

The ponderosa pine-dominated forests of the Colorado Front Range are characterized by rugged, dry, and hot conditions. Annual precipitation is only 500 mm and average annual temperature is 6°C; most precipitation falls as snow but there is a monsoonal influence in the south. As shown in Figure 16.5, the vegetation is characterized by tall conifers with ground cover of herbs or shrubs (Peet 1981). Productivity of forests is fairly low: heights of typical 100-year-old ponderosa pines range from 8 to 24 m. At lower elevations (1700–2000 m) where the foothills meet the Great Plains, scattered Rocky Mountain ponderosa pines (*P. ponderosa* var. *scopulorum*) mix with shrubs or graminoids, depending on the soil type (Figure 16.5). Graminoids dominate the ground cover on finer textured soils and Rocky Mountain juniper (*Juniperus scopulorum*) is often present. An increase in elevation (2000–2200 m) or a northern aspect (or both) provides some additional moisture, allowing Rocky Mountain Douglas-fir (*Pseudotsuga menziesii* var. *glauca*) to become codominant with ponderosa pine. At elevations from 2200 to 2350 m and on southern aspects,

(a)

(b)

FIGURE 16.5 Forests of the Colorado Front Range: (a) Low-elevation ponderosa pine stands on a rocky site at 1900–2200 m elevation; (b) ponderosa pine and Douglas-fir growing on coarse-textured soils. *(Continued)*

(c)

(d)

FIGURE 16.5 (*Continued*) Forests of the Colorado Front Range: (c) Fire-excluded ponderosa pine/ Douglas-fir forest with a dense forest floor of Douglas-fir seedlings; and (d) restored ponderosa pine forest showing individuals, groups of trees, and openings. (Photographs courtesy of Michael Battaglia.)

the xeric foothill woodland consists of sparse ponderosa pine with some Rocky Mountain juniper (Peet 1981). At 2450–2850 m, ponderosa pine woodlands occur on fine-textured soils with a ground cover of mountain muhly (*Muhlenbergia montana*). On coarse-textured soils, the tree density is higher and ponderosa pine and Douglas-fir grow together (Figure 16.5).

Fire Regime

The historical (before 1860) fire regime for the Colorado Front Range was influenced by latitude, elevation, and aspect. Average fire-return intervals ranged from 12 to 59 years with more frequent fires at lower elevations (Veblen et al. 2000; Hunter et al. 2007). Fires occurred throughout the growing season but slightly earlier in the south (Brown et al. 1999; Brown and Shepperd 2001). A low-severity, frequent surface fire regime prevailed at lower elevations (Sherriff and Veblen 2006; Sherriff et al. 2014; Brown et al. 2015). At higher, steeper elevations, with mixtures of Douglas-fir and ponderosa pine, fire return-intervals were longer (>35 years) and historical fire regimes were likely of mixed severity.

Forests under mixed-severity fire regimes contain areas of low, moderate, and high severity burns. Low severity fires might kill seedlings and sometimes saplings, but the high severity fires would kill trees of all sizes, leaving patches of high mortality. Despite ongoing debate about the scale of high severity patches in these forest types and how high severity is defined (Fulé et al. 2014; Odion et al. 2014; Sherriff et al. 2014), the steep complex topography, variable weather conditions, and mixture of forest types clearly had a strong effect on the fire regime.

Changes Resulting from Fire Exclusion, Grazing, and Logging

Ponderosa pine and mixed ponderosa-pine/Douglas-fir forests of the Colorado Front Range have increased in density since settlers began arriving from the East after gold was discovered in the mountains in 1859 (Kaufmann et al. 2000; Ehle and Baker 2003; Sherriff and Veblen 2006). In the early 20th century, fire frequency was substantially reduced with increases in timber harvesting, livestock grazing, and mining (Sherriff and Veblen 2007; Brown et al. 2015). These activities promoted new tree establishment of Douglas-fir in particular, contributing to denser forests and increased canopy continuity across the landscape (Figure 16.5). The prevalence of openings, especially small ones (<50 m in diameter) has decreased (Dickinson 2014). Present-day low-elevation ponderosa pine forests have more regular, homogeneous spacing and age structure than historical forests (Brown et al. 2015). As a result of these changes, contemporary forested landscapes have become vulnerable to high-severity crown fire; over recent decades, several wildfires have burned at high intensity across large, contiguous areas (Graham 2003) resulting in complete overstory mortality at spatial scales that limit ponderosa pine regeneration, especially at lower elevations (Chambers et al. 2016).

Restoration, Management, and Conservation

The Colorado Front Range urban corridor—an area mostly on the Great Plains but also adjacent to forests—is already densely populated and development into the wildland-urban interface is projected to increase (Theobald and Romme 2007). By 2010, the increase in negative ecological, social, and economic impacts from large, high-severity fires led to funding under the U.S. Forest Service Collaborative Forest Landscape Restoration Program, whose goal is to reduce the threat of uncharacteristic fire while increasing forest resilience to fire, insects, disease, drought, and climate change (Haas et al. 2015). The Front Range program uses collaborative, science-based ecosystem restoration and has developed desired conditions for ponderosa-pine dominated forests that follow the principles articulated by Hessburg et al. (2015). The intent is to establish forests that are spatially heterogeneous across plots, stands, watersheds, and landscapes by creating canopy openings and groups of trees while retaining individual trees; the proportion of openings, clumps, and individual trees is determined by topography (Figure 16.5). Old, large-diameter trees are protected. Monitoring of implementation, effectiveness, and ecological impacts is paramount to the adaptive management approach of the program.

Implementation of restoration treatments has revealed some obstacles to whole-landscape restoration efforts (Underhill et al. 2014). Early on, most restoration prescriptions were traditional thin-from-below fuels-reduction treatments in which contractors selected the trees to remove in a process called "designation by prescription" (Dickinson and Cadry 2016). The switch to variable tree spacing using the individuals-clumps-openings approach (Churchill et al. 2013) has been challenging, and for now, leave-tree marking in demonstration areas is being used to guide operators and tree-marking crews. Similarly, the introduction of prescribed fire has been stymied by concerns about smoke and escapes. Prescription burning windows have been narrowed by the proximity to major urban areas, travel corridors, and private lands (Ryan et al. 2013). Heavy surface and canopy fuel loads are a major concern because of the potential for high tree mortality or escapes during prescribed burning (Dether and Black 2006).

Mixed-Conifer and Ponderosa Pine Forests of the Sierra Nevada Range

The mixed-conifer forest of California grows in the rugged mountain ranges of the Sierra Nevada, Cascades, and parts of the coastal ranges. The mixed-conifer forest occurs at mid-elevations, roughly 1000–2000 m. The climate is Mediterranean in seasonality, with most precipitation falling as snow in winter or as rain in spring and autumn; summer droughts can persist from April through September. Average precipitation is about 1000 mm, and average monthly temperatures range from -5°C in January to 30°C in July. The asynchrony between the arrival of moisture and the availability of energy (from increases in air temperature) in the growing season profoundly affects the character of the vegetation (Stephenson 1998).

The vegetation is characterized by tall conifers with a ground cover of shrubs and herbs (Figure 16.6); shrub cover is often larger than that of forbs and grasses. The shrub cover is diverse and includes California lilacs (*Ceanothus* spp.) and manzanita (*Arctostaphylos* spp.). Both conifers and shrubs are sclerophyllous, meaning that they have thick tough leaves that are characteristic of drought-stressed environments. The dominant trees of the mixed-conifer forest (Fites-Kaufman et al. 2007) include California black oak (*Q. kelloggii*), ponderosa pine

FIGURE 16.6 Mixed-conifer forests of the northern Sierra Nevada in California and Nevada: (a) Large Jeffrey and sugar pines, left middle ground, with dense forest floor of small-diameter white fir and Douglas-fir, background. *(Continued)*

(b)

(c)

FIGURE 16.6 (*Continued*) Mixed-conifer forests of the northern Sierra Nevada in California and Nevada: (b) Ground cover of mountain whitethorn shrubs with overstory of incense cedar; (c) stand of white fir and Douglas-fir after mastication of small-diameter stems. *(Continued)*

(d)

FIGURE 16.6 (*Continued*) Mixed-conifer forests of the northern Sierra Nevada in California and Nevada: (d) Smoke from slash pile burn after fuels-reduction thinning, Lake Tahoe Basin. (Photographs courtesy of Seth Bigelow.)

(ssp. *critchfieldiana*), Jeffrey pine (*P. jeffreyi*), sugar pine (*P. lambertiana*), white fir (*Abies concolor*), incense cedar (*Calocedrus decurrens*), and Douglas-fir (*Pseudotsuga menziesii* var. *menziesii*). The conifers grow readily to diameters of >1 m; the maximum height of the tallest of these trees, the sugar pine, is >60 m.

Fire Regime

Estimates of median point fire-return interval before settlement from eastern states (before 1849) for mixed-conifer and ponderosa pine forests in the Sierra Nevada and southern Cascades range from 10 to 20 years (Taylor 2000; Stephens and Collins 2004; North et al. 2005; Scholl and Taylor 2010). The historical fire regime was mixed severity, consisting of mostly low-intensity fires that spread slowly through the undergrowth; flare-ups during which flames passed from the surface to the canopy and then spread from tree to tree—resulting in patches of dead trees that eventually became openings or shrub fields—were likely a normal feature of the mixed-severity fire regime (Collins and Stephens 2010; Collins et al. 2015). Estimates of patch size range from 0.2 ha (Scholl and Taylor 2010) to <4 ha, with occasional patches reaching 60 ha (Collins and Stephens 2010).

Changes Resulting from Fire Exclusion, Grazing, and Logging

The structure of the mixed-conifer forest has changed dramatically from the onset of settlement to the present day. Historical accounts (Muir 1894) and reconstructions of stand structure from early timber surveys and other sources suggest groves of immense trees with open ground cover, characterized by low tree densities (2–315 per hectare). Canopy tree cover ranged from 20% to 30% or 45% (Collins et al. 2011; Lydersen et al. 2013; Barth et al. 2015; Stephens et al. 2015), compared to the present-day range of about 50%–80%. Increases in canopy tree cover have been accompanied by decreases in ground cover; for example, a site in the central Sierra Nevada has undergone a decrease in shrub cover since 1929, down from 30% to 2.5% (E. Knapp et al. 2013).

The higher-density canopy cover of the present-day forest is accompanied by a change in canopy structure—with many more branches and much more leaf area in the midstory. The shade-tolerant white fir, incense cedar, and Douglas-fir have become more common. Because frequent surface fires are no longer a part of the disturbance regimes in most of the forest, regeneration is no longer regulated by fire; consequently, a density-dependent, competition-driven mortality regime prevails. Diameter-frequency distributions more closely resemble the negative-exponential J-shaped curve of closed forests than the flatter curve of open forests, because of numerous small-diameter, shade-tolerant trees (North et al. 2005; Youngblood 2010).

Most dramatic among recent forest changes is the prevalence of large high-severity fires known as "megafires" (Miller et al. 2008). Increasingly, evidence points to a warming climate as the major driver for increased fire extent, intensity, and severity. The proximate mechanisms include a longer fire season owing to the earlier disappearance of the snowpack in spring and later appearance of rains in autumn (Westerling et al. 2006). If the new, high-severity fire regime persists and burns over areas that have not recovered since the last severe burn, the forest may be driven toward a lower-stature, shrubbier, open woodland condition (Collins et al. 2009).

Restoration, Management, and Conservation

Forest restoration management, driven by the need to reduce fire hazard, has been underway at large scales in publicly owned forests of the Sierra Nevada for more than a decade. Fuels reduction generally involves thinning trees from the midstory to remove ladder fuels and removal of some codominant trees from the canopy (Agee and Skinner 2005). The resulting forest has lower canopy cover, lower density of foliage in the upper canopy (lower canopy bulk density), increased spacing among tree crowns, and—because shade-tolerant species are targeted—fewer firs (Figure 16.6). Post fuels-reduction forests still have higher stem density, canopy cover, and abundance of shade-tolerant trees than historical reconstructions suggest they ought to have, which could limit recruitment of shade-intolerant trees (North et al. 2007; Bigelow et al. 2011).

Prescribed fire is required for reestablishing a diverse ground cover of herbs and shrubs (Keeley and Fotheringham 2000; Wayman and North 2007); indeed many of these species have seeds that require smoke for germination, meaning that they will not establish under surrogate treatments. Prescribed fire is integral to reduction of surface fuels (live or dead flammable material within 2 m of the forest floor). Although prescribed fire is a common treatment in the national forests of the Sierra Nevada, benefits could be achieved by applying it—along with managed wildfire—over much larger areas. Constraints on wider use of prescribed fire and managed wildfire on public lands in California have been identified as air quality (Figure 16.6), rural house density, and a risk-averse agency culture (North et al. 2012).

IMPLICATIONS FOR FREQUENT-FIRE FOREST RESTORATION

Restoration and management practices in the frequent-fire longleaf pine forests of the Southeast and other frequent-fire forests have the potential to inform each other. Fire exclusion, heavy logging, and grazing have left their mark everywhere, and degraded frequent-fire forests share functional similarities regardless of location. Potential impacts from climate change are omnipresent, and although novel ecosystems may be created as a result, a better understanding of the historical range of variation is needed in all forests. Although some restoration objectives may be similar among frequent-fire forests degraded by fire suppression, basic differences in the ecological processes of individual systems may require strategic variations from established approaches for changing successional trajectories. Following are common themes and landscape-based variations on needs for restoration research.

Thinning

Thinning is frequently used to restore tree species composition, diameter class distribution, density, and spacing in degraded frequent-fire forests. Midstory trees are conspicuously absent from most frequent-fire forests that have active fire regimes, and prescribing the removal by harvesting of such trees is common for fire-excluded forests (Agee and Skinner 2005). But the results of decades of fire exclusion cannot be undone with a single treatment. Structural and species composition targets can be challenging to meet in stands where minor-component species have become significant parts of the stand. Removing all undesired trees—often shade-tolerant species—can result in a canopy cover that is too sparse to support some wildlife species (Stephens et al. 2014) or provides too little flammable litter to perpetuate a continuous fire regime (Jack, Mitchell, et al. 2006).

When less-desired tree species are retained to maintain a continuous source of fuels, regeneration from these trees can be abundant, overwhelming the seed production of large-seeded pine species. For example, after a restoration thinning experiment in a Sierra Nevada mixed-conifer forest, seed rain of residual shade-tolerant white fir and incense cedar was 5–26 times larger than Jeffrey and sugar pine, compromising efforts to shift stands toward increased pine abundance (Zald et al. 2008). In shortleaf and longleaf pine forests, midstory hardwood removal can leave behind rootstocks that sprout readily, requiring the application of herbicides or a rigorous burn schedule to keep pruning back the new growth by top-killing the sprouts, preventing them from getting large enough to become fire resistant (Jack, Mitchell, et al. 2006).

In forests that have very long-lived trees such as longleaf pine and ponderosa pine, stable populations can be maintained with highly infrequent and episodic reproduction and size distributions in which large trees frequently outnumber small trees (O'Hara 2009). This broad irregular distribution of tree sizes represents a striking contrast to the negative exponential curve that is commonly used as the target stand structure in uneven-aged management systems such as the BDq system in which stocking is controlled by a basal area level, maximum diameter, and a q factor (Guldin 1991; Guldin and Baker 1998). These observations suggest that cutting/thinning to create an exponentially decreasing diameter distribution may not be an appropriate approach for restoring many frequent-fire forests (Franklin et al. 2007). Of particular concern is that the choice of model parameters for the BDq system can result in the cutting of large legacy trees that are both irreplaceable (Hessburg et al. 2015) and critically needed for wildlife habitat, carbon sequestration, and other ecosystem services (Brockway et al. 2014).

Restoring Spatial Pattern

Longleaf pine ecosystems do not have the aridity and topographic complexity that would predispose them to a mixed-severity fire regime, nor do they possess the characteristic patch size of even-aged trees that is sometimes found in conifer forests of semi-arid western landscapes (Palik and Pederson 1996; Pederson et al. 2008; Collins and Stephens 2010). Historical reconstructions are useful for developing silvicultural prescriptions, establishing tree-marking guidelines, and monitoring benchmarks when restoring fine-scale stand structure and tree spatial patterns (Larson and Churchill 2012; Clyatt et al. 2016). As mentioned in the vignette on Ponderosa Pine and Mixed-Conifer Forests of the Interior Northwest, spatial patterns of trees ought to reflect expected fine-scale heterogeneity given natural disturbance regimes and biophysical settings (Hessburg et al. 2015). Intensive exploitation of the easily accessible forests of the southeastern Coastal Plain have left few intact old-growth longleaf pine forests available for study (Varner and Kush 2004; Mitchell, Engstrom, et al. 2009), but spatial analysis of one of the remaining few suggested clumps of seedlings and saplings in a mosaic that is superimposed on a matrix of widely spaced (loosely aggregated) mature trees (Platt, Evans, and Rathbun 1988; Noel et al. 1998).

In most present-day fire-excluded forests, small- and medium-diameter trees occur in larger, denser clumps than they did under historical conditions. Producing more small clumps and small openings (for example, openings with a 25 m radius) is a common treatment goal, which is often achieved by breaking up large clumps (Taylor 2004; Dickinson 2014). In arid western forests, spatial targets include increasing within-stand heterogeneity by creating tree patterns that are composed of local tree clumps, openings, and widely spaced single trees (Knapp et al. 2012; Larson and Churchill 2012; Underhill et al. 2014). Approaches for restoring a historical spatial structure involve increasing stand heterogeneity at several spatial scales (for example, from clumps of trees to stands with gaps of varying sizes). In contrast, the usual fuels-reduction thinning approach strives to maximize space around each residual tree and thereby minimize the risk of crown fire and maximize growth. This approach, although of demonstrated effectiveness in changing fire behavior, deters regeneration of shade-intolerant, fire-tolerant tree species and fails to provide high-quality habitat for wildlife (Bigelow et al. 2011; Stephens et al. 2014). In the western forests, where fire models in current use cannot adequately simulate within-stand heterogeneity, more stand-level research is needed on fire-behavior responses to within-stand heterogeneous spacing treatments. Models that can simulate within-stand heterogeneity and fire behavior are under development (Parsons et al. 2011; Hoffman et al. 2016) but their effectiveness is not yet known (Alexander and Cruz 2013).

Research on the historical structure and disturbance regime of frequent-fire forests suggests that gaps are a pervasive structural element (Palik and Pederson 1996; Noel et al. 1998; Larson and Churchill 2012; Dickinson 2014) that may be necessary for the successful regeneration of light-demanding fire-tolerant species (Stambaugh et al. 2002; Palik et al. 2003; Bigelow et al. 2011); considerations specific to longleaf pine are discussed in Chapters 4 and 7. Gap-based, group-selection silviculture has been suggested as a management system for longleaf pine forests, but results have been mixed. The optimal opening size is unclear despite substantial research on the question (Palik et al. 1997; Pecot et al. 2007; McGuire et al. 2001), and securing consistent regeneration and fire behavior in experimental openings has been difficult (Jack, Mitchell, et al. 2006; Mitchell, Engstrom, et al. 2009). Some have suggested possible gap sizes (Grace and Platt 1995b; Brockway and Outcalt 1998; McGuire et al. 2001; Mitchell et al. 2006) but there is not agreement among the different studies (Pecot et al. 2007). It is probable that there is no one consistent group opening size that can be uniformly applied across the natural range of longleaf pine, and appropriate group opening size should be based on specific site and vegetation conditions. An alternative to gap-based approaches—single-tree selection—has been suggested as an effective silvicultural system that ensures adequate regeneration to sustainably manage multiaged longleaf forests with modest timber yields (Pecot et al. 2007; Neel et al. 2010).

Tree marking is increasingly recognized as an important element of forest management that is worthy of study in its own right (Vitková et al. 2016). Variable-retention or multiscale thinning designs require adjustments to traditional tree-marking systems (Churchill et al. 2013; Brockway et al. 2014). The individuals-clumps-openings (ICO) system (Churchill et al. 2013) is a practical method to manage and monitor heterogeneous tree patterns within patches. Substantial training is required for tree markers to faithfully translate complex prescriptions into marked stands (Underhill et al. 2014). Rapid assessment of tree marking, timely monitoring, and providing feedback to markers are crucial. A team of two markers can be more effective than a larger team that is deployed in the traditional manner of a straight-line formation; markers are encouraged to collaborate on which natural clumps of trees should be retained (Knapp et al. 2012). Designation by prescription may save the time of foresters, but this approach risks becoming a zero-sum game if time saved by foresters becomes time lost by loggers (Dickinson and Cadry 2016). Finally, although research approaches exist for assessing the effectiveness of silvicultural interventions to restore historical spatial structure (North et al. 2007; Churchill et al. 2013), managers lack practical methods for assessing whether the application of spatially heterogeneous prescriptions has succeeded.

Reintroducing Fire to Fire-Excluded Landscapes

Reintroduction of fire is a universally advocated measure for restoration of frequent-fire forests, but the effort is not without many associated challenges (Brown et al. 2004; Taylor 2004; Dey and Hartman 2005; Mitchell et al. 2006; Ryan et al. 2013). Successful reintroduction often requires management of fuel amount and configuration and careful consideration of weather conditions and firing techniques to avoid undesired fire behavior and consequent damage to vegetation. Silvicultural measures such as eliminating ladder fuels and decreasing canopy bulk density (Agee and Skinner 2005) are more likely to be necessary in arid western forests than in central or eastern ones. The humid climate of the Southeast means that extreme fire behavior and escaped fires are lesser concerns than in western forests (Dether and Black 2006); but in all situations, restoration and management of frequent-fire forests benefit from close coordination of fire and vegetation-management specialists.

Successful reintroduction of fire in the Southeast and elsewhere may require that initial applications be conducted under cool-burning conditions to gradually reduce fuels, particularly duff, without causing excessive tree mortality (Stephens and Finney 2002; Varner et al. 2005). Mortality of legacy trees is a common concern when fire is reintroduced to ecosystems that have a long history of fire-exclusion (Maloney et al. 2008; Varner et al. 2009; Harrington 2012). Raking duff away from tree bases is an effective yet time-consuming method of forestalling mortality of large trees; a better understanding of the controls on duff moisture would likely provide more efficient ways of managing tree mortality that results from smoldering duff (Banwell et al. 2013).

Understanding public perceptions—and correcting misperceptions—is vital for successful fire reintroduction. Smoke management, the problem of decreasing burn windows that are enforced to address air quality concerns, and a changing climate (Mitchell et al. 2014) are widespread challenges (see Chapter 13). Research indicates that public approval increases as people become more familiar with the goals of prescribed fire application (Jacobson et al. 2001; Winter et al. 2004). Southeastern states are fortunate in this regard because their long history of forest burning and large area that is burned annually contribute to public acceptance (McCaffrey 2009; Way 2011; Melvin 2015). Effective communication about the goals and procedures of prescribed burning becomes even more important in regions that do not have this history.

Because no pine seedling is fire-tolerant immediately after it germinates, understanding the developmental schedule of fire tolerance is essential for creating prescribed fire regimens that foster pine regeneration and recruitment while selectively reducing competing plants. Prescribed fire that is applied either too often or not often enough for the target species can result in regeneration failure and can sometimes cause pine forests to be replaced by grassland (Myers and Rodríguez-Trejo 2009) or other vegetation types. The development of fire tolerance in longleaf pine is well known (see Chapter 4): newly germinated seedlings become fire tolerant after a year by entering a grass stage in which their long needles insulate the apical meristem; after 5–15 years, the juveniles rapidly grow taller (known as bolting) for several years and are again vulnerable to fire (Grace and Platt 1995a). With continued growth, they eventually become much less vulnerable to fire-induced mortality. The same developmental sequence occurs in tropical pine (de las Heras et al. 2005). Shortleaf pine does not have a grass stage and—depending on site conditions—can take up to 7 years to develop significant fire tolerance (Weddell and Ware 1935). The long fire-free period required for shortleaf pine establishment and recruitment can provide an advantage to competing hardwoods, which can become impossible to control without applications of herbicides. Thus, even with knowledge of the fire-tolerance developmental schedule for a species, achieving management goals using only fire as a tool may still be difficult.

Ground Cover Restoration

In frequent-fire forests, fire is carried by leaves and other ground fuels, by surface fuels such as grasses and shrubs, and sometimes by canopy fuels. The characteristic vegetation that carries fire changes in different areas according to climate and plant life form distribution. In the humid Southeast, the interaction between ground fuels (such as needles) and surface fuels takes on particular importance because high decomposition rates can quickly render needles inflammable (Mitchell, Hiers, et al. 2009; Platt et al. 2016). For example, wiregrass (*Aristida stricta*), with its spreading crown of rigid leaves, provides perches that allow fallen leaves to dry (Hendricks et al. 2002; Nelson and Hiers 2008). When fire is excluded, species richness of the ground cover declines (Moore et al. 2006; Peterson and Reich 2008; Kirkman et al. 2016) and the growth of wiregrass and other surface fuels is curtailed.

Spatial continuity between ground and surface fuels is required for prescribed fire to carry (Miller and Urban 2000). Anything that disrupts fuels continuity (such as logging, roads, or a change in species composition) changes fire behavior and the effectiveness of prescribed fire (Loudermilk et al. 2012). Longleaf pine restoration guidelines emphasize the need for evenly distributed overstory pines to provide the needle cast that is necessary for the spatial continuity of surface fuels (Mitchell, Hiers, et al. 2009). The complete removal of longleaf pine from extensive parts of its former range has provided an opportunity to study the establishment of the species in agricultural fields from the ground up (without recourse to remnant stands or trees). Such restoration involves challenges of establishing a scaffolding of trees, suppressing old-field vegetation, introducing native ground cover, and applying fire (Addington, Greene, et al. 2015). If longleaf pines and native ground cover are planted at the same time, frequent thinning is required to prevent the development of a high, uniform canopy cover that can shade out the ground cover (Harrington and Edwards 1999). Mulligan et al. (2002) propose an alternative: establish longleaf pines first, wait until a dense canopy develops that suppresses old-field plants, harvest some trees for economic returns, and then plant native ground cover (see Chapter 11). Although an even thinning (usually every third to fifth row) would typically be applied, the creation of larger openings—as would be done in a group selection—might be preferable when forest floor restoration is the primary consideration (Sharma et al. 2012). When larger openings are created, however, less needle cast reaches the center of openings, prescribed fire intensity diminishes, and control of hardwoods becomes tenuous (Jack, Mitchell, et al. 2006).

Landscape and Topographic Approaches to Restoration

The incorporation of landscape topography into planning is a frontier for restoration research (Palik et al. 2000; Reynolds et al. 2013; Hessburg et al. 2015). Geomorphology, topography, and soil characteristics can be used to guide restoration of the longleaf pine forests and wet depressions in the southeastern Coastal Plain: because fire frequency is linked to geomorphological characteristics, maintaining longleaf pine structure and regeneration may require a shorter fire interval in more fertile or mesic landscapes than in the deep sand of xeric sandhills (Gilliam and Platt 1999; Kirkman, Goebel, et al. 2004). In western landscapes, topography can guide restoration of successional and habitat patchworks when patch sizes and stand structures are tailored to a particular ridge, valley, or aspect (Underwood et al. 2010; Hessburg et al. 2015). These topographic settings give rise to contrasting forest communities and fire regimes: sparse pine-dominated forests with lower levels of clumping and more openings on southern aspects and ridges compared to mixed-species stands containing multiple canopy layers and large tree clumps on north aspects and valley bottoms (Heyerdahl et al. 2001; Lydersen and North 2012). Regardless of biophysical setting, independent stand-level restoration treatments need to be developed into integrated, multiscale landscape restoration plans (Hessburg et al. 2015).

Wetlands are topographic landscape features of frequent-fire forests that have a much higher conservation importance than their size would suggest (Cohen et al. 2016). Such wetlands include the wet depressions of longleaf pine forests and the meadows of western mountain forests. Wet depressions and meadows have high plant diversity and provide foraging and breeding grounds for many aquatic and terrestrial organisms (Kaeser and Kirkman 2009; Erwin et al. 2016); they are highly vulnerable to degradation resulting from alterations in fire regimes, hydrological period, and grazing regimes (Miller and Halpern 1998; Dull 1999). When the fire regime is interrupted in the wet depressions of longleaf pine forests in southwest Georgia, mesic oaks become established, displacing wetland grass species and drastically altering wetland hydrology and habitat value (Kirkman et al. 2000; Martin and Kirkman 2009). When shrubs and conifers invade meadows of western mixed-conifer and ponderosa pine forests, the culprit is often an alteration in hydrology but changes in grazing and fire frequency can also be responsible (Weixelman et al. 1997; Berlow et al. 2003; Haugo and Halpern 2007). Identifying and correcting sources of wetland degradation is essential for maintaining biodiversity of frequent-fire ecosystems. Fire often burns into wetlands under the natural fire regime, although perhaps not with the same frequency as the rest of the forest burns; this should be anticipated and planned for under prescribed fire regimens.

The Fire Next Time: Managing in the Anthropocene

Human-caused global climate change has accelerated the frequency and scale of stand-replacing wildfires in frequent-fire forests of western landscapes (Abatzoglou and Williams, 2016); these effects are projected to increase and spread throughout other U.S. regions over the 21st century (Pechony and Shindell 2010; Liu, Goodrick, et al. 2013). The increase in uncharacteristically high-severity fire has created a need for better landscape-scale restoration strategies in the wake of large wildfires: how do we manage previously fire-excluded, unrestored sites that have experienced uncharacteristically high severity fire? At present, management actions are often limited to short-term emergency response activities (such as culvert replacement) and controversial salvage proposals. Landowners and forest managers critically need models and tools to help them determine where fires have achieved restoration objectives and where additional postfire restoration or climate change adaptation treatments are indicated. The millions of trees killed by bark beetle outbreaks and historic weather events such as the California drought of 2013–2015 (Potter 2016) creates a related need: this raises questions about how to manage the resulting dead biomass and how to foster the development of an ecologically appropriate, climate-resilient forest community. Historic droughts and fires of 2016 in the Southeast suggest that this region is not immune from the cataclysmic ecosystem changes that are occurring in western landscapes.

CONCLUSIONS

The preservation of a culture that favors prescribed burning and the development of a fire-dependent silvicultural system were key contributions of the Southeast to the restoration and management of frequent-fire forests. The science of frequent-fire forest management has rapidly spread throughout the United States and, despite markedly different landscapes, all regions can benefit from sharing information and analytic approaches. Understanding how regional restoration and fire management practices fit into frameworks for addressing issues—such as land use, climate, and topography—forms the basis for comparing and analyzing different approaches. What must be demonstrated are silvicultural and restoration practices that can be refined to serve specific goals, meet long-term management objectives for specific forest and wildlife habitats, and ensure continuation of ecosystem services such as watershed protection.

ACKNOWLEDGMENTS

We thank Lindsay Boring, Mike Conner, Yvette Dickinson, and William T. Flatley for reviews that led to improvements in the manuscript.

17 Longleaf Pine Ecosystems
The Path Forward

Steven B. Jack and L. Katherine Kirkman

CONTENTS

LINKING BASIC AND APPLIED SCIENCE

The preface described the goal for this book: to report what is known about the basic ecology of longleaf pine (*Pinus palustris*) and to find connections between this knowledge and its application for successful management and restoration. Developing sustainable restoration trajectories and management approaches for any ecosystem requires both an understanding of its complex ecological processes and unique natural history and a clear recognition that time is a major factor during the developmental process. The transfer of such basic information into practice requires successful application at operational scales, followed by iterative monitoring of change toward long-term goals; when monitoring indicates divergence from the desired trajectory of structural and functional development, adjustment of management practices may be necessary.

This chapter describes how the scientific understanding of ecological processes and the socioeconomic concepts and realities presented in previous chapters shed new light on established paradigms and their application for restoration and management. We also discuss some issues that are larger in scale and focus than the individual chapters. The Preface presented several areas of study for which a more complete understanding about basic ecological processes could improve ecological management or restoration. The reviews and syntheses provided in this book addressed many of the basic and applied questions posed. Here, we present the major findings organized into four areas: plant community dynamics, forest structure and functional relationships, the landscape context of wildlife habitat, and ecosystem services of the future. For each of these areas, we summarize key ecological findings and highlight their implications for restoration and management.

Plant Community Dynamics

Key Ecological Findings

- Species richness is driven by a combination of dispersal-related and environmental filtering factors. A large species pool serves as a lottery-based seed supply, which is sorted by environmental conditions that fluctuate through time; this means that high species richness at fine scales is the result of a very lengthy process, especially for species with a low frequency of occurrence.
- In most sandy Coastal Plain soils, two factors promote high species richness: a large species pool of fire-adapted species and an assemblage of dominant grasses which do not increase in competitiveness with frequent fires or with increased resources. Thus, with frequent fire, dominant grasses are not strong regulators of species richness through competitive exclusion, even at resource-rich sites.
- Although recruitment of some species may be limited by seed dispersal ability, available soil moisture is of paramount importance to species establishment in many localities. Interwoven with recruitment microsite availability and establishment is the stochastically determined mortality of seeds and seedlings that results from fine-scale, heterogeneous fire behavior and intensity.
- Because seed of most ground cover species, including dominant matrix grasses, do not persist in the seed bank, ground cover plant communities are not able to reassemble from the seed bank after a severe disturbance and rely on seed dispersal for recruitment.
- Season-of-burn does not have a direct effect on species richness of the ground cover. Annual summer burning eventually results in the decline of small hardwoods and shrubs.
- In resource-poor sites such as sandhills, the presence of pyrogenic midstory hardwoods has little influence on species richness. In more resource-rich sites, however, fire-intolerant hardwoods can rapidly encroach if they become well established during fire-free periods, competing with other plant species and changing fire behavior.

Implications for Restoration and Management

- Establishment of fine fuels in the ground cover is critical to successful restoration; because C_4 grasses—warm-season grasses that collectively contribute large volumes of fine fuels—do not competitively exclude the establishment of other ground cover species, reintroduction of these grass species needs to be a top priority.
- Ground cover diversity in extremely xeric fire-excluded sites is best enhanced when efforts are directed to the application of frequent fire rather than the extensive removal of pyrogenic hardwoods.
- Establishing fine-scale species richness of the ground cover is a very long process; even if numerous species are reintroduced to a restoration site, the conditions for species establishment are often episodic and the accumulation of species at fine scales depends on environmental stochasticity.
- Assessments of ground cover restoration results would be more informative if changes in reference sites were monitored over time to provide more robust benchmarks than those that are based on static-reference assumptions.
- The most appropriate fire regime for ground cover restoration and management depends on the immediate burn objectives—such as need for hardwood or saw palmetto (*Serenoa*

repens) reduction, promotion of wiregrass (*Aristida stricta*) seed production, or reduction of accumulated fuels—which can change from one burn event to the next. In general, ground cover richness is maintained when longleaf pine sites are burned as frequently as fuel accumulation permits.

- At mesic sites that require aggressive control of midstory shrub encroachment, ground cover restoration is more problematic with fire alone because hardwood growth is extremely rapid in these highly fertile soils; such situations often require supplementary mechanical or chemical intervention.
- When specific information about seed performance or genetics is unavailable, the safest approach is to acquire seed from sources that have similar habitat conditions and are as close to the restoration site as possible. Seed sources from the Southeast are preferable to those from outside the natural range of longleaf pine.

Forest Structure and Functional Relationships: The Critical Pieces of the Restoration Puzzle

Key Ecological Findings

- In longleaf pine forests, natural canopy disturbance processes act over long time spans, resulting in a multiaged forest structure. Small natural canopy gaps—the result of mortality of one to several trees (frequently in response to lightning strikes)—provide regeneration sites for longleaf pine seedlings.
- Aboveground and belowground competition both affect hardwood encroachment and establishment of longleaf pine regeneration. The size of canopy gaps influences the balance between these two kinds of competition by altering soil resource availability, light availability for longleaf pine regeneration, and distribution of fine fuels for application of prescribed fire.
- Productivity can vary twofold or more across a soil moisture gradient; in the absence of fire, site fertility influences both the abundance of fuels and the rate of successional change.
- Longleaf pine forests are limited by water, nitrogen, and phosphorus. Aboveground net primary production is positively correlated with soil moisture availability and negatively correlated with nitrogen availability.
- Individual species and groups of species have important functional roles. Where present in the ground cover, wiregrass serves as a particularly effective fuel for carrying fire; compared to sites in which wiregrass is absent, the presence of dense wiregrass acts to retard woody plant establishment directly (also known as holding the site). The role of nitrogen fixation by legumes is significant in maintaining nitrogen that is volatilized by frequent fire. Fine root turnover of pines and C_4 grasses and the presence of black carbon resulting from frequent fire both play important roles in carbon accumulation in the soil.
- Fine-scale heterogeneity in fuels creates heterogeneous fire behavior. Compared to other U.S. pine species, longleaf pines produce needles that are among the most flammable and cones that smolder longer; thus the distribution of these fuel types can strongly influence fine-scale, localized fire behavior and fire effects.
- Long-term frequent fire regime effects can dramatically shift fuelbed properties toward a condition that promotes positive feedback loops and reinforces the capacity to apply fire over time.
- Despite their small size, embedded wetlands often play a large role as biogeochemical "hot spots" of ecosystem processes across the longleaf pine landscape. The hydrological variability among wetlands promotes spatial and temporal heterogeneity in biogeochemical cycling.

- Land use legacies of agriculture include large reductions of soil carbon and significant changes in soil-available nitrogen and phosphorus, effects that remain for decades after agriculture ceases.

Implications for Restoration and Management

- Ecological restoration and sustainable management of longleaf pine forests for biodiversity and other ecological values is best achieved through a strategy that maintains a perpetual forest canopy and a regime of frequent fire.
- The choice of silvicultural approaches for a multiaged forest structure is affected by the trade-offs associated with canopy gap sizes: gaps must be large enough to provide adequate light for successful longleaf pine regeneration and small enough to allow the fuel continuity needed to limit the establishment and growth of hardwood competitors.
- Because longleaf pine seedlings can survive in the grass stage for many years and then grow out of the grass stage when resources become more available, establishing advanced regeneration that can be released by harvesting provides a competitive advantage in newly created canopy gaps.
- In sites that lack native ground cover, including C_4 grasses in reintroduction projects will ensure the presence of fine fuels and increase organic matter turnover; including nitrogen-fixing legumes will ameliorate the nitrogen lost through repeated fires.
- The role of frequent fire, and thus the role of fine fuels to sustain the application of prescribed fire, is the most critical factor in longleaf pine management, affecting decisions about overstory stocking rates, midstory management (usually mechanical or chemical treatments), ground cover restoration, and wildlife management.

The Landscape Context of Wildlife Habitat

Key Ecological Findings

- Several important herbivore and predator species that were once a component of the longleaf pine ecosystem are no longer present in the Southeast, but were replaced by many nonnative animals. The resulting change in predator-prey relationships has the potential to influence plant communities through trophic cascades.
- In longleaf pine ecosystems, viable populations of semiaquatic animals are sustained by a landscape that includes: a mosaic of embedded wetlands that are spatially distributed so as to accommodate animal dispersal; wetlands that have appropriate vegetation structures, hydroperiods, and other conditions for aquatic habitat; and being embedded within a matrix of frequently burned longleaf pine uplands for suitable terrestrial and dispersal habitat.
- Nearly two-thirds of the vertebrates that inhabit longleaf pine forests either excavate burrows for themselves, use existing burrows of other animals, or inhabit stump holes and belowground root systems.
- Regulatory policies tend to focus management on individual species, sometimes at the expense of other target species or broader ecosystem objectives.

Implications for Restoration and Management

- Predator-prey interactions alter behavior and population dynamics on both sides of the interaction. Even though the interactions between contemporary apex predators and their large herbivore prey are frequent, the two rarely rely on each other to persist. Thus, direct management of herbivore populations may be needed to avoid negative impacts on plant communities.

- Restoration that involves the direct management of large herbivore populations may be aided by the presence of apex carnivores, which directly reduce herbivore population size, stimulate antipredator behaviors that result in decreased herbivore reproduction, and reduce overconsumption of preferred browse species.
- Growing a mosaic of stands that are maintained by frequent fire and have the structural characteristics of old-growth forests is one of the best strategies for increasing and enhancing wildlife habitat. Desirable attributes for such stands are old trees, coarse woody debris, stump holes and snags, and native ground cover.
- Landscape connectivity—including tracts with significant acreage and corridors that connect them—is a key consideration in prioritizing habitat restoration, especially when embedded fire-maintained wetlands are also present in the landscape.
- Because many imperiled wildlife species evolved in fire-maintained longleaf pine systems and are adapted to the resulting structural attributes, successful restoration or management at broad scales requires the presence of the vegetative attributes that are necessary for the persistence of these rarer species (at a landscape scale, albeit not on every acre).
- Because restoration of longleaf pine forest structure will not ensure restoration of all native wildlife, many species will require translocation to establish populations in more favorable conditions. The key factors in translocating species are moving organisms the shortest possible distance, taking a region-wide perspective, using as many individuals as possible, and ensuring a commitment to long-term habitat management and monitoring.
- Long-term monitoring of restoration sites is necessary to determine whether desired habitat alterations are attracting targeted wildlife species or whether species reintroductions are warranted.

Ecosystem Services of the Future

Key Ecological Findings

- Climate change is predicted to cause longer, more frequent, and more intense droughts, while increasing the probability of catastrophic wildfires in the Southeast, thereby threatening key ecosystem services from forests. Water yield—the difference between incoming precipitation and the water that exits the system as evapotranspiration—has changed drastically in the Southeast since European settlement. Compared to other southeastern forests, frequently burned longleaf pine stands have low annual evapotranspiration rates.
- The relatively open canopy of longleaf pine reduces both transpiration by dominant trees and water loss by direct interception of rainfall. The ground cover plants of longleaf pine forests use relatively little water because of their high water-use efficiency—the quantity of carbon fixed per unit of water consumed—and the low leaf area of grasses and herbs. When carbon budgets incorporate the effects of fire-induced carbon loss, assessments indicate that longleaf pine stands managed with frequent prescribed fire tend to be small carbon sinks and can become short-term carbon sources under certain conditions.
- When combusted, a large proportion of fuel on the forest floor is converted to CO_2, and a substantial proportion of biomass carbon is transformed into black carbon by incomplete combustion; because of its stability, black carbon can represent a major carbon sink when it becomes integrated in the soil matrix. Some fractions of the soil black carbon pool are soluble, meaning that the transport of this carbon in longleaf pine dominated watersheds would likely have an impact on soils, wetlands, and streams.

Implications for Restoration and Management

- Although prescribed fire results in an immediate loss of biomass carbon as CO_2, actively managing longleaf pine stands with prescribed fire greatly reduces the risk of massive and long-term carbon loss from catastrophic stand-replacing fires.
- Broad-scale restoration of the drought-tolerant, low-density, frequently burned longleaf pine forest structure could benefit water resources by reducing evapotranspiration; this reduction could result in increased stream runoff or groundwater recharge (or both), while also contributing modestly and sustainably to long-term carbon sequestration.
- An emphasis on carbon accounting could interfere with restoration efforts if loss of forest carbon storage is construed as a primary effect of longleaf pine restoration activities. Carbon accounting principles that use current southeastern forests as the baseline for carbon sequestration incentive programs may need to be modified to reflect the fact that longleaf pine is the historical reference condition for much of the region.
- Using silvicultural thinning operations to lower tree densities could increase resilience to drought by reducing water stress on the trees that remain after harvesting.

These key findings provide many new insights into how the longleaf pine ecosystem functions and, in some instances, they contradict concepts that were assumed to apply to all longleaf pine ecosystems. Incorporating these new insights into management actions, especially on a site-specific basis, should improve the success of attempts to restore and sustain longleaf pine forests.

SOME PERSPECTIVES ON PRESCRIBED FIRE

One process that we address primarily from a management perspective is prescribed fire. Frequent fire is recognized as essential for the establishment and maintenance of the longleaf pine ecosystem, and discussion of natural and prescribed fire is threaded throughout the book. Unlike the nearly universal recognition that fire is required to sustain the longleaf pine ecosystem, the appropriate season in which to apply prescribed fire is still in debate. The controversy stems partly from recognition that lightning-ignited fire tends to be more frequent in warmer months under dry fuel conditions and that frequent fires originating from lightning have contributed to the evolutionary history of this ecosystem (Komarek 1964; Noss 2013; Noss et al. 2015). Debate also continues about the historical role of fires ignited by native tribes, which may have also influenced the timing of fire on the landscape for thousands of years (Delcourt and Delcourt 2004; Fowler and Konopik 2007). Thus, the question is whether prescribed fire should mimic the season-of-burn under which selective forces presumably operated, or whether season-of-burn has been historically variable and less important than the establishment of a frequent prescribed fire regime. Both sides of this debate are supported by valid arguments; rather than rehashing them, we focus on the management implications that this philosophical dispute engenders.

The season in which fire is applied is one of many factors that influence the outcome of prescribed fire. Other elements—such as fuel type and amount, wind speed, relative humidity, air temperature, and soil moisture—collectively contribute to the ecological effects of fire. Because these factors vary from one season to the next, the timing of fire can be an effective additional tool for achieving specific burn objectives. Individual plant species also differ in their responses to fire at different times of the year, providing options for meeting restoration and management objectives that can include hardwood midstory reduction, duff fuel reduction, stimulation of wiregrass flowering, wildlife habitat enhancement, promotion of herbaceous ground cover, or longleaf pine regeneration (see Chapters 4, 10, 11, and 13).

In today's world of forest fragmentation and wildland-urban interface expansion, the only effective way to establish a frequent-fire regime is to burn in a socially acceptable manner. However, prescribed fire regulations ensure that permitted fires are safe, contained, and conducted under appropriate conditions for adequate smoke dispersion, and the subsequent fires may not burn with the same intensities or over the same acreages as historical "natural" fires. If prescribed fire is further restricted to only growing-season months, the number of suitable burn days that meet regulatory parameters will be significantly reduced; the resulting failure to apply fire to all areas that need to be burned will likely jeopardize management restoration objectives.

Ultimately, a policy advocating prescribed burning exclusively in growing-season months would result in fewer suitable burn days, fewer acres burned, and longer fire-return intervals. Research shows that the most important practice for longleaf pine restoration and management is to employ an objective-driven regime of frequent fire with a return interval that is appropriate for the vegetation type, and consciously varying the season-of-burn over the long term. This approach provides the flexibility needed to develop fire prescriptions that are most suitable for achieving management objectives for a particular site.

THE SOCIOECONOMIC CONTEXT

The real-world context in which longleaf pine is restored can be as important as the ecological considerations and challenges encountered. Social, economic, and policy concerns play extremely important roles in the management of this ecosystem, and such issues will likely have even more influence in the future. These factors can be important drivers of where, when, how, and even if longleaf pine restoration and management are appropriate. Available resources for longleaf pine restoration are usually limited and difficult decisions must be made to leverage these resources. True restoration of longleaf pine ecosystems unfolds over long time scales, requiring a long-term commitment and site stability for ecological values to be realized.

Land use change can affect forest restoration and management in many different ways and—because the Southeast is one of the most rapidly growing U.S. regions—is especially important for longleaf pine across its historical range. For example:

- Increasing urbanization and wildland-urban interface can restrict burning in nearby forests, even though restoring longleaf pine will lead to a requirement for increased use of prescribed fire.
- Growth of urban and exurban footprints drives up land prices, and land can become too highly valued to justify forest timber management. Similarly, economic values for alternative products—such as agricultural commodities—can cause conversion from forests to other land uses.
- The demographics of family forest owners are changing, with present owners aging and intergenerational transfers leading to parcelization, fragmentation, and conversion to other land uses.

Management costs and economic considerations greatly impact the interest in and ability to restore and manage longleaf pine forests, especially for private landowners and especially given the long-term nature of the commitment. Some examples include:

- Choosing to restore and manage longleaf pine, especially for ecological objectives, represents opportunity costs compared to managing other forest types, including higher establishment costs, higher cost of ongoing management (such as establishing a frequent fire regime), lower growth rates and economic returns, and the extended time scales that are inherent in managing for mature, multiaged forest structures.
- Incentive programs, particularly for establishment costs, are available for family forest owners and can offset some opportunity costs. However, these incentive programs are generally not available to private owners with large landholdings or to corporate owners, even though these owners could have substantial impacts on longleaf pine restoration.

- Regulatory uncertainty and perceived disincentives, such as the management restrictions for red-cockaded woodpeckers (*Picoides borealis*) under the Endangered Species Act, can cause some landowners to avoid specific management approaches that would create mature forest structure due to a concern that they will be unable to utilize the resources in the future if certain species are present.
- Restoration and long-term management of longleaf pine forests produce many societal goods and services—carbon storage, soil maintenance, regulation of water quality and quantity, wildlife habitat, scenic beauty—but private landowners are not compensated for these public benefits even though they bear the costs of their "production."
- Local markets for wood products (such as solid wood, pulpwood, and chips), emerging markets for new products such as bioenergy, and the increasing globalization of the forest products industry all affect the economic viability of long-term longleaf pine restoration and management, and the ability to offset operational management costs.

THE FUTURE OF LONGLEAF PINE ECOSYSTEMS

The chapters in this book identified several potential areas of basic and applied research that could have major impacts on the capacity to carry out effective restoration and sustainable management of the longleaf pine ecosystem. For example, ecosystem process models that incorporate climate predictions would provide insights into how the ecosystems would respond to changes in climate and how climate would alter fire behavior. Such models would also be useful in addressing landscape-level ecohydrologic questions that are tied to carbon sequestration and water yield. Most planted stands of longleaf pine are at the present time too young for thinning; consequently, very little management guidance is available to inform questions about optimal stand densities or timber quality of future harvests, or how best to transition these plantations into multiaged forests. We also do not know the best time to establish desired ground cover plant communities in new plantations, whether at the time the trees are planted or at some later point during stand development. Increased understanding of the ecology of saw palmetto on flatwood sites would significantly contribute to approaches for regenerating longleaf pine and maximizing biodiversity in stands of off-site pine species (those adapted to different site conditions than the historically resident species) that have high densities of saw palmetto. Finally, better assessment of fire behavior in ecotonal sites (particularly transition zones between uplands and wetlands) would help advance tactical approaches for restoring ecologically important upland-wetland linkages.

We acknowledge that longleaf pine will never dominate the forests of the southeastern Coastal Plain as it once did. However, given the expanding recognition of the ecological and societal values associated with the species, we remain optimistic about the future of this imperiled ecosystem. The challenges of increasing sustainable multiaged stands of frequently burned longleaf pine as habitat for associated species will continue. Developing innovative and viable ecological restoration options for the future depends on collaborative efforts between scientists and managers to incorporate rapidly emerging new information into operational practices.

Progress will also require identification of realistic goals that integrate the socioeconomic conditions of the region into restoration strategies. In addition to examining how incentives for private sector programs can be improved, landscape-level analyses would help to focus resources and efforts in spatially explicit priority areas, with both long-term and broad-scale outcomes. Priority areas need to have high connectivity for wildlife populations, freedom to use prescribed fire, economic viability of forest management in relation to development, and expertise or partnerships to carry out the required management. Areas with stable long-term land use potential—including public lands, private lands under conservation easement, and potentially large corporate ownerships—are strong candidates for restoration efforts. Shifting some resources away from new pine forest establishment and toward maintaining existing longleaf pine forests, even those that are degraded, could more quickly provide desired forest conditions; in many instances, these existing forests immediately provide some desired characteristics (such as habitat structure and fine fuels for prescribed fire). Altering existing incentive policies could facilitate this change in focus.

Finally, understanding that both science and societal values evolve over time is an important perspective (Jackson 2016). Management inherently adapts to new discoveries, changes in the environment, and changes in the forest characteristics that are considered important and desirable. Consequently, restoration practices evolve; many of the ideas and practices that seemed radical or impractical just a few years ago now are acceptable alternatives that are applied widely. An example from the Joseph W. Jones Ecological Research Center is the concept of a gradual conversion of overstory species—the strategy of preserving existing forest structure and using underplanting to introduce desired species over time. This once novel and somewhat dubiously viewed experiment is now regarded as a fairly common restoration approach for off-site pine conversion to longleaf pine. The concept has been implemented extensively with numerous variations in the pattern of underplanted pine seedlings (see Chapter 10). Although the origin of this perpetual forest restoration concept is not widely known (nor is the original source particularly important), its acceptance as a viable restoration technique has advanced restoration efforts for longleaf pine.

Interest in longleaf pine has grown dramatically over the last 20 years, and enthusiasm for restoration and management of the species has never been higher. One certainty is that restoration and sustainability of the longleaf pine ecosystem will require active management to be successful. In the future, ecological restoration concepts and practices for longleaf pine forests will continue to evolve as they are more widely applied in an operational context, are attempted in more forest types, and respond to a rapidly changing natural and socioeconomic environment (Richardson et al. 2007). Effective conservation of the longleaf pine ecosystem in the future will require multifaceted actions that robustly integrate forest and wildlife conservation at landscape scales, and targeted efforts to merge ecological with socioeconomic values.

ACKNOWLEDGMENTS

We thank Steven Brantley, Mike Conner, Kevin McIntyre, and Lora Smith for their thoughtful contributions to the content of this chapter and reviews of an early draft.

Literature Cited

Abatzoglou, J. T. and A. P. Williams. 2016. Impact of anthropogenic climate change on wildfire across western U.S. forests. *Proceedings of the National Academy of Sciences* 113:11770–11775.

Aber, J. D. and J. M. Melillo. 2001. *Terrestrial Ecosystems*. San Diego, CA: Academic Press.

Abrahamson, W. G. and D. C. Hartnett. 1990. Pine flatwoods and dry prairies. Pages 103–149 *in* R. L. Myers, and J. J. Ewel, editors. *Ecosystems of Florida*. Gainesville, FL: University Press of Florida.

Achtemeier, G. L. 2003. On the origins of "superfog"—A combination of smoke and water vapor that produces zero visibility over roadways. Pages 1–4 *in Second International Wildland Fire Ecology and Fire Management Congress and Fifth Symposium on Fire and Forest Meteorology*, November 16–20, Orlando, FL.

Adams, J. C. and T. R. Clason. 2002. Loblolly pruning and growth characteristics at different planting spacings. Pages 153–155 *in* K. W. Outcalt, editor. *Proceedings of the 11th Biennial Southern Silvicultural Research Conference*. General Technical Report SRS-48. USDA Forest Service, Southern Research Station, Asheville, NC.

Adams, A. S. and L. K. Rieske. 2001. Herbivory and fire influence white oak seedling vigor. *Forest Science* 47:331–337.

Addington, R. N., L. A. Donovan, R. J. Mitchell, J. M. Vose, S. D. Pecot, S. B. Jack, U. G. Hacke, J. S. Sperry, and R. Oren. 2006. Adjustments in hydraulic architecture of *Pinus palustris* maintain similar stomatal conductance in xeric and mesic habitats. *Plant, Cell, and Environment* 29:535–545.

Addington, R. N., T. A. Greene, M. L. Elmore, C. E. Prior, and W. C. Harrison. 2012. Influence of herbicide site preparation on longleaf pine ecosystem development and fire management. *Southern Journal of Applied Forestry* 36:173–180.

Addington, R. N., T. A. Greene, W. C. Harrison, G. G. Sorrell, M. L. Elmore, and S. M. Hermann. 2015. Restoring longleaf pine: Effects of seasonal prescribed fire and overstory density on vegetation structure of a young longleaf pine plantation. *Forest Science* 61:135–149.

Addington, R. N., B. O. Knapp, G. G. Sorrel, M. L. Elmore, G. G. Wang, and J. L. Walker. 2015. Factors affecting broadleaf woody vegetation in upland pine forests managed for longleaf pine restoration. *Forest Ecology and Management* 354:130–138.

Adler, P. B., J. HilleRisLambers, and J. M. Levine. 2007. A niche for neutrality. *Ecology Letters* 10:95–104.

Agee, J. K. 1993. *Fire Ecology of Pacific Northwest Forests*. Washington, DC: Island Press.

Agee, J. K. and M. R. Lolley. 2006. Thinning and prescribed fire effects on fuels and potential fire behavior in an Eastern Cascades Forest, Washington, USA. *Fire Ecology* 2:3–19.

Agee, J. K. and C. N. Skinner. 2005. Basic principles of forest fuel reduction treatments. *Forest Ecology and Management* 211:83–96.

Alavalapati, J. R., G. A. Stainback, and D. R. Carter. 2002. Restoration of the longleaf pine ecosystem on private lands in the U.S. South: An ecological economic analysis. *Ecological Economics* 40:411–419.

Albin, L. T. 2014. Restoring the longleaf pine (*Pinus palustris*) forests using pineywoods cattle grazing in conjunction with prescribed burning, MS thesis. University of Southern Mississippi, Hattiesburg, MS.

Albini, F. A. and R. G. Baughmann. 1979. Estimating windspeeds for predicting wildland fire behavior. Research Paper INT-RP-221. USDA Forest Service, Intermountain Forest and Range Experiment Station, Ogden, UT.

Aleric, K. M. and L. K. Kirkman. 2005. Growth and photosynthetic responses of the federally endangered shrub, *Lindera melissifolia* (Lauraceae) to varied light environments. *American Journal of Botany* 92:682–689.

Alexander, M. E. and M. G. Cruz. 2013. Are the applications of wildland fire models getting ahead of their evaluation again? *Environmental Modelling and Software* 41:65–71.

Allen, R. M. 1954. Shade may improve longleaf survival. Research Note SO-90. USDA Forest Service, Southern Forest Experiment Station, New Orleans, LA.

Allen, C. R., S. Demarais, and R. S. Lutz. 1997. Effects of red imported fire ants on recruitment of white-tailed deer fawns. *Journal of Wildlife Management* 61:911–916.

Allen, H. L., T. R. Fox, and R. G. Campbell. 2005. What is ahead for intensive pine plantation silviculture in the south? *Southern Journal of Applied Forestry* 29:62–69.

Amazega, J. M., L. Santamaría, and A. J. Green. 2002. Biotic wetland connectivity-supporting a new approach for wetland policy. *Acta Oecologica* 23:213–222.

America's Longleaf Restoration Initiative (ALRI). 2009. Rangewide conservation plan for longleaf pine. http://www.americaslongleaf.org/media/86/conservation_plan.pdf (accessed 4 February 2017).

Andersen, H. E., R. J. McGaughey, and S. E. Reutebuch. 2005. Estimating forest canopy fuel parameters using LIDAR data. *Remote Sensing of Environment* 94:441–449.

Anderson, H. E. 1982. Aids to determining fuel models for estimating fire behavior grass, shrub, timber, and slash, photographic examples, danger ratings. General Technical Report INT-122. USDA Forest Service, Intermountain Forest and Range Experiment Station, Ogden, UT.

Anderson, M. G., M. Clark, and A. O. Sheldon. 2014. Estimating climate resilience for conservation across geophysical settings. *Conservation Biology* 28:959–970.

Anderson, H. W., M. D. Hoover, and K. G. Reinhart. 1976. Forests and water: Effects of forest management on floods, sedimentation, and water supply. General Technical Report PSW-18. USDA Forest Service, Pacific Southwest Research Station, Berkeley, CA.

Andreu, M. G., C. W. Hedman, M. H. Friedman, and A. G. Andreu. 2009. Can managers bank on seed banks when restoring *Pinus taeda* L. plantations in southwest Georgia? *Restoration Ecology* 17:586–596.

Andrews, P., C. Bevins, and R. Seli. 2004. *BehavePlus Fire Modeling System, Version 3.0: User's Guide.* Ogden, UT: USDA Forest Service, Rocky Mountain Research Station.

Andrews, P. L. and C. H. Chase. 1989. BEHAVE: Fire behavior prediction and fuel modeling system—BURN subsystem, Part 2. General Technical Report INT-260. USDA Forest Service, Intermountain Research Station, Ogden, UT.

Andrews, P. L. and L. P. Queen. 2001. Fire modeling and information system technology. *International Journal of Wildland Fire* 10:343–352.

Ansley, R. J., T. W. Boutton, and J. O. Skjemstad. 2006. Soil organic carbon and black carbon storage and dynamics under different fire regimes in temperate grass-savanna. *Global Biogeochemical Cycles* 20:GB3006.

Anttila, C. K., C. C. Daehler, N. E. Rank, and D. R. Strong. 1998. Greater male fitness of a rare invader (*Spartina alterniflora*, Poaceae) threatens a common native (*Spartina foliosa*) with hybridization. *American Journal of Botany* 85:1597–1601.

Archibald, S., C. E. R. Lehmann, J. L. Gómez-Dans, and R. A. Bradstock. 2013. Defining pyromes and global syndromes of fire regimes. *Proceedings of the National Academy of Science* 110:6442–6447.

Aresco, M. J. and C. Guyer. 1999. Burrow abandonment by gopher tortoises in slash pine plantations of the Conecuh National Forest. *Journal of Wildlife Management* 63:26–35.

Arno, S. F. 1980. Forest fire history in the northern Rockies. *Journal of Forestry* 78:460–465.

Arno, S. F., J. H. Scott, and M. G. Hartwell. 1995. Age-class structure of old growth ponderosa pine/Douglas-fir stands and its relationship to fire history. Research Paper INT-RP-481. USDA Forest Service, Intermountain Research Station, Ogden, UT.

Association of Official Seed Analysts (AOSA). 2002. *Seed Vigor Testing Handbook. Contribution No. 32 to the Handbook on Seed Testing*. Las Cruces, New Mexico: The Association of Official Seed Analysts.

Atkinson, C. L. and C. C. Vaughn. 2015. Biogeochemical hotspots: Temporal and spatial scaling of the impact of freshwater mussels on ecosystem function. *Freshwater Biology* 60:563–574.

Aubrey, D. P., B. Mortazavi, J. J. O'Brien, J. D. McGee, J. J. Hendricks, K. A. Kuehn, R. O. Teskey, and R. J. Mitchell. 2012. Influence of repeated canopy scorching on soil CO_2 efflux. *Forest Ecology and Management* 282:142–148.

Auffenberg, W. and R. Franz. 1982. The status and distribution of the gopher tortoise (*Gopherus polyphemus*). Pages 95–126 *in* R. B. Bury, editor. *North American Tortoises: Conservation and Ecology*. Wildlife Research Report 12. Washington, DC: U.S. Fish and Wildlife Service.

Auld, T. D. 1987. Post-fire demography in the resprouting shrub *Angophora hispida* (Sm.) Blaxell: Flowering, seed production, dispersal, seedling establishment, and survival. *Proceedings of the Linnean Society* 109:259–269.

Avery, C., R. S. Cohen, K. C. Parker, and J. S. Kush. 2004. Spatial patterns of longleaf pine (*Pinus palustris*) seedling establishment on the Croatan National Forest, North Carolina. *Journal of the North Carolina Academy of Science* 120:131–142.

Avnery, S., D. L. Mauzerall, J. Liu, and L. W. Horowitz. 2011. Global crop yield reductions due to surface ozone exposure: 2. Year 2030 potential crop production losses and economic damage under two scenarios of O_3 pollution. *Atmospheric Environment* 45:2297–2309.

Babbitt, K. J., M. J. Baber, and L. A. Brandt. 2006. The effect of woodland proximity and wetland characteristics on larval anuran assemblages in an agricultural landscape. *Canadian Journal of Zoology* 84:510–519.

Babbitt, K. J. and G. W. Tanner. 1997. Effects of cover and predator identity on predation of *Hyla squirella* tadpoles. *Journal of Herpetology* 31:128–130.

Baer, S. G., S. L. Collins, J. M. Blair, and A. K. Knapp. 2005. Soil heterogeneity effects on tallgrass prairie community heterogeneity: An application of ecological theory to restoration ecology. *Restoration Ecology* 13:413–424.

Bailes, J. C. and S. Wendell. 1979. Capital budgeting practices in the forest products industry. *Management Accounting* 61:46–51.

Bailey, R. G. 1998. *Ecoregions*. New York, NY: Springer.

Bailey, M. A. 2015. *Pine Straw Management and Harvest in Longleaf Pine Forest: Best Management Guidelines to Achieve Sustainability of Wildlife Resources*. Washington, DC: National Wildlife Federation.

Bailey, J. D. and W. W. Covington. 2002. Evaluating ponderosa pine regeneration rates following ecological restoration treatments in northern Arizona, USA. *Forest Ecology and Management* 155:271–278.

Bailian, L., S. E. McKeand, and R. J. Weir. 1999. Tree improvement and sustainable forestry–impact of two cycles of loblolly pine breeding in the USA. *Forest Genetics* 6:229–234.

Baker, J. B., M. D. Cain, J. M. Guldin, P. A. Murphy, and M. G. Shelton. 1996. Uneven-aged silviculture for the loblolly and shortleaf pine forest cover types. General Technical Report SO-118. USDA Forest Service, Southern Research Station, Asheville, NC.

Bakker, E. S., J. L. Gill, C. N. Johnson, F. C. Vera, C. J. Sandom, G. P. Asner, and J. C. Svenning. 2015. Combining paleo data and modern exclosure experiments to assess the impact of megafauna extinctions on woody vegetation. *Proceedings of the National Academy of Sciences* 113:847–855.

Bakker, E. S., M. E. Ritchie, H. Olff, D. G. Milchunas, and J. M. H. Knops. 2006. Herbivore impact on grassland plant diversity depends on habitat productivity and herbivore size. *Ecology Letters* 9:780–788.

Bale, A. M. 2009. Fire effects and litter accumulation dynamics in a montane longleaf pine ecosystem, MS thesis. University of Missouri, Columbia, MO.

Band, L. E., P. Patterson, R. R. Nemani, and S. W. Running. 1993. Forest ecosystem processes at the watershed scale: Incorporating hillslope hydrology. *Agricultural and Forest Meteorology* 63:93–126.

Banwell, E. M., J. M. Varner, E. E. Knapp, and R. W. van Kirk. 2013. Spatial, seasonal, and diel forest floor moisture dynamics in Jeffrey pine-white fir forests of the Lake Tahoe Basin, USA. *Forest Ecology and Management* 305:11–20.

Barnard, E. L., E. C. Ash, and W. N. Dixon. 1995. Benomyl root dip and scalping improve~performance of longleaf pine on pest-infested agricultural croplands. *Tree Planters' Notes* 46:93–96.

Barnett, J. P. and J. C. Brissette. 1986. Producing southern pine seedlings in containers. General Technical Report SO-59. USDA Forest Service, Southern Forest Experiment Station, New Orleans, LA.

Barnett, J. P. and J. M. McGilvray. 1997. Practical guidelines for producing longleaf pine seedlings in containers. General Technical Report SRS-14. USDA Forest Service, Southern Research Station, Asheville, NC.

Barnes, M. 2001. Effects of large herbivores and fire on the regeneration of *Acacia erioloba* woodlands in Chobe National Park, Botswana. *African Journal of Ecology* 39:340–350.

Barnes, B. V., D. R. Zak, S. R. Denton, and S. H. Spurr. 1998. *Forest Ecology*, 4th edition. New York, NY: John Wiley & Sons.

Barot, S. 2004. Mechanisms promoting plant coexistence: Can all the proposed processes be reconciled? *Oikos* 106:185–192.

Barth, M. A. F., A. J. Larson, and J. A. Lutz. 2015. A forest reconstruction model to assess changes to Sierra Nevada mixed-conifer forest during the fire suppression era. *Forest Ecology and Management* 354:104–118.

Bartram, W. 1791. Travels through North and South Carolina, Georgia, East and West Florida. *In* F. Harper, editor. *The Travels of William Bartram*. New Haven, CT: Yale University Press.

Basey, A. C., J. B. Fant, and A. T. Kramer. 2015. Producing native plant materials for restoration: 10 rules to collect and maintain genetic diversity. *Native Plants Journal* 16:37–53.

Baskin, C. C. and J. M. Baskin. 1998. *Seeds: Ecology, Biogeography, and Evolution of Dormancy and Germination*. San Diego, CA: Academic Press.

Battaglia, M. A., R. J. Mitchell, P. P. Mou, and S. D. Pecot. 2003. Light transmittance estimates in a longleaf pine woodland. *Forest Science* 49:752–762.

Battaglia, M. A., P. Mou, B. Palik, and R. J. Mitchell. 2002. The effect of spatially variable overstory on the understory light environment of an open-canopied longleaf pine forest. *Canadian Journal of Forest Research* 32:1984–1991.

Battle, J. and S. W. Golladay. 2001a. Hydroperiod influence on breakdown of leaf litter in cypress-gum wetlands. *American Midland Naturalist* 146:128–145.

Battle, J. and S. W. Golladay. 2001b. Water quality and macroinvertebrate assemblages in three types of seasonally inundated limesink wetlands in southwest Georgia. *Journal of Freshwater Ecology* 16:189–207.

Battle, J. M. and S. W. Golladay. 2002. Aquatic invertebrates in hardwood depressions of southwest Georgia. *Southeastern Naturalist* 1:149–158.

Battle, J. and S. W. Golladay. 2003. Prescribed fire's impact on water quality of depressional wetlands in southwestern Georgia. *American Midland Naturalist* 150:15–25.

Battle, J. M. and S. W. Golladay. 2007. How hydrology, habitat type, and litter quality affect leaf breakdown in wetlands on the gulf coastal plain of Georgia. *Wetlands* 27:251–260.

Battle, J. M., S. W. Golladay, and B. Clayton. 2001. Aquatic macroinvertebrates and water quality characteristics in five wetland types: Preliminary results on biomonitoring. Pages 333–336 *in* K. J. Hatcher, editor. *Proceedings of the 2001 Georgia Water Resources Conference*, 26–27 March 2001, University of Georgia, Athens, GA.

Bauder, J. M., C. Castellano, J. B. Jensen, D. J. Stevenson, and C. L. Jenkins. 2014. Comparison of movements, body weight, and habitat selection between translocated and resident gopher tortoises. *Journal of Wildlife Management* 78:1444–1455.

Baum, K. A., K. J. Haynes, F. P. Dillemuth, and J. T. Cronin. 2004. The matrix enhances the effectiveness of corridors and stepping stones. *Ecology* 85:2671–2676.

Beckage, B. and C. Ellingwood. 2008. Fire feedbacks with vegetation and alternative stable states. *Complex Systems* 18:159–173.

Beckage, B., L. J. Gross, and W. J. Platt. 2011. Grass feedbacks on fire stabilize savannas. *Ecological Modelling* 222:2227–2233.

Beckerman, A. P., M. Uriarte, and O. J. Schmitz. 1997. Experimental evidence for a behavior-mediated trophic cascade in a terrestrial food chain. *Proceedings of the National Academy of Sciences* 94:10735–10738.

Becknell, J. M., A. R. Desai, M. C. Dietze, C. A. Schultz, G. Starr, P. A. Duffy, J. F. Franklin, A. Pourmokhtarian, J. Hall, P. C. Stoy, M. W. Binford, L. R. Boring, and C. L. Staudhammer. 2015. Assessing interactions among changing climate, management, and disturbance in forests: A macrosystems approach. *Bioscience* 65:263–274.

Bell, G. 2005. The co-distribution of species in relation to the neutral theory of community ecology. *Ecology* 86:1757–1770.

Bendel, P. R. and G. D. Therres. 1994. Movements, site fidelity and survival of Delmarva fox squirrels following translocation. *American Midland Naturalist* 132:227–233.

Bengston, G., J. DuPre, W. Twomey, and R. Hooper. 1993. Longleaf ecosystem restoration in the wake of Hurricane Hugo. Pages 339–347 *in* S. M. Hermann, editor. *Proceedings of the 18th Tall Timbers Fire Ecology Conference*, 30 May–2 June 1991, Tallahassee, Florida, USA. Tall Timbers Research Station, Tallahassee, FL.

Bennett, F. A. 1980. Growth and yield stands of slash pine and suggested management alternatives. Research Paper SE-21. USDA Forest Service, Southeasten Forest Experiment Station, Asheville, NC.

Berish, J. E., L. D. Wendland, and C. A. Gates. 2000. Distribution and prevalence of upper respiratory tract disease in gopher tortoises in Florida. *Journal of Herpetology* 34:5–12.

Berlow, E. L., C. M. D'Antonio, and H. Swartz. 2003. Response of herbs to shrub removal across natural and experimental variation in soil moisture. *Ecological Applications* 13:1357–1387.

Beven, K. J. and M. J. Kirkby. 1979. A physically based, variable contributing area model of basin hydrology. *Hydrological Sciences Bulletin* 24:43–69.

Bhuta, A. A. R., L. M. Kennedy, and N. Pederson. 2009. Climate-radial growth relationships of northern latitudinal range margin longleaf pine (*Pinus palustris* P. Mill) in the Atlantic coastal plain of southeastern Virginia. *Tree-Ring Research* 65:105–115.

Bigelow, S. W., M. P. North, and C. F. Salk. 2011. Using light to predict fuels-reduction and group-selection effects on succession in Sierran mixed-conifer forest. *Canadian Journal of Forest Research* 41:2051–2063.

Binkley, C. S. 2007. *The Rise and Fall of the Timber Investment Management Organizations: Ownership Changes in U.S. Forestlands*. Washington, DC: Pinchot Distinguished Lecture, Pinchot Institute for Conservation.

Binkley, C. S., C. F. Raper, and C. L. Washburn. 1996. Institutional ownership of U.S. timberland. *Journal of Forestry* 94:21–28.

Birch, T. W. 1997. Private forest-land owners of the southern United States, 1994. Resource Bulletin NE-138. USDA Forest Service, Northeastern Forest Experiment Station, Randor, PA.

Birch, T. W., D. G. Lewis, and H. F. Kaiser. 1982. The private forest-land owners of the United States. Resource Bulletin WO-1. USDA Forest Service, Washington, DC.

Birkhead, R. and T. D. Tuberville. 2008. Gopher tortoise, *Gopherus polyphemus*. Pages 514–516 *in* B. Jensen, C. D. Camp, W. Gibbons, and M. J. Elliot, editors. *The Amphibians and Reptiles of Georgia*. Athens, GA: University of Georgia Press.

Biswell, H. 1989. *Prescribed Burning in California Wildlands Vegetation Management*. Berkeley, CA: University of California Press.

Biswell, H. H. and P. C. Lemon. 1943. Effect of fire upon seed-stalk production of range grasses. *Journal of Forestry* 41:844.

Bizzari, L. E., C. D. Collins, L. A. Brudvig, and E. I. Damschen. 2015. Historical agriculture and contemporary fire frequency alter soil properties in longleaf pine woodlands. *Forest Ecology and Management* 349:45–54.

Blanc, L. A. and J. R. Walters. 2008. Cavity excavation and enlargement as mechanisms for indirect interactions in an avian community. *Ecology* 89:506–514.

Blaustein, A. R., S. C. Walls, B. A. Bancroft, J. J. Lawler, C. L. Searle, and S. S. Gervasi. 2010. Direct and indirect effects of climate change on amphibian populations. *Diversity* 2:281–313.

Blihovde, W. B. 2006. Terrestrial movements and upland habitat use of gopher frogs in central Florida. *Southeastern Naturalist* 5:265–276.

Bonan, G. B. 2008. Forests and climate change: Forcings, feedbacks, and the climate benefits of forests. *Science* 320:1444–1449.

Bond, W. J. and J. E. Keeley. 2005. Fire as a global herbivore: The ecology and evolution of flammable ecosystems. *Trends in Ecology and Evolution* 20:387–394.

Bond, W. J., G. F. Midgley, and F. I. Woodward. 2003. What controls South African vegetation—climate or fire? *South African Journal of Botany* 69:79–91.

Boone, M. D. and S. M. James. 2003. Interactions of an insecticide, herbicide, and natural stressors in amphibian community mesocosms. *Ecological Applications* 13:829–841.

Borders, B. E. and R. L. Bailey. 2001. Loblolly pine—pushing the limits of growth. *Southern Journal of Applied Forestry* 25:69–74.

Boring, L. R., J. J. Hendricks, and M. B. Edwards. 1991. Loss, retention, and replacement of nitrogen associated with site preparation burning in southern pine-hardwood forests. Pages 145–153 *in* S. C. Nodvin, and T. A. Waldrop, editors. *Fire and the Environment: Ecological and Cultural Perspectives. Proceedings of an International Symposium*, 20–24 March 1990, Knoxville, Tennessee, USA. General Technical Report SE-69. USDA Forest Service, Southeastern Forest Experiment Station, Asheville, NC.

Boring, L. R., C. A. Wilson, J. J. Hendricks, and R. J. Mitchell. 2004. Season of burn and nutrient losses in a longleaf pine ecosystem. *International Journal of Wildland Fire* 13:443–453.

Bosch, J. M. and J. D. Hewlett. 1983. A review of catchment experiments to determine the effect of vegetation changes on water yield and evapotranspiration. *Journal of Hydrology* 55: 3–23.

Bower, A. D., J. B. St. Clair, and V. Erickson. 2014. Generalized provisional seed zones for native plants. *Ecological Applications* 24:913–919.

Boyd, W. 2015. *The Slain Wood: Papermaking and its Environmental Consequences in the American South*. Baltimore, MD: Johns Hopkins University Press.

Boyer, W. D. 1963. Development of longleaf pine seedlings under parent trees. USDA Forest Service Research Paper SO-4. USDA Forest Service, Southern Forest Experiment Station, New Orleans, LA.

Boyer, W. D. 1983. Variations in height-over-age curves for young longleaf pine plantations. *Forest Science* 29:15–27.

Boyer, W. D. 1985. Timing of longleaf pine seedling release from overtopping hardwoods: A look 30 years later. *Southern Journal of Applied Forestry* 9:114–116.

Boyer, W. D. 1988. Effects of site preparation and release on the survival and growth of planted bare-root and container-grown longleaf pine. Georgia Forest Research Paper 76.

Boyer, W. D. 1990. *Pinus palustris* Mill. (longleaf pine). Pages 405–412 *in* R. M. Burns, and B. H. Honkala, editors. *Silvics of North America. Volume I: Conifers*. Agricultural Handbook 654. Washington, DC: USDA Forest Service.

Boyer, W. D. 1993. Long-term development of regeneration under longleaf pine seedtree and shelterwood stands. *Southern Journal of Applied Forestry* 17:10–15.

Boyer, W. D. 1996. Longleaf pine can catch up. Pages 28–29 *in* J. S. Kush, editor. Longleaf pine: A regional perspective of challenges and opportunities. *Proceedings of the First Longleaf Alliance Conference*, 17–19 September 1996, Mobile, Alabama, USA. Longleaf Alliance Report No. 1. The Longleaf Alliance, Solon Dixon Forestry Center, Andalusia, AL.

Boyer, W. D. 1998. Long-term changes in flowering and cone production by longleaf pine. Pages 92–98 *in* T. A. Waldrop, editor. *Proceedings of the 9th Biennial Southern Silvicultural Research Conference*, 25–27 February 1997, Clemson, South Carolina, USA. USDA Forest Service, Southern Research Station, Asheville, NC.

Boyer, W. D. and R. M. Farrar. 1981. Thirty years of management on a small longleaf pine forest. *Southern Journal of Applied Forestry* 5:73–77.

Boyer, W.D. and D.W. Peterson. 1983. Longleaf pine. Pages 153–156 *in* R. M. Burns, editor. *Silvicultural Systems for the Major Forest Types of the United States*. Agriculture Handbook 445. New Orleans, LA: USDA Forest Service, Southern Research Station.

Bozeman, J. R. 1971. A sociologic and geographic study of the sand ridge vegetation in the coastal plain of Georgia, PhD dissertation. University of North Carolina, Chapel Hill, NC.

Bracho, R., G. Starr, H. L. Gholz, T. A. Martin, W. P. Cropper, and H. W. Loescher. 2013. Controls on carbon dynamics by ecosystem structure and climate for southeastern U.S. slash pine plantations. *Ecological Monographs* 82:101–128.

Brantley, S. T., C. R. Ford, and J. M. Vose. 2013. Future species composition will affect forest water use after loss of hemlock from southern Appalachian forests. *Ecological Applications* 23:777–790.

Brantley, S. T., C. F. Miniat, K. J. Elliott, S. H. Laseter, and J. M. Vose. 2015. Changes to southern Appalachian water yield and stormflow after loss of a foundation species. *Ecohydrology* 8:518–528.

Bratton, S. P. 1975. The effect of the European wild boar, *Sus scrofa*, on gray beech forest in the Great Smoky Mountains. *Ecology* 56:1356–1366.

Breckheimer, I., N. M. Haddad, W. F. Morris, A. M. Trainor, W. R. Fields, R. Jobe, B. R. Hudgens, A. Moody, and J. R. Walters. 2014. Defining and evaluating the umbrella species concept for conserving and restoring landscape connectivity. *Conservation Biology* 28:1584–1593.

Brennan, L. A., R. T. Engstrom, W. E. Palmer, S. M. Hermann, G. A. Hurst, L. W. Burger, and C. L. Hardy. 1998. Whither wildlife without fire. *Transactions of the North American Natural Resources Conference* 63:402–414.

Brenner, J. 2002. Measuring live fuel moistures in Florida: Standard methods and procedures. Florida Division of Forestry, Department of Agriculture and Consumer Services. Tallahassee, FL. http://www.freshfromflorida.com/content/download/4761/30337/procedures.pdf (accessed 24 May 2017).

Brenner, J. and D. Wade. 2003. Florida's revised prescribed fire law: Protection for responsible burners. Pages 132–136 *in* K. E. M. Galley, R. C. Klinger, and N. G. Sugihara, editors. *Proceedings of Fire Conference 2000: The First National Congress on Fire Ecology, Prevention, and Management*, 27 November-1 December 2000, San Diego, California, USA. Miscellaneous Publication 13. Tall Timbers Research Station, Tallahassee, FL.

Brewer, J. S. and W. J. Platt. 1994. Effects of fire season and herbivory on reproductive success in a clonal forb, *Pityopsis graminifolia*. *Journal of Ecology* 81:665–675.

Brewer, J. S., W. J. Platt, J. S. Glitzenstein, and D. R. Streng. 1996. Effects of fire-generated gaps on growth and reproduction of golden aster (*Pityopsis graminifolia*). *Bulletin of the Torrey Botanical Club* 123:295–303.

Bright, B., E. L. Loudermilk, S. Pokswinski, A. Hudak, and J. J. O'Brien. 2016. Introducing photogrammetry for characterizing 3D surface fuel structure and plant diversity at fine scales. *Canadian Journal of Remote Sensing* 42:460–472.

Brockerhoff, E. G., H. Jactel, J. A. Parrotta, C. P. Quine, and J. Sayer. 2008. Plantation forests and biodiversity: Oxymoron or opportunity? *Biodiversity and Conservation* 17:925–951.

Brockway, D. G. and C. E. Lewis. 1997. Long-term effects of dormant-season prescribed fire on plant community diversity, structure and productivity in a longleaf pine wiregrass ecosystem. *Forest Ecology and Management* 96:167–183.

Brockway, D. G., E. F. Loewenstein, and K. W. Outcalt. 2014. Proportional basal area method for implementing selection silviculture systems in longleaf pine forests. *Canadian Journal of Forest Research* 44:977–985.

Brockway, D. G. and K. W. Outcalt. 1998. Gap-phase regeneration in longleaf pine wiregrass ecosystems. *Forest Ecology and Management* 106:125–139.

Brockway, D. G. and K. W. Outcalt. 2000. Restoring longleaf pine wiregrass ecosystems: Hexazinone application enhances effects of prescribed fire. *Forest Ecology and Management* 137:121–138.

Brockway, D. G. and K. W. Outcalt. 2015. Influence of selection systems and shelterwood methods on understory plant communities of longleaf pine forests in flatwoods and uplands. *Forest Ecology and Management* 357:138–150.

Brockway, D. G., K. W. Outcalt, and W. D. Boyer. 2006. Longleaf pine regeneration ecology and methods. Pages 95–133 *in* S. Jose, E. J. Jokela, and D. L. Miller, editors. *The Longleaf Pine Ecosystem: Ecology, Silviculture, and Restoration*. New York, NY: Springer.

Brockway, D. G., K. W. Outcalt, B. L. Estes, and R. B. Rummer. 2009. Vegetation response to midstorey mulching and prescribed burning for wildfire hazard reduction and longleaf pine (*Pinus palustris* Mill.) ecosystem restoration. *Forestry* 82:299–314.

Brockway, D. G., K. W. Outcalt, J. M. Guldin, W. D. Boyer, J. L. Walker, D. C. Rudolph, R. B. Rummer, J. P. Barnett, S. Jose, and J. Nowak. 2005. Uneven-aged management of longleaf pine forests: A scientist and manager dialogue. General Technical Report SRS-78. USDA Forest Service, Southern Research Station, Asheville, NC.

Brockway, D. G., K. W. Outcalt, D. J. Tomczak, and E. E. Johnson. 2005. Restoration of longleaf pine ecosystems. General Technical Report SRS-83. USDA Forest Service, Southern Research Station, Asheville, NC.

Brockway, D. G., K. W. Outcalt, and R. N. Wilkins. 1998. Restoring longleaf pine wiregrass ecosystems: Plant cover, diversity and biomass following low-rate hexazinone application on Florida sandhills. Forest Ecology and Management 103:159–175.

Brooks, R. T. 2009. Potential impacts of global climate change on the hydrology and ecology of ephemeral freshwater systems of the forests of the northeastern United States. *Climatic Change* 95:469–483.

Brooks, M. L., C. M. D'Antonio, and D. M. Richardson. 2004. Effects of invasive alien plants on fire regimes. *BioScience* 54:677–688.

Brooks, J. R. and S. B. Jack. 2016. An updated whole stand growth and yield system for planted longleaf pine in southwest Georgia. Pages 258–263 *in* C. Schweitzer, W. K. Clatterbuck, and C. M. Oswalt, editors. *Proceedings of the 18th Biennial Southern Silvicultural Research Conference*. e-General Technical Report SRS-212. USDA Forest Service, Southern Research Station, Asheville, NC.

Brose, P. H., D. C. Dey, R. P. Guyette, J. M. Marschall, and M. C. Stambaugh. 2013. The influences of drought and humans on the fire regimes of northern Pennsylvania, USA. *Canadian Journal of Forest Research* 43:757–767.

Brown, J. K. 1974. Handbook for inventorying downed woody material. General Technical Report GTR-INT-16. USDA Forest Service, Intermountain Research Station, Ogden, UT.

Brown, J. K. 1981. Bulk densities of nonuniform surface fuels and their application to fire modeling. *Forest Science* 27:667–683.

Brown, J. S. 1999. Vigilance, patch use and habitat selection: Foraging under risk of predation. *Evolutionary Ecology Research* 1:49–71.

Brown, S. K. 2012. Movements, home range and habitat selection of Bachman's sparrows (*Peucaea aestivalis*) on a longleaf sandhill forest: Implications for fire management, MS thesis. University of Georgia, Athens, GA.

Brown, R. T., J. K. Agee, and J. F. Franklin. 2004. Forest restoration and fire: Principles in the context of place. *Conservation Biology* 18:903–912.

Brown, P. M., M. A. Battaglia, P. J. Fornwalt, B. Gannon, L. S. Huckaby, C. Julian, and A. S. Cheng. 2015. Historical (1860) forest structure in ponderosa pine forests of the northern Front Range, Colorado. *Canadian Journal of Forest Research* 45:1462–1473.

Brown, J. K. and C. D. Bevins. 1986. Surface fuel loadings and predicted fire behavior for vegetation types in the northern Rocky Mountains. Research Note INT-358. USDA Forest Service, Intermountain Research Station, Ogden, UT.

Brown, P. M., M. R. Kaufmann, and W. D. Shepperd. 1999. Long-term, landscape patterns of past fire events in a montane ponderosa pine forest of central Colorado. *Landscape Ecology* 14:513–532.

Brown, J. S. and B. P. Kotler. 2004. Hazardous duty pay and the foraging cost of predation. *Ecology Letters* 7:999–1014.

Brown, P. M. and W. D. Shepperd. 2001. Fire history and fire climatology along a five degree gradient in latitude in Colorado and Wyoming, USA. *Paleobotanist* 50:133–140.

Brown, M. T. and M. B. Vivas. 2005. Landscape development intensity index. *Environmental Monitoring and Assessment* 101:289–309.

Brown, A. E., L. Zhang, T. A. McMahon, A. W. Western, and R. A. Vertessy. 2005. A review of paired catchment studies for determining changes in water yield resulting from alterations in vegetation. *Journal of Hydrology* 310:28–61.

Bruce, D. 1951. Fire resistance of longleaf pine seedlings. *Journal of Forestry* 49:739–740.

Bruce, D. and C. A. Bickford. 1950. Use of fire in natural regeneration of longleaf pine. *Journal of Forestry* 48:114–117.

Brudvig, L. A. and E. I. Damschen. 2011. Land-use history, historical connectivity, and land management interact to determine longleaf pine woodland understory richness and composition. *Ecography* 34:257–266.

Brudvig, L. A., E. I. Damschen, J. J. Tewksbury, N. M. Haddad, and D. J. Levey. 2009. Landscape connectivity promotes plant biodiversity spillover into non-target habitats. *Proceedings of the National Academy of Sciences* 106:9328–9332.

Brudvig, L. A., E. Grman, C. W. Habeck, J. L. Orrock, and J. A. Ledvina. 2013. Strong legacy of agricultural land use on soils and understory plant communities in longleaf pine woodlands. *Forest Ecology and Management* 310:944–955.

Brudvig, L. A., J. L. Orrock, E. I. Damschen, C. D. Collins, P. G. Hahn, W. B. Mattingly, J. W. Veldman, and J. L. Walker. 2014. Land-use history and contemporary management inform an ecological reference model for longleaf pine woodland understory plant communities. *PLOS ONE* 9:e86604.

Brudvig, L. A., S. A. Wagner, and E. I. Damschen. 2012. Corridors promote fire via connectivity and edge effects. *Ecological Applications* 22:937–946.

Brühl, C. A., T. Schmidt, S. Pieper, and A. Alscher. 2013. Terrestrial pesticide exposure of amphibians: An underestimated cause of global decline? *Scientific Reports* 3: 1135. doi: 1.1038/srep01135.

Bruno, J. F., J. J Stachowicz, and M. D. Bertness. 2003. Inclusion of facilitation into ecological theory. *Trends in Ecology and Evolution* 18:119–125.

Bryan, A. L., Jr. 2005. Wood stork. Pages 289–294 *in* J. C. Kilgo, and J. I. Blake, editors. *Ecology and Management of a Forested Landscape: Fifty Years on the Savannah River Site*. Washington, DC: Island Press.

Buckland, S. T., D. R. Anderson, K. P. Burnham, J. L. Laake, D. L. Borchers, and L. Thomas. 2001. *Introduction to Distance Sampling*. New York, NY: Oxford University Press.

Buckner, J. L. and J. L. Landers. 1979. Fire and disking effects on herbaceous food plants and seed supplies. *Journal of Wildlife Management* 43:807–811.

Buhlmann, K. A. and J. W. Gibbons. 2001. Terrestrial habitat use by aquatic turtles from a seasonally fluctuating wetland: Implications for wetland conservation boundaries. *Chelonian Conservation and Biology* 4:115–127.

Bullard, S. H., J. E. Gunter, M. L. Doolittle, and K. G. Arano. 2002. Discount rates for nonindustrial private forest landowners in Mississippi: How high a hurdle? *Southern Journal of Applied Forestry* 26:26–31.

Bureau of Land Management (BLM). 2015. Technical protocol for the collection, study, and conservation of seeds from native plant species for Seeds of Success. Bureau of Land Management, National Coordinating Office of Seeds of Success, Washington, DC. http://www.blm.gov/style/medialib/blm/wo/Planning_and_Renewable_Resources/fish__wildlife_and/plants/sos0.Par.67406.File.dat/protocol.pdf (accessed 4 February 2017).

Burgan, R. E. and R. C. Rothermel. 1984. BEHAVE: Fire behavior prediction and fuel modeling system-FUEL subsystem. General Technical Report INT-GTR-167. USDA Forest Service, Intermountain Research Station, Ogden, UT.

Burger, L. W., D. McKenzie, R. Thackston, and S. J. Demaso. 2006. The role of farm policy in achieving large-scale conservation: Bobwhite and buffers. *Wildlife Society Bulletin* 34:986–993.

Burger, L. W., D. A. Miller, and R. I. Southwick. 1999. Economic impact of northern bobwhite hunting in the southeastern United States. *Wildlife Society Bulletin* 27:1010–1018.

Burkle, L. A., J. A. Myers, and R. T. Belote. 2015. Wildfire disturbance and productivity as drivers of plant species diversity across spatial scales. *Ecosphere* 6:1–4.

Burney, D. A. and T. F. Flannery. 2005. Fifty millennia of catastrophic extinctions after human contact. *Trends in Ecology and Evolution* 20:395–401.

Busby, R. L., C. E. Thomas, and R. E. Lohrey. 1993. Potential product values from thinned longleaf pine plantations in Louisiana. Pages 645–650 *in* J. C. Brissette, editor. *Proceedings of the Seventh Biennial Southern Silvicultural Research Conference*, 17–19 November 1992, Mobile, Alabama, USA. General Technical Report SO-93. USDA Forest Service, Southern Forest Experiment Station, New Orleans, LA.

Butler, B. J. 2008. Family forest owners of the United States, 2006. General Technical Report NRS-27. USDA Forest Service, Northern Research Station, Newtown Square, PA.

Butler, B. J., J. H. Hewes, B. J. Dickinson, K. Andrejczyk, S. M. Butler, and M. Markowski-Lindsay. 2016. USDA Forest Service national woodland owner survey: National, regional, and state statistics for family forest and woodland ownerships with 10+ acres, 2011–2013. Resource Bulletin NRS-99. USDA Forest Service, Northern Research Station, Newtown Square, PA.

Butler, B. J. and E. C. Leatherberry. 2004. America's family forest owners. *Journal of Forestry* 102:4–14.

Butler, B. J. and D. N. Wear. 2013. Forest ownership dynamics of southern forests. Pages 103–121 *in* D. N. Wear, and J. G. Greis, editors. *The Southern Forest Futures Project: Technical Report*. General Technical Report SRS-GTR-178. USDA Forest Service, Southern Research Station, Asheville, NC.

Butterfield, B. J. 2009. Effects of facilitation on community stability and dynamics: Synthesis and future directions. *Journal of Ecology* 97:1192–1201.

Cabral, A. C., J. M. Miguel, A. J. Rescia, M. F. Schmitz, and F. D. Pineda. 2003. Shrub encroachment in Argentinean savannas. *Journal of Vegetation Science* 14:145–152.

Cain, M. D. 1991. The influence of woody and herbaceous competition on early growth of naturally regenerated loblolly and shortleaf pines. *Southern Journal of Applied Forestry* 15:179–185.

Caldwell, P. V., C. F. Miniat, K. J. Elliott, W. T. Swank, S. T. Brantley, and S. H. Laseter. 2016. Declining water yield from forested mountain watersheds in response to climate change and forest mesophication. *Global Change Biology* 22: 2997–3012.

Caldwell, P. V., C. Muldoon, C. Ford-Miniat, E. Cohen, S. Krieger, G. Sun, S. McNulty, and P. V. Bolstad. 2014. Quantifying the role of National Forest system lands in providing surface drinking water supply for the southern United States. General Technical Report SRS-197. USDA Forest Service, Southern Research Station, Asheville, NC.

Calhoun, A. J. K., N. A. Miller, and M. W. Klemens. 2005. Conserving pool-breeding amphibians in human-dominated landscapes through local implementation of Best Development Practices. *Wetlands Ecology and Management* 13:291–304.

Cameron, G. N. and S. R. Spencer. 1981. Sigmodon hispidus. *Mammalian Species Supplement* 158:1–9.

Canadell, J. G. and M. R. Raupach. 2008. Managing forests for climate change mitigation. *Science* 320:1456–1457.

Cannon, J. B., J. J. O'Brien, E. L. Loudermilk, M. B. Dickinson, and C. J. Peterson. 2014. The influence of experimental wind disturbance on forest fuels and fire characteristics. *Forest Ecology and Management* 330:294–303.

Cano Carmona, E. and A. Cano Ortiz. 2012. Establishment of biogeographic areas by distributing endemic flora and habitats (Dominican Republic, Haiti r.). Pages 99–118 *in* L. Stevens, editor. *Global Advances in Biogeography*. Rijeka: Intech Europe.

Cansler, C. A. and D. McKenzie. 2014. Climate, fire size, and biophysical setting control fire severity and spatial pattern in the northern Cascade Range, USA. *Ecological Applications* 24:1037–1056.

Capps, K. A., K. A. Berven, and S. D. Tiegs. 2015. Modelling nutrient transport and transformation by pool-breeding amphibians in forested landscapes using a 21-year dataset. *Freshwater Biology* 60:500–511.

Carpenter, S. R., H. A. Mooney, J. Agard, D. Capistrano, R. S. Defries, S. Diaz, T. Dietz, A. K. Duraiappah, A. Oteng-Yeboah, H. M. Pereira, C. Perrings, W. V. Reid, J. Sarukhan, R. J. Scholes, and A. Whyte. 2009. Science for managing ecosystem services: Beyond the millennium ecosystem assessment. *Proceedings of the National Academy of Sciences* 106:1305–1312.

Carr, S. C., K. M. Robertson, W. J. Platt, and R. K. Peet. 2009. A model of geographical, environmental and regional variation in vegetation composition of pyrogenic grasslands of Florida. *Journal of Biogeography* 36:1600–1612.

Carter, L. M., J. W. Jones, L. Berry, V. Burkett, J. F. Murley, J. Obeysekera, P. J. Schramm, and D. Wear. 2014. Southeast and the Caribbean. Pages 396–417 *in* J. M. Melillo, T. C. Richmond, and G. W. Yohe, editors. *Climate Change Impacts in the United States: The Third National Climate Assessment*, Chap. 17. U.S. Global Change Research Program. doi: 10.7930/JON-P22CB U.S. Government Printing Office: Washington, DC.

Carter, M. C. and C. D. Foster. 2004. Prescribed burning and productivity in southern pine forests: A review. *Forest Ecology and Management* 191:93–109.

Carter, M. C., R. C. Kellson, and R. S. Wallinger. 2015. *Forestry in the U.S. South: A History*. Baton Rouge, LA: Louisiana State University Press.

Casey, W. P. and K. C. Ewel. 2006. Patterns of succession in forested depressional wetlands in North Florida, USA. *Wetlands* 26:147–160.

Casper, B. B. and R. B. Jackson. 1997. Plant competition underground. *Annual Review of Ecology and Systematics* 28:545–570.

Catano, C. P., T. L. Dickson, and J. A. Myers. 2017. Dispersal and neutral sampling mediate contingent effects of disturbance on beta-diversity: A meta-analysis. Ecology Letters. doi: 10.1111/ele.12733.

Catesby, M. 1771. *The Natural History of Carolina, Florida and the Bahama Islands*. Chapel Hill, NC: University of North Carolina Press.

Cathey, S. E., L. R. Boring, and T. R. Sinclair. 2010. Assessment of N_2 fixation capability of native legumes from the longleaf pine-wiregrass ecosystem. *Experimental Botany* 67:444–450.

Cavender-Bares, J., K. Kitajima, and F. A. Bazzaz. 2004. Multiple trait associations in relation to habitat differentiation among 17 Floridian oak species. *Ecological Monographs* 74:635–662.

Cerame, B., J. A. Cox, R. T. Brumfield, J. W. Tucker, and S. S. Taylor. 2014. Adaptation to ephemeral habitat may overcome natural barriers and severe habitat fragmentation in a fire-dependent species, the Bachman's Sparrow (*Peucaea aestivalis*). *PLOS ONE* 9:e105782. doi:10.1371/journal.pone.0105782.

Chamberlain, M. J., B. D. Leopold, and L. M. Conner. 2003. Space use, movements and habitat selection of adult bobcats (*Lynx rufus*) in central Mississippi. *American Midland Naturalist* 146:395–405.

Chambers, M. E., P. J. Fornwalt, S. L. Malone, and M. A. Battaglia. 2016. Patterns of conifer regeneration following high severity wildfire in ponderosa pine-dominated forests of the Colorado Front Range. *Forest Ecology and Management* 378:57–67.

Chanda, D., D. L. Herrighty, P.M. Castelli, A. W. Burnett, and J. R. Garris. 2009. New Jersey Northern Bobwhite action plan. Report to the New Jersey Fish and Game Council, NJ.

Chandler, J. L., J. L. Orrock, and J. Resasco. 2016. Invasive ants generate heterogeneity in patterns of seed survival. *American Midland Naturalist* 176:289–297.

Changnon, S. A. and T. R. Karl. 2003. Temporal and spatial variations of freezing rain in the contiguous United States: 1948–2000. *Journal of Applied Meteorology* 42:1302–1315.

Channell, R. and M. V. Lomolino. 2000. Dynamic biogeography and conservation of endangered species. *Nature* 403:84–86.

Chapman, C. S. 1905. *A Working Plan for Forest Lands in Berkeley County, South Carolina*. USDA Bureau of Forestry Bulletin 56 Washington, DC.

Chapman, H. H. 1932. Is the longleaf type a climax? *Ecology* 13:328–334.

Chapman, H. H. 1942. *Management of Loblolly Pine in the Pine-Hardwood Region in Arkansas and in Louisiana West of the Mississippi River*. Forestry Bulletin No. 49. School of Forestry, Yale University, New Haven, CT.

Chapuis-Lardy, L., S. Vanderhoeven, N. Dassonville, L. S. Koutika, and P. Meerts. 2006. Effects of the exotic invasive plant *Solidago gigantea* on soil phosphorus status. *Biology and Fertility of Soils* 42:481–489.

Chase, J. M. 2003. Community assembly: When should history matter? *Oecologia* 136:489–498.

Chase, J. M., K. C. Amarasekare, A. Gonzalez, R. D. Holt, M. Holyoak, M. F. Hoopes, M. A. Leipold, M. Loreau, N. Mouquet, J. B. Shurin, and D. Tilman. 2005. Competing theories for competitive metacommunities. Pages 335–354 *in* M. Holyoak, M. A. Leipold, and R. D. Holt, editors. *Metacommunities: Spatial Dynamics and Ecological Communities*. Chicago, IL: University of Chicago Press.

Chase, J. M. and M. A. Leibold. 2003. *Ecological Niches: Linking Classical and Contemporary Approaches*. Chicago, IL: University of Chicago Press.

Chase, J. M. and J. A. Myers. 2011. Disentangling the importance of ecological niches from stochastic processes across scales. *Philosophical Transactions of the Royal Society B-Biological Sciences* 366:2351–2363.

Chen, D. and H. W. Chen. 2013. Using the Köppen classification to quantify climate variation and change: An example for 1901–2010. *Environmental Development* 6:69–79.

Cherry, M. J., K. L. Turner, M. B. Howze, D. S. Cohen, L. M. Conner, and R. J. Warren. 2016. Coyote diets in a longleaf pine ecosystem. *Wildlife Biology* 22:64–70.

Cherry, M. J, R. J. Warren, and L. M. Conner. 2015. Effects of predation risk and group dynamics on white-tailed deer foraging behavior in a longleaf pine savanna. *Behavioral Ecology* 26: 1091–1099. doi: 10.1093/beheco/arv054.

Cherry, M. J., R. J. Warren, and L. M. Conner. 2016. Fear, fire, and behaviorally mediated trophic cascades. *Forest Ecology and Management* 368:133–139.

Cherry, M.J., Warren, R.J. and Conner, L.M., 2017. Fire-mediated foraging tradeoffs in white-tailed deer. Ecosphere, 8(4):e01784. doi: 0.1002/ecs2.1784.

Chesson, P. 2000. Mechanisms of maintenance of species diversity. *Annual Review of Ecology and Systematics* 31:343–366.

Chitwood, M. C., M. A. Lashley, J. C. Kilgo, C. E. Moorman, and C. S. Deperno. 2015. White-tailed deer population dynamics and adult female survival in the presence of a novel predator. *Journal of Wildlife Management* 79:211–219.

Christensen, N. L. 1981. Fire regimes in southeastern ecosystems. Pages 112–136 *in* H. T. Mooney, T. M. Bonnicksen, N. L. Christensen, J. E. Lotan, and W. A. Reiners, editors. *Fire Regimes and Ecosystem Process*. General Technical Report WO-26. USDA Forest Service, Washington, DC.

Christensen, N. L. 1988. Vegetation of the southeastern coastal plain. Pages 317–364 *in* M. G. Barbour, and W. D. Billings, editors. *North American Terrestrial Vegetation*. New York, NY: Cambridge University Press.

Christensen, N. L., R. B. Burchell, A. Liggett, and E. L. Simms. 1981. The structure and development of pocosin vegetation. Pages 43–61 *in* C. J. Richardson, editor. *Pocosin Wetlands*. Stroudsburg, PA: Hutchinson Ross Publishing Company.

Christianson, D. and S. Creel. 2008. Risk effects in elk: Sex-specific responses in grazing and browsing due to predation risk from wolves. *Behavioral Ecology* 19:1258–1266.

Churchill, D. J., A. J. Larson, M. C. Dahlgren, J. F. Franklin, P. F. Hessburg, and J. A. Lutz. 2013. Restoring forest resilience: From reference spatial patterns to silvicultural prescriptions and monitoring. *Forest Ecology and Management* 291:442–457.

Chuvieco, E., D. Riano, I. Aguado, and D. Cocero. 2002. Estimation of fuel moisture content from multitemporal analysis of Landsat Thematic Mapper reflectance data: Applications in fire danger assessment. *International Journal of Remote Sensing* 23:2145–2162.

Cieslar, A. 1909. Licht- und Schattholzarten, Lichtgenuss und Bodenfeuchtigkeit. *Centralblatt für das gesamte Forstwese (Austrian Journal of Forest Science)* 35:4–22.

Cipollini, M. L., J. Culberson, C. Strippelhoff, T. Baldvins, and K. Miller. 2012. Herbaceous plants and grasses in a mountain longleaf pine forest undergoing restoration: A survey and comparative study. *Southeastern Naturalist* 11:637–668.

Clark County. 2015. Boulder City conservation easement management plan, Version 3.2. January 2015. http://www.clarkcountynv.gov/airquality/dcp/Documents/mitigation/bcce/BCCE%20Management%20 Plan_Version%203.2_01062015.pdf (accessed 4 February 2017).

Clark County. 2016. 2016 Clark County multiple species habitat conservation plan. Adaptive Management Report prepared for Desert Conservation Program by Enduring Conservation Outcomes. Clark County, NV. http://www.clarkcountynv.gov/airquality/dcp/Documents/Library/dcp%20reports/2016/2016%20% 20Adaptive%20Management%20Report.pdf (accessed 4 February 2017).

Clark, J. S. 2009. Beyond neutral science. *Trends in Ecology and Evolution* 24:8–15.

Clark, J. P., J. E. Dunn, and K. G. Smith. 1993. A multivariate model of female black bear habitat for a Geographic Information System. *Journal of Wildlife Management* 57:519–526.

Clark, K. L., H. L. Gholz, and M. S. Castro. 2004. Carbon dynamics along a chronosequence of slash pine plantations in North Florida. *Ecological Applications* 14:1154–1171.

Clark, J. S., E. Macklin, and L. Wood. 1998. Stages and spatial scales of recruitment limitation in southern Appalachian forests. *Ecological Monographs* 68:213–235.

Clark, M. A., J. Siegrist, and P. A. Keddy. 2008. Patterns of frequency in species-rich vegetation in pine savannas: Effects of soil moisture and scale. *Ecoscience* 15:529–535.

Clarke, P. J., M. J. Lawes, and J. J. Midgley. 2010. Resprouting as a key functional trait in woody plants – challenges to developing new organizing principles. *New Phytologist* 188:651–654.

Clayton, B. and D. W. Hicks. 2007. Hydrologic monitoring of a hardwood encroached, isolated depressional wetland, southwest Georgia. Pages 596–598 *in* T. C. Rasmussen, G. D. Carroll, and A. P. Georgakakos, editors. *Proceedings of the 2007 Georgia Water Conference*, 27–29 March 2007, Institute of Ecology, University of Georgia, Athens, GA.

Clayton, L.A., E.C. Moore, and V.J. Knight Jr. (editors). 1995. *The DeSoto Chroniles: The Expedition of Hernando de Soto to North America in 1539–1543*. University of Alabama Press. 1200 p.

Clean Air Act (CAA) of 1970, as amended August 1977; 42 U.S.C. 7401 et seq. Clean Air Act Amendments of 1990. 104 Stat. 2399.

Clewell, A. F. 1989. Natural history of wiregrass (*Aristida stricta* Michx., Gramineae). *Natural Areas Journal* 9:223–233.

Clewell, A. F. and J. Aronson. 2013. *Ecological Restoration: Principles, Values, and Structure of an Emerging Profession*. Washington, DC: Island Press.

Clinton, B. D., C. A. Maier, C. R. Ford, and R. J. Mitchell. 2011. Transient changes in transpiration, and stem and soil CO_2 efflux in longleaf pine (*Pinus palustris* Mill.) following fire-induced leaf area reduction. *Trees* 25:997–1007.

Clutter, M. L., B. Mendell, D. Newman, D. Wear, and J. Greis. 2005. Strategic factors driving timberland ownership changes in the U.S. South. http://www.iatp.org/files/181_2_78129.pdf (accessed 4 February 2017).

Clyatt, K. A., J. S. Crotteau, M. S. Schaedel, H. L. Wiggins, H. Kelley, D. J. Churchill, and A. J. Larson. 2016. Historical spatial patterns and contemporary tree mortality in dry mixed-conifer forests. *Forest Ecology and Management* 361:23–37.

Coates, K. D. and P. J. Burton. 1997. A gap-based approach for development of silvicultural systems to address ecosystem management objectives. *Forest Ecology and Management* 99:337–354.

Coble, H. D., F. M. Williams, and R. L. Ritter. 1981. Common ragweed (*Ambrosia artemisiifolia*) interference in soybeans (*Glycine max*). *Weed Science* 29:339–342.

Cohen, M. J., I. Creed, L. Alexander, N. Basu, A. J. K. Calhoun, C. Craft, E. D'Amico, E. DeKeyser, L. Fowler, H. E. Golden, J. W. Jawitz, P. Kalla, L. K. Kirkman, C. R. Lane, M. Lang, S. G. Leibowitz, D. B. Lewis,

J. Marton, D. L. McLaughlin, D. Mushet, H. Raanan-Kiperwas, M. C. Rains, L. Smith, and S. C. Walls. 2016. Do geographically isolated wetlands influence landscape functions? *Proceedings of the National Academy of Sciences* 113:1978–1986.

Cohen, M. J., E. J. Dunne, and G. L. Bruland. 2008. Spatial variability of soil properties in cypress domes surrounded by different land uses. *Wetlands* 28:411–422.

Coile, T. S. 1936. The effect of rainfall and temperature on the annual growth of pine in the southern United States. *Ecological Monographs* 6:533–562.

Collins, B. M., R. G. Everett, and S. L. Stephens. 2011. Impacts of fire exclusion and recent managed fire on forest structure in old growth Sierra Nevada mixed-conifer forest. *Ecosphere* 2: Article 51.

Collins, B. M., J. M. Lydersen, R. G. Everett, D. L. Fry, and S. L. Stephens. 2015. Novel characterization of landscape-level variability in historical vegetation structure. *Ecological Applications* 25:1167–1174.

Collins, B. M., J. D. Miller, A. E. Thode, M. Kelly, J. W. van Wagtendonk, and S. L. Stephens. 2009. Interactions among wildland fires in a long-established Sierra Nevada natural fire area. *Ecosystems* 12:114–128.

Collins, S. L. and M. D. Smith. 2006. Scale-dependent interaction of fire and grazing on community heterogeneity in tallgrass prairie. *Ecology* 87:2058–2067.

Collins, B. M. and S. L. Stephens. 2010. Stand-replacing patches within a 'mixed severity' fire regime: Quantitative characterization using recent fires in a long-established natural fire area. *Landscape Ecology* 25:927–939.

Collins, W. B. and P. J. Urness. 1983. Feeding behavior and habitat selection of mule deer and elk on northern Utah summer range. *Journal of Wildlife Management* 47:646–663.

Connell, J. H. 1978. Diversity in tropical rain forests and coral reefs. *Science* 199:1302–1310.

Connell, J. H. 1980. Diversity and the coevolution of competitors, or the ghost of competition past. *Oikos* 35:131–138.

Conner, L. M. 2000. Home range sizes of fox squirrels in southwest Georgia. *Proceedings of the Annual Conference of the Southeastern Association of Fish and Wildlife Agencies* 54:400–406.

Conner, L. M., S. B. Castleberry, and A. M. Derrick. 2011. Effects of mesopredators and prescribed fire on hispid cotton rat survival and cause-specific mortality. *Journal of Wildlife Management* 75:938–944.

Conner, L. M., M. J. Cherry, B. T. Rutledge, C. H. Killmaster, G. Morris, and L. L. Smith. 2016. Predator exclusion as a management option for increasing white-tailed deer recruitment. *Journal of Wildlife Management* 80:162–170.

Conner, R. C. and A. J. Hartsell. 2002. Forest area and conditions. Pages 357–402 *in* D. N. Wear, and J. G. Greis, editors. In *Southern Forest Resource Assessment*. General Technical Report SRS-53. USDA Forest Service, Southern Research Station, Asheville, NC.

Conner, L. M. and G. Morris. 2015. Impacts of mesopredator control on conservation of mesopredators and their prey. *PLOS ONE* 10:e0137169. doi:10.1371/journal.pone.0137169.

Conner, R. N., D. C. Rudolph, D. Saenz, and R. R. Schaefer. 1996. Red-cockaded woodpecker nesting success, forest structure, and southern flying squirrels in Texas. *The Wilson Bulletin* 1996:697–711.

Conner, L. M., J. C. Rutledge, and L. L. Smith. 2010. Effects of mesopredators on nest survival of shrub-nesting songbirds. *Journal of Wildlife Management* 74:73–80.

Cook, E. R., D. M. Meko, D. W. Stahle, and M. K. Cleaveland. 2004. North American Summer PDSI Reconstructions. World Data Center for Paleoclimatology Data Contribution Series #2004-045. NOAA/NGDC Paleoclimatology Program, Boulder, CO. http://www.ncdc.noaa.gov/paleo/newpdsi.html (accessed 4 February 2017).

Cosentino, B. J., R. L. Schooley, and C. A. Phillips. 2011. Spatial connectivity moderates the effect of predatory fish on salamander metapopulation dynamics. *Ecosphere* 2:1–14.

Costa, R. and R. S. DeLotelle. 2006. Reintroduction of fauna to longleaf pine ecosystems: Opportunities and challenges. Pages 335–376 *in* S. Jose, E. J. Jokela, and D. L. Miller, editors. *The Longleaf Pine Ecosystem: Ecology, Silviculture, and Restoration*. New York, NY: Springer Science.

Côté, S. D., T. P. Rooney, J. P. Tremblay, C. Dussault, and D. M. Waller. 2004. Ecological impacts of deer overabundance. *Annual Review of Ecology, Evolution and Systematics* 35:113–147.

Couch, C. A., E. H. Hopkins, and P. S. Hardy. 1996. Influences of environmental settings on aquatic ecosystems in the Apalachicola-Chattahoochee-Flint River basin. USGS Water-Resources Investigations Report 95-4278. U.S. Geological Survey National Water-Quality Assessment Program, Atlanta, GA.

Coulston, J. W., D. N. Wear, and J. M. Vose. 2015. Complex forest dynamics indicate potential for slowing carbon accumulation in the southeastern United States. *Scientific Reports* 5:Article 8002. doi:10.1038/srep08002.

Covey, C. (translator and editor). 1990. *Cabeza de Vaca's Adventures in the Unknown Interior of America*. University of New Mexico Press. 160 p.

Cox, J. A., W. W. Baker, and R. T. Engstrom. 2001. Red-cockaded woodpeckers in the red hills region: A GIS-based assessment. *Wildlife Society Bulletin* 29:1278–1288.

Cox, J. A. and R. T. Engstrom, 2001. Influence of the spatial pattern of conserved lands on the persistence of a large population of red-cockaded woodpeckers. *Biological Conservation* 100:137–150.

Cox, A. C., D. R. Gordon, J. L. Slapcinsky, and G. S. Seamon. 2004. Overstory restoration in longleaf pine sandhills. *Natural Areas Journal* 24:4–14.

Cox, J. A., L. C. Scopel, and M. R. Klostermann. 2012. Brown-headed nuthatch occupancy in central Florida and its relationship to forest type, forest structure, and the presence of red-cockaded woodpeckers. *The Condor* 114:622–628.

Cox, J. A. and G. L. Slater. 2007. Cooperative breeding in the brown-headed nuthatch. *The Wilson Journal of Ornithology* 119:1–8.

Cox, J. A. and B. Widener. 2008. *Lightning-Season Burning: Friend or Foe of Breeding Birds*. Miscellaneous Publication 17. Tallahassee, FL: Tall Timbers Research Station.

Craft, C. B. and W. P. Casey. 2000. Sediment and nutrient accumulation in floodplain and depressional freshwater wetlands of Georgia, USA. *Wetlands* 20:323–332.

Craft, C. B. and C. Chiang. 2002. Forms and amounts of soil nitrogen and phosphorus across a longleaf pine-depressional wetland landscape. *Soil Science Society of America Journal* 66:1713–1721.

Craib, I. J. 1929. *Some Aspects of Soil Moisture in the Forest. Yale University School Research Bulletin 25*. New Haven, CT: Yale University

Cram, M. M., K. W. Outcalt, and S. J. Zarnoch. 2010. Growth of longleaf and loblolly pine planted on South Carolina sandhill sites. *Southern Journal of Applied Forestry* 34:79–83.

Craul, P. J., J. S. Kush, and W. D. Boyer. 2005. Longleaf pine site zones. General Technical Report SRS-89. USDA Forest Service, Southern Research Station, Asheville, NC.

Creech M. N.., L. K. Kirkman, and L. A. Morris. 2012. Alteration and recovery of slash pile burn sites in the restoration of a fire-maintained ecosystem. *Restoration Ecology* 20:505–516.

Creel, S. and D. Christianson. 2009. Wolf presence and increased willow consumption by Yellowstone elk: Implications for trophic cascades. *Ecology* 90:2454–2466.

Creel, S., J. Winnie, Jr., B. Maxwell, K. Hamlin, and M. Creel. 2005. Elk alter habitat selection as an antipredator response to wolves. *Ecology* 86:3387–3397.

Crespi, E. J. and R. W. Warne. 2013. Environmental conditions experienced during the tadpole stage alter post-metamorphic glucocorticoid response to stress in an amphibian. *Integrative and Comparative Biology* 53:989–1001.

Crofton, E. W. 2001. The fire prophet: Leon Neel's forest management plan leaves something for future generations. J. R. Wilson, editor, *In the Fire Forest: Longleaf Pine Wiregrass Ecosystem*. Covington, GA: Georgia Wildlife Press. http://www.sherpaguides.com/georgia/fire_forest/leon_neel/index.html (accessed 4 February 2017).

Croker, T. C., Jr. 1968. Longleaf pine: An annotated bibliography, 1946 through 1967. Research Paper SO-35. USDA Forest Service, Southern Forest Experiment Station, New Orleans, LA.

Croker, T. C., Jr. 1975. Seedbed preparation aids natural regeneration of longleaf pine. Forest Service Research Paper SO-112. USDA Forest Service, Southern Forest Experiment Station, New Orleans, LA.

Croker, T. C., Jr. 1979. Longleaf: The longleaf pine story. *Journal of Forest History* 23:32–43.

Croker, T. C., Jr. 1987. Longleaf pine, a history of man and a forest. Forestry Report R8-FR7. USDA Forest Service, Southern Forest Experiment Station, New Orleans, LA.

Croker, T. C., Jr. 1990. Longleaf pine-myths and facts. Pages 2–10 *in* R. M. Farrar, editor. *Proceedings of the Symposium on the Management of Longleaf Pine*, 4–6 April 1989, Long Beach, Mississippi, USA. General Technical Report SO-75. USDA Forest Service, Southern Forest Experiment Station, New Orleans, LA.

Croker, T. C., Jr. and W. D. Boyer. 1975. Regenerating longleaf pine naturally. Research Paper SO-105. USDA Forest Service, Southern Forest Experiment Station, New Orleans, LA.

Cronan, J. B., C. S. Wright, and M. Petrova. 2015. Effects of dormant and growing season burning on surface fuels and potential fire behavior in northern Florida longleaf pine (*Pinus palustris*) flatwoods. *Forest Ecology and Management* 354:318–333.

Crooks, K. R. and M. E. Soulé. 1999. Mesopredator release and avifaunal extinctions in a fragmented system. *Nature* 400:563–566.

Crossa, J. and R. Vencovsky. 2011. Basic sampling strategies: Theory and practice. Pages 1–28 *in* L. Guarino, R. Rao, and E. Goldberg, editors. *Collecting Plant Genetic Diversity: Technical Guidelines*. Rome: Bioversity International. http://cropgenebank.sgrp.cgiar.org/images/file/procedures/collecting2011/Chapter5-2011.pdf (accessed 4 February 2017).

Cubbage, F. W. 1983. Economics of forest tract size: Theory and literature. General Technical Report SO-41. USDA Forest Service, Southern Forest Experiment Station, New Orleans, LA.

Cubbage, F. W. and D. Hodges. 1989. *The economics of managing longleaf pine. Proceedings of the Symposium on the Management of Longleaf Pine*. General Technical Report SO-75. USDA Forest Service, Southern Forest Experiment Station, New Orleans, LA.

Cubbage, F. W., P. Mac Donagh, J. S. Júnior, R. Rubilar, P. Donoso, A. Ferreira, V. Hoeflich, V. M. Olmos, G. Ferreira, and G. Balmelli. 2007. Timber investment returns for selected plantations and native forests in South America and the Southern United States. *New Forests* 33:237–255.

Cubbage F. W. and C. H. Redmond. 1985. Capital budgeting practices in forest products industry. *Journal of Forest Products* 35:55–60

Cumberland, M. S. and L. K. Kirkman. 2013. The effects of the red imported fire ant on seed fate in the longleaf pine ecosystem. *Plant Ecology* 214:717–724.

Currie, K., J. M. Varner, J. Kush, and M. Cipollini. 2006. A survey of the herbaceous vegetation found in the Berry Longleaf Pine Management Area. Pages 35–38 *in* M. L. Cipollini, compiler. *Proceedings of the Second Montane Longleaf Pine Conference Workshop*, 18–19 November 2005, Berry College, Mount Berry, GA. Longleaf Alliance Report No. 9.

Cushwa, C. T., E. Czuhai, R. W. Cooper, and W. H. Julian. 1969. Burning clearcut openings in loblolly pine to improve wildlife habitat. Georgia Forest Research Paper 61.

Cushwa, C. T., M. Hopkins, and B. S. McGinnes. 1970. Response of legumes to prescribed burns in loblolly pine stands of the South Carolina Piedmont. U.S. Forest Service Research Note SE-140. USDA Forest Service, Southeastern Forest Experiment Station, Asheville, NC.

Czapka, S. J. and J. C. Kilgo. 2011. Importance of Carolina bays to the avifauna of pinelands in the southeastern United States. *Southeastern Naturalist* 10:321–332.

Czimczik, C. I. and C. A. Masiello. 2007. Controls on black carbon storage in soils. *Global Biogeochemical Cycles* 21:GCB3005.

Dahl, T. E. 1990. Wetlands losses in the United States, 1780s to 1980s. Report to the Congress, No. PB-91-169284/XAB. National Wetlands Inventory, St. Petersburg, FL.

Daily, G. C. and P. A. Matson. 2008. Ecosystem services: From theory to implementation. *Proceedings of the National Academy of Sciences* 105:9455–9456.

Dale, V. H., S. C. Beyeler, and B. Jackson. 2002. Understory vegetation indicators of anthropogenic disturbance in longleaf pine forests at Fort Benning, Georgia, USA. *Ecological Indicators* 1:155–170.

Daly, C., M. Halbleib, J. I. Smith, W. P. Gibson, M. K. Doggett, G. H. Taylor, J. Curtis, and P. P. Pasteris. 2008. Physiographically sensitive mapping of temperature and precipitation across the conterminous United States. *International Journal of Climatology* 28:2031–2064.

Damschen, E. I., D. V. Baker, G. Bohrer, R. Nathan, J. L. Orrock, J. R. Turner, L. A. Brudvig, N. M. Haddad, D. J. Levey, and J. J. Tewksbury. 2014. How fragmentation and corridors affect wind dynamics and seed dispersal in open habitats. *Proceedings of the National Academy of Sciences* 111:3484–3489.

Damschen, E. I. and L. A. Brudvig. 2012. Landscape connectivity strengthens local-regional richness relationships in successional plant communities. *Ecology* 93:704–710.

Dandois, J. P. and E. C. Ellis. 2013. High spatial resolution three-dimensional mapping of vegetation spectral dynamics using computer vision. *Remote Sensing of Environment* 136:259–276.

Daniel, T. W., J. A. Helms, and F. S. Baker. 1979. *Principles of Silviculture*. New York, NY: McGraw-Hill.

Darley-Hill, S. and W. C. Johnson. 1981. Acorn dispersal by the blue jay. *Oecologia* 50:231–232.

Darracq, A. K., L. M. Conner, J. S. Brown, and R. A. McCleery. 2016. Cotton rats alter foraging in response to an invasive ant. *PLOS ONE* 11:e0163220.

Dassot, M., T. Constant, and M. Fournier. 2011. The use of terrestrial LiDAR technology in forest science: Application fields, benefits and challenges. *Annals of Forest Science* 68:959–974.

Davidson, E. A. and I. L. Ackerman. 1993. Changes in soil inventories following cultivation of previously untilled soils. *Biogeochemistry* 20:161–193.

Davies, M. G., A. A. Smith, A. J. MacDonald, J. D. Bakker, and C. J. Legg. 2010. Fire intensity, fire severity and ecosystem response in heathlands: Factors affecting the regeneration of *Calluna vulgaris*. *Journal of Applied Ecology* 47:356–365.

Davis, M. A., S. G. Pritchard, R. J. Mitchell, S. A. Prior, H. H. Rogers, and G. B. Runion. 2002. Elevated atmospheric CO_2 affects structure of a model regenerating longleaf pine community. *Journal of Ecology* 90:130–140.

Dawson, J. R., J. C. Lee, D. A. Osborn, and K. V. Miller. 2009. Survival, movements and habitat use of translocated southern fox squirrels. *American Midland Naturalist* 162:335–345.

Dean, S., E. C. Farrer, and E. S. Menges. 2015. Fire effects on soil biogeochemistry in Florida scrubby flatwoods. *American Midland Naturalist* 174:49–64.

DeBano, L. F., D. G. Neary, and P. F. Folliott. 1998. *Fires Effects on Ecosystems*. New York, NY: John Wiley & Sons.

De las Heras, J., M. Bonilla, and L. W. Martinez. 2005. Early vegetation dynamics of *Pinus tropicalis* Morelet forests after experimental fire (W. Cuba). *Annals of Forest Science* 62:771–777.

Delcourt, H. R. and P. A. Delcourt. 1977. Presettlement magnolia-beech climax of the Gulf coastal plain: Quantitative evidence from the Apalachicola River Bluffs, north-central Florida. *Ecology* 58:1085–1093.

Delcourt, P. A. and H. R. Delcourt. 2004. *Prehistoric Native Americans and Ecological Change: A Panarchical Perspective on the Evolution of Human Ecosystems in Eastern North America Since the Pleistocene.* Cambridge: Cambridge University Press.

Delis, P. R. and H. R. Mushinsky. 2005. Human disturbance and Florida anurans. Pages 15–22 *in* W. E. Meshaka, Jr. and K .J. Babbitt, editors. *Amphibians and Reptiles: Status and Conservation in Florida.* Malabar, FL: Krieger Publishing Company.

Delis, P. R., H. R. Mushinsky, and E. D. McCoy. 1996. Decline of some west-central Florida anuran populations in response to habitat degradation. *Biodiversity and Conservation* 5:1579–1595.

DeLuca, T. H. and G. H. Aplet. 2008. Charcoal and carbon storage in forest soils of the Rocky Mountain West. *Frontiers in Ecology and the Environment* 6:18–24.

DeLuca, T. H., M. D. McKenzie, M. J. Gundale, and W. E. Holben. 2006. Wildfire produced charcoal directly influences N cycling in Ponderosa pine forests. *Soil Science Society of America Journal* 70:448–453.

Del-Val, E. K. and M. K. Crawley. 2005. What limits herb biomass in grasslands: Competition or herbivory? *Oecologia* 142:202–211.

Delworth, T. L., A. J. Broccoli, A. Rosati, R. J. Stouffer, V. Balaji, J. A. Beesley, W. F. Cooke, K. W. Dixon, J. Dunne, K. A. Dunne, J. W. Durachta, K. L. Findell, P. Ginoux, A. Gnanadesikan, C. T. Gordon, S. M. Griffies, R. Gudgel, M. J. Harrison, I. M. Held, R. S. Hemler, L. W. Horowitz, S. A. Klein, T. R. Knutson, P. J. Kushner, A. R. Langenhorst, H. C. Lee, S. J. Lin, J. Lu, S. L. Malyshev, P. C. D. Milly, V. Ramaswamy, J. Russell, M. D. Schwarzkopf, E. Shevliakova, J. J. Sirutis, M. J. Spelman, W. F. Stern, M. Winton, A. T. Wittenberg, B. Wyman, F. Zeng, and R. Zhang. 2006. GFDL's CM2 global coupled climate models. Part 1. Formulation and simulation characteristics. *Journal of Climate* 19:643–674.

DeMaynadier, P. G. and M. L. Hunter, Jr. 1995. The relationship between forest management and amphibian ecology: A review of the North American literature. *Environmental Reviews* 3:230–261.

Denevan, W. 1992. *The Native Population of the Americas in 1492*, 2nd edition. Madison, WI: University of Wisconsin Press.

De Steven, D. D. and J. M. Gramling. 2012. Diverse characteristics of wetlands restored under the Wetlands Reserve Program in the southeastern United States. *Wetlands* 32:593–604.

De Steven, D. D. and R. Lowrance. 2011. Agricultural conservation practices and wetland ecosystem services in the wetland-rich Piedmont-Coastal Plain region. *Ecological Applications* 21:S3–S17.

De Steven, D. D. and M. M. Toner. 2004. Vegetation of upper coastal plain depression wetlands: Environmental templates and wetland dynamics within a landscape framework. *Wetlands* 24:34–42.

Dether, D. and A. Black. 2006. Learning from escaped prescribed fires—lessons for high reliability. *Fire Management Today* 66:50–66.

Devall, M. S., J. Myers Grender, and J. Koretz. 1991. Dendroecological analysis of a longleaf pine *Pinus palustris* forest in Mississippi. *Vegetatio* 93:1–8.

Dey, D. C. and G. Hartman. 2005. Returning fire to Ozark Highland forest ecosystems: Effects on advance regeneration. *Forest Ecology and Management* 217:37–53.

Diaz, S., M. Cabido, and F. Casanoves. 1998. Plant functional traits and environmental filters at a regional scale. *Journal of Vegetation Science* 9:113–122.

Dickens, E. D., C. W. Dangerfield Jr., and D. J. Moorhead. 2006. An economic comparison of slash and loblolly pine under various levels of management in the lower Atlantic and Gulf Coastal Plain. Pages 351–356 in K. F. Connor, editor. *Proceedings of the 13th Biennial Southern Silvicultural Research Conference*, 28 February–4 March 2005, Memphis, TN, USA. General Technical Report SRS-92. USDA Forest Service, Southern Research Station, Asheville, NC.

Dickens, E. D., D. J. Moorhead, C. T. Bargeron, and B. C. McElvany. 2011. Pine straw yields and economic benefits when added to traditional wood products in loblolly, longleaf and slash pine stands. Pages 264–271 *in* S. F. Ashton, S. W. Workman, W. G. Hubbard, and D. J. Moorhead, editors. *Agroforestry: A Profitable Land Use. Proceedings of the 12th North American Agroforestry Conference*, 4–9 June 2011, Athens, GA.

Dickens, E. D., D. J. Moorhead, C. T. Bergeron, L. A. Morris, L. A. Ogden, and B. C. McElvany. 2012. A summary of pine straw yields and economic benefits in loblolly, longleaf and slash pine stands. *Agroforestry Systems* 86:315–321.

Dickens, E. D., D. J. Moorhead, L. A. Morris, and D. E. Kissel. 2012. Fertilizer response expectations, application levels, and material needs for loblolly, longleaf, and slash pine. Warnell School of Forest Resources, University of Georgia, Athens, GA. https://www.warnell.uga.edu/sites/default/files/publications/Fertilizer_response_expectations_Nov_2012_0.pdf (accessed 3 February 2017).

Dickens, E. D., J. Sunday, and D. J. Moorhead. 2014. Economics of growing loblolly, longleaf, and slash pine to various rotation ages with three stumpage price sets, four establishment cost sets, four discount rates, with and without pine straw—soil expectation value. Series paper #8. Warnell School of Forest Resources, University of Georgia, Athens, GA. https://www.warnell.uga.edu/sites/default/files/publications/Series%20Paper_8.pdf (accessed 3 February 2017).

Dickinson, Y. 2014. Landscape restoration of a forest with a historically mixed-severity fire regime: What was the historical landscape pattern of forest and openings? *Forest Ecology and Management* 331:264–271.

Dickinson, Y. L. and J. D. Cadry. 2016. An evaluation of tree marking methods for implementing spatially heterogeneous restoration. *Journal of Sustainable Forestry* 36:47–64. doi: 10.1080/10549811.2016.1251325.

Dickinson, M. B., A. T. Hudak, T. Zajkowski, E. L. Loudermilk, W. Schroeder, L. Ellison, R. L. Kremens, W. Holley, O. Martinez, and A. Paxton. 2016. Measuring radiant emissions from entire prescribed fires with ground, airborne and satellite sensors—RxCADRE 2012. *International Journal of Wildland Fire* 25:48–61.

Dickinson, M. B. and K. C. Ryan. 2010. Introduction: Strengthening the foundation of wildland fire effects prediction for research and management. *Fire Ecology* 6:1–12. doi: 10.4996/fireecology0601001.

Diefenderfer, H. L., G. E. Johnson, R. M. Thom, K. E. Buenau, L. A. Weitkamp, C. M. Woodley, A. B. Borde, and R. K. Kropp. 2016. Evidence-based evaluation of the cumulative effects of ecosystem restoration. *Ecosphere* 7:1–33 e01242.10.1002/ecs2.1242.

Diemer, J. E. 1986. The ecology and management of the gopher tortoise in the southeastern United States. *Herpetologica* 42:125–133.

Dilcher, D. L. 1973. A paleoclimatic interpretation of the Eocene floras of southeastern North America. Pages 39–59 *in* A. Graham, editor. *Vegetation and Vegetational History of Northern Latin America*. Amsterdam: Elsevier Science Publishers.

Dilcher, D. L. 2000. Geological history of the vegetation in southeast United States. *Sida Botanical Miscellany* 18:1–21.

Diop, A., E. Palola, A. Staudt, and B. B. Stein. 2009. Standing tall: How restoring the longleaf pine can help prepare the southeast for global warming. National Wildlife Federation, Reston, VA. http://www.nwf.org/pdf/Reports/LongleafPineReport.pdf (accessed 4 February 2017).

Dittmar T., C. E. Rezende, M. Manecki, J. Niggemann, A. R. C. Ovalle, A. Stubbins, and M. C. Bernardes. 2012. Continuous flux of dissolved black carbon from a vanished tropical forest biome. *Nature Geoscience* 5:618–622.

Dixit, A. K. and R. S. Pyndyck. 1994. *Investment under Uncertainty*. Princeton, NJ: Princeton University Press.

Dixon, W. N., J. A. Corneil, R. C. Wilkinson, and J. L. Foltz. 1984. Using stem char to predict mortality and insect infestation of fire-damaged slash pines. *Southern Journal of Applied Forestry* 8:85–88.

Dixon, R. O. D.. and C. T. Wheeler. 1983. Biochemical, physiological and environmental aspects of symbiotic nitrogen fixation. Pages 107–171 *in* J. C. Gordon, and C. T. Wheeler, editors. *Biological Nitrogen Fixation in Forest Ecosystems: Foundations and applications*. The Hague: Martinus Nijhoff.

Dobrowski, S. Z., A. K. Swanson, J. T. Abatzoglou, Z. A. Holden, H. D. Safford, M. K. Schwartz, and D. G. Gavin. 2015. Forest structure and species traits mediate projected recruitment declines in western U.S. tree species. *Global Ecology and Biogeography* 24:917–927.

Dodd, C. K., Jr. 1992. Biological diversity of a temporary pond herpetofauna in north Florida sandhills. *Biodiversity and Conservation* 1:125–142.

Dodd, C. K., Jr. 1993. Cost of living in an unpredictable environment: The ecology of striped newts (*Notophthalmus perstriatus*) during a prolonged drought. *Copeia* 1993:605–614.

Dodd, C. K., Jr. and B. S. Cade. 1998. Movement patterns and the conservation of amphibians breeding in small, temporary wetlands. *Conservation Biology* 12:331–339.

Dodd, C. K., Jr. and L. L. Smith. 2003. Habitat destruction and alteration: Historical trends and future prospects for amphibians. Pages 94–112 *in* R. D. Semlitsch, editor. *Amphibian Conservation*. Washington, DC: Smithsonian Institution.

Doliner, L. H. and P. A. Jolliffe. 1979. Ecological evidence concerning the adaptive significance of the C_4 dicarboxylic acid pathway of photosynthesis. *Oecologica* 38:23–24.

Douglass, J. E. 1983. The potential for water yield augmentation from forest management in the eastern United States. *Water Resources Bulletin* 19:351–358.

Drever C. R., G. Peterson, C. Messier, Y. Bergeron, and M. Flannigan. 2006. Can forest management based on natural disturbances maintain ecological resilience? *Canadian Journal of Forest Research* 36:2285–2299.

Drew, M. B., L. K. Kirkman, and A. K. Gholson, Jr. 1998. The vascular flora of Ichauway, Baker County, Georgia: A remnant longleaf pine/wiregrass ecosystem. *Castanea* 63:1–24.

Drewa, P. B., W. Platt, and E. B. Moser. 2002. Fire effects on resprouting of shrubs in headwaters of southeastern longleaf pine savannas. *Ecology* 83:755–767.

Driver, H. E. 1969. *Indians of North America*. Chicago, IL: University of Chicago Press.

Duerr, W. A, and J. B. Duerr. 1975. The role of faith in forest resource management. Pages 30–41 *in* F. Rumsey, and W. A. Duerr, editors. *Social Sciences in Forestry: A Book of Readings*. Philadelphia, PA: W. B. Saunders Company.

Duever, L. C. 2011. *Ecology and Management of Saw Palmetto*. Micanopy, FL: Conway Conservation LLC.

Dull, R. A. 1999. Palynological evidence for 19th century grazing-induced vegetation change in the southern Sierra Nevada, California, USA. *Journal of Biogeography* 26:899–912.

Duncan, B. W. and P. A. Schmalzer. 2004. Anthropogenic influences on potential fire spread in a pyrogenic ecosystem of Florida, USA. *Landscape Ecology* 19:153–165.

Dunne, E. J., J. Smith, D. B. Perkins, M. W. Clark, J. W. Jawitz, and K. R. Reddy. 2007. Phosphorus storages in historically isolated wetland ecosystems and surrounding pasture uplands. *Ecological Engineering* 31:16–28.

Dziadzio, M. C., A. K. Long, L. L. Smith, R. D. Chandler, and S. B. Castleberry. 2016. Presence of red imported fire ants at gopher tortoise nests. *Wildlife Society Bulletin* 40:202–206.

Dziadzio, M. C., L. L. Smith, R. B. Chandler, and S. B. Castleberry. 2016. Effect of nest location on gopher tortoise nest survival. *Journal of Wildlife Management* 80:1314–1322.

Earl, J. E. and R. D. Semlitsch. 2012. Reciprocal subsidies in ponds: Does leaf input increase frog biomass export? *Oecologia* 170:1077–1087.

Earley, L. S. 2004. *Looking for Longleaf: The Fall and Rise of an American Forest*. Chapel Hill, NC: University of North Carolina Press.

Eckert, A. J. and B. D. Hall. 2006. Phylogeny, historical biogeography, and patterns of diversification for *Pinus* (Pinaceae): Phylogenetic tests of fossil-based hypotheses. *Molecular Phylogenetics and Evolution* 40:166–182.

Edmisten, J. A. 1963. The ecology of the Florida pine flatwoods, PhD dissertation. University of Florida, Gainesville, FL.

Edwards, J. 1983. Diet shifts in moose due to predator avoidance. *Oecologia* 60:185–189.

Edwards, L., J. Ambrose, and L. K. Kirkman. 2013. *The Natural Communities of Georgia*. Athens, GA: University of Georgia Press.

Edwards, A. E. and A. S. Weakly. 2001. Population biology and management of rare plants in depression wetlands of the southeastern coastal plain, USA. *Natural Areas Journal* 21:12–35.

Ehle, D. S. and W. L. Baker. 2003. Disturbance and stand dynamics in ponderosa pine forests in Rocky Mountain National Park, USA. *Ecological Monographs* 73:543–566.

Eisenberg, J. F. 1983. The gopher tortoise as a keystone species. Pages 1–4 *in* R. J. Bryant and R. Franz, editors. *Proceedings of the Fourth Annual Meeting of the Gopher Tortoise Council*, Valdosta, GA, Florida State Museum, Gainesville, Florida.

Eisenbies, M. H., E. D. Vance, W. M. Aust, and J. R. Seiler. 2009. Intensive utilization of harvest residues in southern pine plantations: Quantities available and implications for nutrient budgets and sustainable site productivity. *Bioenergy Research* 2:90–98.

Ekblad, A., H. Wallander, D. L. Godbold, C. Cruz, D. Johnson, P. Baldrian, R. G. Bjork, D. Epron, B. Kieliszewska-Rokicka, R. Kjoller, H. Kraigher, E. Matzner, J. Neumann, and C. Plassard. 2013. The production and turnover of extramatrical mycelium of ectomycorrhizal fungi in forest soils: Role in carbon cycling. *Plant and Soil* 366:1–27.

Ellair, D. P. and W. J. Platt. 2013. Fuel composition influences fire characteristics and understorey hardwoods in pine savanna. *Journal of Ecology* 101:192–201.

Elliott, M. 2015. Multistate sandhills/uplands longleaf ecological restoration project (Phase 2): Alabama, Florida, Goergia, Louisiana, and Mississippi. Final Performance Resport. U.S. FWS State Wildlife Grant Program, Grant No.: U2-4-$-1, F11AP00769.

Ellis-Felege S. N., M. J. Conroy, W. E. Palmer, and J. P. Carroll. 2012. Predator reduction results in compensatory shifts in losses of avian ground nests. *Journal of Applied Ecology* 3:661–669.

Elton, C. 1927. *Animal Ecology*. New York, NY: The Macmillan Company.

Elzinga, C. L., D. W. Salzer, J. W. Willoughby, and J. P. Gibbs. 2001. *Monitoring Plant and Animal Populations*. Oxford: Blackwell Science.

Emanuel, R. E., H. E. Epstein, B. L. McGlynn, D. L. Welsch, D. J. Muth, and P. D'Odorico. 2010. Spatial and temporal controls on watershed ecohydrology in the northern Rocky Mountains. *Water Resources Research* 46:W11553

Engstrom, R. T. 1993. Characteristic mammals and birds of longleaf pine forests. Pages 127–138 *in* S. M. Hermann, editor. *Proceedings of the Tall Timbers Fire Ecology Conference*, no. 18, Tall Timbers Research Station, Tallahassee, FL.

Engstrom, R. T. 2010. First-order fire effects on animals: Review and recommendations. *Fire Ecology* 6:115–130.

Engstrom, R. T., D. B. McNair, and W. W. Baker. 1984. Breeding bird populations in relation to changing forest structure following fire exclusion: A 15-year study. *The Wilson Bulletin* 96:437–450.

Engstrom, R. T. and F. J. Sanders. 1997. Red-cockaded woodpecker foraging ecology in an old-growth longleaf pine forest. *The Wilson Bulletin* 1997:203–217.

Eno, A., L. M. Dover, and W. Dyche. 2006. State of the Land 2006, a brief inventory of public and private land in the United States. Resources First Foundation, Yarmouth, ME. https://www.privatelandownernetwork.org/pdfs/StateoftheLand.pdf (accessed 4 February 2017).

Eom, A. H., D. C. Hartnett, G. W. T. Wilson, and D. A. H. Figge. 1999. The effect of fire, mowing and fertilizer amendment on arbuscular mycorrhizas in tallgrass prairie. *American Midland Naturalist* 142:55–70.

Erickson, V. J., N. L. Mandel, and F. C. Sorensen. 2004. Landscape patterns of phenotypic variation and population structuring in a selfing grass, *Elymus glaucus* (blue wildrye). *Canadian Journal of Botany* 82:1776–1789.

Erwin, K. J., H. C. Chandler, J. G. Palis, T. A. Gorman, and C. A. Haas. 2016. Herpetofaunal communities in ephemeral wetlands embedded within longleaf pine flatwoods of the Gulf coastal plain. *Southeastern Naturalist* 15:421–447.

Eskew, E. A., J. D. Willson, and C. T. Winne. 2009. Ambush site selection and ontogenetic shifts in foraging strategy in a semi-aquatic pit viper, the Eastern cottonmouth. *Journal of Zoology* 277:179–186.

Estes, J. A., J. Terborgh, J. S. Branshares, M. E. Power, J. Berger, W. J. Bond, S. R. Carpenter, T. E. Essington, R. D. Holt, J. B. C. Jackson, R. J. Marquis, L. Oksanen, T. Oksanen, R. T. Paine, E. K. Pikitch, W. J. Ripple, S. A. Sandin, M. Scheffer, T. W. Schoener, B. Shurin, A. R. E.. Sinclair, M. E. Soule, R. Virtanen, and D. A. Wardle. 2011. Trophic downgrading of planet earth. *Science* 33:301–306.

Ewel, K. C. 1998. Pondcypress swamps. Pages 405–420 *in* M. Messina, and W. Conner, editors. *Southern Forested Wetlands: Ecology and Management*. Boca Raton, FL: Lewis Publishers.

Fagan, W. F. 1997. Omnivory as a stabilizing feature of natural communities. *American Naturalist* 150:554–567.

Falk, D. A. 2006. Process-centered restoration in a fire-adapted ponderosa pine forest. *Journal for Nature Conservation* 14:140–151.

Fancy, S. G. and R. E. Bennetts. 2012. Institutionalizing an effective long-term monitoring program in the U.S. National Park Service. Pages 481–497 *in* R. A. Gitzen, J. J. Millspaugh, A. B. Cooper, and D. S. Licht, editors. *Design and Analysis of Long-Term Ecological Monitoring Studies*. Cambridge: Cambridge University Press.

Farjon, A. 2013. *Pinus palustris*. The IUCN Red List of Threatened Species 2013: e.T39068A2886222. http://dx.doi.org/10.2305/IUCN.UK.2013-1.RLTS.T39068A2886222.en (accessed 2 February 2017).

Farrar, R. M., Jr. 1985. Volume and growth predictions for thinned even-aged natural longleaf pine stands in the East Gulf area. Research Paper SO-220. USDA Forest Service, Southern Forest Experiment Station, New Orleans, LA.

Farrar, R. M., Jr. 1996. *Fundamentals of Uneven-Aged Management in Southern Pine*. Miscellaneous Publication 9. Tallahassee, FL: Tall Timbers Research Station.

Farrar, R. M. and W. D. Boyer. 1990. Managing longleaf pine under the selection system—promises and problems. USDA Forest Service General Technical Report S0-70.

Faurby, S. and J. C. Svenning. 2015. Historic and prehistoric human-driven extinctions have reshaped global mammal diversity patterns. *Diversity and Distributions* 21:1155–1166.

Fenneman, N. 1938. *Physiography of Western United States*. New York, NY: McGraw-Hill.

Ferguson, S. A., J. E. Ruthford, S. J. McKay, D. Wright, C. Wright, and R. D. Ottmar. 2002. Measuring moisture dynamics to predict fire severity in longleaf pine forests. *International Journal of Wildland Fire* 11:267–279.

Fesenmyer, K. A. and N. L. Christensen. 2010. Reconstructing Holocene fire history in a southern Appalachian forest using soil charcoal. *Ecology* 91:662–670.

Fickle, J. E. 2001. *Mississippi Forests and Forestry*. Jackson, MS: University Press of Mississippi.

Fill, J. M., J. S. Glitzenstein, D. R. Streng, and J. Stowe. 2017. Wiregrass (*Aristida beyrichiana*) may limit woody plant encroachment in longleaf pine (*Pinus palustris*) ecosystems. *American Midland Naturalist* 217:331–342.

Fill, J. M., W. J. Platt, S. M. Welch, J. L. Waldron, and T. A. Mousseau. 2015. Updating models for restoration and management of fiery ecosystems. *Forest Ecology and Management* 356:54–63.

Fill, J. M., W. J. Platt, S. M. Welch, J. L. Waldron, and T. A. Mousseau. 2016. Flammability of the keystone savanna bunchgrass *Aristida stricta*. *Plant Ecology* 217:331–342.

Finney, M. A. 2001. Design of regular landscape fuel treatment patterns for modifying fire growth and behavior. *Forest Science* 47:219–228.

Finney, M. A. 2003. Calculation of fire spread rates across random landscapes. *International Journal of Wildland Fire* 12:167–174.

Firbank, L. G. and A. R. Watkinson. 1985. On the analysis of competition within two-species mixtures of plants. *Journal of Applied Ecology* 22:503–517.

Fitch, H. S., P. Goodrum, and C. Newman. 1952. The armadillo in the southeastern United States. *Journal of Mammalogy* 33:21–37.

Fites-Kaufman, J. A., P. W. Rundel, N. Stephenson, and D. A. Weixelman. 2007. Montane and subalpine vegetation of the Sierra Nevada and Cascade ranges. Pages 456–501 *in* M. G. Barbour, T. Keeler-Wolf, and A. A. Schoenherr, editors. *Terrestrial Vegetation of California*. Berkeley, CA: University of California Press.

Fitzhugh, T. W. and B. D. Richter. 2004. Quenching urban thirst: Growing cities and their impacts on freshwater ecosystems. *BioScience* 54:741–754.

Fleckenstein, N. 2013. The economic impact of the Red Hills region of southwest Georgia and North Florida. http://talltimbers.org/wp-content/uploads/2014/03/Economic-Impact-Report_RedHills_Opt.pdf (accessed 4 February 2017).

Fleckenstein, N. 2014. The economic impact of the quail hunting lands of Georgia's greater Albany Region. Tall Timbers Research Station and Land Conservancy, Tallahassee, FL. http://talltimbers.org/wp-content/uploads/2014/03/Economic-Impact-Report_Albany_Opt.pdf (accessed 4 February 2017).

Fletcher, P. W. and R. E. McDermott. 1957. *Influence of Geologic Parent Material and Climate on Distribution of Shortleaf Pine in Missouri*. Research Bulletin 625. Columbia, MO: University of Missouri.

Flinn, K. M. and P. L. Marks. 2007. Agricultural legacies in forest environments: Tree communities, soil properties, and light availability. *Ecological Applications* 17:452–463.

Florida Fish and Wildlife Conservation Commission (FFWCC). 2012. Gopher tortoise management plan, *Gopherus polyphemus*. Unpublished Report. Florida Fish and Wildlife Conservation Commission, Tallahassee, FL.

Flower, B. P., D. W. Hastings, H. W. Hill, and T. M. Quinn. 2004. Phasing of deglacial warming and Laurentide Ice Sheet meltwater in the Gulf of Mexico. *Geology* 32:597–600.

Focardi, S., D. Capizzi, and D. Monett. 2000. Competition for acorns among wild boar (*Sus scrofa*) and small mammals in a Mediterranean woodland. *Journal of Zoology* 250:329–334.

Folkerts, G. W. 1997. Citronelle ponds: Little-known wetlands of the central Gulf coastal plain, USA. *Natural Areas Journal* 17:6–16.

Fonda, R. W. 2001. Burning characteristics of needles from eight pine species. *Forest Science* 47:390–396.

Fonda, R. W. and J. M. Varner. 2004. Burning characteristics of cones from eight pine species. *Northwest Science* 78:322–333.

Forbes, M. S., R. J. Raison, and J. O. Skjemstad. 2006. Formation, transformation, and transport of black carbon (charcoal) in terrestrial and aquatic ecosystems. *Science of the Total Environment* 370:190–206.

Ford, R. I. 1985. Prehistoric food production in North America. Pages 341–364 *in* R. I. Ford, editor. *Prehistoric Food Production in North America. Anthropological Paper No. 75. Museum of Anthropology*, Ann Arbor, MI: University of Michigan.

Ford, C. R., S. H. Laseter, W. T. Swank, and J. M. Vose. 2011. Can forest management be used to sustain water-based ecosystem services in the face of climate change? *Ecological Applications* 21:2049–2067.

Ford, W. M., J. M. Menzel, M. A. Menzel, J. W. Edwards, and J. C. Kilgo. 2006. Presence and absence of bats across habitat scales in the upper coastal plain of South Carolina. *Journal of Wildlife Management* 70:1200–1209.

Ford, C. R., R. J. Mitchell, and R. O. Teskey. 2008. Water table depth affects productivity, water use, and the response to nitrogen addition in a savanna system. *Canadian Journal of Forest Research* 38:2118–2127.

Fordham, D., A. Georges, B. Corey, and B. W. Brook. 2006. Feral pig predation threatens the indigenous harvest and local persistence of snake-necked turtles in northern Australia. *Biological Conservation* 133:379–388.

Fortin, D., H. L. Beyer, M. S. Boyce, D. W. Smith, T. Duchesne, and J. S. Mao. 2005. Wolves influence elk movements: Behavior shapes a trophic cascade in Yellowstone National Park. *Ecology* 86:1320–1330.

Foster, B. L. 2001. Constraints on colonization and species richness along a grassland productivity gradient: The role of propagule availability. *Ecology Letters* 4:530–535.

Foster, T. E., II, B. Black, and M. D. Abrams. 2004. A witness tree analysis of the effects of Native American Indians on the Pre-European settlement forest in east-central Alabama. *Human Ecology* 32:27–47.

Foster, D., F. Swanson, J. Aber, I. Burke, N. Brokaw, D. Tilman, and A. Knapp. 2003. The importance of land-use legacies to ecology and conservation. *BioScience* 53:77–88.

Fowler, C. and E. Konopik. 2007. The history of fire in the southern United States. *Human Ecology Review* 14:165–176.

Fox, M. D. and B. J. Fox. 1986. The effects of fire frequency on the structure and floristic composition of a woodland understory. *Australian Journal of Ecology* 11:77–85.

Fox, T. R., E. J. Jokela, and H. L. Allen. 2007. The development of pine plantation silviculture in the southern United States. *Journal of Forestry* 105:337–347.

Franklin, J. F. 1993. Preserving biodiversity: Species, ecosystems, or landscapes? *Ecological Applications* 3:202–205.

Franklin, J. F., D. E. Berg, D. A. Thornburgh, and J. C. Tappenier. 1997. Alternative silvicultural approaches to timber harvesting: Variable retention harvest systems. Pages 111–139 *in* K. A. Kohm, and J. F. Franklin, editors. *Creating a Forestry for the 21st Century: The Science of Ecosystem Management*. Washington, DC: Island Press.

Franklin, J. F., K. N. Johnson, and D. Johnson. 2017. *Ecological Forest Management*. Long Gove, IL: Waveland Press.

Franklin, J. F., R. J. Mitchell, and B. J. Palik. 2007. Natural disturbance and stand development principles for ecological forestry. General Technical Report NRS-GTR-19. USDA Forest Service, Northern Research Station, Newtown Square, PA.

Franz, R. 2005. Up close and personal: A glimpse into the life of the Florida Pine Snake in a north Florida sand hill. Pages 120–131 *in* W. E. Meshaka, and K. J. Babbitt, editors. *Amphibians and Reptiles: Status and Conservation in Florida*. Malabar, FL: Kreiger Publishing Company.

Freckleton, R. P. and A.R. Watkinson. 2001. Asymmetric competition between plant species. *Functional Ecology* 15:615–623.

Frederiksen, N. O. 1980. Mid-tertiary climate of southeastern United States: The sporomorph evidence. *Journal of Paleontology* 54:728–739.

Free, A. and L. Smith. 2012. Gopher tortoise survey, population evaluation, and habitat suitability. Final Report to Georgia Department of Natural Resources, Social Circle, GA.

Freeman, J. E. and S. Jose. 2009. The role of herbicide in savanna restoration: Effects of shrub reduction treatments on the understory and overstory of a longleaf pine flatwoods. *Forest Ecology and Management* 257:978–986.

Freestone, A. L. 2006. Facilitation drives local abundance and regional distribution of a rare plant in a harsh environment. *Ecology* 87:2728–2735.

Fricke, K. 1904. "Licht- und Schattenholzarten," ein wissenschaftlich nicht begründetes Dogma. *Centralblatt für das gesamte Forstwese (Austrian Journal of Forest Science)* 30:315–325.

Fritz, H. and A. Loison. 2006. Large herbivores across biomes. Pages 19–49 *in* K. Danell, R. Bergstrom, P. Duncan, and J. Pastor, editors. *Large Herbivore Ecology, Ecosystem Dynamics and Conservation*. Cambridge: Cambridge University.

Frost, C. C. 1993. Four centuries of changing landscape patterns in the longleaf pine ecosystem. Pages 17–43 *in* S. M. Hermann, editor. *Proceedings of the Tall Timbers Fire Ecology Conference*, No. 18. Tall Timbers Research Station, Tallahassee, FL.

Frost, C. C. 1998. Presettlement fire frequency regimes of the United States: A first approximation. Pages 70–81 *in* T. L. Pruden, and L. A. Brennan, editors. *Proceedings of the 20th Tall Timbers Fire Ecology Conference*, 7–10 May 1996, Boise, Idaho, USA. Tall Timbers Research Station, Tallahassee, FL.

Frost, C. C. 2006. History and future of the longleaf pine ecosystem. Pages 9–42 *in* S. Jose, E. J. Jokela, and D. L. Miller, editors. *The Longleaf Pine Ecosystem: Ecology, Silviculture, and Restoration*. New York, NY: Springer.

Frost, C. C. and L. J. Musselman. 1987. History and vegetation of the Blackwater Ecologic Preserve. *Castanea* 52:16–46.

Fukami, T. 2015. Historical contingency in community assembly: Integrating niches, species pools and priority effects. *Annual Review of Ecology, Evolution and Systematics* 46:1–23.

Fulé, P. Z., T. W. Swetnam, P. M. Brown, D. A. Falk, D. L. Peterson, C. D. Allen, G. H. Aplet, M. A. Battaglia, D. Binkley, C. Farris, R. E. Keane, E. Q. Margolis, H. Grissino-Mayer, C. Miller, C. H. Sieg, C. Skinner, S. L. Stephens, and A. Taylor. 2014. Unsupported inferences of high-severity fire in historical dry forests of the western United States: Response to Williams and Baker. *Global Ecology and Biogeography* 23:825–830.

Funderburg, J. B. and D. S. Lee. 1968. The amphibian and reptile fauna of pocket gopher (*Geomys*) mounds in central Florida. *Journal of Herpetology* 1:99–100.

Gagnon, P. R., K. E. Harms, W. J. Platt, H. A. Passmore, and J. A. Myers. 2012. Small-scale variation in fuel loads differentially affects two co-dominant bunchgrasses in a species-rich pine savanna. *PLOS ONE* 7:e29674.

Gagnon, J. L., E. J. Jokela, W. K. Moser, and D. A. Huber. 2003. Dynamics of artificial regeneration in gaps within a longleaf pine flatwoods ecosystem. *Forest Ecology and Management* 172:133–144.

Gagnon, P. R., H. A. Passmore, W. J. Platt, J. A. Myers, C. E. T. Paine, and K. E. Harms. 2010. Does pyrogenicity protect burning plants? *Ecology* 91:3481–3486.

Gagnon, P. R., H. A. Passmore, M. Slocum, J. A. Myers, K. E. Harms, W. J. Platt, and C. E. T. Paine. 2015. Fuels and fires influence vegetation via above and below-ground pathways in a high-diversity plant community. *Journal of Ecology* 103:1009–1019.

Gaines, E. M., R. S. Campbell, and J. J. Brasington. 1954. Forage production on longleaf pine lands of southern Alabama. *Ecology* 35:59–62.

Gamble, L. R., K. McGarigal, and B. W. Compton. 2007. Fidelity and dispersal in the pond-breeding amphibian, *Ambystoma opacum*: Implications for spatio-temporal population dynamics and conservation. *Biological Conservation* 139:247–257.

Ganey, J. L. and S. C. Vojta. 2011. Tree mortality in drought-stressed mixed-conifer and ponderosa pine forests, Arizona, USA. *Forest Ecology and Management* 261:162–168.

Gano, L. 1917. A study in physiographic ecology in northern Florida. *Botanical Gazette* 63:337–372.

Garren, K. H. 1943. Effects of fire on vegetation of the southeastern United States. *The Botanical Review* 9:617–654.

Gawlik, D. E. 2002. The effects of prey availability on the numerical response of wading birds. *Ecological Monographs* 72:329–346.

Georgia Forestry Commission (GFC). 2010. Longleaf pine mortality—Rhizoctonia blight. Georgia Forestry Commission, Dry Branch, GA. http://www.gatrees.org/forest-management/forest-health/alerts-and-updates/LongleafPineMortalityRhizoctoniablightAug2010.pdf (accessed 4 February 2017).

Georgia Forestry Commission (GFC). 2015. Statewide Assessment of Forest Resources. Georgia Forestry Commission, Dry Branch, GA. http://www.gfc.state.ga.us/about-us/strategic-plan/GAStatewideAssessmentofForestResources.pdf (accessed 30 May 2017).

Gernandt, D. S., G. Gaeda López, and S. Ortiz García. 2005. Phylogeny and classification of *Pinus*. *Taxon* 54:29–42.

Gibbons, J. W. 2003. Terrestrial habitat: A vital component for herpetofauna of isolated wetlands. *Wetlands* 23:630–635.

Gibbons, J. W., C. T. Winne, D. E. Scott, J. D. Willson, X. Glaudas, K. M. Andrews, B. D. Todd, L. A. Fedewa, L. Wilkinson, R. N. Tsaliagos, and S. J. Harper. 2006. Remarkable amphibian biomass and abundance in an isolated wetland: Implications for wetland conservation. *Conservation Biology* 20:1457–1465.

Gilbert, J. C., J. S. Kush, and J. P. McGuire. 2005. Uprooting the wiregrass myth. Pages 32–36 *in* J. S. Kush, editor. *Proceedings of the Fifth Longleaf Alliance Regional Conference*, 12–15 October 2004, Hattiesburg, Mississippi, USA. Longleaf Alliance Report no. 8. The Longleaf Alliance, Solon Dixon Forestry Center, Andalusia, AL.

Gilbert, S. L., M. S. Lindberg, K. J. Hundermakr, and D. K. Person. 2014. Dead before detection: Addressing the effects of left truncation on survival estimation and ecological inference for neonates. *Methods in Ecology and Evolution* 5:992–1001.

Gill, J. L., J. W. Williams, S. T. Jackson, K. B. Lininger, and G. S. Robinson. 2009. Pleistocene megafaunal collapse, novel plant communities, and enhanced fire regimes in North America. *Science* 326:1100–1103.

Gilliam, F. S. 1988. Interactions of fire with nutrients in the herbaceous layer of a nutrient-poor coastal plain forest. *Bulletin of the Torrey Botanical Club* 115:265–271.

Gilliam, F. S. and W. J. Platt. 1999. Effects of long-term fire exclusion on tree species composition and stand structure in an old-growth *Pinus palustris* (longleaf pine) forest. *Plant Ecology* 140:5–26.

Gilliam, F. S., W. J. Platt, and R. K. Peet. 2006. Natural disturbances and the physiognomy of pine savannas: A phenomenological model. *Applied Vegetation Science* 9:83–96.

Gillon, D., V. Gomendy, C. Houssard, J. Marechal, and J. C. Valette. 1995. Combustion and nutrient loss during laboratory burns. *International Journal of Wildland Fire* 5:1–12.

Glasser, S. P. 2005. History of watershed management in the U.S. Forest Service: 1897–2005. *Journal of Forestry* 103:253–358.

Glazebrook, H. S. and A. I. Robertson. 1999. The effect of flooding and flood timing on leaf litter breakdown rates and nutrient dynamics in a river red gum (*Eucalyptus camaldulensis*) forest. *Australian Journal of Ecology* 24:625–635.

Gleim, E. R., L. M. Conner, R. D. Bergaus, M. L. Levin, G. E. Zemstova, and M. J. Yabsley. 2014. The phenology of ticks and the effects of long-term prescribed burning on tick population dynamics in southwestern Georgia and northwestern Florida. *PLOS ONE* 9:e112174.

Glenn, J. V. 2012. Economic assessment of landowner incentives: Analyses in North Carolina and Malawi, MS thesis. North Carolina State University, Raleigh, NC. http://repository.lib.ncsu.edu/ir/bitstream/1840.16/8176/1/etd.pdf (accessed 4 February 2017).

Glitzenstein, J. S., W. J. Platt, and D. R. Streng. 1995. Effects of fire regime and habitat on tree dynamics in north Florida longleaf pine savannas. *Ecological Monographs* 65:441–476.

Glitzenstein, J. S., D. R. Streng, R. E. Masters, and W. J. Platt. 2008. Clarifying long-term impacts of fire frequency and fire season in southeastern pine savannas. Pages 14–24 *in* W. C. Stringer, J. Andre, and G. Yarrow, editors. *Proceedings of the Sixth Eastern Native Grass Symposium*, 7–10 October 2008, Columbia, South Carolina, USA. Agriculture, Forestry and Life Sciences, Clemson University, Clemson, SC.

Glitzenstein, J. S., D. R. Streng, R. E. Masters, K. M. Robertson, and S. M. Hermann. 2012. Fire frequency effects on vegetation in north Florida pinelands: Another look at the long-term Stoddard fire research plots at Tall Timbers Research Station. *Forest Ecology and Management* 264:197–209.

Glitzenstein, J. S., D. R. Streng, and D. D. Wade. 2003. Fire frequency effects on longleaf pine (*Pinus palustris* P. Miller) vegetation in South Carolina and northeast Florida, USA. *Natural Areas Journal* 23:22–37.

Glitzenstein, J. S., D. R. Streng, D. D. Wade, and J. Brubaker. 2001. Starting new populations of longleaf pine ground-layer plants in the outer coastal plain of South Carolina, USA. *Natural Areas Journal* 21:89–110.

Godbois, I. A., L. M. Conner, and R. J. Warren. 2003. Bobcat diet on an area managed for northern bobwhite. *Proceedings of the Southeastern Association of Fish and Wildlife Agencies* 57:222–227.

Goebel, P. C., B. J. Palik, L. K. Kirkman, M. B. Drew, L. West, and D. C. Pederson. 2001. Forest ecosystems of a lower Gulf coastal plain landscape: Multifactor classification and analysis. *Journal of the Torrey Botanical Society* 128:47–75.

Goldberg, E. 1985. *Black Carbon in the Environment*. New York, NY: Wiley and Sons.

Golladay, S. W. and D. W. Hicks. 2013. Indicators of long-term hydrologic change in the Flint River. *Proceedings of the 2013 Georgia Water Resources Conference*, 10–11 April 2013, University of Georgia, Athens, GA. https://smartech.gatech.edu/bitstream/handle/1853/48531/3.4.3_Golladay.pdf?sequence=1 (accessed 4 February 2017).

Golladay, S. W. and D. W. Hicks. 2015. Using the sustainable boundary approach (SBA) to assess and develop flow guidelines: The Flint River, Georgia. *Proceedings of the 2015 Georgia Water Resources Conference*, 28–29 April 2015, University of Georgia, Athens, GA. http://gwri.gatech.edu/sites/default/files/files/docs/2015/6.4.3golladay.pdf (accessed 4 February 2017).

Golladay, S. W., D. W. Hicks, and T. K. Muenz. 2007. Stream flow changes associated with water use and climatic variation in the lower Flint River Basin, southwest Georgia. *Proceedings of the 2007 Georgia Water Conference*, 27–29 March 2007, University of Georgia, Athens, GA. http://gwri.gatech.edu/sites/default/files/files/docs/2007/6.6.3.pdf. (accessed 4 February 2017).

Golladay, S. W., K. L. Martin, J. M. Vose, D. N. Wear, A. P. Covich, R. J. Hobbs, K. D. Klepzig, G. E. Likens, R. J. Naiman, and A. W. Shearer. 2016. Achievable future conditions as a framework for guiding forest conservation and management. *Forest Ecology and Management* 360:80–96.

Gonzalez-Benecke, C. A., T. A. Martin, and W. P. Cropper, Jr. 2011. Whole-tree water relations of co-occurring mature *Pinus palustris* and *Pinus elliottii* var. *elliottii*. *Canadian Journal of Forest Research* 41:509–523.

Gonzalez-Benecke, C. A., L. J. Samuelson, T. A. Martin, W. P. Cropper, Jr., K. H. Johnsen, T. A. Stokes, J. R. Butnor, and P. H. Anderson. 2015. Modeling the effects of forest management on *in situ* and *ex situ* longleaf pine forest carbon stocks. *Forest Ecology and Management* 355:24–36.

Good, S. P., D. Noone, and G. Bowen. 2015. Hydrologic connectivity constrains partitioning of global terrestrial water fluxes. *Science* 349:175–177.

Goodrick, S. L., D. Shea, and J. Blake. 2010. Estimating fuel consumption for the upper coastal plain of South Carolina. *Southern Journal of Applied Forestry* 34:5–12.

Goolsby, D., L. Boring, and M. Wimberly. 2005. Land use legacies and upland forest overstory structural change in southwest Georgia. Pages 41–43 *in* J. S. Kush, editor. *Longleaf Pine: Making Dollar$ and Sense. Proceedings of the Fifth Longleaf Alliance Regional Conference*, 12–15 October 2004, Hattiesburg, Mississippi, USA. Longleaf Alliance Report no. 8. The Longleaf Alliance, Solon Dixon Forestry Center, Andalusia, AL.

Gordon, I. J., A. J. Hester, and M. Festa-Bianchet. 2004. The management of wild large herbivores to meet economic, conservation, and environmental objectives. *Journal of Applied Ecology* 41:1021–1031.

Gordon, D. R., L. Provencher, and J. L. Hardesty. 1997. Measurement scales and ecosystem management. Pages 262–273 *in* S. T. A.. Pickett, R. S. Ostfeld, and G. E. Likens, editors. *The Ecological Basis of Conservation, Heterogeneity, Ecosystems and Biodiversity*. New York, NY: Chapman and Hall.

Gorman, T. A., C. A. Haas, and D. C. Bishop. 2009. Factors related to occupancy of breeding wetlands by flatwoods salamander larvae. *Wetlands* 29:323–329.

Gorman, T. A., C. A. Haas, and J. G. Himes. 2013. Evaluating methods to restore amphibian habitat in fire-suppressed pine flatwoods wetlands. *Fire Ecology* 9:96–109.

Grace, S. L. and W. J. Platt. 1995a. Effects of adult tree density and fire on the demography of pregrass state juvenile longleaf pine (*Pinus palustris* Mill.). *Journal of Ecology* 83:75–86.

Grace, S. L. and W. J. Platt. 1995b. Neighborhood effects on juveniles in an old-growth stand of longleaf pine, *Pinus palustris*. *Oikos* 72:99–105.

Graham, A. 1999. *Late Cretaceous and Cenozoic history of North American Vegetation North of Mexico*. Oxford: Oxford University Press.

Graham, R. 2003. Hayman fire case study: Summary. Pages 1–32 *in* R. Graham, editor. *Hayman Fire Case Study*. General Technical Report RMRS-GTR-114. USDA Forest Service, Intermountain Research Station, Ogden, UT.

Graney, D. L. and E. R. Ferguson. 1972. Shortleaf pine site-index relationship in the Ozark highlands. *Soil Science Society of America Proceedings* 36:495–500.

Grano, C. X. 1949. Is litter a barrier to the initial establishment of shortleaf and loblolly pine reproduction? *Journal of Forestry* 47:544–548.

Gravel, D., C. D. Canham, M. Beaudet, and C. Messier. 2006. Reconciling niche and neutrality: The continuum hypothesis. *Ecology Letters* 9:399–409.

Graves, H. S. 1919. *Private Forestry*, Vol. 129. Washington, DC: USDA Forest Service, Office of the Secretary.

Gray, J. 1988. Growth of southern higher education forestry programs and their impact on the South's timber resources and industries. Supplement to The South's Fourth Forest: Alternatives for the Future. USDA-FS Forest Resource Report 24. USDA Forest Service, Washington, DC.

Greaves, H. E., L. A. Vierling, J. U. Eitel, N. T. Boelman, T. S. Magney, C. M. Prager, and K. L. Griffin. 2015. Estimating aboveground biomass and leaf area of low-stature Arctic shrubs with terrestrial LiDAR. *Remote Sensing of Environment* 164:26–35.

Greenberg C. H. and R. W. Simons. 1999. Age, composition, and stand structure of old-growth oak sites in the Florida high pine landscape: Implications for ecosystem management and restoration. *Natural Areas Journal* 19:30–40.

Greenberg, C. H. and G. W. Tanner. 2004. Breeding pond selection and movement patterns by eastern spadefoot toads (*Scaphiopus holbrookii*) in relation to weather and edaphic conditions. *Journal of Herpetology* 38:569–577.

Greenberg, C. H. and G. W. Tanner. 2005. Spatial and temporal ecology of eastern spadefoot toads on a Florida landscape. *Herpetologica* 61:20–28.

Greene, S. W. 1931. The first forest that fire made. *America Forests* October:53–54.

Greene, S. W. 1935. Relation between winter grass fires and cattle grazing in the longleaf pine belt. *Journal of Forestry* 33:339–341.

Grelen, H. E. 1962. Plant succession on cleared sandhills in northwest Florida. *American Midland Naturalist* 67:36–44.

Grelen, H. E. 1978. Forest grazing in the South. *Journal of Range Management* 31:244–250.

Grelen, H. E. 1983. May burning favors survival and early height growth of longleaf pine seedlings. *Southern Journal of Applied Forestry* 7:16–20.

Gresham, C. A. 1982. Litterfall patterns in mature loblolly and longleaf pine stands in Coastal South Carolina. *Forest Science* 28:223–231.

Griffith, G. E., J. M. Omernik, J. A. Comstock, S. Lawrence, G. Martin, A. Goddard, V. J. Hulcher, and T. Foster. 2001. *Ecoregions of Alabama and Georgia (color poster with map, descriptive text, summary tables, and photographs)*. Reston, VA: U.S. Geological Survey.

Grime, J. P. 1979. *Plant Strategies and Vegetation Processes*. New York, NY: John Wiley & Sons.

Grimm, E. C., G. L. Jacobson, Jr., W. A. Watts, B. C. S. Hansen, and K. A. Maasch. 1993. A 50,000-year record of climate oscillations from Florida and its temporal correlation with the Heinrich events. *Science* 261:198–200.

Grimm, E. C., W. A. Watts, G. L. Jacobson, Jr., B. C. S. Hansen, H. R. Almquist, and A. C. Dieffenbacher-Krall. 2006. Evidence for warm wet Heinrich events in Florida. *Quaternary Science Reviews* 25:2197–2211.

Grman, E., J. L. Orrock, C. W. Habeck, J. A. Ledvina, and L. A. Brudvig. 2015. Altered beta diversity in post-agricultural woodlands: Two hypotheses and role of scale. *Ecography* 38:614–621.

Grossmann, E. B. and D. J. Mladenoff. 2008. Farms, fires, and forestry: Disturbance legacies in the soils of the northwest Wisconsin (USA) Sand Plain. *Forest Ecology and Management* 256:827–836.

Grubb, P. J. 1986. The ecology of establishment. Pages 83–97 *in* D. Bradshaw, D.A. Goode, and E. Thorp, editors. *Ecology and Design in Landscape*. Oxford: Blackwell Scientific Publications.

Guerin, D. N. 1993. Oak dome clonal structure and fire ecology in a Florida longleaf pine dominated community. *Bulletin of the Torrey Botanical Club* 120:107–114.

Guldin, J. M. 1991. Uneven-aged BDq regulation of Sierra Nevada mixed conifers. *Western Journal of Applied Forestry* 6:27–32.

Guldin, J. M. 2006. Uneven-aged silviculture of longleaf pine. Pages 217–241 *in* S. Jose, E. J. Jokela, and D. L. Miller, editors. *The Longleaf Pine Ecosystem: Ecology, Silviculture, and Restoration*. New York, NY: Springer Science+Business Media, LLC.

Guldin, J. M. and J. B. Baker. 1998. Uneven-aged silviculture, southern style. *Journal of Forestry* 96:22–26.

Guldin, J. M., J. F. Rosson, Jr. and C. D. Nelson. 2016. Restoration of longleaf pine—the status of our knowledge. Pages 323–331 *in* C. J. Schweitzer, W. K. Clatterbuck, and C. M. Oswalt, editors. *Proceedings of the 18th Biennial Southern Silvicultural Research Conference*, 2–5 March, 2015, Knoxville, Tennessee, USA. General Technical Report SRS-212. USDA Forest Service, Southern Research Station, Asheville, NC.

Gulf Coastal Plains and Ozarks Landscape Conservation Cooperative (GCPO LCC). 2013. Integrated science agenda, Version 4. http://gcpolcc.org/group-hub/asmt/asmt-pages/the-gcpo-lcc-science-agenda-development-process (accessed 4 February 2017).

Gulsby, W. D., C. H. Killmaster, J. W. Bowers, J. D. Kelly, B. N. Sacks, M. J. Stratham, and K. V. Miller. 2015. White-tailed deer fawn recruitment before and after experimental coyote removals in central Georgia. *Wildlife Society Bulletin* 39:248–255.

Guo, D., R. J. Mitchell, and J. J. Hendricks. 2004. Fine root branch orders respond differently to carbon source-sink manipulations in a longleaf pine forest. *Oecologia* 140:450–457.

Guo, D., R. J. Mitchell, J. M. Withington, P. P. Fan, and J. J. Hendricks. 2008. Endogenous and exogenous controls of root life span, mortality, and nitrogen flux in a longleaf pine forest: Root branch order predominates. *Journal of Ecology* 96:737–745.

Gupta, V., K. J. Reinke, S. D. Jones, L. Wallace, and L. Holden. 2015. Assessing metrics for estimating fire induced change in the forest understorey structure using terrestrial laser scanning. *Remote Sensing* 7:8180–8201.

Guyer, C. and M. A. Bailey. 1993. Amphibians and reptiles of longleaf pine communities. Pages 139–158 *in* S. M. Hermann, editor. *Proceedings of the Tall Timbers Fire Ecology Conference*, no. 18. Tall Timbers Research Station, Tallahassee, FL.

Guyette, R. P., R. M. Muzika, and D. C. Dey. 2002. Dynamics of an anthropogenic fire regime. *Ecosystems* 5:472–486.

Guyette, R. P., M. A. Spetich, and M. C. Stambaugh. 2006. Historic fire regime dynamics and forcing factors in the Boston Mountains, Arkansas, USA. *Forest Ecology and Management* 234:293–304.

Guyette, R. P., M. C. Stambaugh, D. C. Dey, and R. Muzika. 2012. Predicting fire frequency with chemistry and climate. *Ecosystems* 15:322–335.

Guyette, R. P., F. R. Thompson, J. Whittier, M. C. Stambaugh, and D. C. Dey. 2014. Future fire probability modeling with climate change data and physical chemistry. *Forest Science* 60:862–870.

Gwayze, D., R. Melick, L. McClure, C. Studyvin, and D. Massengele. 2007. Effects of site preparation subsoiling and prescribed burning on survival and growth of shortleaf pine in the Mark Twain National Forest: Results after 20 growing seasons. Pages 129–133 *in* J. M. Kabrick, D. C. Dey, and D. Gwayze, editors. *Shortleaf Pine Restoration and Ecology in the Ozarks. Proceedings of a Symposium*, 7–9 November 2006, Springfield, Missouri, USA. General Technical Report NRS-P-15. USDA Forest Service, Northern Research Station, Newtown Square, PA.

Haas, J. R., D. E. Calkin, and M. P. Thompson. 2015. Wildfire risk transmission in the Colorado Front Range, USA. *Risk Analysis* 35:226–240.

Habeck, J. R. and R. W. Mutch. 1973. Fire dependent forests in the northern Rocky Mountains. *Quaternary Research* 3:408–424.

Haggstrom, D. A. and D. G. Kelleyhouse. 1996. Silviculture and wildlife relationships in the boreal forest of interior Alaska. *The Forestry Chronicle* 72:59–62.

Hagmann, R. K., J. F. Franklin, and K. N. Johnson. 2014. Historical conditions in mixed-conifer forests on the eastern slopes of the northern Oregon Cascade Range, USA. *Forest Ecology and Management* 330:158–170.

Hainds, M. J., R. J. Mitchell, B. J. Palik, L. R. Boring, and D. H. Gjerstad. 1999. Distribution of native legumes (Leguminoseae) in frequently burned longleaf pine (Pinaceae)-wiregrass (Poaceae) ecosystems. *American Journal of Botany* 86:1606–1614.

Hall, R. J. and E. Kolbe. 1980. Bioconcentration of organophosphorus pesticides to hazardous levels by amphibians. *Journal of Toxicology and Environmental Health, Part A Current Issues* 6:853–860.

Hallett, L. M., R. J. Standish, K. B. Hulvey, M. R. Gardener, K. N. Suding, B. M. Starzomski, S. D. Murphy, and J. A. Harris. 2013. Towards a conceptual framework for novel ecosystems. Pages 16–28 *in* R. J. Hobbs, E. S. Higgs, and C. Hall, editors. *Novel Ecosystems: Intervening in the New Ecological World Order*. Somerset, NJ: Wiley-Blackwell.

Halligan, J. J., M. R. Waters, A. Perrotti, I. J. Owens, J. M. Feinberg, M. D. Bourne, B. Fenerty, B. Winsborough, D. Carlson, D. C. Fisher, T. W. Stafford, Jr., and J. S. Dunbar. 2016. Pre-Clovis occupation 14,550 years ago at the Page-Ladson site, Florida, and the peopling of the Americas. *Science Advances* 2:e1600375.

Halsted, M. and J. Lynch. 1996. Phosphorus response of C_3 and C_4 species. *Journal of Experimental Botany* 47:497–505.

Hamada, S., T. Ohta, T. Hiyama, T. Kuwada, A. Takahashi, and T. C. Maximov. 2004. Hydrometeorological behavior of pine and larch forests in eastern Siberia. *Hydrological Processes* 18:23–39.

Hamilton, R. A. 1998. Longleaf pine plantations: Economics or crystal balling with a cloudy ball. Pages 16–19 *in* J. S. Kush, editor. *Proceedings of the Second Longleaf Alliance Conference*, 17–19 November 1998, Charleston, South Carolina, USA. Longleaf Alliance Report No. 4. The Longleaf Alliance, Solon Dixon Forestry Center, Andalusia, AL.

Hammond, D. H., J. M. Varner, J. S. Kush, and Z. Fan. 2015. Contrasting sapling bark allocation of five southeastern USA hardwood tree species in a fire prone ecosystem. *Ecosphere* 6:Article 112.

Hanberry, B. B., D. T. Jones-Farrand, and J. M. Kabrick. 2014. Historical open forest ecosystems in the Missouri Ozarks: Reconstruction and restoration targets. *Ecological Restoration* 32:407–416.

Hanberry, B. B., R. F. Noss, H. D. Safford, S. K. Allison, and D. C. Dey. 2015. Restoration is preparation for the future. *Journal of Forestry* 113:425–429.

Hannah, T. I. 2015. Influence of landscape-and stand-scale factors on avian communities in open pine ecosystems, MS thesis. Mississippi State University, Starkville, MS.

Hansen, A. J. and R. DeFries. 2007. Ecological mechanisms linking protected areas to surrounding lands. *Ecological Applications* 17:974–988.

Hansen, B. C. S., E. C. Grimm, and W. A. Watts. 2001. Palynology of the Peace Creek site, Polk County, Florida. *Bulletin of the Geological Society of America* 113:682–692.

Hansen, L., J. Hoffman, C. Drews, and E. Mielbrecht. 2010. Designing climate-smart conservation: Guidance and case studies. *Conservation Biology* 24:63–69.

Hanski, I. 1999. Habitat connectivity, habitat continuity, and metapopulations in dynamic landscapes. *Oikos* 87:209–219.

Hanski, I. and E. Ranta. 1983. Coexistence in a patchy environment: Three species of *Daphnia* in rock pools. *Journal of Animal Ecology* 52:263–279.

Hardesty, J. L., K. E. Gault, and H. F. Percival. 1997. Ecological correlates of red-cockaded woodpecker (*Picoides borealis*) foraging preference, habitat use, and home range size in northwest Florida (Eglin Air Force Base). Final Report Research Work Order 99. Florida Cooperative Fish and Wildlife Unit, University of Florida, Gainesville, FL.

Hardin, D. E. and D. L. White. 1989. Rare vascular plant taxa associated with wiregrass (*Aristida stricta*) in the southeastern United States. *Natural Areas Journal* 9:234–245.

Harley, G. L., H. D. Grissino-Mayer, and S. P. Horn. 2013. Fire history and forest structure of an endangered subtropical ecosystem in the Florida Keys, USA. *International Journal of Wildland Fire* 22:394–404.

Harmon, M. E., J. F. Franklin, F. J. Swanson, P. Sollins, S. V. Gergory, J. D. Lattin, N. H. Anderson, S. P. Cline, N. G. Aumen, J. R. Sedell, G. W. Lienkaemper, K. Cromack, and K. W. Cummins. 1986. Ecology of coarse woody debris in temperate ecosystems. *Advances in Ecological Research* 15:133–263.

Harper, R. M. 1905. Some noteworthy stations for *Pinus palustris*. *Torreya* 5:55–60.

Harper, R. M. 1907. A midsummer journey through the coastal plain of the Carolinas and Virginia. *Bulletin of the Torrey Botanical Club* 34:351–377.

Harper, R. M. 1911. The relation of climax vegetation to islands and peninsulas. *Bulletin of the Torrey Botanical Club* 38:515–525.

Harper, R. M. 1913. *Economic Botany of Alabama. Part 1. Geologic Survey of Alabama, Monograph 8.* Tuscaloosa, AL: University of Alabama.

Harper, R. M. 1914. A superficial study of the pine-barren vegetation of Mississippi. *Bulletin of the Torrey Botanical Club* 41:551–567.

Harper, R. M. 1928. *Economic Botany of Alabama. Part 2. Geological Survey of Alabama, Monograph 9.* Tuscaloosa, AL: University of Alabama.

Harper, R. M. 1943. *Forests of Alabama. Geological Survey of Alabama, Monograph 10.* Tuscaloosa, AL: University of Alabama.

Harper, R. M. 1956. Lowering of groundwater in the coastal plain of Georgia. *Memorandum Folio of the Association of American Geographers Southeastern Division* 8:33–34.

Harper, E. B., D. A. Patrick, and J. P. Gibbs. 2015. Impact of forestry practices at a landscape scale on the dynamics of amphibian populations. *Ecological Applications* 25:2271–2284.

Harpole, D. N. and C. A. Haas. 1999. Effects of seven silvicultural treatments on terrestrial salamanders. *Forest Ecology and Management* 114:349–356.

Harrington, T. B. 2006. Plant competition, facilitation, and other overstory-understory interactions in longleaf pine ecosystems. Pages 135–156 *in* S. Jose, E. J. Jokela, and D. L. Miller, editors. *The Longleaf Pine Ecosystem: Ecology, Silviculture, and Restoration.* New York, NY: Springer Science+Business Media, LLC.

Harrington, T. B. 2011. Overstory and understory relationships in longleaf pine plantations 14 years after thinning and woody control. *Canadian Journal of Forest Research* 41:2301–2314.

Harrington, M. G. 2012. Duff mound consumption and cambium injury for centuries-old western larch from prescribed burning in western Montana. *International Journal of Wildland Fire* 22:359–367.

Harrington, T. B., C. M. Dagley, and M. B. Edwards. 2003. Above- and belowground competition from longleaf pine plantations limits performance of reintroduced herbaceous species. *Forest Science* 49:681–695.

Harrington, T. B. and M. B. Edwards. 1999. Understory vegetation, resource availability, and litterfall responses to pine thinning and woody vegetation control in longleaf pine plantations. *Canadian Journal of Forest Research* 29:1055–1064.

Harris, A. H., J. S. Kush, R. S. Meldahl, and J. M. Varner. 2000. And the winner is . . . longleaf pine: Response of planted pines to various cultural treatments 39 years later. Pages 129–131 *in* J. S. Kush, editor. *Proceedings of the Third Longleaf Alliance Regional Conference*, 16–18 October 2000, Alexandria, Louisiana, USA. Longleaf Alliance Report No. 5. The Longleaf Alliance, Solon Dixon Forestry Center, Andalusia, AL.

Harris, L. and A. H. Taylor. 2015. Topography, fuels, and fire exclusion drive fire severity of the Rim Fire in an old-growth mixed-conifer forest, Yosemite National Park, USA. *Ecosystems* 18:1192–1208.

Harrison, S. and H. Cornell. 2008. Toward a better understanding of the regional causes of local community richness. *Ecology Letters* 11:969–979.

Hartnett, D. C. 1987. Effects of fire on clonal growth and dynamics of *Pityopsis graminifolia* (Asteraceae). *American Journal of Botany* 74:1737–1743.

Hartnett, D. C. and D. M. Krofta. 1989. Fifty-five years of post-fire succession in a southern mixed hardwood forest. *Bulletin of the Torrey Botanical Club* 116:107–113.

Haskins, K. E. and C. A. Gehring. 2004. Long-term effects of burning slash on plant communities and arbuscular mycorrhizae in a semi-arid woodland. *Journal of Applied Ecology* 41:379–388.

Hatcher, J. E., T. J. Straka, and J. L. Greene. 2013. The size of forest holding/parcelization problem in forestry: A literature review. *Resources* 2:39–57.

Haugo, R. D. and C. B. Halpern. 2007. Vegetation responses to conifer encroachment in a western Cascade meadow: A chronosequence approach. *Canadian Journal of Botany* 3:285–298.

Havens, K. 2015. Seeds sourcing for restoration in an era of climate change. *Natural Areas Journal* 35:122–133.

Hawbaker, T. J., V. C. Radeloff, S. I. Stewart, R. B. Hammer, N. S. Keuler, and M. K. Clayton. 2013. Human and biophysical influences on fire occurrence in the United States. *Ecological Applications* 23:565–582.

Hawthorne, D. W. 1994. Armadillos. *In Prevention and Control of Wildlife Damage*. Lincoln, NE: Institute of Agriculture and Natural Resources, Cooperative Extension Office, University of Nebraska.

Hayes, T. B., L. L. Anderson, V. R. Beasley, S. R. De Solla, T. Iguchi, H. Ingraham, P. Kestemont, J. Kniewald, Z. Kniewald, V. S. Langlois, and E. H. Luque. 2011. Demasculinization and feminization of male gonads by atrazine: Consistent effects across vertebrate classes. *Journal of Steroid Biochemistry and Molecular Biology* 127:64–73.

Hayes, T. B., P. Falso, S. Gallipeau, and M. Stice. 2010. The cause of global amphibian declines: A developmental endocrinologist's perspective. *Journal of Experimental Biology* 213:921–933.

Haymes, K. L. and G. A. Fox. 2012. Variation among individuals in cone production in *Pinus palustris* (Pinaceae). *American Journal of Botany* 99:640–645.

Haywood, J. D. 2002. Delayed prescribed burning in a seedling and sapling longleaf pine plantation in Louisiana. Pages 103–108 *in* K. W. Outcalt, editor. *Proceedings of the 11th Biennial Southern Silvicultural Research Conference*. General Technical Report SRS-48. USDA Forest Service, Southern Research Station, Asheville, NC.

Haywood, J. D. 2007. Restoring fire-adapted forested ecosystems-research in longleaf pine on the Kisatchie National Forest. Pages 87–105 *in* R. F. Powers, editor. *Restoring fire-adapted ecosystems. Proceedings of the 2005 National Silviculture Workshop*. General Technical Report PSW-GTR-203. USDA Forest Service, Pacific Southwest Research Station, Albany, CA.

Haywood, J. D. 2009. Eight years of seasonal burning and herbicidal brush control influence sapling longleaf pine growth, understory vegetation, and the outcome of an ensuing wildfire. *Forest Ecology and Management* 258:295–305.

Haywood, J. D. 2011. Influence of herbicides and felling, fertilization, and prescribed fire on longleaf pine growth and understory vegetation through ten growing seasons and the outcome of an ensuing wildfire. *New Forests* 41:55–73.

Haywood, J. D. 2015. Influence of herbicides and improvement cutting, fertilization, and prescribed fire on planted longleaf pine development. *Forest Science* 61:363–369.

Hecnar, S. J. and R. T. M'Closkey. 1997. The effects of predatory fish on amphibian species richness and distribution. *Biological Conservation* 79:123–131.

Hedman, C. W., S. L. Grace, and S. E. King. 2000. Vegetation composition and structure of southern coastal plain pine forests: An ecological comparison. *Forest Ecology and Management* 134:233–247.

Hedrick, L. D., G. A. Bukenhofer, W. G. Montague, W. F. Pell, and J. M. Guldin. 2007. Shortleaf pine-bluestem restoration in the Ouachita National Forest. Pages 206–213 *in* J. M. Kabrick, D. C. Dey, and D. Gwayze, editors. *Shortleaf Pine Restoration and Ecology in the Ozarks. Proceedings of a Symposium*, 7–9 November 2006, Springfield, MO. General Technical Report NRS-P-15. USDA Forest Service, Northern Research Station, Newtown Square, PA.

Heinsch, F. A., R. J. Innes, C. C. Hardy, K. M. Lee, editors. 2015. Fire, Fuel, and Smoke Program: 2014 Research Accomplishments. Publication R1-15-10. Missoula, MT: U.S. Department of Agriculture, Forest Service, Rocky Mountain Research Station, Fort Collins, CO.

Henderson, J. P. 2006. Dendroclimatological analysis and fire history of longleaf pine (*Pinus palustris* Mill.) in the Atlantic and Gulf coastal plain, PhD dissertation. University of Tennessee, Knoxville, TN.

Henderson, J. P. and H. D. Grissino-Mayer. 2009. Climate-tree growth relationships of longleaf pine (*Pinus palustris* Mill.) in the southeastern coastal plain, USA. *Dendrochronologia* 27:31–43.

Hendricks, E. L. 1954. Some notes on the relation of ground-water levels to pond levels in limestone sinks of southwestern Georgia. *Transactions, American Geophysical Union* 35:796–804.

Hendricks, J. J. and L. R. Boring. 1992. Litter quality of native herbaceous legumes in a burned pine forest of the Georgia Piedmont. *Canadian Journal of Forest Research* 22:2007–2010.

Hendricks, J. J. and L. R. Boring. 1999. N_2-fixation by native herbaceous legumes in burned pine ecosystems of the southeastern United States. *Forest Ecology and Management* 113:167–177.

Hendricks, E. L. and M. H. Goodwin. 1952. Water-level fluctuations in limestone sinks in southwestern Georgia. Water-Supply Paper 1110-E. U.S. Geological Survey, Atlanta, GA.

Hendricks, J. J., R. L. Hendrick, C. A. Wilson, R. J. Mitchell, S. D. Pecot, and D. Guo. 2006. Assessing the patterns and controls of fine root dynamics: An empirical test and methodological review. *Journal of Ecology* 94:40–57.

Hendricks, J. J., R. J. Mitchell, K. A. Kuehn, and S. D. Pecot. 2016. Ectomycorrhizal fungal mycelia turnover in a longleaf pine forest. *New Phytologist* 209:1693–1704.

Hendricks, J. J., C. A. Wilson, and L. R. Boring. 2002. Foliar litter position and decomposition in a fire-maintained longleaf pine-wiregrass ecosystem. *Canadian Journal of Forest Research* 32:928–941.

Henke, S. E. and F. C. Bryant. 1999. Effects of coyote removal on the faunal community in western Texas. *Journal of Wildlife Management* 63:1066–1081.

Hermann, S. M., C. Guyer, J. H. Waddle, and M. G. Nelms. 2002. Sampling on private property to evaluate population status and effects of land use practices on the gopher tortoise, *Gopherus polyphemus*. *Biological Conservation* 108:289–298.

Hermann, S. M., T. Van Hook, R. W. Flowers, L. A. Brennan, J. S. Glitzenstein, D. R. Streng, J. L. Walker, and R. L. Meyers. 1998. Fire and biodiversity: Studies of vegetation and arthropods. Pages 384–401.

Transactions of the 63rd North American Wildlife and Natural Resources Conference, Vol. 63, Wildlife Management Institute, Orlando, FL.

Hernández, L. and J. W. Laundré. 2005. Foraging in the "landscape of fear" and its implications for habitat use and diet quality of elk *Cervus elaphus* and bison *Bison bison*. *Wildlife Biology* 11:215–220.

Hess, C. A. 2014. Restoration of longleaf pine in slash pine plantations: Using fire to avoid the landscape trap, PhD dissertation. Florida State University, Tallahassee, FL.

Hessburg, P. F., J. K. Agee, and J. F. Franklin. 2005. Dry forests and wildland fires of the inland Northwest USA: Contrasting the landscape ecology of the pre-settlement and modern eras. *Forest Ecology and Management* 211:117–139.

Hessburg, P. F., D. J. Churchill, A. J. Larson, R. D. Haugo, C. Miller, T. A. Spies, M. North, N. Povak, T. R. Belote, P. H. Singleton, W. L. Gaines, R. E. Keane, G. H. Aplet, S. L. Stephens, P. Morgan, P. A. Bisson, B. E. Rieman, R. B. Slater, and G. H. Reeves. 2015. Restoring fire-prone inland Pacific landscapes: Seven core principles. *Landscape Ecology* 30:1805–1835.

Heyerdahl, E. K., L. B. Brubaker, and J. K. Agee. 2001. Spatial controls of historical fire regimes: A multiscale example from the interior west USA. *Ecology* 82:660–678.

Heyerdahl, E. K., D. McKenzie, L. D. Daniels, A. E. Hessl, J. S. Littell, and N. J. Mantua. 2008. Climate drivers of regionally synchronous fires in the inland Northwest (1651–1900). *International Journal of Wildland Fire* 17:40–49.

Heyward, F. 1933. The root system of the longleaf pine on the deep sands of western Florida. *Ecology* 14:136–148.

Heyward, F. 1939. The relation of fire to stand composition of longleaf pine forests. *Ecology* 20:287–304.

Hiers, J. K., S. C. Laine, J. J. Bachant, J. H. Furman, W. W. Greene, and V. Compton. 2003. Simple spatial modeling tool for prioritizing prescribed burning activities at the landscape scale. *Conservation Biology* 17:571–1578.

Hiers, J. K. and R. J. Mitchell. 2007. The influence of burning and light availability on N_2-fixation of native legumes in longleaf pine woodlands. *Journal of the Torrey Botanical Society* 134:398–409.

Hiers, J. K., R. J. Mitchell, A. Barnett, J. R. Walters, M. Mack, B. Williams, and R. Sutter. 2012. The dynamic reference concept: Measuring restoration success in a rapidly changing no-analogue future. *Ecological Restoration* 30:27–36.

Hiers, J. K., R. J. Mitchell, L. R. Boring, J. J. Hendricks, and R. Wyatt. 2003. Legumes native to longleaf pine savannahs exhibit capacity for high N_2-fixation rates and negligible impacts due to timing of fire. *New Phytologist* 157:327–338.

Hiers, J. K., J. J. O'Brien, R. J. Mitchell, J. M. Grego, and E. L. Loudermilk. 2009. The wildland fuel cell concept: An approach to characterize fine-scale variation in fuels and fire in frequently burned longleaf pine forests. *International Journal of Wildland Fire* 18:315–325.

Hiers, J. K., J. J. O'Brien, R. E. Will, and R. J. Mitchell. 2007. Forest floor depth mediates understory vigor in xeric *Pinus palustris* ecosystems. *Ecological Applications* 17:806–814.

Hiers, J. K., J. R. Walters, R. J. Mitchell, J. M. Varner, L. M. Conner, L. A. Blanc, and J. Stowe. 2014. Ecological value of retaining pyrophytic oaks in longleaf pine ecosystems. *Journal of Wildlife Management* 78:383–393.

Hiers, J. K., R. Wyatt, and R. J. Mitchell. 2000. The effects of fire regime on legume production in longleaf pine savannas: Is season selective? *Oecologia* 125:521–530.

Higgins, S. I., W. J. Bond, and W. S. W. Trollope. 2000. Fire, resprouting, and variability: A recipe for grass–tree coexistence in savanna. *Journal of Ecology* 80:213–229.

Hillman, S. S., R. C. Drewes, M. S. Hedrick, and T. V. Hancock. 2014. Physiological vagility and its relationship to dispersal and neutral genetic heterogeneity in vertebrates. *Journal of Experimental Biology* 217:3356–3364.

Hinton, J. W. and M. J. Chamberlain. 2010. Space and habitat use by a red wolf pack and their pups during pup-rearing. *Journal of Wildlife Management* 74:55–58.

Hinton, J. W., M. J. Chamberlain, and D. R. Rabon. 2013. Red wolf recovery: A review with suggestions for future research. *Animals* 3:722–744.

Hobbs, R. J. 2007. Setting effective and realistic restoration goals: Key directions for research. *Restoration Ecology* 15:354–357.

Hobbs, R. J. and L. Atkins. 1988. Spatial variability of experimental fires in south-west Western Australia. *Australian Journal of Ecology* 13:295–299.

Hobbs, R. J., E. Higgs, and J. A. Harris. 2009. Novel ecosystems: Implications for conservation and restoration. *Trends in Ecology and Evolution* 24:599–605.

Hobbs, R. J. and D. A. Norton. 1996. Towards a conceptual framework for restoration ecology. *Restoration Ecology* 4:93–110.

Hoctor, T. 2013. Updating the Florida Ecological Greenways Network. FWC Final Report. Department of Landscape Architecture, University of Florida, Gainesville, FL. http://conservation.dcp.ufl.edu/Downloads/FEGN%20Update%20Final%20Report.pdf (accessed 4 February 2017).

Hodges, J. D., W. W. Elam, W. F. Watson, and T. E. Nebeker. 1979. Oleoresin characteristics and susceptibility of four southern pines to southern pine beetle (Coleoptera: Scolytidae) attacks. *The Canadian Entomologist* 111:889–896.

Hoffman, C. M., J. Canfield, R. R. Linn, W. Mell, C. H. Sieg, F. Pimont, and J. Ziegler. 2016. Evaluating crown fire rate of spread predictions from physics-based models. *Fire Technology* 52:221–237.

Hogaboam, L. S. and S. R. Shook. 2004. Capital budgeting practices in the U.S. forest products industry: A reappraisal. *Forest Products Journal* 54:149.

Hohausová, E., R. J. Lavoy, and M. S. Allen. 2010. Fish dispersal in a seasonal wetland: Influence of anthropogenic structures. *Marine and Freshwater Research* 61:682–694.

Homer, C. G., J. A. Dewitz, L. Yang, S. Jin, P. Danielson, G. Xian, J. Coulston, N. D. Herold, J. D. Wickham, and K. Megown. 2015. Completion of the 2011 National Land Cover Database for the conterminous United States—representing a decade of land cover change information. *Photogrammetric Engineering and Remote Sensing* 81:345–354.

Hopkins, T., A. J. Larson, and R. T. Belote. 2014. Contrasting effects of wildfire and ecological restoration in old-growth western larch forests. *Forest Science* 60:1005–1013.

Hopkinson, C., L. Chasmer, Y. P. Colin, and P. Treitz. 2004. Assessing forest metrics with a ground-based scanning lidar. *Canadian Journal of Forest Research* 34:573–583.

Horn, H. S. and R. H. MacArthur. 1972. Competition among fugitive species in a harlequin environment. *Ecology* 53:749–752.

Hoss, S. K., C. Guyer, L. L. Smith, and G. W. Schuett. 2010. Multiscale influences of landscape composition and configuration on the spatial ecology of eastern diamond-backed rattlesnakes (*Crotalus adamanteus*). *Journal of Herpetology* 44:110–123.

Hough, F. B. 1878. *Report Upon Forestry, Volume 1*. Washington, DC: U.S. Department of Agriculture.

Hough, W. A. 1982. Phytomass and nutrients in the understory and forest floor of slash/longleaf pine stands. *Forest Science* 28:359–372.

Howe, G. E., R. Gillis, and R. C. Mowbray. 1998. Effect of chemical synergy and larval stage on the toxicity of atrazine and alachlor to amphibian larvae. *Environmental Toxicology and Chemistry* 17:519–525.

Howze, B. M., L. M. Conner, R. J. Warren, and K. V. Miller. 2009. Predator removal and white-tailed deer recruitment in southwestern Georgia. *Proceedings of the Southeastern Association of Fish and Wildlife Agencies* 63:17–20.

Howze, J. M. and L. L. Smith. 2015. Spatial ecology and habitat use of the coachwhip. *Southeastern Naturalist* 14:242–250.

Hu, H. 2011. Restoring longleaf pine (*Pinus palustris*) in loblolly pine (*P. taeda*) stands on the coastal plain of North Carolina, PhD dissertation. Clemson University, Clemson, SC.

Hu, H., B. O. Knapp, G. G. Wang, and J. L. Walker. 2016. Silvicultural treatments for converting loblolly pine to longleaf pine dominance: Effects on ground layer and midstorey vegetation. *Applied Vegetation Science* 19:280–290.

Hubbell, S. P. 2001. *The Unified Neutral Theory of Biodiversity and Biogeography*. Princeton, NJ: Princeton University Press.

Hubbell, S. P. and R. B. Foster. 1986. Biology, chance, and history and the structure of tropical rain forest tree communities. Pages 314–329 *in* J. Diamond, and T. J. Case, editors. *Community Ecology*. New York, NY: Harper and Row.

Hudak, A. T., M. B. Dickinson, B. C. Bright, R. L. Kremens, E. L. Loudermilk, J. J. O'Brien, B. S. Hornsby, and R. D. Ottmar. 2016. Measurements relating fire radiative energy density and surface fuel consumption–RxCADRE 2011 and 2012. *International Journal of Wildland Fire* 25:25–37.

Hudak, A. T., J. S. Evans, and A. M. Stuart Smith. 2009. LiDAR utility for natural resource managers. *Remote Sensing* 1:934–951.

Hudson, C. 1976. *The Southeastern Indians*. Knoxville, TN: University of Tennessee Press.

Hudson, J. C. 1988. North American origins of middlewestern frontier populations. *Annals of the Association of American Geographers* 78:395–413.

Huffman, J. M. 2006. Historical fire regimes in southeastern pine savannas, PhD dissertation. University of Florida, Gainesville, FL.

Huffman, J. M., W. J. Platt, H. Grissino-Mayer, and C. J. Boyce. 2004. Fire history of a barrier island slash pine (*Pinus elliottii*) savanna. *Natural Areas Journal* 24:258–268.

Hufford, K. M. and S. J. Mazer. 2003. Plant ecotypes: Genetic differentiation in the age of ecological restoration. *Trends in Ecology and Evolution* 18:147–155.

Huggett, R., D. N. Wear, R. Li, J. Coulston, and S. Liu. 2013. Forecasts of forest conditions. Pages 73–101 *in* D. N. Wear, and J. G. Greis, editors. 2013. The southern forest futures project: Technical Report. General Technical Report SRS-GTR-178. USDA Forest Service, Southern Research Station, Asheville, NC.

Humphries, W. J. and M. A. Sisson. 2012. Long distance migrations, landscape use, and vulnerability to prescribed fire of the gopher frog (*Lithobates capito*). *Journal of Herpetology* 46:665–670.

Hunter, M. L., Jr. 1990. *Wildlife, Forests, and Forestry. Principles of Managing Forests for Biological Diversity.* Englewood Cliffs, NJ: Prentice Hall.

Hunter, W. C., A. J. Mueller, and C. L. Hardy. 1994. Managing for red-cockaded woodpeckers and neotropical migrants–is there a conflict? *Proceedings of the Annual Conference of the Southeastern Association of Fish and Wildlife Agencies* 48:383–394.

Hunter, M. E., W. D. Shepperd, J. E. Lentile, J. E. Lundquist, M. G. Andreu, J. L. Butler, and F. W. Smith. 2007. A comprehensive guide to fuels treatment practices for ponderosa pine in the Black Hills, Colorado Front Range, and Southwest. General Technical Report RMRS-GTR-198. USDA Forest Service, Rocky Mountain Research Station, Fort Collins, CO.

Hurteau, M. and M. North. 2009. Fuel treatment effects on tree-based forest carbon storage and emissions under modeled wildfire scenarios. *Frontiers in Ecology and the Environment* 8:409–414.

Hurteau, M. D., M. T. Stoddard, and P. Z. Fule. 2011. The carbon costs of mitigating high severity wildfire in southwestern ponderosa pine. *Frontiers in Ecology and the Environment* 8:409–414.

Hurtt, G. C. and S. W. Pacala. 1995. The consequences of recruitment limitation: Reconciling chance history and competitive differences between plants. *Journal of Theoretical Biology* 176:1–12.

Hussey, T. C. 1993. A 20,000-year history of vegetation and climate at Clear Pond, northeastern South Carolina, MS thesis. University of Maine, Orono, ME.

Huston, M. 1979. A general hypothesis of species diversity. *American Naturalist* 113:81–101.

Huston, M. A. 2014. Disturbance, productivity, and species diversity: Empiricism vs. logic in ecological theory. *Ecology* 95:2382–2396.

Huston, M. A. and D. L. DeAngelis. 1994. Competition and coexistence: The effects of resource transport and supply rates. *American Naturalist* 144:954–977.

Hwang, T., L. Band, and T. C. Hales. 2009. Ecosystem processes at the watershed scale: Extending optimality theory from plot to catchment. *Water Resources Research* 45:1–20.

Hyslop, N. L. 2007. Movements, habitat use, and survival of the threatened eastern indigo snake *(Drymarchon couperi)* in Georgia, PhD dissertation. University of Georgia, Athens, GA.

Iacona, G. D. 2008. Seedling recruitment as a driver of species richness in the understory of the longleaf pine savanna, MS thesis. University of Florida, Gainesville, FL.

Iacona, G. D., L. K. Kirkman, and E. M. Bruna. 2010. Effects of resource availability on seedling recruitment in a fire-maintained savanna. *Oecologia* 163:171–180.

Iacona, G. D., L. K. Kirkman, and E. M. Bruna. 2012. Experimental test for facilitation of seedling recruitment by the dominant bunchgrass in a fire-maintained savanna. *PLOS ONE* 7:e39108.

Ike, J. C. 2010. Spatial variability and land use change: Effects on total soil carbon content in the coastal plain of Georgia, MS thesis. University of Georgia, Athens, GA.

Ince, P., A. Schuler, H. Spelter, and W. Luppold. 2007. Globalization and structural change in the U.S. forest sector: An evolving context for sustainable forest management. General Technical Report FPL-GTR-170. USDA Forest Service, Forest Products Laboratory, Madison, WI.

Ingram, K., K. Dow, L. Carter, and J. Anderson. 2013. *Climate of the Southeast United States: Variability, Change, Impacts, and Vulnerability*. Washington, DC: Island Press.

Inkley, M. D., S. A. Wissinger, and B. L. Baros. 2008. Effects of drying regime on microbial colonization and shredder preference in seasonal woodland wetlands. *Freshwater Biology* 53:435–445.

Intergovernmental Panel on Climate Change (IPCC). 2014. Climate change 2014: Synthesis Report. R. K. Pachauri, and L. A. Meyer, editors. Contribution of Working Groups I, II and III to the Fifth Assessment Report of the Intergovernmental Panel on Climate Change, Geneva, Switzerland.

International Forest Company (IFCO). 2016. Shop seedlings. http://www.internationalforest.co/seedlings.php (accessed 4 February 2017).

Iverson, L. R., A. M. Prasad, S. N. Matthews, and M. Peters. 2008. Estimating potential habitat for 134 eastern U.S. tree species under six climate scenarios. *Forest Ecology and Management* 254:390–406.

Iverson, L. R., D. A. Yaussy, J. Rebbeck, T. F. Hutchinson, R. P. Long, and A. M. Prasad. 2004. A comparison of thermocouples and temperature paints to monitor spatial and temporal characteristics of landscape-scale prescribed fires. *International Journal of Wildland Fire* 13:311–322.

Jack, S. B., J. L. Gagnon, and M. A. Melvin. 2005. Chemical control of woody sprouts following mechanical removal of large hardwoods within a longleaf pine matrix. Pages 62–64 *in* J. S. Kush, editor. *Proceedings of the Fifth Longleaf Alliance Regional Conference*, 12–15 October 2004, Hattiesburg, Mississippi, USA. Longleaf Alliance Report no. 8. The Longleaf Alliance, Solon Dixon Forestry Center, Andalusia, AL.

Jack, S. B., J. K. Hiers, R. J. Mitchell, and J. L. Gagnon. 2010. Fuel loading and fire intensity effects on longleaf pine seedling survival. Pages 275–279 *in* J. A. Stanturf, editor. *Proceedings of the 14th Biennial Southern Silvicultural Research Conference*. General Technical Report SRS-121. USDA Forest Service, Southern Research Station, Asheville, NC.

Jack, S. B., R. J. Mitchell, and S. D. Pecot. 2006. Silvicultural alternatives in a longleaf pine/wiregrass woodland in southwest Georgia: Understory hardwood response to harvest-created gaps. Pages 85–89 *in* K. F. Connor, editor. *Proceedings of the 13th Biennial Southern Silvicultual Research Conference*, 28 February–4 March 2005, Memphis, Tennessee, USA. General Technical Report SRS-92. USDA Forest Service, Southern Research Station, Asheville, NC.

Jack, S. B., W. L. Neel, and R. J. Mitchell. 2006. The Stoddard-Neel approach: A conservation-oriented approach. Pages 242–245 in S. Jose, E. J. Jokela, and D. L. Miller, editors. *The Longleaf Pine Ecosystem: Ecology, Silviculture, and Restoration*. New York, NY: Springer Science+Business Media, LLC.

Jackson, J. A. 1994. Red-cockaded woodpecker *(Picoides borealis)*. In A. Poole and F. Gill (eds), *The Birds of North America*. Philadelphia, PA: National Academy of Sciences.

Jackson, L. W. R. 1968. Growth and specific gravity of unthinned longleaf, slash pine (*Pinus palustris, P. elliottii*) in Georgia Piedmont. *Naval Stores Review and Terpene Chemicals* 78:10–11.

Jackson, S. T. 2012. Conservation and resource management in a changing world: Extending historical range of variation beyond the baseline. Pages 92–109 in J. A. Wiens, G. D. Hayward, H. D. Safford, and C. Giffen, editors. *Historical Environmental Variation in Conservation and Natural Resource Management*. Chichester, UK: Wiley-Blackwell.

Jackson, S. T. 2016. Reinventing conservation—Again. *Frontiers in Ecology and the Environment* 14:519.

Jackson, R. B., E. G. Jobbagy, R. Avissar, S. B. Roy, D. J. Barrett, C. W. Cook, K. A. Farley, D. C. le Maitre, B. A. McCarl, and B. C. Murray. 2005. Trading water for carbon with biological carbon sequestration. *Science* 310:1944–1947.

Jackson, D. R. and E. G. Milstrey. 1989. The fauna of gopher tortoise burrows. Pages 86–98 *in* J. E. Diemer, D. R. Jackson, J. L. Landers, J. N. Layne, and D. A. Wood, editors. *Proceedings of the Gopher Tortoise Relocation Symposium*. Technical Report No. 5. Florida Game and Fresh Water Fish Commission, Nongame Wildlife Program, Tallahassee, FL.

Jackson, S. T., J. T. Overpeck, T. Webb III, S. E. Keattch, and K. H. Anderson. 1997. Mapped plant macrofossil and pollen records of Late Quaternary vegetation change in eastern North America. *Quaternary Science Reviews* 16:1–70.

Jackson, C. R., G. Sun, D. Amatya, W. T. Swank, M. Riedel, J. Patric, T. Williams, J. M. Vose, C. Trettin, W. M. Aust, R. S. Beasley, H. Williston, and G. G. Ice. 2004. Fifty years of forest hydrology in the Southeast. Pages 33–112 *in* G. G. Ice, and J. D. Stednick, editors. *A Century of Forested and Wildland Watershed Lessons*. Bethesda, MD: The Society of American Foresters.

Jackson, S. T., R. S. Webb, K. H. Anderson, J. T. Overpeck, T. Webb III, J. W. Williams, and B. C. S. Hansen. 2000. Vegetation and environment in Eastern North America during the Last Glacial Maximum. *Quaternary Science Reviews* 19:489–508.

Jackson, S. T. and C. Weng. 1999. Late Quaternary extinction of a tree species in eastern North America. *Proceedings of the National Academy of Sciences* 96:13847–13852.

Jacobson, S. K., M. C. Monroe, and S. Marynowski. 2001. Fire at the wildland interface: The influence of experience and mass media on public knowledge, attitudes, and behavioral intentions. *Wildlife Society Bulletin* 29:929–937.

Jacqmain, E. I., R. H. Jones, and R. J. Mitchell. 1999. Influences of frequent cool-season burning across a soil moisture gradient on oak community structure in longleaf pine ecosystems. *American Midland Naturalist* 141:85–100.

Jaffe, R., Y. Ding, J. Niggemann, A. V. Vähätalo, A. Stubbins, R. G. M. Spencer, J. Campbell, and T. Dittmar. 2013. Global charcoal mobilization from soils via dissolution and riverine transport to oceans. *Science* 340:345–347.

James, F. C., C. A. Hess, B. C. Kicklighter, and R. A. Thum. 2001. Ecosystem management and the niche gestalt of the red-cockaded woodpecker in longleaf pine forests. *Ecological Applications* 11:854–870.

Janzen, F. J. 1994. Climate change and temperature-dependent sex determination in reptiles. *Proceedings of the National Academy of Sciences* 91:7487–7490.

Jasechko, S., Z. D. Sharp, J. J. Gibson, S. J. Birks, Y. Yi, and P. J. Fawcett. 2013. Terrestrial water fluxes dominated by transpiration. *Nature* 496:347–350.

Jodice, P. G. and A. R. Humphrey. 1992. Activity and diet of an urban population of Big Cypress fox squirrels. *Journal of Wildlife Management* 56:685–692.

Jofré, J. M., M. R. Warn, and C. J. Reading. 2016. The role of managed coniferous forest in the conservation of reptiles. *Forest Ecology and Management* 362:69–78.

Johnsen, K. H., J. R Butnor, J. S. Kush, R. C. Schmidtling, and C. D. Nelson. 2009. Hurricane Katrina winds damaged longleaf pine less than loblolly pine. *Southern Journal of Applied Forestry* 33:178–181.

Johnson, S. A. 2002. Life history of the striped newt at a north-central Florida breeding pond. *Southeastern Naturalist* 1:381–402.

Johnson, S. A. 2005. Conservation and life history of the striped newt: The importance of habitat connectivity. Pages 91–98 *in* W. E. Meshaka, and K. J. Babbitt, editors. *Amphibians and Reptiles: Status and Conservation in Florida*. Malabar, FL: Kreiger Publishing Company.

Johnson, R. 2011. The economics of longleaf pine management: A road to making dollars and sense. http://www.americaslongleaf.org/media/8580/The%20Economics%20of%20Longleaf%20Pine%20Management%20Final%20(Longleaf%20Alliance).pdf (accessed 4 February 2017).

Johnson, W. C., S. E. Boettcher, K. A. Poiani, and G. Guntenspergen. 2004. Influence of weather extremes on the water levels of glaciated prairie wetlands. *Wetlands* 24:385–398.

Johnson, R. and D. Gjerstad. 2006. Restoring the overstory of longleaf pine ecosystems. Pages 271–295 *in* S. Jose, E. J. Jokela, and D. L. Miller, editors. *The Longleaf Pine Ecosystem: Ecology, Silviculture, and Restoration*. New York, NY: Springer Science+Business Media, LLC.

Johnson, E. A. and J. L. Kovner. 1956. Effect on streamflow of cutting forest understory. *Forest Science* 2:82–91.

Johnson, E. and K. Miyanishi. 2001. Strengthening fire ecology's roots. Pages 1–9 *in* E. Johnson, and K. Miyanishi, editors. *Forest Fires: Behavior and Ecological Effects*. San Diego, CA: Academic Press.

Johnson, E. A., K. Miyanishi, and S. R. J. Bridge. 2001. Wildfire regime in the boreal forest and the idea of suppression and fuel buildup. *Conservation Biology* 15:1554–1557.

Jones, C. D. 2013. Landscape-level investigations of avian species at Fort Benning, Georgia, PhD dissertation. University of Georgia, Athens, GA.

Jones, R. A. 2014. The role of Native American fire regimes in the maintenance of longleaf pine savanna, PhD dissertation. University of Wyoming, Laramie, WY.

Jones, D. D., L. M. Conner, T. H. Storey, and R. J. Warren. 2004. Prescribed fire and raccoon use of longleaf pine forests: Implications for managing nest predation? *Wildlife Society Bulletin* 32:1255–1259.

Jones, C. D., J. A. Cox, E. Toriani-Moura, and R. J. Cooper. 2013. Nest-site characteristics of Bachman's Sparrows and their relationship to plant succession following prescribed burns. *The Wilson Journal of Ornithology* 125:293–300.

Jones, J. A., I. F. Creed, K. L. Hatcher, R. J. Warren, M. B. Adams, M. H. Benson, E. Boose, W. A. Brown, J. L. Campbell, A. Covich, D. W. Clow, C. N. Dahm, K. Elder, C. R. Ford, N. B. Grimm, D. L. Henshaw, K. L. Larson, E. S. Miles, K. M. Miles, S. D. Sebestyen, A. T. Spargo, A. B. Stone, J. M. Vose, and M. W. Williams. 2012. Ecosystem processes and human influences regulate streamflow response to climate change at long-term ecological research sites. *Bioscience* 62:390–404.

Jones, P. D., B. B. Hanberry, and S. Demarais. 2010. Managing the southern pine forest-retained wetland interface for wildlife diversity: Research priorities. *Wetlands* 30:381–391.

Jones, K. C., P. Hill, T. A. Gorman, and C. A. Haas. 2012. Climbing behavior of flatwoods salamanders (*Ambystoma bishopi/A. cingulatum*). *Southeastern Naturalist* 11:537–542.

Jones, C. G., J. H. Lawton, and M. Shachak. 1994. Organisms as ecosystem engineers. *Oikos* 69:373–386.

Jones, R. H., R. J. Mitchell, G. N. Stevens, and S. D. Pecot. 2003. Controls of fine root dynamics across a gradient of gap sizes in a pine woodland. *Oecologia* 134:132–143.

Jose, S., E. J. Jokela, and D. L. Miller. 2006. *The Longleaf Pine Ecosystem: Ecology, Silviculture, and Restoration*. New York, NY: Springer.

Jose, S., S. Ranasinghe, and C. L. Ramsey. 2010. Longleaf pine (*Pinus palustris* P. Mill.) restoration using herbicides: Overstory and understory vegetation responses on a coastal plain flatwoods site in Florida, USA. *Restoration Ecology* 18:244–251.

Joseph W. Jones Ecological Research Center (JWJERC). 2015. Workshop summary: Challenges and priorities for managing frequent fire conifer ecosystems. *Workshop*, 7–9 January 2015, Joseph W. Jones Ecological Research Center, Newton, GA. http://www.jonesctr.org/research/FFCE_Workshop_Summary_2015.pdf (accessed 4 February 2017).

Joyce, J. E., J. M. Prutzman, and L. R. C. Tjalsma. 1990. High-resolution planktic stable isotope record and spectral analysis for the last 5.35 m.y.: Ocean drilling program site 625 northeast Gulf of Mexico. *Paleoceanography* 5:507–529.

Joyce, J. E., L. R. C. Tjalsma, and J. M. Prutzman. 1993. North American glacial meltwater history for the past 2.3 m.y.: Oxygen isotope evidence from the Gulf of Mexico. *Geology* 21:483–486.

Jurney, D., R. Evans, J. Ippolito, and V. Bergstrom. 2004. The role of wildland fire in portions of southeastern North America. Pages 95–116 *in* R. T. Engstrom, K. E. M. Galley, W. J. de Groot, editors. Fire in temperate, boreal, and montane ecosystems. *Proceedings of the 22nd Tall Timbers Fire Ecology Conference*, 15–18 October 2001, Kananaskis, Alberta. Tall Timbers Research Station, Tallahassee, FL.

Kaczor, S. A. and D. C. Hartnett. 1990. Gopher tortoise (*Gopherus polyphemus*) effects on soils and vegetation in a Florida sandhill community. *American Midland Naturalist* 123:100–111.

Kaeser M. J. and L. K. Kirkman. 2009. Estimating total plant species richness in seasonally inundated depressional wetlands in the longleaf pine ecosystem. *Wetlands* 20:373–85.

Kaeser, M. J. and L. K. Kirkman. 2010. The effects of pre- and post-emergent herbicides on non-target native plant species of the longleaf pine ecosystem. *Journal of the Torrey Botanical Society* 137:420–430.

Kaeser, M. J. and L. K. Kirkman. 2012. Seed longevity of 12 native herbaceous species in a fire-maintained pine savanna after 8 years of burial. *Forest Ecology and Management* 281:68–74.

Kale, H. W., II, B. S. Pranty, B. S. Stith, and C. W. Biggs. 1992. An atlas of Florida's breeding birds. Nongame Wildlife Program Final Report. Florida Game and Fresh Water Fish Commission, Tallahassee, FL.

Kalisz, P. J. and E. L. Stone. 1984a. Soil mixing by scarab beetles and pocket gophers in North-Central Florida. *Soil Science Society of America Journal* 48:169–172.

Kalisz, P. J. and E. L. Stone. 1984b. The longleaf pine islands of the Ocala National Forest, Florida: A soil study. *Ecology* 65:1743–1754.

Kane, J. M., J. M. Varner, and J. K. Hiers. 2008. The burning characteristics of southeastern oaks: Discriminating fire facilitators from fire impeders. *Forest Ecology and Management* 256:2039–2045.

Kantola, A. T. and S. R. Humphrey. 1990. Habitat use by Sherman's fox squirrel (*Sciurus niger shermani*) in Florida. *Journal of Mammalogy* 71:411–419.

Kauffeld, C. 1957. *Snakes and Snake Hunting*. Garden City, NY: Hanover House.

Kaufmann, M. R., C. M. Regan, and P. M. Brown. 2000. Heterogeneity in ponderosa pine/Douglas-fir forests: Age and size structure in unlogged and logged landscapes of central Colorado. *Canadian Journal of Forest Research* 30:698–711.

Keane, R. E. and L. J. Dickinson. 2007. The photoload sampling technique: Estimating surface fuel loadings from downward-looking photographs of synthetic fuelbeds. General Technical Report RMRS-GTR-190. USDA Forest Service, Rocky Mountain Research Station, Fort Collins, CO.

Keane, R. E., M. Rollins, and Z. L. Zhu. 2007. Using simulated historical time series to prioritize fuel treatments on landscapes across the United States: The LANDFIRE prototype project. *Ecological Modeling* 204:485–502.

Keddy, P. A. 1992. Assembly and response rules: Two goals for predictive community ecology. *Journal of Vegetation Science* 3:157–164.

Keddy, P. A., L. Smith, D. R. Campbell, M. Clark, and G. Montz. 2006. Patterns of herbaceous plant diversity in southeastern Louisiana pine savannas. *Applied Vegetation Science* 9:17–26.

Keeley, J. E. 2009. Fire intensity, fire severity and burn severity: A brief review and suggested usage. *International Journal of Wildland Fire* 18:116–126.

Keeley, J. E. and C. J. Fotheringham. 2000. Role of fire in regeneration from seed. Pages 311–330 *in* M. Fenner, editor. *Seeds: The Ecology of Regeneration in Plant Communities*. Oxon: CAB International.

Keeley, J. E. and P. H. Zedler. 1998. Evolution of life histories in *Pinus*. Pages 219–251 *in* D. M. Richardson, editor. *Ecology and Biogeography of Pinus*. Cambridge: Cambridge University Press.

Keetch, J. J. and G. M. Byram. 1968. A drought index for forest fire control. Research Paper SE-38. USDA Forest Service, Southeastern Forest Experiment Station, Asheville, NC.

Keever, C. 1950. Causes of succession on old fields of the Piedmont, North Carolina. *Ecological Monographs* 20:229–250.

Keith, D. A., T. G. Martin, E. McDonald-Madden, and C. Walters. 2011. Uncertainty and adaptive management for biodiversity conservation. *Biological Conservation* 144:1175–1178.

Kellman, M. 1984. Synergistic relationships between fire and low soil fertility in neotropical savannas: A hypothesis. *Biotropica* 16:158–160.

Kelly, J. F. and W. A. Bechtold. 1990. The longleaf pine resource. Pages 11–22 *in* R. M. Farrar, Jr., editor. *Proceedings of the Symposium on the Management of Longleaf Pine*, 4–6 April 1989, Long Beach, Mississippi, USA. General Technical Report SO-75. USDA Forest Service, Southern Research Station, New Orleans, LA.

Kennamer, R. A. and G. R. Hepp. 2000. Integration of research with long-term monitoring: Breeding wood ducks on the Savannah River Site. *Studies in Avian Biology* 21:39–49.

Kennard, D. K., K. W. Outcalt, D. Jones, and J. J. O'Brien. 2005. Comparing techniques for estimating flame temperature of prescribed fires. *Fire Ecology* 1:75–84.

Kesler, D. C. and J. R. Walters. 2012. Social composition of destination territories and matrix habitat affect red-cockaded woodpecker dispersal. *Journal of Wildlife Management* 76:1028–1035.

Kessler, G. and T. Straka. 1991. Economics of longleaf management. Presented at Establishing and Managing Longleaf Pine Workshop, Sumter, SC.

Kiflawi, M., L. Blaustein, and M. Mangel. 2003. Predation-dependent oviposition habitat selection by the mosquito *Culiseta longiareolata*: A test of competing hypotheses. *Ecology Letters* 6:35–40.

Kilgo, J. C. and A. L. Bryan, Jr. 2005. Nongame birds. Pages 223–252 *in* J. C. Kilgo, and J. I. Blake, editors. *Ecology and Management of a Forested Landscape: Fifty Years on the Savannah River Site*. Washington, DC: Island Press.

Kilgo J. C., H. S. Ray, C. Ruth, and K. V. Miller. 2010. Can coyotes affect deer populations in southeastern North America? *Journal of Wildlife Management* 75:929–933.

Kilgo, J. C., H. S. Ray, M. Vukovich, M. J. Goode, and C. Ruth. 2012. Predation by coyotes on white-tailed deer neonates in South Carolina. *Journal of Wildlife Management* 76:1420–1430.

King, J. S., R. Ceulemans, J. M. Albaugh, S. Y. Dillen, J. Domec, R. Fichot, M. Fischer, Z. Leggett, E. Sucre, M. Trnka, and T. Zenone. 2013. The challenge of lignocellulosic bioenergy in a water-limited world. *Bioscience* 63:102–117.

Kinlaw, A. and M. Grasmueck. 2012. Evidence for and geomorphologic consequences of a reptilian ecosystem engineer: The burrowing cascade initiated by the gopher tortoise. *Geomorphology* 157:108–121.

Kirkman, L. K., A. Barnett, B. W. Williams, J. K. Hiers, S. M. Pokswinski, and R. J. Mitchell. 2013. A dynamic reference model: A framework for assessing biodiversity restoration goals in a fire-dependent ecosystem. *Ecological Applications* 23:1574–1587.

Kirkman, L. K., C. L. Brown, and D. J. Leopold. 2007. *Native Trees of the Southeast—An Identification Guide*. Pages 268–270. Portland, OR: Timber Press.

Kirkman, L. K., K. L. Coffey, R. J. Mitchell, and E. B. Moser. 2004. Ground cover recovery patterns and life-history traits: Implications for restoration obstacles and opportunities in a species-rich savanna. *Journal of Ecology* 92:409–421.

Kirkman, L. K., M. B. Drew, L. T. West, and E. R. Blood. 1998. Ecotone characterization between upland longleaf pine/wiregrass stands and seasonally ponded isolated wetlands. *Wetlands* 18:346–364.

Kirkman, L. K., L. M. Giencke, R. S. Taylor, L. R. Boring, C. L. Staudhammer, and R. J. Mitchell. 2016. Productivity and species richness in longleaf pine woodlands: Resource-disturbance influences across an edaphic gradient. *Ecology* 97:2259–2271.

Kirkman, L. K., P. C. Goebel, B. J. Palik, and L. T. West. 2004. Predicting plant species diversity in a longleaf pine landscape. *Ecoscience* 11:80–93.

Kirkman. L. K., P. C. Goebel, L. T. West, M. B. Drew, and B. J. Palik. 2000. Depressional wetland vegetation types: A question of plant community development. *Wetlands* 20:373–385.

Kirkman, L. K., S. W. Golladay, L. Laclaire, and R. Sutter. 1999. Biodiversity in southeastern, seasonally ponded, isolated wetlands: Management and policy perspectives for research and conservation. *Journal of the North American Benthological Society* 18:553–562.

Kirkman, L. K. and R. J. Mitchell. 2002. A forest gap approach to restoring longleaf pine-wiregrass ecosystems (Georgia and Florida). *Ecological Restoration* 20:50–51.

Kirkman, L. K. and R. J. Mitchell. 2006. Conservation management of *Pinus palustris* ecosystems from a landscape perspective. *Applied Vegetation Science* 9:67–74.

Kirkman, L. K., R. J. Mitchell, R. C. Helton, and M. B. Drew. 2001. Productivity and species richness across an environmental gradient in a fire-dependent ecosystem. *American Journal of Botany* 88:2119–2128.

Kirkman, L. K., R. J. Mitchell, M. J. Kaeser, S. D. Pecot, and K. L. Coffey. 2007. The perpetual forest: Using undesirable species to bridge restoration. *Journal of Applied Ecology* 44:604–614.

Kirkman, L. K., L. L. Smith, P. F. Quintana-Ascencio, M. J. Kaeser, S. W. Golladay, and A. L. Farmer. 2012. Is species richness congruent among taxa? Surrogacy, complementarity, and environmental correlates among three disparate taxa in geographically isolated wetlands. *Ecological Indicators* 18:131–139.

Kittredge, D. B. 2005. The cooperation of private forest owners on scales larger than one individual property: International examples and potential application in the United States. *Forest Policy and Economics* 7:671–688.

Klaus, J. M. and R. F. Noss. 2016. Specialist and generalist amphibians respond to wetland restoration treatments. *Journal of Wildlife Management* 80:1106–1119.

Klepzig, K., R. Shelfer, and Z. Choice. 2014. Outlook for coastal plain forests: A subregional report from the Southern Forest Futures Project. General Technical Report SRS-GTR-196. USDA Forest Service, Southern Research Station, Asheville, NC.

Knapp, E. E. 2015. Long-term dead wood changes in a Sierra Nevada mixed conifer forest: Habitat and fire hazard implications. *Forest Ecology and Management* 339:87–95.

Knapp, A. K., J. M. Briggs, S. L. Collins, S. R. Archer, M. S. Bret-Harte, B. E. Ewers, D. P. Peters, D. R. Young, G. R. Shaver, E. Pendall, and M. B. Cleary. 2008. Shrub encroachment in North American grasslands: Shifts in growth form dominance rapidly alters control of ecosystem carbon inputs. *Global Change Biology* 14:615–623.

Knapp, E. E., B. L. Estes, and C. N. Skinner. 2009. Ecological effects of prescribed fire season: A literature review and synthesis for managers. General Technical Report PSW-GTR-224. USDA Forest Service, Pacific Southwest Research Station, Redding, CA.

Knapp, E. E., M. North, M. Benech, and B. Estes. 2012. The variable-density thinning study at Stanislaus-Tuolomne experimental forest. Pages 127–139 in M. North, editor. Managing Sierra Nevada Forests. General Technical Report PSW-GTR-237. USDA Forest Service, Pacific Southwest Research Station, Albany, CA.

Knapp, E. E. and K. J. Rice. 1994. Starting from seed: Genetic issues in using native grasses for restoration. *Restoration & Management Notes* 12:40–45.

Knapp, E. E., C. N. Skinner, M. P. North, and B. Estes. 2013. Long-term overstory and understory change following logging and fire exclusion in a Sierra Nevada mixed-conifer forest. *Forest Ecology and Management* 310:904–914.

Knapp, A. K. and M. D. Smith. 2001. Variation among biomes in temporal dynamics of aboveground primary production. *Science* 291:481–484.

Knapp, B. O., J. L. Walker, G. G. Wang, H. Hu, and R. N. Addington. 2014. Effects of overstory retention, herbicides, and fertilization on sub-canopy vegetation structure and functional group composition in loblolly pine forests restored to longleaf pine. *Forest Ecology and Management* 320:149–160.

Knapp, B. O., G. G. Wang, H. Hu, J. L. Walker, and C. Tennant. 2011. Restoring longleaf pine (*Pinus palustris* Mill.) in loblolly pine (*Pinus taeda* L.) stands: Effects of restoration treatments on natural loblolly pine regeneration. *Forest Ecology and Management* 262:1157–1167.

Knapp, B. O., G. G. Wang, and J. L. Walker. 2013. Effects of canopy structure and cultural treatments on the survival and growth of *Pinus palustris* Mill. seedlings underplanted in *Pinus taeda* L. stands. *Ecological Engineering* 57:46–56.

Knapp, B. O., G. G. Wang, J. L. Walker, and H. Hu. 2015. From loblolly to longleaf: Fifth-year results of a longleaf pine restoration study at two ecologically distinct sites. Pages 470–472 *in* A. G. Holley, K. F. Connor, and J. D. Haywood, editors. *Proceedings of the 17th Biennial Silvicultural Research Conference*. General Technical Report SRS-203. USDA Forest Service, Southern Research Station, Asheville, NC.

Knight, T. 2004. Reconstruction of Flint River streamflow using tree-rings. Water Policy Working Paper # 2004–005, Georgia State University, Atlanta, GA.

Knight, T. 2006. Dendrochronological investigations of longleaf pine on Lavender Mountain. Pages 18–24 *in* M. L. Cipollini, compiler. *Proceedings of the Second Montane Longleaf Pine Conference Workshop*, 18–19 November 2005, Berry College, Mount Berry, GA. Longleaf Alliance Report No. 9.

Knutson, M. G., W. B. Richardson, D. M. Reineke, B. R. Gray, J. R. Parmelee, and S. E. Weick. 2004. Agricultural ponds support amphibian populations. *Ecological Applications* 14:669–684.

Knutson, M. G., J. R. Sauer, D. A. Olsen, M. J. Mossman, L. M. Hemesath, and M. J. Lannoo. 1999. Effects of landscape composition and wetland fragmentation on frog and toad abundance and species richness in Iowa and Wisconsin, USA. *Conservation Biology* 13:1437–1446.

Kobziar, L. N., D. L. Godwin, L. Taylor, and A. C. Watts. 2015. Perspectives on trends, effectiveness, and impediments to prescribed burning in the southern U.S. *Forests* 6:561–580.

Kobziar, L. N., M. E. Rocca, C. A. Dicus, C. Hoffman, N. Sugihara, A. E. Thode, J. M. Varner, and P. Morgan. 2009. Challenges to educating the next generation of wildland fire professionals in the United States. *Journal of Forestry* 107:339–345.

Kohm, K. A. and J. F. Franklin. 1997. *Creating a Forestry for the 21st Century: The Science of Ecosystem Management*. Washington, DC: Island Press.

Komarek, E. V. 1964. The natural history of lightning. Pages 139–183 *in Proceedings of the Third Tall Timbers Fire Ecology Conference*. Tall Timbers Research Station, Tallahassee, FL.

Komarek, E. V. 1967. Fire and animal behavior. Pages 161–207 *in Proceedings of the Seventh Tall Timbers Fire Ecology Conference*. Tall Timbers Research Station, Tallahassee, FL.

Komarek, E. V. 1968. Lightning and lightning fires as ecological forces. *Proceedings of the Eighth Tall Timbers Fire Ecology Conference*. Tall Timbers Research Station, Tallahassee, FL.

Komarek, E.V., Sr. 1974. Effects of fire on temperate forests and related ecosystems: Southeastern United States. Pages 251–278 *in* T. T. Kozlowski, and C.E. Ahlgren, editors. *Fire and Ecosystems*. New York, NY: Academic Press.

Korb, J. E., N. C. Johnson, and W. W. Kovington. 2004. Slash pile burning effects on soil biotic and chemical properties and plant establishment. *Restoration Ecology* 12:52–62.

Korstian, C. F. and T.S. Coile. 1938. Plant competition in forest stands. *Duke University School of Forestry Bulletin 3*. Durham, NC: Duke University.

Kosola, K. R., D. I. Dickman, E. A. Paul, and D. Parry. 2001. Repeated insect defoliation effects on growth, nitrogen acquisition, carbohydrates, and root demography of poplars. *Oecologia* 129:65–74.

Kraft, N. J. B., L. S. Comita, J. M. Chase, N. J. Sanders, N. G. Swenson, T. O. Crist, J. C. Stegen, M. Vellend, B. Boyle, M. J. Anderson, H. V. Cornell, K. F. Davies, A. L. Freestone, B. D. Inouye, S. P. Harrison, and J. A. Myers. 2011. Disentangling the drivers of β-diversity along latitudinal and elevational gradients. *Science* 333:1755–1758.

Kramer, A. T., D. J. Larkin, and J. B. Fant. 2015. Assessing potential seed transfer zones for five forb species from the Great Basin floristic region, USA. *Natural Areas Journal* 35:174–188.

Kremens, R., M. Dickinson, and A. Bova. 2012. Radiant flux density, energy density and fuel consumption in mixed-oak forest surface fires. *International Journal of Wildland Fire* 21:722–730.

Kreye, J. K., J. M. Varner, J. K. Hiers, and J. Mola. 2013. Toward a mechanism for eastern North American forest mesophication: Differential litter drying across 17 species. *Ecological Applications* 23:1976–1986.

Krysko, K. L., L. P. Nuñez, C. A. Lippi, D. J. Smith, and M. C. Granatosky. 2016. Pliocene-Pleistocene lineage diversifications in the Eastern Indigo Snake (*Drymarchon couperi*) in the southeastern United States. *Molecular Phylogenetics and Evolution* 98:111–122.

Kush, J. S. 2016. The longleaf pines are few and far between. *Ecological Restoration* 34:15–16.

Kush, J. S. and R. S. Meldahl. 2000. Composition of a virgin stand of longleaf pine in south Alabama. *Castanea* 65:56–63.

Kush, J. S., R. S. Meldahl, and W. D. Boyer. 1999. Understory plant community response after 23 years of hardwood control treatments in natural longleaf pine (*Pinus palustris*) forests. *Canadian Journal of Forest Research* 29:1047–1054.

Kush, J. S., R. S. Meldahl, C. K. McMahon, and W. D. Boyer. 2004. Longleaf pine: A sustainable approach for increasing terrestrial carbon in the southern United States. *Environmental Management* 33:S138–S147.

Labisky, R. F. and M. C. Boulay. 1998. Behaviors of bobcats preying on white-tailed deer in the Everglades. *American Midland Naturalist* 139:275–281.

Lajeunesse, S. D., J. J. Dilustro, R. R. Sharitz, and B. S. Collins. 2006. Ground layer carbon and nitrogen cycling and legume nitrogen inputs following fire in mixed pine forests. *American Journal of Botany* 93:84–93.

Laliberte, L., J. O. Luken, J. J. Hutchens, Jr., and K. S. Godwin. 2007. The ecological boundaries of six Carolina bays: Community composition and ecotone distribution. *Wetlands* 27:873–883.

Laliberte, A. S. and W. J. Ripple. 2004. Range contractions of North American carnivores and ungulates. *BioScience* 54:123–138.

Lancia, R. A., J. P. Roise, D. A. Adams, and M. R. Lennartz. 1989. Opportunity costs of red-cockaded woodpecker foraging habitat. *Southern Journal of Applied Forestry* 13:81–85.

Landers, J. L. 1987. Prescribed burning for managing wildlife in southeastern pine forests. Pages 19–27 *in* J. G. Dickson, and O. E. Maughan, editors. *Managing Southern Forests for Wildlife and Fish*. General Technical Report SO-65. USDA Forest Service, Southern Forest Experiment Station, New Orleans, LA.

Landers, J. L. 1991. Disturbance influences on pine traits in the southeastern United States. Pages 61–98 *in* S. M. Hermann, editor. *Proceedings of the 17th Tall Timbers Fire Ecology Conference*, 18–21 May 1989, Tallahassee, Florida, USA. Tall Timbers Research Station, Tallahassee, FL.

Landers, J. L., N. A. Byrd, and R. Komarek. 1990. A holistic approach to managing longleaf pine communities. *Proceedings of the Symposium on the Management of Longleaf Pine*. General Technical Report SO-75. USDA Forest Service, Washington, DC.

Landers, J. L., D. H. Van Lear, and W. D. Boyer. 1995. The longleaf pine forests of the Southeast: Requiem or renaissance? *Journal of Forestry* 93:39–44.

LANDFIRE. 2008. LANDFIRE national fire regime condition class (FRCC). https://www.conservationgateway.org/PublishingImages/LF_1%200_FRCC_map.jpg (accessed 4 February 2017).

Lane, C. R., E. D'Amico, and B. Autrey. 2012. Isolated wetlands of the southeastern United States: Abundance and expected condition. *Wetlands* 32:753–767.

Langley, J. A., B. G. Drake, and B. A. Hungate. 2002. Extensive belowground carbon storage supports roots and mycorrhizae in regenerating scrub oaks. *Oecologia* 13:542–548.

Langpap, C. 2006. Conservation of endangered species: Can incentives work for private landowners? *Ecological Economics* 57:558–572.

Larkin, J. L., D. S. Maehr, T. S. Hoctor, M. A. Orlando, and K. Whitney. 2004. Landscape linkages and conservation planning for the black bear in west-central Florida. *Animal Conservation* 7:23–34.

Larson, K. 2004. Family forests—The bigger picture. Response to "America's family forest owners." *Journal of Forestry* 102:13–14.

Larson, A. J., R. T. Belote, C. A. Cansler, S. A. Parks, and M. S. Dietz. 2013. Latent resilience in ponderosa pine forest: Effects of resumed frequent fire. *Ecological Applications* 23:1243–1249.

Larson, A. J., R. T. Belote, M. A. Williamson, and G. H. Aplet. 2013. Making monitoring count: Project design for active adaptive management. *Journal of Forestry* 111:348–356.

Larson, A. J. and D. Churchill. 2012. Tree spatial patterns in fire-frequent forests of western North America, including mechanisms of pattern formation and implications for designing fuel reduction and restoration treatments. *Forest Ecology and Management* 267:74–92.

Lashley, M. A., M. C. Chitwood, R. Kays, C. A. Harper, C. S. DePerno, and C. W. Moorman. 2015. Prescribed fire affects female white-tailed deer habitat use during summer lactation. *Forest Ecology and Management* 348:220–225.

Lashley, M. A., M. C. Coltwood, A. Price, M. B. Elfelt, E. L. Kilburg, C. S. DePerno, and C. E. Moorman. 2014. Subtle effects of a managed fire regime: A case study in the longleaf pine ecosystem. *Ecological Indicators* 38:212–217.

Laughlin, D. C., M. M. Moore, and P. Z. Fule. 2011. A century of increasing pine density and associated shifts in understory plant strategies. *Ecology* 92:556–561.

Lavoie, M., M. C. Mack, J. K. Heirs, S. Pokswinski, A. Barnett, and L. Provencher. 2014. Effects of restoration techniques on soil carbon and nitrogen dynamics in Florida longleaf pine (*Pinus palustris*) sandhill forests. *Forests* 5:498–517.

Lavoie, M., G. Starr, T. A. Martin, and H. L. Gholz. 2010. Effects of a prescribed fire on understory vegetation, carbon pools, and soil nutrients in a longleaf pine-slash pine forest in Florida. *Natural Areas Journal* 30:82–94.

Law, B. E. 2015. Regional analysis of drought and heat impacts on forests: Current and future science directions. *Global Change Biology* 20:3595–3599.

Lawson, J. 1860. *History of Carolina, Containing the Exact Description and Natural History of that Country.* Raleigh, NC: Strother and Marcom.

Lazarus, E. and D. Schaible. 2015. A hidden scale dependency in conserving working woodlands. *PeerJ PrePrints* 3:e1333v1.

Leeper, D. A. and B. E. Taylor. 1998. Abundance, biomass and production of aquatic invertebrates in Rainbow Bay, a temporary wetland in South Carolina, USA. *Archiv für Hydrobiologie* 143:335–362.

Leibowitz, S. G. 2003. Isolated wetlands and their functions: An ecological perspective. *Wetlands* 23:517–531.

Lemon, P. C. 1949. Successional responses of herbs in the longleaf-slash pine forest after fire. *Ecology* 30:135–145.

Lemon, P. C. 1967. Effects of fire on herbs of the southeastern United States and central Africa. Pages 112–127. *Proceedings of the Seventh Tall Timbers Fire Ecology Conference*. Tall Timbers Research Station, Tallahassee, FL.

Lentile, L. B., Z. A. Holden, A. M. S. Smith, M. J. Falkowski, A. T. Hudak, P. Morgan, S. A. Lewis, P. E. Gessler, and N. C. Benson. 2006. Remote sensing techniques to assess active fire characteristics and post-fire effects. *International Journal of Wildland Fire* 15:319–345.

Lenton, T. M. 2011. Early warning of climate tipping points. *Nature Climate Change* 1:201–209.

Leopold, A. 1949. *A Sand County Almanac and Sketches Here and There*. New York, NY: Oxford University Press.

Leroux, S. J. and M. Loreau. 2008. Subsidy hypothesis and strength of trophic cascades across ecosystems. *Ecology Letters* 11:1147–1156.

Lesica, P. and F. W. Allendorf. 1999. Ecological genetics and the restoration of plant communities: Mix or match? *Restoration Ecology* 7:42–50.

Letnic, M., E. G. Ritchie, and C. R. Dickman. 2012. Top predators as biodiversity regulators: The dingo *Canis lupus dingo* as a case study. *Biological Reviews* 87:390–413.

Levins, R. 1969. Some demographic and genetic consequences of environmental heterogeneity for biological control. *Bulletin of the Entomological Society of America* 15:237–240.

Lewis, C. E. and R. H. Hart. 1972. Some herbage responses to fire on pine-wiregrass range. *Journal of Range Management* 25:209–213.

Li, B., S. E. McKeand, A. V. Hatcher, and R. J. Weir. 1997. Genetic gains of second generation selections from the North Carolina State University-Industry Cooperative Tree Improvement Program. Pages 234–238 *in* T. White, D. Huber, and G. Powell, compilers. *Proceedings of the 24th Southern Tree Improvement Conference*, 9–12 June 1997, Orlando, Florida, USA. University of Florida, Gainesville, FL.

Liang, B., J. Lehmann, D. Solomon, S. Sohi, J. E. Thies, J. O. Skjemstad, F. J. Luizao, M. H. Engelhard, E. G. Neves, and S. Wirick. 2008. Stability of biomass-derived black carbon in soils. *Geochimica et Cosmochimica Acta* 72:6069–6078.

Lide, R. F., V. G. Meentemeyer, J. E. Pinder III, and L. M. Beatty. 1995. Hydrology of a Carolina bay located on the upper coastal plain of western South Carolina. *Wetlands* 15:47–57.

Lidicker, W. Z., Jr., J. O. Wolff, L. N. Lidicker, and M. H. Smith. 1992. Utilization of a habitat mosaic by cotton rats during a population decline. *Landscape Ecology* 6:259–268.

Liechty, H. O., K. R. Luckow, and J. M. Guldin. 2005. Soil chemistry and nutrient regimes following 17–21 years of shortleaf pine-bluestem restoration in the Ouachita Mountains of Arkansas. *Forest Ecology and Management* 204:345–357.

Lima, S. L. and L. M. Dill. 1990. Behavioral decisions made under the risk of predation—a review and prosepectus. *Canadian Journal of Zoology* 68:619–640.

Lindenmayer, D. B. and J. F. Franklin. 2002. *Conserving Forest Biodiversity: A Comprehensive Multiscaled Approach.* Washington, DC: Island Press.

Lindenmayer, D. B., K. B. Hulvey, R. J. Hobbs, M. Colyvan, A. Felton, H. Possingham, W. Steffen, K. Wilson, J. Youngentob, and P. Gibbons. 2012. Avoiding bio-perversity from carbon sequestration solutions. *Conservation Letters* 5:28–36.

Lindenmayer, D. B., W. F. Laurance, and J. F. Franklin. 2012. Global decline in large old trees. *Science* 338:1305–1306.

Lindenmayer, D. B. and G. E. Likens. 2010. The science and application of ecological monitoring. *Biological Conservation* 143:1317–1328.

Liner, A. E., L. L. Smith, S. B. Castleberry, S.W. Golladay, and J. W. Gibbons. 2008. Amphibian distributions within three types of isolated wetlands in southwest Georgia. *American Midland Naturalist* 160:69–81.

Linn, R., J. Reisner, J. J. Colman, and J. Winterkamp. 2002. Studying wildfire behavior using FIRETEC. *International Journal of Wildland Fire* 11:233–246.

Linnartz, N. E., C. Y. Hse, and V. L. Duvall. 1966. Grazing impairs physical properties of a forest soil in central Louisiana. *Journal of Forestry* 64:239–243.

Lipps, E. L. 1966. Plant communities of a portion of Floyd County, Georgia—Especially the Marshall Forest, PhD dissertation. University of Tennessee, Knoxville, TN.

Lips, K. R., J. Diffendorfer, J. R. Mendelson, and M. W. Sears. 2008. Riding the wave: Reconciling the roles of disease and climate change in amphibian declines. *PLOS Biology* 6:72.

Lipscomb, D. J. 1989. Impacts of feral hogs on longleaf pine regeneration. *Southern Journal of Applied Forestry* 13:177–181.

Litt, A. R. and D. R. Gordon. 2003. Predictors of species richness in northwest Florida longleaf pine sandhills. *Conservation Biology* 17:1660–1671.

Litt, A. R., B. J. Herring, and L. Provencher. 2001. Herbicide effects on ground-layer vegetation in southern pinelands, USA: A review. *Natural Areas Journal* 21:177–188.

Littell, J. S., D. McKenzie, D. L. Peterson, and A. L. Westerling. 2009. Climate and wildfire area burned in western U.S. ecoprovinces, 1916–2003. *Ecological Applications* 19:1003–1021.

Little, E. L., Jr. 1971. *Atlas of United States Trees, Volume 1, Conifers and Important Hardwoods*. Miscellaneous Publication 1146. Washington, DC: USDA Forest Service.

Little, E. L., Jr., and W. B. Critchfield. 1969. *Subdivisions of the Genus Pinus*. Miscellaneous Publication 1144. Washington, DC: USDA Forest Service.

Little, A. R., S. Demarais, K. L. Gee, S. L. Webb, S. M. Riffell, J. A. Gaskamp, and J. L. Belant. 2014. Does human predation risk affect harvest susceptibility of white-tailed deer during hunting season? *Wildlife Society Bulletin* 38:797–805.

Little, A. R., M. M. Streich, M. J. Chamberlain, L. M. Conner, and R. J. Warren. 2014. Eastern wild turkey reproductive ecology in frequently-burned longleaf pine savannas. *Forest Ecology and Management* 331:180–187.

Liu, Y., J. J. Andersen, J. W. Williams, and S. T. Jackson. 2013. Vegetation history in central Kentucky and Tennessee during the last glacial and deglacial period. *Quaternary Research* 79:189–198.

Liu, Y., S. L. Goodrick, and J. A. Stanturf. 2013. Future U.S. wildfire potential trends projected using a dynamically downscaled climate change scenario. *Forest Ecology and Management* 294:120–135.

Liu, Y. Q., J. Prestemon, S. L. Goodrick, T. P. Holmes, J. A. Stanturf, J. M. Vose, and G. Sun. 2013. Future wildfire trends, impacts and mitigation options in the southern United States. Pages 85–126 in J. M. Vose, and K. Klepzig, editors. *Climate Change Adaptation and Mitigation Management Options*. Boca Raton, FL: CRC Press.

Livingston, R. J. 1991. Historical relationships between research and resource management in the Apalachicola River estuary. *Ecological Applications* 1:361–382.

Lockaby, G., C. Nagy, J. M. Vose, C. R. Ford, G. Sun, S. McNulty, P. Caldwell, E. Cohen, and J. Moore Myers. 2013. Forests and water. Pages 309–339 *in* D. N. Wear and J. G. Greis, editors. *The Southern Forest Futures Project*, Chap. 13. General Technical Report SRS-GTR-178. USDA Forest Service, Southern Research Station, Asheville, NC.

Lodewick, J. E. 1930. Effect of certain climatic factors on the diameter growth of longleaf pine in western Florida. *Journal of Agricultural Research* 41:349–363.

Loeb, S. C. 1996. The role of coarse woody debris in the ecology of southeastern mammals. Pages 108–118 *in* J. W. McMinn, and D. A. Crossley, editors. *Biodiversity and Coarse Woody Debris in Southern Forests: Effects on Biodiversity*. General Technical Report GTR-SE-94:108–118. USDA Forest Service, Southern Research Station, Asheville, NC.

Long, A. K. 2015. Effects of an invasive ant on invertebrates in the southeastern United States, PhD dissertation. University of Florida, Gainesville, FL.

Long, A. K, L. M. Conner, L. L. Smith, and R. A. McCleery. 2015. Effects of an invasive ant and native predators on cotton rat recruitment and survival. *Journal of Mammalogy* 96:1135–1141.

Long, A. K., D. D. Knapp, L. Mccullough, L. L. Smith, L. M. Conner, and R. A. Mccleery. 2015. Southern toads alter their behavior in response to red-imported fire ants. *Biological Invasions* 17:2179–2186.

Longleaf Partnership Council. 2013. Longleaf Partnership Council white paper—longleaf pine planting density. http://www.americaslongleaf.org/media/10886/lpc-longleaf_pine_density_white-paper-final_053013_logo.pdf (accessed 4 February 2017).

Longleaf Partnership Council. 2014. 2013 Range-wide Accomplishment Report. Longleaf Partnership Council. February 2014. http://www.americaslongleaf.org/resources/2013-range-wide-accomplishment-report-and-executive-summary/ (accessed 2 February 2017).

Longleaf Partnership Council. 2015. America's Longleaf Restoration Initiative. 2014 Range-wide Accomplishment Report. Longleaf Partnership Council. February 2015. http://www.americaslongleaf.org/resources/2014-range-wide-accomplishment-report-and-executive-summary/ (accessed 4 February 2017).

Longleaf Partnership Council. 2016. America's Longleaf Restoration Initiative. 2015 Range-wide Accomplishment Report. Longleaf Partnership Council. March 2016. http://www.americaslongleaf.org/resources/2015-range-wide-accomplishment-report/ (accessed 4 February 2017).

Loreau, M., N. Mouquet, and A. Gonzalez. 2003. Biodiversity as spatial insurance in heterogeneous landscapes. *Proceedings of the National Academy of Sciences* 100:12765–12770.

Loudermilk, E. L., G. L. Achtemeier, J. J. O'Brien, J. K. Hiers, and B. S. Hornsby. 2014. High-resolution observations of combustion in heterogeneous surface fuels. *International Journal of Wildland Fire* 23:1016–1026.

Loudermilk, E. L., W. P. Cropper, Jr., R. J. Mitchell, and H. Lee. 2011. Longleaf pine (*Pinus palustris*) and hardwood dynamics in a fire-maintained ecosystem: A simulation approach. *Ecological Modelling* 222:2733–2750.

Loudermilk, E. L., J. K. Hiers, J. J. O'Brien, R. J. Mitchell, A. Singhania, J. C. Fernandez, W. P. Cropper Jr., and K. C. Slatton. 2009. Ground-based LIDAR: A novel approach to quantify fine-scale fuelbed characteristics. *International Journal of Wildland Fire* 18:676–685.

Loudermilk, E. L., J. K. Hiers, S. Pokswinski, J. J. O'Brien, A. Barnett, and R. J. Mitchell. 2016. The path back: Oaks (*Quercus* spp.) facilitate longleaf pine (*Pinus palustris*) seedling establishment in xeric sites. *Ecosphere* 7:e01361. doi:10.1002/ecs2.1361.

Loudermilk, E. L., J. J. O'Brien, R. J. Mitchell, W. P. Cropper, J. K. Hiers, S. Grunwald, J. Grego, and J. C. Fernandez-Diaz. 2012. Linking complex forest fuel structure and fire behaviour at fine scales. *International Journal of Wildland Fire* 21:882–893.

Lubowski, R. N., A. J. Plantinga, and R. N. Stavins. 2006. Land-use change and carbon sinks: Econometric estimation of the carbon sequestration supply function. *Journal of Environmental Economics and Management* 51:135–152.

Lunan, J. S. and J. R. Habeck. 1973. The effects of fire exclusion on ponderosa pine communities in Glacier National Park, Montana. *Canadian Journal of Forest Research* 3:574–579.

Lundberg, J. and F. Moberg. 2003. Mobile link organisms and ecosystem functioning: Implications for ecosystem resilience and management. *Ecosystems* 6:87–98.

Luo, Y., D. Gerten, G. L. Maire, W. J. Parton, E. Weng, X. Zhou, C. Keough, C. Beier, P. Ciais, W. Cramer, J. S. Dukes, B. Emmett, P. J. Hanson, A. Knapp, S. Linder, D. Nepstad, and L. Rustad. 2008. Modeled interactive effects of precipitation, temperature, and [CO_2] on ecosystem carbon and water dynamics in different climatic zone. *Global Change Biology* 14:1–14.

Lupardus, J. L., L. Muller, and J. L. Kindall. 2011. Seasonal forage availability and diet for reintroduced elk in the Cumberland Mountains, Tennessee. *Southeastern Naturalist* 10:53–74.

Lutz, J. A., A. J. Larson, M. E. Swanson, and J. A. Freund. 2012. Ecological importance of large-diameter trees in a temperate mixed-conifer forest. *PLOS ONE* 7:e36131.

Lydersen, J. M. and M. P. North. 2012. Topographic variation in structure of mixed-conifer forests under an active-fire regime. *Ecosystems* 15:1134–1146.

Lydersen, J. M., M. P. North, E. E. Knapp, and B. M. Collins. 2013. Quantifying spatial patterns of tree groups and gaps in mixed-conifer forests: Reference conditions and long-term changes following fire suppression and logging. *Forest Ecology and Management* 304:370–382.

MacArthur, R. H. and E. O. Wilson. 1967. *The Theory of Island Biogeography*. Princeton, NJ: Princeton University Press.

MacDonald, L. A. and H. R. Mushinsky. 1988. Foraging ecology of the gopher tortoise in a sandhill habitat. *Herpetologica* 44:345–353.

Macie, E. A. and L. A. Hermansen. 2002. Human influences on forest ecosystems: The southern wildland-urban interface assessment. General Technical Report SRS-55. USDA Forest Service, Southern Research Station, Asheville, NC.

Mack, M. C. and C. M. D'Antonio. 1998. Impacts of biological invasions on disturbance regimes. *Trends in Ecology and Evolution* 13:195–198.

Maldague, X. 2001. *Theory and Practice of Infrared Technology for Nondestructive Testing*. New York, NY: John Wiley & Sons.

Maloney, P. E., T. F. Smith, C. E. Jensen, J. Innes, D. Rizzo, and M. P. North. 2008. Initial tree mortality and insect and pathogen response to fire and thinning restoration treatments in an old-growth mixed-conifer forest of the Sierra Nevada, California. *Canadian Journal of Forest Research* 38:3011–3020.

Malumphy, C., M. A. Hamilton, B. N. Manco, P. W. C. Green, M. D. Sanchez, M. Corcoran, and E. Salamanca. 2012. *Toumeyella parvicornis* (Hemiptera: Coccidae), causing severe decline of *Pinus caribaea* var. *bahamensis* in the Turks and Caicos islands. *Florida Entomologist* 95:113–119.

Mamo, L. B. and E. G. Bolen. 1999. Effects of area, isolation, and landscape on the avifauna of Carolina bays. *Journal of Field Ornithology* 70:310–320.

Mann, C. C. 2005. 1491: *New Revelations of the Americas Before Columbus*. New York, NY: Knopf.

Mann, C. J. and R. G. Wetzel. 1996. Loading and utilization of dissolved organic carbon from emergent macrophytes. *Aquatic Botany* 53:61–72.

Maple, W. R. 1969. Shaded longleaf pine seedlings can survive prescribed burns. *Forest Farmer* 29:13.

Maple, W. R. 1975. Spring burn aids longleaf pine seedling height growth. Report SO-228. USDA Forest Service, Southern Forest Experiment Station, New Orleans, LA.

Marcoux, H. M., L. D. Daniels, S. E. Gergel, E. Da Silva, Z. Gedalof, and P. F. Hessburg. 2015. Differentiating mixed- and high-severity fire regimes in mixed-conifer forests of the Canadian Cordillera. *Forest Ecology and Management* 341:45–58.

Marion, D. A., G. Sun, P. V. Caldwell, C. F. Miniat, Y. Ouyang, D. M. Amatya, B. D. Clinton, P. A. Conrads, S. Gull-Laird, Z. Dai, J. A. Clingenpeel, Y. Liu, E. A. Roehl Jr., J. A. Moore Meyers, and C. Trettin. 2014. Pages 249–306 *in* J. M. Vose, and K. D. Klepzig, editors. *Climate Change Adaption and Mitigation Management Options: A Guide for Natural Resource Managers in Southern Forest Ecosystems*. Boca Raton, FL: CRC Press.

Markewitz, D., F. Sartori, and C. Craft. 2002. Soil change and carbon storage in longleaf pine stands planted on marginal agricultural lands. *Ecological Applications* 12:1276–1285.

Markwardt, L. J. and T. R. C. Wilson. 1935. *Strength and related properties of woods grown in the United States*. Technical Bulletin No. 479. Washington, DC: USDA Forest Service, Forest Products Laboratory, Division of Research.

Marlon, J. R., P. J. Bartlein, C. Carcaillet, D. G. Gavin, S. P. Harrison, P. E. Higuera, F. Joos, M. J. Power, and I. C. Prentice. 2008. Climate and human influences on global biomass burning over the past two millennia. *Nature Geoscience* 1:697–702.

Marschner, P. 2012. *Marschner's Mineral Nutrition of Higher Plants*. New York, NY: Academic Press.

Marsh, G. M., V. R. Lane, K. V. Miller, S. B. Castleberry, D. A. Miller, and T. B. Wigley. 2012. Responses of white-tailed deer and northern bobwhite food plants to a gradient of stand establishment intensities in pocosin pine plantations. *Wildlife Society Bulletin* 36:693–703.

Martin, T. A. 2000. Winter season tree sap flow and stand transpiration in an intensively managed loblolly and slash pine plantation. *Journal of Sustainable Forestry* 10:155–163.

Martin, G. I., J. Hepinstall-Cymerman, and L. K. Kirkman. 2013. Six decades (1948–2007) of landscape change in the Dougherty plain of southwest Georgia, USA. *Southeastern Geographer* 53:28–49.

Martin, K. L., M. D. Hurteau, B. A. Hungate, G. W. Koch, and M. P. North. 2015. Carbon trade-offs of restoration and provision of endangered species habitat in a fire-maintained forest. *Ecosystems* 18:76–88.

Martin, K. L. and L. K. Kirkman. 2009. Management of ecological thresholds to re-establish disturbance-maintained herbaceous wetlands of the south-eastern USA. *Journal of Applied Ecology* 46:906–914.

Martin, G. I., L. K. Kirkman, and J. Hepinstall-Cymerman. 2012. Mapping geographically isolated wetlands in the Dougherty Plain, Georgia, USA. *Wetlands* 32:149–160.

Martinuzzi, S., S. I. Stewart, D. P. Helmers, M. H. Mockrin, R. B. Hammer, and V. C. Radeloff. 2015. The 2010 wildland-urban interface of the conterminous United States. Research Map NRS-8. USDA Forest Service, Northern Research Station, Newtown Square, PA.

Martinuzzi, S., L. A. Vierling, W. A. Gould, M. J. Falkowski, J. S. Evans, A. T. Hudak, and K. T. Vierling. 2009. Mapping snags and understory shrubs for a LiDAR-based assessment of wildlife habitat suitability. *Remote Sensing of Environment* 113:2533–2546.

Maschinski, J., M. S. Ross, H. Liu, J. O'Brien, E. J. von Wettberg, and K. E. Haskins. 2011. Sinking ships: Conservation options for endemic taxa threatened by sea level rise. *Climatic Change* 107:147–167.

Masters, R. E., R. L. Lochmiller, and D. M. Engle. 1993. Effects of timber harvest and prescribed fire on white-tailed deer forage production. *Wildlife Society Bulletin* 21:401–411.

Matthews, S. 2014. Dead fuel moisture research: 1991–2012. *International Journal of Wildland Fire* 23:78–92.

Mathews, N. E. and W. F. Porter. 1988. Black bear predation of white-tailed deer neonates in the central Adirondacks. *Canadian Journal of Zoology* 66:1241–1242.

Mazzotti, F. J. and L. A. Brandt. 1994. Ecology of the American alligator in a seasonally fluctuating environment. Pages 485–505 *in* J. C. Ogden, and S. M. Davis, editors. *Everglades: The Ecosystem and Its Restoration*. Delray Beach, FL: St. Lucie Press.

McCaffrey, S. M. 2006. Prescribed fire: What influences public approval? Pages 192–198 *in* M. B. Dickinson, editor. *Fire in Eastern Oak Forests: Delivering Science to Land Managers. Proceedings of a Conference*, 15–17 November 2005, Ohio State University, Columbus, Ohio, USA. General Technical Report NRS-P-1. USDA Forest Service, Northern Research Station, Newtown Square, PA.

McCaffrey, S. 2009. Crucial factors influencing public acceptance of fuels treatments. *Fire Management Notes* 69:9–12.

McCain, K. N. S., S. G. Baer, J. M. Blair, and G. W. Wilson. 2010. Dominant grasses suppress local diversity in restored tallgrass prairie. *Restoration Ecology* 18:40–49.

McCay, D. H. 2000. Effects of chronic human activities on invasion of longleaf pine forests by sand pine. *Ecosystems* 3:283–292.

McClain, M. E., E. W. Boyer, C. L. Dent, S. E. Gergel, N. B. Grimm, P. M. Groffman, S. C. Hart, J. W. Harvey, C. A. Johnston, E. Mayorga, W. H. McDowell, and G. Pinay. 2003. Biogeochemical hot spots and hot moments at the interface of terrestrial and aquatic ecosystems. *Ecosystems* 6:301–312.

McCleery, R. A., C. E. Moorman, and M. N. Peterson. 2015. Introduction. Pages 1–10 *in* R. A. McCleery, C. E. Moorman, and M. N. Peterson, editors. *Urban Wildlife Conservation: Theory and Practice*. New York, NY: Springer.

McCleery, R. A., C. E. Moorman, M. C. Wallace, and D. Drake. 2012. Managing urban environments for wildlife. Pages 169–191 *in* N. J. Silvy, editor. *The Wildlife Techniques Manual: Management*. The Wildlife Society, Bethesda, MD.

McConville, D. J., S. B. Jack, and R. J. Mitchell. 1999. A quantitative assessment of slash pine crown structure in trees marked for gap creation. Pages 539–541 *in* J. D. Haywood, editor. *Proceedings of the Tenth Biennial Southern Silvicultural Research Conference*, 16–18 February 1999, Shreveport, Louisiana, USA. General Technical Report SRS-30. USDA Forest Service, Southern Research Station. Asheville, NC.

McCormick, P. V. and L. C. Baron. 2015. Effects of reduced summertime stream flows on instream habitat in the lower Flint River Basin, Georgia, USA. *Proceedings of the 2015 Georgia Water Resources Conference*, 28–29 April 2015, Athens, GA. http://gwri.gatech.edu/sites/default/files/files/docs/2015/6.1.2mccormick.pdf (accessed 4 February 2017).

McDonald-Madden, E., P. W. J. Baxter, R. A. Fuller, T. G. Martin, E. T. Game, J. Montambault, and H. P. Possingham. 2010. Monitoring does not always count. *Trends in Ecology and Evolution* 25:547–550.

McDowell, N. G. and C. D. Allen. 2015. Darcy's law predicts widespread forest mortality under climate warming. *Nature Climate Change* 5:669–672.

McEachern, K. and R. Sutter. 2010. Assessment of eleven years of rare plant monitoring data from the San Diego Multiple Species Conservation Plan. Administrative Report 2010-01. USGS Western Ecological Research Center, Channel Islands Field Station, Ventura, CA.

McGill, B. J., B. A. Maurer, and M. D. Weiser. 2006. Empirical evaluation of neutral theory. *Ecology* 87:1411–1423.

McGraw, R. L., F. W. Shockley, J. F. Thompson, and C. A. Roberts. 2004. Evaluation of native legume species for forage yield, quality, and seed production. *Native Plants Journal* 5:152–159.

McGuire, J. P., R. J. Mitchell, E. B. Moser, S. D. Pecot, D. H. Gjerstad, and C. W. Hedman. 2001. Gaps in a gappy forest: Plant resources, longleaf pine regeneration, and understory response to tree removal in longleaf pine savannas. *Canadian Journal of Forest Research* 31:765–778.

McGuire, J. L., L. L. Smith, C. Guyer, J. M. Lockhart, G. W. Lee, and M. J. Yabsley. 2014. Surveillance for upper respiratory tract disease and Mycoplasma in free-ranging gopher tortoises (*Gopherus polyphemus*) in Georgia, USA. *Journal of Wildlife Diseases* 50:733–744.

McIntyre, R. K. 2012. Longleaf pine restoration assessment: Conservation outcomes and performance metrics. Final Report to National Fish and Wildlife Foundation, Contract nr NFWF 1903.12.034011. Joseph W. Jones Ecological Research Center, Newton, GA.

McIntyre, P. B., A. S. Flecker, M. J. Vanni, J. M. Hood, B. W. Taylor, and S. A. Thomas. 2008. Fish distributions and nutrient cycling in streams: Can fish create biogeochemical hotspots? Ecology 89:2335–2346.

McIntyre, R. K., S. B. Jack, B. B. McCall, and R. J. Mitchell. 2010. Financial feasibility of selection-based multiple-value management on private lands in the South: A heuristic case study approach. *Journal of Forestry* 108:230–237.

McIntyre, R. K., S. B. Jack, R. J. Mitchell, J. K. Hiers, and W. L. Neel. 2008. *Multiple Value Management: The Stoddard-Neel Approach to Ecological Forestry in Longleaf Pine Grasslands*. Miscellaneous Publication. Newton, GA: Joseph W. Jones Ecological Research Center.

McIntyre, R. K. and B. B. McCall. 2014. Longleaf pine economics: Large-acreage landowner perspectives and opportunities. Report to National Fish and Wildlife Foundation, Contract nr NFWF 1903.13.040686. Joseph W. Jones Ecological Research Center, Newton, GA.

McKay, J. K., C. E. Christian, S. Harrison, and K. J. Rice. 2005. "How local is local?"—a review of practical and conceptual issues in the genetics of restoration. *Restoration Ecology* 13:432–440.

McKee, A. M. 2012. Correlations between community diversity and within-species genetic diversity in an amphibian assemblage: Potential processes and implications for conservation management, PhD dissertation. University of Georgia, Athens, GA.

McLachlan, J. S., J. J. Hellman, and M. W. Schwartz. 2007. A framework for debate of assisted migration in an era of climate change. *Conservation Biology* 21:297–302.

McLauchlan, K. 2006. The nature and longevity of agricultural impacts on soil carbon and nutrients: A review. *Ecosystems* 9:1364–1382.

McLauchlan, K. K., S. E. Hobbie, and W. M. Post. 2006. Conversion from agriculture to grassland builds soil organic matter on decadal timescales. *Ecological Applications* 16:143–153.

McLaughlin, D. L., D. A. Kaplan, and M. J. Cohen. 2013. Managing forests for increased regional water yield in the southeastern U.S. coastal plain. *Journal of the American Water Resources Association* 49:953–965.

McNab, W. H., M. B. Edwards, and W. A. Hough. 1978. Estimating fuel weights in slash pine-palmetto stands. *Forest Science* 24:345–358.

McNair, D. B. 1998. Henslow's sparrow and sedge wren response to a dormant-season prescribed burn in a pine savanna. *Florida Field Naturalist* 26:46–47.

McVey, J. M., D. T. Cobb, R. A. Powell, M. K. Stoskpof, J. H. Bohling, L. P. Waits, and C. E. Moorman. 2013. Diets of sympatric red wolves and coyotes in northeastern North Carolina. *Journal of Mammalogy* 94:1141–1148.

Means, D. B. 1985. Radio-tracking the eastern diamondback rattlesnake. *National Geographic Society Research Reports* 18:529–536.

Means, D. B. 1996. Longleaf pine forest, going, going… Pages 210–219 *in* M.B. Davis, editor. *Eastern Old-Growth Forests: Prospects for Rediscovery and Recovery*. Washington, DC: Island Press.

Means, D. B. 1997. Wiregrass restoration: Probable shading effects in a slash pine plantation. *Restoration and Management Notes* 15:52–54.

Means, D. B. 2005. The value of dead trees bases and stumpholes as habitat for wildlife. Pages 74–78 *in* W. E. Meshaka, and K. J. Babbitt, editors. *Amphibians and Reptiles: Status and Conservation in Florida*. Malabar, FL: Kreiger Publishing Company.

Means, D. B. 2006. Vertebrate faunal diversity of longleaf pine ecosystems. Pages 157–213 *in* S. Jose, E. J. Jokela, and D. L. Miller, editors. *The Longleaf Pine Ecosystem: Ecology, Silviculture, and Restoration*. New York, NY: Springer.

Means, R. C., R. P. M. Means, M. Beshel, R. Mendyk, S. Reichling, and B. Summerford. 2015. A conservation strategy for the imperiled striped newt (*Notophthalmus perstriatus*) in the Apalachicola National Forest, Florida. Fifth Annual Report, USDA Forest Service.

Means, D. B., J. G. Palis, and M. Baggett. 1996. Effects of slash pine silviculture on a Florida population of flatwoods salamander. *Conservation Biology* 10:426–437.

Medley, K. S., E. H. Boughton, D. G. Jenkins, J. E. Fauth, P. J. Bohlen, and P. F. Quintana-Ascencio. 2015. Intense ranchland management tips the balance of regional and local factors affecting wetland community structure. *Agriculture, Ecosystems and Environment* 212:207–244.

Meehan, K. and P. G. Jodice. 2010. Landscape scale correlates of fox squirrel presence on golf courses in coastal South Carolina. *Southeastern Naturalist* 9:573–586.

Meeks. 2016. Meeks farm and nursery website. http://www.meeksfarms-nurserys.com/productsservices.html (accessed 4 February 2017).

Meentemeyer, V. 1978. Macroclimate and lignin control of litter decomposition rates. *Ecology* 59:465–472.

Meier, E. 2016. The wood database. http://www.wood-database.com/ (accessed 4 February 2017).

Meldahl, R. S., N. Pederson, J. S. Kush, and J. M. Varner. 1999. Dendrochronological investigations of climate and competitive effects on longleaf pine growth. Pages 265–285 *in* R. Wimmer, and R. E. Vetter, editors. *Tree Ring Analysis: Biological, Methodological and Environmental Aspects*. Wallingford: CABI Publishing.

Meléndez, J., A. Foronda, J. Aranda, F. Lopez, and F. L. Del Cerro. 2010. Infrared thermography of solid surfaces in a fire. *Measurement Science and Technology* 21:105504.

Melville, H. I. A. S, W. C. Conway, M. L. Morrison, C. E Comer, and J. B. Hardin. 2014. Artificial nests identify possible predators of eastern wild turkeys. *Southeastern Naturalist* 13:80–91.

Melvin, M. 2012. National prescribed fire use survey report. Technical Report 01-12. Newton, GA: Coalition of Prescribed Fire Councils.

Melvin, M. 2015. National prescribed fire use survey report. Technical Report 02–15. Newton, GA: Coalition of Prescribed Fire Councils.

Menges, E. S. 1999. Ecology and conservation of Florida scrubs. Pages 7–22 *in* R. C. Anderson, J. S. Fralish, and J. M. Baskin, editors. *Savannas, Barrens, and Rock Outcrop Plant Communities of North America*. Cambridge: Cambridge University Press.

Menges, E. S. and D. R. Gordon. 2010. Should mechanical treatments and herbicides be used as fire surrogates to manage Florida's uplands? A review. *Florida Scientist* 73:147–174.

Menges, E. S. and P. L. Marks. 2008. Fire and flood: Why are south-central Florida seasonal ponds treeless? *American Midland Naturalist* 159:8–20.

Menzel, J. M., M. A. Menzel, J. C. Kilgo, W. M. Ford, and J. W. Edwards. 2005. Bat response to Carolina bays and wetland restoration in the southeastern U.S. coastal plain. *Wetlands* 25:542–550.

Merschel, A. G., T. A. Spies, and E. K. Heyerdahl. 2014. Mixed-conifer forests of central Oregon: Effects of logging and fire exclusion vary with environment. *Ecological Applications* 24:1670–1688.

Michener, W. K., J. W. Brunt, J. J. Helly, T. B. Kirchner, and S. G. Stafford. 1997. Nongeospatial metadata for the ecological sciences. *Ecological Applications* 7:330–342.

Miles, P. D. 2016. Forest Inventory EVALIDator web-application, Version 1.6.0.03. USDA Forest Service, Northern Research Station, St. Paul, MN. http://apps.fs.fed.us/Evalidator/evalidator.jsp (accessed 4 February 2017).

Miles, A. C., S. B. Castleberry, D. A. Miller, and L. M. Conner. 2006. Multi-scale roost-site selection by evening bats on pine-dominated landscapes in southwest Georgia. *Journal of Wildlife Management* 70:1191–1199.

Millar, C. I. 1998. Early evolution of pines. Pages 69–91 *in* D. M. Richardson, editor. *Ecology and Biogeography of Pinus*. Cambridge: Cambridge University Press.

Miller, S. A., A. Bartow, M. Gisler, K. Ward, A. S. Young, and T. N. Kaye. 2011. Can an ecoregion serve as a seed transfer zone? Evidence from a common garden study with five native species. *Restoration Ecology* 19:268–276.

Miller, E. A. and C. B. Halpern. 1998. Effects of environment and grazing disturbance on tree establishment in meadows of the Central Cascade range, Oregon, USA. *Journal of Vegetation Science* 9:265–282.

Miller, J. H. and K. V. Miller. 1999. *Forest Plants of the Southeast and Their Wildlife Uses*. Auburn, AL: Southern Weed Science Society.

Miller, D. R., C. P. Quine, and W. Hadley. 2000. An investigation of the potential of digital photogrammetry to provide measurements of forest characteristics and abiotic damage. *Forest Ecology and Management* 135:279–288.

Miller, J. D., H. D. Safford, M. Crimmins, and A. E. Thode. 2008. Quantitative evidence for increasing forest fire severity in the Sierra Nevada and southern Cascade Mountains, California and Nevada, USA. *Ecosystems* 12:16–32.

Miller, G. J., L. L. Smith, S. A. Johnson, and R. Franz. 2012. Home range size and habitat selection in the Florida pine snake (*Pituophis melanoleucus mugitus*). *Copeia* 2012:706–713.

Miller, S. D. and D. W. Speake. 1978. Prey utilization on quail plantations in southern Alabama. *Proceedings of the Annual Conference of the Southeastern Association of Fish and Wildlife Agencies* 32:100–111.

Millar, C. I., N. L. Stephenson, and S. L. Stephens. 2007. Climate change and forests of the future: Managing in the face of uncertainty. *Ecological Applications* 17:2145–2151.

Miller, C. and D. L. Urban. 2000. Connectivity of forest fuels and surface fire regimes. *Landscape Ecology* 15:145–154.

Mills, S. D. and C. T. Stiff. 2008. *Financial Performance of Loblolly and Longleaf Pine Plantations*. Summerville, SC: FORSight Resources, LLC. http://www.forsightresources.com/library/MillsStiff2008.pdf (accessed 4 February 2017).

Mills, S. D. and C. T. Stiff. 2013. Financial performance of loblolly and longleaf pine plantations. Pages 183–191 *in* J. M. Guldin, editor. *Proceedings of the 15th Biennial Southern Silvicultural Research Conference*. e-General Technical Report SRS-GTR-175. USDA Forest Service, Southern Research Station, Asheville, NC.

Minogue, P., D. J. Moorhead, and E. D. Dickens. 2012 (Revised). Herbaceous weed control recommendations for planted longleaf sites. http://www.bugwood.org/HWC_longleaf_March_2012.pdf. (accessed 4 February 2017).

Mississippi State University Extension Service. 2016. Prescribed burning in southern pine forests: Fire ecology, techniques, and uses for wildlife management. http://extension.msstate.edu/publications/publications/prescribed-burning-southern-pine-forests-fire-ecology-techniques-and-uses (accessed 4 February 2017).

Mitchell, R., T. Engstrom, R. R. Sharitz, D. De Steven, K. Hiers, and L. K. Kirkman. 2009. Old forests and endangered woodpeckers: Old-growth in the southern coastal plain. *Natural Areas Journal* 29:301–310.

Mitchell, R. J., J. K. Hiers, J. J. O'Brien, S. B. Jack, and R. T. Engstrom. 2006. Silviculture that sustains: The nexus between silviculture, frequent prescribed fire, and conservation of biodiversity in longleaf pine forests of the southeastern United States. Canadian Journal of Forest Research 36:2724–2736.

Mitchell, R. J., J. K. Hiers, J. J. O'Brien, and G. Starr. 2009. Ecological forestry in the Southeast: Understanding the ecology of fuels. *Journal of Forestry* 107:391–397.

Mitchell, R. J., L. K. Kirkman, S. D. Pecot, C. A. Wilson, B. J. Palik, and L. R. Boring. 1999. Patterns and controls of ecosystem function in longleaf pine-wiregrass savannas. I. Aboveground net primary productivity. *Canadian Journal of Forest Research* 29:743–751.

Mitchell, R. J., Y. Liu, J. J. O'Brien, K. J. Elliott, G. Starr, C. F. Miniat, and J. K. Hiers. 2014. Future climate and fire interactions in the southeastern region of the United States. *Forest Ecology and Management* 327:316–326.

Mittelbach, G. G. and D. W. Schemske. 2015. Ecological and evolutionary perspectives on community assembly. *Trends in Ecology and Evolution* 30:241–247.

Mobley, H. E. 1989. Summary of smoke-related accidents in the south from prescribed fire (1979–1988). Technical Release 90-R-11. American Pulpwood Association, Rockville, MD.

Mohr, C. 1896. *The Timber Pines of the Southern United States*. Bulletin No. 13. Washington, DC: USDA Division of Forestry.

Mola, J. M., J. M. Varner, E. S. Jules, and T. Spector. 2014. Altered community flammability in Florida's Apalachicola ravines and implications for the persistence of the endangered conifer *Torreya taxifolia*. *PLOS ONE* 9:e103933.

Moler, P. E. 1992. *Rare and Endangered Biota of Florida: Amphibians and Reptiles*. Gainesville, FL: University Press of Florida.

Moler, P. E. and R. Franz. 1987. Wildlife values of small isolated wetlands in the southeastern coastal plain. Pages 234–241 *in* R. R. Odom, K. A. Riddleberger, and J. C. Ozier, editors. *Third Southeastern Nongame and Endangered Wildlife Symposium*, 8–10 August 1987, Athens, Georgia, USA. Georgia Department of Natural Resources, Game and Fish Division, Athens, GA.

Monk, C. D. 1968. Successional and environmental relationships of the forest vegetation of north central Florida. *American Midland Naturalist* 79:441–457.

Monleon, V. J., K. Cromack Jr., and J. D. Landsberg. 1997. Short- and long-term effects of prescribed underburning on N availability in ponderosa pine stands in central Oregon. *Canadian Journal of Forest Research* 27:369–378.

Montalvo, A. M. 1994. Inbreeding depression and maternal effects in *Aquilegia caerulea*, a partially selfing plant. *Ecology* 75:2395–2409.

Montambault, J. R., S. Wongbusarakum, T. Leberer, E. Joseph, W. Andrew, F. Castro, B. Nevitt, Y. Golbuu, N. W. Oldiais, C. R. Groves, and W. Kostka. 2015. Use of monitoring data to support conservation management and policy decisions in Micronesia. *Conservation Biology* 29:1279–1289.

Mooney, H. A. and S. L. Gulmon. 1982. Constraints on leaf structure and function in relation to herbivory. *Bioscience* 32:198–206.

Moore, P. D. 1990. Soils and ecology: Temperate wetlands. Pages 95–114 *in* M. Williams, editor. *Wetlands: A Threatened Landscape*. Oxford: Basil Blackwell.

Moore, M. M., C. A. Casey, J. D. Bakker, J. D. Springer, P. Z. Fule, W. W. Covington, and D. C. Laughlin. 2006. Herbaceous vegetation responses (1992–2004) to restoration treatments in a ponderosa pine forest. *Rangeland Ecology and Management* 59:135–144.

Morgan, P., E. K. Heyerdahl, and C. E. Gibson. 2008. Multi-season climate synchronized forest fires throughout the 20th century, northern Rockies, USA. *Ecology* 89:717–728.

Morris, G., J. A. Hostetler, L. M. Conner, and M. K. Oli. 2011. Effects of prescribed fire, supplemental feeding, and mammalian predator exclusion on hispid cotton rat populations. *Oecologia* 167:1005–1016.

Morris, G., J. A. Hostetler, M. K. Oli, and L. M. Conner. 2011. Effects of predation, fire, and supplemental feeding on populations of two species of *Peromyscus* mice. *Journal of Mammalogy* 92:934–944.

Morris, L. A., E. J. Jokela, and J. B. O'Connor. 1992. Silvicultural guidelines for pinestraw management in the southeastern United States. Georgia Forest Research Paper No. 88. Research Division, Georgia Forestry Commission.

Moser, W. K. 2006. The Stoddard-Neel system: Case studies. Pages 246–249 *in* S. Jose, E. J. Jokela, and D. L. Miller, editors. *The Longleaf Pine Ecosystem: Ecology, Silviculture, and Restoration*. New York, NY: Springer Science+Business Media, LLC.

Moser, W. K., S. M. Jackson, V. Podrazsky, and D. R. Larsen. 2002. Examination of stand structure on quail plantations in the Red Hills region of Georgia and Florida managed by the Stoddard-Neel system: An example for forest managers. *Forestry* 75:443–449.

Mount, R. H. 1963. The natural history of the red-tailed skink, *Eumeces egregius* Baird. *American Midland Naturalist* 70:356–385.

Mouquet, N. and M. Loreau. 2003. Community patterns in source-sink metacommunities. *American Naturalist* 162:544–557.

Muir, J. 1894. *The Mountains of California*. New York, NY: Penguin Books (original publisher: Century).

Mulligan, M. K. and L. K. Kirkman. 2002. Competition effects on wiregrass (*Aristida beyrichiana*) growth and survival. *Plant Ecology* 163:39–50

Mulligan, M. K., L. K. Kirkman, and R. J. Mitchell. 2002. *Aristida beyrichiana* (wiregrass) establishment and recruitment: Implications for restoration. *Restoration Ecology* 10:68–72.

Murphy, B. P. and D. M. J. S. Bowman. 2012. What controls the distribution of tropical forest and savanna? *Ecology Letters* 15:748–758.

Myers, R. L. 1990. Scrub and high pine. Pages 150–193 *in* R. L. Myers, and J. J. Ewel, editors. *Ecosystems of Florida*. Orlando, FL: University of Central Florida Press.

Myers, J. A. 2010. Ecological assembly of high-diversity plant communities: Dispersal, competition, and environmental filtering in longleaf pine savannas, PhD dissertation. Louisiana State University, Baton Rouge, LA.

Myers, J. A., J. M. Chase, R. M. Crandall, and I. Jimenez. 2015. Disturbance alters beta-diversity but not the relative importance of community assembly mechanisms. *Journal of Ecology* 103:1291–1299.

Myers, J. A., J. M. Chase, I. Jimenez, P. M. Jørgense, A. Araujo-Murakami, N. Paniagua-Zambrana, and R. Seidel. 2013. Beta-diversity in temperate and tropical forests reflects dissimilar mechanisms of community assembly. *Ecology Letters* 16:151–157.

Myers, J. A. and K. E. Harms. 2009a. Local immigration, competition from dominant guilds, and the ecological assembly of high-diversity pine savannas. *Ecology* 90:2745–2754.

Myers, J. A. and K. E. Harms. 2009b. Seed arrival, ecological filters, and plant species richness: A meta-analysis. *Ecology Letters* 12:1250–1260.

Myers, J. A. and K. E. Harms. 2011. Seed arrival and ecological filters interact to assemble high-diversity plant communities. *Ecology* 92:676–686.

Myers, R. L., J. O'Brien, and S. Morrison. 2006. A general description of fire management in Caribbean pine (*Pinus caribaea*) savannas of the Mosquitia, Honduras. Technical Report 2006-la. The Nature Conservancy, Global Fire Initiative, Arlington, VA.

Myers, R. L. and D. A. Rodríguez-Trejo. 2009. Fire in tropical pine ecosystems. Pages 557–605 *in* M. A. Cochrane, editor. *Tropical Fire Ecology: Climate Change, Land Use, and Ecosystem Dynamics.* New York, NY: Springer.

Myers, R. K. and D. H. Van Lear. 1998. Hurricane-fire interactions in coastal forests of the south: A review and hypothesis. *Forest Ecology and Management* 103:265–276.

Naesset, E. 2002. Predicting forest stand characteristics with airborne scanning laser using a practical two-stage procedure and field data. *Remote Sensing of Environment* 80:88–99.

Naficy, C., A. Sala, E. G. Keeling, J. Graham, and T. H. deLuca. 2010. Interactive effects of historical logging and fire exclusion on ponderosa pine forest structure in the northern Rockies. *Ecological Applications* 20:1851–1864.

Nathan, R. and H. C. Muller-Landau. 2000. Spatial patterns of seed dispersal, their determinants and consequences for recruitment. *Trends in Ecology and Evolution* 15:278–285.

National Oceanic and Atmospheric Administration (NOAA)/ National Weather Service. 2015. Lightning strike density product description. http://www.opc.ncep.noaa.gov/lightning/lightning_pdd.php (accessed 4 February 2017).

National Wildfire Coordinating Group (NWCG). 2015. *Interagency Aerial Ignition Guide.* PMS 501. Boise, ID: National Wildfire Coordinating Group.

Neale, J. C. and B. N. Sacks. 2001. Food habits and space use of gray foxes in relation to sympatric coyotes and bobcats. *Canadian Journal of Zoology* 79:1794–1800.

Neel, L., P. S. Sutter, and A. G. Way. 2010. *The Art of Managing Longleaf: A Personal History of the Stoddard-Neel Approach.* Athens, GA: University of Georgia Press.

Nelson, A. S. and D. C. Bragg. 2016. Multidecadal response of naturally regenerated southern pine to early competition control and commercial thinning. *Forest Science* 62:115–124.

Nelson, L. R. and R. L. Cantrell. 2002. *Herbicide Prescription Manual for Southern Pine Management.* Publication EC-659. Clemson, SC: Clemson University Extension and American Forest & Paper Association, Clemson University.

Nelson, M. A., M. J. Cherry, M. B. Howze, R. J. Warren, and L. M. Conner. 2015. Coyote and bobcat predation on white-tailed deer fawns in a longleaf pine ecosystem in southwestern Georgia. *Journal of the Southeastern Association of Fish and Wildlife Agencies* 2:208–213.

Nelson, M. D., G. C. Liknes, and B. J. Butler. 2010. *Forest Ownership in the Conterminous United States: ForestOwn_v1 Geospatial Dataset.* Newtown Square, PA: USDA Forest Service, Northern Research Station. doi:10.2737/RDS-2010-0002 (accessed 3 February 2017).

Nelson, R. M., Jr. and J. K. Hiers. 2008. The influence of fuelbed properties on moisture drying rates and timelags of longleaf pine litter. *Canadian Journal of Forest Research* 38:2394–2404.

Nestler, R. B. 1949. Nutrition of bobwhite quail. *Journal of Wildlife Management* 13:342–358.

Newman, D. H. and D. N. Wear. 1993. Production economics of private forestry: A comparison of industrial and nonindustrial forest owners. *American Journal of Agricultural Economics* 75:674–684.

Newsome, T. M. and W. J. Ripple. 2015. A continental scale trophic cascade from wolves through coyotes to foxes. *Journal of Animal Ecology* 84:49–59.

Nichols, V., N. Verhulst, R. Cox, and B. Govaerts. 2015. Weed dynamics and conservation agriculture principles: A review. *Field Crops Research* 183:56–68.

Nichols, J. D. and B. K. Williams. 2006. Monitoring for conservation. *Trends in Ecology and Evolution* 21:668–673.

Niemi, G. J., J. M. Hanowski, A. R. Lima, T. Nicholls, and N. Weiland. 1997. A critical analysis on the use of indicator species in management. *Journal of Wildlife Management* 61:1240–1252.

Niemi, G. J. and M. E. McDonald. 2004. Application of ecological indicators. *Annual Review of Ecology, Evolution, and Systematics* 35:89–111.

Nieminen, M. F., D. L. Evans, J. M. Varner, S. D. Roberts, H. A. Londo, and S. A. Tweddale. 2014. Spectral separability of longleaf and loblolly pines in southeast Mississippi: Preliminary results. *Proceedings of the Ninth Southern Forestry and Natural Resource Management GIS Conference*, 8–10 December 2013, Athens, GA. http://www.soforgis.net/2013/files/Nieminen_etal_SOFORGIS_2014.pdf (accessed 4 February 2017).

Niinemets, U. and F. Valladares. 2006. Tolerance to shade, drought, and waterlogging of temperate northern hemisphere trees and shrub. *Ecological Monographs* 76:521–547.

Noel, J. M., W. J. Platt, and E. B. Moser. 1998. Structural characteristics of old-and second-growth stands of longleaf pine (*Pinus palustris*) in the Gulf Coastal Region of the USA. *Conservation Biology* 12:533–548.

Norden, A. H. and L. K. Kirkman. 2004. Persistence and prolonged winter dormancy of *Schwalbea americana* L. (Scrophulariaceae) following experimental management techniques. *Natural Areas Journal* 24:129–134.

Nordman, C., R. White, R. Wilson, C. Ware, C. Rideout, M. Pyne, and C. Hunter. 2016. Rapid assessment metrics to enhance wildlife habitat and biodiversity within southern open pine ecosystems, version 1.0. U.S. Fish and Wildlife Service and NatureServe, for the Gulf Coastal Plains and Ozarks Landscape Conservation Cooperative. http://www.lmvjv.org/library/OpenPineMetrics-FinalReport_12May16.pdf (accessed 4 February 2017).

North, M., B. M. Collins, and S. Stephens. 2012. Using fire to increase the scale, benefits, and future maintenance of fuels treatments. *Journal of Forestry* 110:392–401.

North, M., M. Hurteau, R. Fiegener, and M. G. Barbour. 2005. Influence of fire and El Niño on tree recruitment varies by species in Sierran mixed conifer. *Forest Science* 51:187–197.

North, M., J. Innes, and H. Zald. 2007. Comparison of thinning and prescribed fire restoration treatments to Sierran mixed-conifer historic conditions. *Canadian Journal of Forest Research* 37:331–342.

Noss, R. F. 1989. Longleaf pine and wiregrass: Keystone components of an endangered ecosystem. *Natural Areas Journal* 9:234–235.

Noss, R. F. 2013. *Forgotten Grasslands of the South: Natural History and Conservation*. Washington, DC: Island Press.

Noss, R. F., J. F. Franklin, W. L. Baker, T. Schoennagel, and P. B. Moyle. 2006. Managing fire-prone forests in the western United States. *Frontiers in Ecology and the Environment* 4:481–487.

Noss, R. F., E. T. LaRoe III, and J. M. Scott. 1995. *Endangered Ecosystems of the United States: A Preliminary Assessment of Loss and Degradation*. Washington, DC: U.S. Department of the Interior, National Biological Service.

Noss, R. F., W. J. Platt, B. A. Sorrie, A. S. Weakley, D. B. Means, J. Costanza, and R. K. Peet. 2015. How global biodiversity hotspots may go unrecognized: Lessons from the North American coastal plain. *Diversity and Distributions* 21:236–244.

Novick, K. A., A. C. Oishi, E. J. Ward, M. B. S. Siqueira, J. Y. Juang, and P. C. Stoy. 2015. On the difference in the net ecosystem exchange of CO_2 between deciduous and evergreen forests in the southeastern United States. *Global Change Biology* 21:827–842.

Nowacki, G. J. and M. D. Abrams. 2008. The demise of fire and the "mesophication" of forests in the Eastern United States. *BioScience* 58:123–128.

Nyland, R. D. 1996. *Silviculture: Concepts and Applications*. New York, NY: McGraw-Hill Companies.

O'Brien, J. J., J. K. Hiers, M. A. Callaham, Jr., R. J. Mitchell, and S. B. Jack. 2008. Interactions among overstory structure, seedling life-history traits, and fire in frequently burned neotropical pine forests. *Ambio* 37:542–547.

O'Brien, J. J., J. K. Hiers, R. J. Mitchell, J. M. Varner III, and K. Mordecai. 2010. Acute physiological stress and mortality following fire in a long-unburned longleaf pine ecosystem. *Fire Ecology* 6:1–12. doi:4996/fireecology.0602001.

O'Brien, J. J., E. L. Loudermilk, J. K. Hiers, S. Pokswinski, B. Hornsby, A. Hudak, D. Strother, E. Rowell, and B. C. Bright. 2016. Canopy-derived fuels drive patterns of in-fire energy release and understory plant mortality in a longleaf pine (*Pinus palustris*) sandhill in northwest Florida, USA. *Canadian Journal of Remote Sensing* 42:489–500. doi:10.1080/07038992.2016.1199271.

O'Brien, J. J., E. L. Loudermilk, B. Hornsby, A. T. Hudak, B. C. Bright, M. B. Dickinson, J. K. Hiers, C. Teske, and R. D. Ottmar. 2016. High-resolution infrared thermography for capturing wildland fire behaviour: RxCADRE 2012. *International Journal of Wildland Fire* 25:62–75.

Odion, D. C., C. T. Hanson, A. Arsenault, W. L. Baker, D. A. DellaSala, R. L. Hutto, W. Klenner, M. A. Moritz, R. L. Sherriff, T. T. Veblen, and M. A. Williams. 2014. Examining historical and current mixed-severity fire regimes in ponderosa pine and mixed-conifer forests of western North America. *PLOS ONE* 9:e87852.

Office of the Chief Economist (OCE), United States Department of Agriculture. 2015. Major World Crop Areas, North America. USDA Office of the Chief Economist, Washington, DC. http://www.usda.gov/oce/weather/pubs/Other/MWCACP/namerica.htm (accessed 4 February 2017).

O'Hara, K. L. 2009. Multiaged silviculture in North America. *Journal of Forest Science* 55:432–436.

O'Hara, K. L. 2014. *Multiaged Silviculture: Managing for Complex Forest Stand Structures*. Oxford: Oxford University Press.

Oishi, A. C., R. Oren, K. A. Novick, S. Palmroth, and G. G. Katul. 2010. Interannual invariability of forest evapotranspiration and its consequences to water flow downstream. *Ecosystems* 13:421–436.

Ojima, D. S., D. S. Schimel, W. J. Parton, and C. E. Owensby. 1994. Long- and short-term effects of fire on nitrogen cycling in tallgrass prairie. *Biogeochemistry* 24:67–84.

Olff, H. and M. E. Ritchie. 1998. Effects of herbivores on grassland plant diversity. *Trends in Ecology and Evolution* 13:261–265.

Olson, J. S. 1963. Energy storage and the balance of producers and decomposers in ecological systems. *Ecology* 44:322–331.

Olson, M. S. and W. J. Platt. 1995. Effects of habitat and growing season fires on resprouting of shrubs in longleaf pine savannas. *Vegetatio* 119:101–118.

Olsoy, P. J., N. F. Glenn, P. E. Clark, and D. R. Derryberry. 2014. Aboveground total and green biomass of dryland shrub derived from terrestrial laser scanning. *ISPRS Journal of Photogrammetry and Remote Sensing* 88:166–173.

Omernik, J. M. 1987. Ecoregions of the conterminous United States. *Annals of the Association of American Geographers* 77:118–125.

O'Neill, S., P. Lahm, and A. Mathews. 2011. Basic smoke management practices. USDA Forest Service and Natural Resources Conservation Service. http://www.nrcs.usda.gov/Internet/FSE_DOCUMENTS/stelprdb1046311.pdf (accessed 4 February 2017).

Opsahl, S. P. 2005. Organic carbon composition and oxygen metabolism across a gradient of seasonally inundated limesink and riparian wetlands in the southeast coastal plain, USA. *Biogeochemistry* 76:47–68.

Orchard, V. A. and F. Cook. 1983. Relationship between soil respiration and soil moisture. *Soil Biology and Biochemistry* 15:447–453.

Ostertag, T. E. and K. M. Robertson. 2007. A comparison of native versus old-field vegetation in upland pinelands managed with frequent fire, South Georgia, USA. Pages 109–120 *in* R. E. Masters, and K. E. M. Galley, editors. *Fire in Grassland and Shrubland Ecosystems. Proceedings of the 23rd Tall Timbers Fire Ecology Conference*. Tall Timbers Research Station, Tallahassee, FL.

Ostfeld, R. S. and R. D. Holt. 2004. Are predators good for your health? Evaluating evidence for top-down regulation of zoonotic disease reservoirs. *Frontiers in Ecology and the Environment* 2:13–20.

Oswalt, C. M., J. A. Cooper, D. G. Brockway, H. W. Brooks, J. L. Walker, K. F. Connor, S. N. Oswalt, and R. C. Conner. 2012. History and current condition of longleaf pine in the Southern United States. General Technical Report SRS-166. USDA Forest Service, Southern Research Station, Asheville, NC.

Oswalt, S. N., W. B. Smith, P. D. Miles, and S. A. Pugh. 2014. Forest Resources of the United States, 2012: A technical document supporting the Forest Service 2010 update of the RPA Assessment. General Technical Report WO-91. USDA Forest Service, Washington, DC.

Ottmar, R. D. 2014. Wildland fire emissions, carbon, and climate: Modeling fuel consumption. *Forest Ecology and Management* 317:41–50.

Ottmar, R. D., J. K. Hiers, B. W. Butler, C. B. Clements, M. B. Dickinson, A. T. Hudak, J. J. O'Brien, B. E. Potter, E. M. Rowell, T. M. Strand, and T. J. Zajkowski. 2016. Measurements, datasets and preliminary results from the RxCADRE project-2008, 2011 and 2012. *International Journal of Wildland Fire* 25:1–9.

Ottmar, R. D., A. T. Hudak, S. J. Prichard, C. S. Wright, J. C. Restaino, M. C. Kennedy, and R. E. Vihnanek. 2016. Pre-fire and post-fire surface fuel and cover measurements collected in the south-eastern United States for model evaluation and development-RxCADRE 2008, 2011 and 2012. *International Journal of Wildland Fire* 25:10–24.

Ottmar, R. D., A. I. Miranda, and D. V. Sandberg. 2008. Characterizing sources of emissions from wildland fires. *Developments in Environmental Science* 8:61–78.

Ottmar, R. D., D. V. Sandberg, C. L. Riccardi, and S. J. Prichard. 2007. An overview of the Fuel Characteristic Classification System-quantifying, classifying, and creating fuelbeds for resource planning. *Canadian Journal of Forest Research* 37:2383–2393.

Ottmar, R. D., R. E. Vihnanek, and J. W. Mathey. 2003. *Stereo Photo Series for Quantifying Natural Fuels*. Volume VIa: Sand hill, sand pine scrub, and hardwood with white pine types in the Southeast United States with supplemental sites for Volume VI. PMS 838, National Wildfire Coordinating Group, National Interagency Fire Center, Boise, ID.

Outcalt, K. W. 1992. Factors affecting wiregrass (*Aristida stricta* Michx.) cover on uncut and site prepared sandhills areas in central Florida. *Ecological Engineering* 1:245–251.

Outcalt, K. W. 1993. Southern pines performance on sandhills sites in Georgia and South Carolina. *Southern Journal of Applied Forestry* 17:100–102.

Outcalt, K. W. 1994. Seed production of wiregrass in central Florida following growing-season prescribed burns. *International Journal of Wildland Fire* 4:123–125.

Outcalt, K. W. 2000. Occurrence of fire in longleaf pine stands in the southeastern United States. Pages 178–182 *in* W. K. Moser, and C. E. Moser, editors. *Fire and forest ecology: Innovative silviculture and vegetation management. Proceedings of the 21st Tall Timbers Fire Ecology Conference*. Tall Timbers Research Station, Tallahassee, FL.

Outcalt, K. W. 2008. Lightning, fire and longleaf pine: Using natural disturbance to guide management. *Forest Ecology and Management* 255:3351–3359.

Outcalt, K. W. and D. G. Brockway. 2010. Structure and composition changes following restoration treatments of longleaf pine forests on the Gulf coastal plain of Alabama. *Forest Ecology and Management* 259:1615–1623.

Outcalt, K. W. and R. M. Sheffield. 1996. *The Longleaf Pine Forest: Trends and Current Conditions*. Resource Bulletin SRS–9. Asheville, NC: USDA Forest Service, Southern Research Station.

Outcalt, K. W., M. E. Williams, and O. Onokpise. 1999. Restoring *Aristida stricta* to *Pinus palustris* ecosystems on the Atlantic coastal plain, USA. *Restoration Ecology* 7:262–270.

Pachauri, R. K., M. R. Allen, V. R. Barros, J. Broome, W. Cramer, R. Christ, J. A. Church, L. Clarke, Q. Dahe, P. Dasgupta, N. K. Dubash, O. Edenhofer, I. Elgizouli, C. B. Field, P. Forster, P. Friedlingstein, J. Fuglestvedt, L. Gomez-Echeverri, S. Hallegatte, G. Hegerl, M. Howden, K. Jiang, B. Jimenez Cisneroz, V. Kattsov, H. Lee, K. J. Mach, J. Marotzke, M. D. Mastrandrea, L. Meyer, J. Minx, Y. Mulugetta, K. O'Brien, M. Oppenheimer, J. J. Pereira, R. Pichs-Madruga, G. K. Plattner, H. O. Pörtner, S. B. Power, B. Preston, N. H. Ravindranath, A. Reisinger, K. Riahi, M. Rusticucci, R. Scholes, K. Seyboth, Y. Sokona, R. Stavins, T. F. Stocker, P. Tschakert, D. van Vuuren, and J. P. van Ypserle. 2014. Climate change 2014: synthesis report. R. Pachauri, and L. Meyer, editors. *Contribution of Working Groups I, II and III to the Fifth Assessment Report of the Intergovernmental Panel on Climate Change*, ISBN: 978-92-9169-143-2. IPCC, Geneva.

Paley, W. S. 1952. Making the most of timber resources. *in Resources for freedom: A report to the President by the President's Material Policy Commission*, Vol. 1, Chap. 8. Washington, DC: U.S. Government Printing Office.

Palik, B. J., P. C. Goebel, L. K. Kirkman, and L. West. 2000. Using landscape hierarchies to guide restoration of disturbed ecosystems. *Ecological Applications* 10:189–202.

Palik, B. J., R. J. Mitchell, and J. K. Hiers. 2002. Modeling silviculture after natural disturbance to sustain biodiversity in the longleaf pine (*Pinus palustris*) ecosystem: Balancing complexity and implementation. *Forest Ecology and Management* 155:347–356.

Palik, B. J., R. J. Mitchell, G. Houseal, and N. Pederson. 1997. Effects of canopy structure on resource availability and seedling responses in a longleaf pine ecosystem. *Canadian Journal of Forest Research* 27:1458–1464.

Palik, B. J., R. J. Mitchell, S. D. Pecot, M. A. Battaglia, and M. Pu. 2003. Spatial distribution of overstory retention influences resources and growth of longleaf pine seedlings. *Ecological Applications* 13:674–686.

Palik, B. J. and N. Pederson. 1996. Overstory mortality and canopy disturbances in longleaf pine ecosystems. *Canadian Journal of Forest Research* 26:2035–2047.

Palis, J. G. 1997. Breeding migration of *Ambystoma cingulatum* in Florida. *Journal of Herpetology* 31:71–78.

Palmer, M. A., R. F. Ambrose, and N. L. Poff. 1997. Ecological theory and community restoration ecology. *Restoration Ecology* 5:291–300.

Palmquist, K. A., R. K. Peet, and S. R. Mitchell. 2015. Scale-dependent responses of longleaf pine vegetation to fire frequency and environmental context across two decades. *Journal of Ecology* 103:998–1008.

Palmquist, K. A., R. K. Peet, and A. S. Weakley. 2014. Changes in plant species richness following reduced fire frequency and drought in one of the most species-rich savannas in North America. *Journal of Vegetation Science* 25:1426–1437.

Pandey, C. B. and J. S. Singh. 1992. Rainfall and grazing effects on net primary productivity in a tropical savanna, India. *Ecology* 73:2007–2021.

Parks, S. A., L. M. Holsinger, C. Miller, and C. R. Nelson. 2015. Wildland fire as a self-regulating mechanism: The role of previous burns and weather in limiting fire progression. *Ecological Applications* 25:1478–1492.

Parrott, R. T. 1967. A study of wiregrass (*Aristida stricta* Michx.) with particular reference to fire, MS thesis. Duke University, Durham, NC.

Parsons, D. J. and S. H. DeBenedetti. 1979. Impact of fire suppression on a mixed-conifer forest. *Forest Ecology and Management* 2:21–33.

Parsons, R. A., W. E. Mell, and P. McCauley. 2011. Linking 3D models of fuels and fire: Effects of spatial heterogeneity on fire behavior. *Ecological Modelling* 222:679–691.

Pärtel, M., M. Zobel, J. Liira, and K. Zobel. 2000. Species richness limitations in productive and oligotrophic plant communities. *Oikos* 90:191–193.

Parton, W. J. 1996. The CENTURY model. Pages 283–291 *in* D. S. Powlson, P. Smith, and J. U. Smith, editors. *Evaluation of Soil Organic Matter Models*. Berlin: Springer-Verlag.

Parton, W. J., D. S. Schimel, C. V. Cole, and D. S. Ojima. 1987. Analysis of factors controlling soil organic levels of grasslands in the Great Plains. *Soil Science Society of America Journal* 51:1173–1179.

Pate, J. S., R. H. Froend, B. J. Bowen, A. Hansen, and J. Kuo. 1990. Seedling growth and storage characteristics of seeder and resprouter species of Mediterranean-type ecosystems of S.W. Australia. *Annals of Botany* 65:585–601.

Pausas, J. G. 2015. Bark thickness and fire regime. *Functional Ecology* 29:315–327.

Pausas, J. G., R. A. Bradstock, D. A. Keith, and J. E. Keeley. 2004. Plant functional traits in relation to fire in crown-fire ecosystems. *Ecology* 85:1085–1100.

Pausas, J. G. and M. Verdú. 2008. Fire reduces morphospace occupation in plant communities. *Ecology* 89:2181–2186.

Peacock, E. and T. Schauwecker. 2003. *Blackland Prairies of the Gulf Coastal Plain*. Tuscaloosa, AL: University of Alabama Press.

Pechmann, J. H., D. E. Scott, J. W. Gibbons, and R. D. Semlitsch. 1989. Influence of wetland hydroperiod on diversity and abundance of metamorphosing juvenile amphibians. *Wetlands Ecology and Management* 1:3–11.

Pechony, O. and D. T. Shindell. 2010. Driving forces of global wildfires over the past millennium and the forthcoming century. *Proceedings of the National Academy of Sciences* 107:19167–19170.

Pecot, S. D., S. B. Horsley, M. A. Battaglia, and R. J. Mitchell. 2005. The influence of canopy, sky condition, and solar angle on light quality in a longleaf pine woodland. *Canadian Journal of Forest Research* 35:156–1366.

Pecot, S. D., R. J. Mitchell, B. J. Palik, E. B. Moser, and J. K. Hiers. 2007. Competitive responses of seedlings and understory plants in longleaf pine woodlands: Separating canopy influences above and below ground. *Canadian Journal of Forest Research* 37:634–648.

Pederson, N., J. S. Kush, R. S. Meldahl, and W. D. Boyer. 2000. Longleaf pine cone crops and climate: A possible link. Pages 255–258 *in* J. D. Haywood, editor. *Proceedings of the Tenth Biennial Southern Silvicultural Research Conference*, 16–18 February 1999, Shreveport, Louisiana, USA. General Technical Report GTR SRS-30. USDA Forest Service, Southern Research Station, Asheville, NC.

Pederson, N., J. M. Varner III, and B. J. Palik. 2008. Canopy disturbance and tree recruitment over two centuries in a managed longleaf pine landscape. *Forest Ecology and Management* 254:85–95.

Peel, M. C., B. L. Finlayson, and T. A. McMahon. 2007. Updated world map of the Köppen-Geiger climate classification. *Hydrology and Earth Systems Science* 11:1633–1644.

Peet, R. K. 1981. Forest vegetation of the Colorado Front Range-composition and dynamics. *Vegetatio* 45:3–75.

Peet, R. K. 2006. Ecological classification of longleaf pine woodlands. Pages 51–93 *in* S. Jose, E. J. Jokela, and D. L. Miller, editors. *The Longleaf Pine Ecosystem: Ecology, Silviculture and Restoration*. New York, NY: Springer.

Peet, R. K. and D. J. Allard. 1993. Longleaf pine vegetation of the southern Atlantic and eastern gulf coast regions: A preliminary classification. Pages 45–81 *in* S. M. Hermann, editor. *Proceedings of the Tall Timbers Fire Ecology Conference*, no. 18. Tall Timbers Research Station, Tallahassee, FL.

Peet, R. K., K. A. Palmquist, and S. M. Tessel. 2014. Herbaceous layer species richness of southeastern forests and woodlands: Patterns and causes. Pages 255–276 *in* F. S. Gilliam, editor. *The Herbaceous Layer in Forests of Eastern North America*. New York, NY: Oxford University Press.

Perkins, M. W., L. M. Conner, and M. B. Howze. 2008. The importance of hardwood trees in the longleaf pine forest ecosystem for Sherman's fox squirrels. *Forest Ecology and Management* 255:1618–1625.

Pessin, L. J. 1934. Annual ring formation in *Pinus palustris* seedlings. *American Journal of Botany* 21:599–603.

Pessin, L. J. 1938. The effect of vegetation on the growth of longleaf pine seedlings. *Ecological Monographs* 8:115–149.

Pessin, L. J. 1939. Root habits of longleaf pine and associated species. *Ecology* 20:47–57.

Peterson, D. W. and P. B. Reich. 2008. Fire frequency and tree canopy structure influence plant species diversity in a forest-grassland ecotone. *Plant Ecology* 194:5–16.

Petes, L. E., A. J. Brown, and C. R. Knight. 2012. Impacts of upstream drought and water withdrawals on the health and survival of downstream estuarine oyster populations. *Ecology and Evolution* 2:1712–1724.

Petrick, C. and B. Hagedorn. 2004. Population status and trend of red-cockaded woodpeckers on Eglin Air Force Base, Florida. Pages 230–240 *in* R. Costa, and R. J. Daniels, editors. *Red-Cockaded Woodpecker: Road to Recovery*. Blaine, WA: Hancock House Publishers.

Petrides, G. A. and R. B. Nestler. 1943. Age determination in juvenal bob-white quail. *American Midland Naturalist* 30:774–782.

Phillips, R. J. and T. A. Waldrop. 2008. Changes in vegetation structure and composition in response to fuel reduction treatments in the South Carolina piedmont. *Forest Ecology and Management* 255:3107–3116.

Pickett, S. T. A. and P. S. White. 1985. *The Ecology of Natural Disturbance and Patch Dynamics*. Orlando, FL: Academic Press.

Pierce, R. R., N. L. Barber, and H. R. Stiles. 1984. Georgia irrigation, 1970–1980: A decade of growth. USGS Water Resource Investigations Report 83–4177. U.S. Geological Survey, Doraville, GA.

Pittman, S. E., A. L. Jendrek, S. J. Price, and M. E. Dorcas. 2008. Habitat selection and site fidelity of Cope's gray treefrog (*Hyla chrysoscelis*) at the aquatic-terrestrial ecotone. *Journal of Herpetology* 42:378–385.

Plant Conservation Alliance (PCA). 2015. National seed strategy for rehabilitation and restoration. Bureau of Land Management, UT. https://www.fs.fed.us/wildflowers/Native_Plant_Materials/documents/SeedStrategy081215.pdf (accessed 4 February 2017).

Platt, W. J. 1999. Southeastern pine savannas. Pages 23–51 *in* R. C. Anderson, J. S. Fralish, and J. M. Baskin, editors. *Savannas, Barrens, and Rock Outcrop Plant Communities of North America*. Cambridge: Cambridge University Press.

Platt, W. J., S. M. Carr, M. Reilly, and J. Fahr. 2006. Pine savanna overstorey influences on ground-cover biodiversity. *Applied Vegetation Science* 9:37–50.

Platt, W. J., D. P. Ellair, J. M. Huffman, S. E. Potts, and B. Beckage. 2016. Pyrogenic fuels produced by savanna trees can engineer humid savannas. *Ecological Monographs* 86:352–372.

Platt, W. J., A. K. Entrup, E. K. Babl, C. Coryell-Turpin, V. Dao, J. A. Hebert, C. D. LaBarbera, J. F. L. Noto, S. O. Ogundare, L. K. Stamper, and N. Timilsina. 2015. Short-term effects of herbicides and a prescribed fire on restoration of a shrub-encroached pine savanna. *Restoration Ecology* 23:909–917.

Platt, W. J., G. W. Evans, and M. M. Davis. 1988. Effects of fire season on flowering forbs and shrubs in longleaf pine forests. *Oecologia* 76:353–363.

Platt, W. J., G. W. Evans, and S. L. Rathbun. 1988. The population dynamics of a long-lived conifer (Pinus palustris). *American Naturalist* 131:491–525.

Platt, W. J., J. S. Glitzenstein, and D. R. Streng. 1991. Evaluating pyrogenicity and its effects on vegetation in longleaf pine savannas. Pages 143–161 *in* S. M. Hermann, editor. *Proceedings of the 17th Tall Timbers Fire Ecology Conference*, Tall Timbers Research Station, Tallahassee, FL.

Platt, W. J. and S. L. Rathbun. 1993. Dynamics of an old-growth longleaf pine population. Pages 275–297 *in* S. M. Hermann, editor. *The Longleaf Pine Ecosystem: Ecology, Restoration and Management. Proceedings of the 18th Tall Timbers Fire Ecology Conference*, 30 May-2 June 1991, Tallahassee, Florida, USA. Tall Timbers Research Station, Tallahassee, FL.

Plentovich, S., J. W. Tucker, Jr., N. R. Holler, and G. E. Hill. 1998. Enhancing Bachman's sparrow habitat via management of red-cockaded woodpeckers. *Journal of Wildlife Management* 62:347–354.

Polis, G. A., W. B. Anderson, and R. D. Holt. 1997. Toward an integration of landscape and food web ecology: The dynamics of spatially subsidized food webs. *Annual Review of Ecology and Systematics* 28:289–316.

Pope, V. D. 2000. The impact of new physical parameterizations in the Hadley Centre climate model—HadCM3. *Climate Dynamics* 16:123–146.

Potter, C. S. 2016. Landsat image analysis of tree mortaility in the southern Sierra Nevada region of California during the 2013–2014 drought. *Journal of Earth Science and Climatic Change* 7:1–7.

Pounds, J. A., M. R. Bustamante, L. A. Coloma, J. A. Consuegra, M. P. Fogden, P. N. Foster, E. La Marca, K. L. Masters, A. Merino-Viteri, R. Puschendorf, and S. R. Ron. 2006. Widespread amphibian extinctions from epidemic disease driven by global warming. *Nature* 439:161–167.

Powell, T. L., D. R. Galbraith, B. O. Christoffersen, A. Harper, H. M. Imbuzeiro, L. Rowland, S. Almeida, P. M. Brando, A. C. da Costa, M. H. Costa, N. M. Levine, Y. Malhi, S. R. Saleska, E. Sotta, M. Williams, P. Meir, and P. R. Moorcroft. 2013. Confronting model predictions of carbon fluxes with measurements of Amazon forest subjected to experimental drought. *New Phytologist* 200:350–365.

Powell, T. L., G. Starr, K. L. Clark, T. A. Martin, and H. L. Gholz. 2005. Ecosystem and understory water and energy exchange for a mature, naturally regenerated pine flatwoods forest in north Florida. *Canadian Journal of Forest Research* 35:1568–1580.

Power, M. J., J. Marlon, N. Ortiz, J. Bartlein, S. P. Harrison, F. E. Mayle, A. Ballouche, R. H. W. Bradshaw, C. Carcaillet, C. Cordova, S. Mooney, P. I. Moreno, I. C. Prentice, K. Thonicke, W. Tinner, C. Whitlock, Y. Zhang, Y. Zhao, A. A. Ali, R. S. Anderson, R. Beer, H. Behling, C. Briles, K. J. Brown, A. Brunelle, M. Bush, P. Camill, G. Q. Chu, J. Clark, D. Colombaroli, S. Connor, A. L. Daniau, M. Daniels, J. Dodson, E. Doughty, M. E. Edwards, W. Finsinger, D. Foster, J. Frechette, M. J. Gaillard, D. G. Gavin, E. Gobet, S. Haberle, D. J. Hallett, P. Higuera, G. Hope, S. Horn, J. Inoue, P. Kaltenrieder, L. Kennedy, Z. C.

Kong, C. Larsen, C. J. Long, J. Lynch, E. A. Lynch, M. McGlone, S. Meeks, S. Mensing, G. Meyer, T. Minckley, J. Mohr, D. M. Nelson, J. New, R. Newnham, R. Noti, W. Oswald, J. Pierce, P. J. H. Richard, C. Rowe, M. F. Sanchez Goni, B. N. Shuman, H. Takahara, J. Toney, C. Turney, D. H. Urrego-Sanchez, C. Umbanhowar, M. Vandergoes, B. Vanniere, E. Vescovi, M. Walsh, X. Wang, N. Williams, J. Wilmshurst, and J. H. Zhang. 2008. Changes in fire regimes since the Last Glacial Maximum: An assessment based on a global synthesis and analysis of charcoal data. *Climate Dynamics* 30:887–907.

Prasad, A. M., L. R. Iverson, and A. Liaw. 2006. Newer classification and regression tree techniques: Bagging and random forests for ecological prediction. *Ecosystems* 9:181–199.

Prasad, A. M., L. R. Iverson, S. Matthews, and M. Peters. 2007. A climate change atlas for 134 forest tree species of the eastern United States [database]. USDA Forest Service, Northern Research Station, Delaware, OH. http://www.nrs.fs.fed.us/atlas/tree (accessed 4 February 2017).

Predmore, S. A., J. McDaniel, and J. S. Kush. 2007. Presettlement forests and fire in southern Alabama. *Canadian Journal of Forest Research* 37:1723–1736.

Preston, C. R. and R. D. Beane. 1993. Red-tailed hawk (*Buteo jamaicensis*). Pages 1–24 *in* A. Poole, and E. Gill, editors. *The Birds of North America*. Philadelphia, PA; Washington, DC: Academy of Natural Sciences and American Ornithologists' Union.

Price, R. A., A. Liston, and S. H. Strauss. 1998. Phylogeny and systematics of *Pinus*. Pages 49–68 *in* D. M. Richardson, editor. *Ecology and Biogeography of Pinus*. Cambridge: Cambridge University Press.

PRISM Group. 2014. PRISM climate chart. Northwest Alliance for Computational Science & Engineering, Oregon State University, Corvallis, OR.

Provencher, L., N. M. Gobris, L. A. Brennan, and T. Grubb. 2002. Effects of hardwood reduction on winter birds in northwest Florida longleaf pine sandhill forests. *The Auk* 119:71–87.

Provencher, L., D. R. Gordon, H. L. Rodgers, E. M. Galley, G. W. Tanner, J. L. Hardesty, and L. A. Brennan. 2001. Effects of hardwood reduction techniques on longleaf pine sandhill vegetation in northwest Florida. *Restoration Ecology* 9:13–27.

Provencher, L., B. J. Herring, D. R. Gordon, H. L. Rodgers, G. W. Tanner, J. L. Hardesty, L. A. Brennan, and A. R. Litt. 2001. Longleaf pine and oak responses to hardwood reduction techniques in fire-suppressed sandhills in northwest Florida. *Forest Ecology and Management* 148:1–15.

Provencher, L., A. R. Litt, D. R. Gordon, H. Rodgers, B. J. Herring, K. E. Galley, J. P. McAdoo, S. J. McAdoo, N. M. Gobris, and J. L. Hardesty. 2001. Restoration fire and hurricanes in longleaf pine sandhills. *Ecological Restoration* 19:92–98.

Prugh, L. R., C. J. Stoner, C. W. Epps, W. T. Bean, W. J. Ripple, A. S. Laliberte, and J. S. Brashares. 2009. The rise of the mesopredator. *Bioscience* 59:779–791.

Puettmann, K. J., K. D. Coates, and C. Messier. 2009. *A Critique of Silviculture: Managing for Complexity*. Washington, DC: Island Press.

Pyne, S. J. 1982. *Fire in America: A Cultural History of Wildland and Rural Fire*. Seattle, WA: University of Washington Press.

Pyzoha J. E., J. J. Callahan, G. Sun, C. C. Trettin, and M. Miwa. 2008. A conceptual hydrologic model for a forested Carolina bay depressional wetland on the coastal plain of South Carolina, USA. *Hydrological Processes* 22:1689–2698.

Raison, R. J., P. K. Khanna, and P. V. Woods. 1985. Mechanisms of element transfer to the atmosphere during vegetation fires. *Canadian Journal of Forest Research* 15:132–140.

Randall, W. K. and P. Berrang. 2002. *Washington Tree Seed Transfer Zones*. Olympia, WA: Washington State Department of Natural Resources.

Rasser, M. K. 2003. Monitoring and assessing longleaf pine ecosystem restoration: A case study in North-Central Florida, MS thesis. University of Florida, Gainesville, FL.

Ravenel, R., M. Tyrrell, and R. Mendelsohn. 2002. Institutional timberland investment: A summary of a forum exploring changing ownership patterns and the implications for conservation of environmental values. School of Forestry and Environmental Studies, Yale University, New Haven, CT. www.yale.edu/gisf (accessed 4 February 2017).

Rebertus, A. J., G. B. Williamson, and E. B. Moser. 1989. Longleaf pine pyrogenicity and turkey oak mortality in Florida xeric sandhills. *Ecology* 70:60–70.

Reed, F. W. 1905. *A Working Plan for Forest Lands in Central Alabama*. Forest Service Bulletin 68. Washington, DC: USDA Forest Service.

Refsnider, J. M. and F. J. Janzen. 2012. Behavioural plasticity may compensate for climate change in a long-lived reptile with temperature-dependent sex determination. *Biological Conservation* 152:90–95.

Regelbrugge, J. C. and S. G. Conard. 1993. Modeling tree mortality following wildfire in *Pinus ponderosa* forests in the central Sierra Nevada of California. *International Journal of Wildland Fire* 3:139–148.

Regester, K. J., K. R. Lips, and M. R. Whiles. 2006. Energy flow and subsidies associated with the complex life cycle of ambystomatid salamanders in ponds and adjacent forest in southern Illinois. *Oecologia* 147:303–314.

Regester, K. J., M. R. Whiles, and K. R. Lips. 2008. Variation in the trophic basis of production and energy flow associated with emergence of larval salamander assemblages from forest ponds. *Freshwater Biology* 53:1754–1767.

Reid, A. M. and K. M. Robertson. 2012. Energy content of common fuels in upland pine savannas of the southeastern U.S. and their application to fire behaviour modelling. *International Journal of Wildland Fire* 21:591–595.

Reid, A. M., K. M. Robertson, and T. L. Hmielowski. 2012. Predicting litter and live herb fuel consumption during prescribed fires in native and old-field upland pine communities of the southeastern United States. *Canadian Journal of Forest Research—Revue Canadienne De Recherche Forestiere* 42:1611–1622.

Reinhardt, E. and R. E. Keane. 1998. FOFEM—A first order fire effects model. *Fire Management Notes* 58:25–28.

Remucal, J. M., J. D. McGee, M. F. Fehrenbacher, C. Best, and R. J. Mitchell. 2013. Application of the Climate Action Reserve's forest project protocol to a longleaf pine forest under restoration management. *Journal of Forestry* 111:59–66.

Reynolds, R. T., A. J. Sánchez Meador, J. A. Youtz, T. Nicolet, M. S. Matonis, P. L. Jackson, D. G. DeLorenzo, and A. D. Graves. 2013. Restoring composition and structure in Southwestern frequent-fire forests: A science-based framework for improving ecosystem resiliency. General Technical Report RMRS-GTR-310. USDA Forest Service, Rocky Mountain Research Station, Fort Collins, CO.

Richardson, C. J. 2003. Pocosins: Hydrologically isolated or integrated wetlands on the landscape? *Wetlands* 23:563–76.

Richardson, C. J. and J. W. Gibbons. 1993. Pocosins, Carolina bays and mountain bogs. Pages 257–310 *in* W. H. Martin, S. G. Boyce, and A. C. Echternacht, editors. *Biodiversity of the Southeastern United States*. New York, NY: Wiley & Sons.

Richardson, D. M. and P. W. Rundel. 1998. Ecology and biogeography of *Pinus*: An introduction. Pages 3–48, *in* D. M. Richardson, editor. *Ecology and Biogeography of Pinus*. Cape Town: Cambridge University Press.

Richardson, D. M., P. W. Rundel, S. T. Jackson, R. O. Teskey, J. Aronson, A. Bytnerwoicz, M. J. Wingfield, and S. Proches. 2007. Human impacts in pine forests: Past, present, and future. *Annual Review of Ecology, Evolution, and Systematics* 38:275–297.

Richter, B. 2014. *Chasing Water: A Guide for Moving from Scarcity to Sustainability*. Washington, DC: Island Press.

Richter, D. D., Jr. and D. Markewitz. 2001. *Understanding Soil Change: Soil Sustainability Over Millennia, Centuries, and Decades*. New York, NY: Cambridge University Press.

Richter, D. D., D. Markewitz, S. E. Trumbore, and C. G. Wells. 1999. Rapid accumulation and turnover of soil carbon in a re-establishing forest. *Nature* 400:56–58.

Ricklefs, R. E. 1987. Community diversity: Relative roles of local and regional processes. *Science* 235:167–171.

Riley, S. J., D. J. Decker, J. W. Enck, P. D. Curtis, T. B. Lauber, and T. L. Brown. 2003. Deer populations up, hunter populations down: Implications of interdependence of deer and hunter population dynamics on management. *Ecoscience* 10:455–461.

Ripple, W. J., J. A. Estes, R. L. Beschta, C. C., Wilmers, E. G. Ritchie, M. Hebblewhite, J. Berger, B. Elmhagen, M. Letnic, M. P. Nelson, O. J. Schmitz, D. W. Smith, A. D. Wallach, and A. J. Wirsing. 2014. Status and ecological effects of the world's largest carnivores. *Science* 343:151–163. doi:10.1126/science.1241484.

Ripple, W. J. and E. J. Larsen. 2000. Historic aspen recruitment, elk, and wolves in Northern Yellowstone National Park, USA. *Biological Conservation* 95:361–370.

Ripple, W. J., E. J. Larsen, R. A. Renkin, and D. W. Smith. 2001. Trophic cascades among wolves, elk, and aspen in Yellowstone National Park's northern range. *Biological Conservation* 102:227–234.

Ritchie, E. G., B. Elmhagen, A. S. Glen, M. Letnic, G. Ludwig, and R. A. McDonald. 2012. Ecosystem restoration with teeth: What role for predators? *Trends in Ecology and Evolution* 27:265–271.

Ritchie, E. G. and C. N. Johnson. 2009. Predator interactions, mesopredator release and biodiversity conservation. *Ecology Letters* 12:982–998.

Rittenhouse, T. A., E. B. Harper, L. R. Rehard, and R. D. Semlitsch. 2008. The role of microhabitats in the desiccation and survival of anurans in recently harvested oak-hickory forest. *Copeia* 2008:807–814.

Rivals, F., N. Solounias, and M. C. Mihlbachler. 2007. Evidence for geographic variation in the diets of late Pleistocene and early Holocene bison in North America, and differences from the diets of recent bison. *Quaternary Research* 68:338–346.

Robbins, L. E. and R. L. Myers. 1992. *Seasonal Effects of Prescribed Burning in Florida: A Review*. Miscellaneous Publication no. 8. Tallahassee, FL: Tall Timbers Research Station.

Robertson, K. M. and T. L. Hmielowski. 2014. Effects of fire frequency and season on resprouting of woody plants in southeastern U.S. pine-grassland communities. *Oecologia* 174:765–776.

Robertson, K. M., Y. P. Hsieh, and G. C. Bugna. 2014. Fire environment effects on particulate matter emission factors in southeastern U.S. pine-grasslands. *Atmospheric Environment* 99:104–111.

Rodrìguez-Trejo, D. A., M. L. Duryea, T. L. White, J. R. English, and J. McGuire. 2003. Artificially regenerating longleaf pine in canopy gaps: Initial survival and growth during a year of drought. *Forest Ecology and Management* 180:25–36.

Roe, J. H. and A. Georges. 2007. Heterogeneous wetland complexes, buffer zones, and travel corridors: Landscape management for freshwater reptiles. *Biological Conservation* 135:67–76.

Roe, J. H., B. A. Kingsbury, and N. R. Herbert. 2004. Comparative water snake ecology: Conservation of mobile animals that use temporally dynamic resources. *Biological Conservation* 118:79–89.

Roemer, G. W., C. J. Donlan, and F. Courchamp. 2002. Golden eagles, feral pigs, and insular carnivores: How exotic species turn native predators into prey. *Proceedings of the National Academy of Sciences* 99:791–796.

Rogalski, A. and K. Chrzanowski. 2002. Infrared devices and techniques. *Optoelectronics Review* 10:111–136.

Rohr, J. R. and K. A. McCoy. 2010. A qualitative meta-analysis reveals consistent effects of atrazine on freshwater fish and amphibians. *Environmental Health Perspectives* 118:20–32.

Rohr, J. R., T. R. Raffel, J. M. Romansic, H. McCallum, and P. J. Hudson. 2008. Evaluating the links between climate, disease spread, and amphibian declines. *Proceedings of the National Academy of Sciences* 105:17436–17441.

Roise, J. P., J. Chung, and R. Lancia. 1991. Red-cockaded woodpecker habitat management and longleaf pine straw production: An economic analysis. *Southern Journal of Applied Forestry* 15:88–92.

Roques, K. G., T. G. O'Connor, and A. R. Watkinson. 2001. Dynamics of shrub encroachment in an African savanna: Relative influences of fire, herbivory, rainfall and density dependence. *Journal of Applied Ecology* 38:268–280.

Rosene, W. 1984. *The Bobwhite Quail Its Life and Management*. Augusta, GA: Morris Communications Corporation.

Rosenzweig, M. L. 1973. Exploitation in three trophic levels. *American Naturalist* 107:275–294.

Ross, M. S., J. J. O'Brien, R. G. Ford, K. Zhang, and A. Morkill. 2009. Disturbance and the rising tide: The challenge of biodiversity management on low-island ecosystems. *Frontiers in Ecology and the Environment* 7:471–478.

Ross, W. G., D. L. Kulhavy, and R. N. Conner. 1997. Stand conditions and tree characteristics affect quality of longleaf pine for red-cockaded woodpecker cavity trees. *Forest Ecology and Management* 91:145–154.

Rostlund, E. 1960. The geographic range of the historic bison in the southeast. *Annals of the Association of American Geographers* 50:395–400.

Roth, A., D. Campbell, P. Keddy, H. Dozier, and G. Montz. 2008. How important is competition in a species-rich grassland? A two-year removal experiment in a pine savanna. *Ecoscience* 15:94–100.

Rothermel, R. C. 1972. A mathematical model for predicting fire spread in wildland fuels. Research Paper INT-115. USDA Forest Service, Intermountain Forest and Range Experiment Station, Odgen, UT.

Rothermel, R. C. 1983. How to predict the spread and intensity of forest and range fires. General Technical Report INT-143. USDA Forest Service, Intermountain Forest and Range Experiment Station, Odgen, UT.

Rothermel, B. B. and R. D. Semlitsch. 2006. Consequences of forest fragmentation for juvenile survival in spotted (*Ambystoma maculatum*) and marbled (*Ambystoma opacum*) salamanders. *Canadian Journal of Zoology* 84:797–807.

Row, C. 1973. Probabilities of financial returns from southern pine timber growing, PhD dissertation. Tulane University, New Orleans, LA.

Row, C. 1978. Economies of tract size in timber growing. *Journal of Forestry* 76:576–582.

Rowe, H. I. 2010. Tricks of the trade: Techniques and opinions from 38 experts in tallgrass prairie restoration. *Restoration Ecology* 18:253–262.

Rowell, E., E. L. Loudermilk, C. Seielstad, and J. J. O'Brien. 2016. Using simulated 3D surface fuelbeds and terrestrial laser scan data to develop inputs to fire behavior model. *Canadian Journal of Remote Sensing* 42:443–459. doi:10.1080/07038992.2016.1220827.

Rowell, E. and C. Seielstad. 2012. Characterizing grass, litter, and shrub fuels in longleaf pine forest pre- and post-fire using terrestrial lidar. *Proceedings of SilviLaser*, 16–19 September 2012, Vancouver, British Columbia, Canada.

Rowell, E. M., C. A. Seielstad, and R. D. Ottmar. 2015. Development and validation of fuel height models for terrestrial lidar-RxCADRE 2012. *International Journal of Wildland Fire* 25:38–47.

Roznik, E. A. and S. A. Johnson. 2009. Canopy closure and emigration by juvenile gopher frogs. *Journal of Wildlife Management* 73:260–268.

Rüegg, J., C. Gries, B. Bond-Lamberty, G. J. Bowen, B. S. Felzer, N. E. McIntyre, P. A. Soranno, K. L. Vanderbilt, and K. C. Weathers. 2014. Completing the data life cycle: Using information management in macrosystems ecology research. *Frontiers in Ecology and the Environment* 12:24–30.

Rugel, K., C. R. Jackson, J. J. Romeis, S. W. Golladay, D. W. Hicks, and J. F. Dowd. 2012. Effects of irrigation withdrawals on streamflows in a karst environment: Lower Flint River Basin, Georgia, USA. *Hydrological Processes* 26:523–534.

Ruhl, J. B. 2005. Water wars, eastern style: Divvying up the Apalachicola-Chattahoochee-Flint River Basin. *Journal of Contemporary Water Research and Education* 131:47–54.

Runge M. C. and M. G. Knutson. 2012. When is adaptive management appropriate? Chapter 3 in adaptive management: Structured decision-making for recurrent decisions. U.S. Fish and Wildlife Service, National Conservation Training Center, Shepherdstown, WV. http://nctc.fws.gov/courses/ALC/ALC3176/resources/index.html (accessed 4 February 2017).

Runion, G. B., M. A. Davis, S. G. Pritchard, S. A. Prior, R. J. Mitchell, H. A. Torbert, H. H. Rogers, and R. R. Dute. 2006. Effects of elevated atmospheric carbon dioxide on biomass and carbon accumulation in a model regenerating longleaf pine community. *Journal of Environmental Quality* 35:1478–1486.

Running, S. W. and J. C. Coughlan. 1988. A general model of forest ecosystem processes for regional applications. *Ecological Modelling* 42:125–154.

Running, S. W. and S. T. Gower. 1991. FOREST-BGC, a general model of forest ecosystem processes for regional applications. II. Dynamic carbon allocation and nitrogen budgets. *Tree Physiology* 9:147–160.

Ruth, A. D., S. Jose, and D. L. Miller. 2008. Seed bank dynamics of sand pine scrub and longleaf pine flatwoods. *Ecological Restoration* 26:19–22.

Ryan, K. C., E. E. Knapp, and J. M. Varner. 2013. Prescribed fire in North American forests and woodlands: History, current practice, and challenges. *Frontiers in Ecology & the Environment* 11:e15–e24.

Ryan, K. C. and T. S. Opperman. 2013. LANDFIRE-a national vegetation/fuels data base for use in fuels treatment, restoration, and suppression planning. *Forest Ecology and Management* 294:208–216.

Saenz, D., K. A. Baum, R. N. Conner, D. C. Rudolph, and R. Costa. 2002. Large-scale translocation strategies for reintroducing red-cockaded woodpeckers. *Journal of Wildlife Management* 66:212–221.

Saenz, D., R. N. Conner, C. E. Shackelford, and D. C. Rudolph. 1998. Pileated woodpecker damage to red-cockaded woodpecker cavity trees in eastern Texas. *Wilson Bulletin* 110:362–367.

Sale, P. 1977. Maintenance of high diversity in coral reef fish communities. *American Naturalist* 111:337–359.

Sampson, N. and L. DeCoster. 2000. Forest fragmentation: Implications for sustainable private forests. *Journal of Forestry* 98:4–8.

Samuelson, L. J. and T. A. Stokes. 2012. Leaf physiological and morphological responses to shade in grass-stage seedlings and young trees of longleaf pine. *Forests* 3:684–699.

Samuelson, L. J., T. A. Stokes, J. R. Butnor, K. H. Johnsen, C. A. Gonzalez-Benecke, P. Anderson, J. Jackson, L. Ferrari, T. A. Martin, and W. P. Cropper, Jr. 2014. Ecosystem carbon stocks in *Pinus palustris* forests. *Canadian Journal of Forest Research* 44:476–486.

Samuelson, L. J., T. A. Stokes, J. R. Butnor, K. H. Johnsen, C. A. Gonzalez-Benecke, T. A. Martin, W. P. Cropper, Jr., P. H. Anderson, M. R. Ramirez, and J. C. Lewis. 2017. Ecosystem carbon density and allocation across a chronosequence of longleaf pine forests in the southeastern USA. *Ecological Application* 27: 244–259. doi:10.1002/eap.1439.

Sandberg, D. V., C. L. Riccardi, and M. D. Schaaf. 2007. Reformulation of Rothermel's wildland fire behaviour model for heterogeneous fuelbeds This article is one of a selection of papers published in the Special Forum on the Fuel Characteristic Classification System. *Canadian Journal of Forest Research* 37:2438–2455.

Sanginga, N., S. K. A. Danso, F. Zapata, and G. D. Bowen. 1995. Phosphorus requirements and nitrogen accumulation by N_2-fixing and non-N_2-fixing leguminous trees growing in low P soils. *Biology and Fertility of Soils* 20:205–211.

Sankaran, M., J. Ratnam, and N. P. Hanan. 2004. Tree-grass coexistence in savannas revisited—insights from an examination of assumptions and mechanisms invoked in existing models. *Ecology Letters* 7:480–490.

Sargent, C. S. 1884. *Report on the Forests of North America*. Washington, DC: U.S. Census Office.

Saterson, K. A. and P. M. Vitousek. 1984. Fine-root biomass and nutrient cycling in *Aristida stricta* in a North Carolina coastal plain savanna. *Canadian Journal of Botany* 62:823–829.

Schmidtling, R. 1987. Relative performance of longleaf compared to loblolly and slash pines under different levels of intensive culture. Pages 395–400 *in* D. R. Phillips, compiler. *Proceedings of the Fourth Biennial Southern Silvicultural Research Conference*, 4–6 November 1986, Atlanta, Georgia, USA. General Technical Report SE-42. USDA Forest Service, Southeastern Forest Experiment Station, Asheville, NC.

Schmidtling, R. 2007. Genetic variation in the southern pines: Evolution, migration, and adaptation following the Pleistocene. Pages 28–32 *in* J. M. Kabrick, D. C. Dey, and D. Gwaze, editors. *Shortleaf Pine Restoration and Ecology in the Ozarks. Proceedings of a Symposium*, 7–9 November 2006, Springfield, Missouri, USA. General Technical Report NRS-P-15. USDA Forest Service, Northern Research Station, Newtown Square, PA.

Schmidtling, R. C. and V. Hipkins. 1998. Genetic diversity in longleaf pine (*Pinus palustris*): Influence of historical and prehistorical events. *Canadian Journal of Forest Research* 28:1135–1145.

Schmitz, O. J., V. Krivan, and O. Ovadia. 2004. Trophic cascades: The primacy of trait-mediated indirect interactions. *Ecology Letters* 7:153–163.

Scholes, R. J. and S. R. Archer. 1997. Tree-grass interactions in savannas. *Annual Review of Ecology and Systematics* 28:517–544.

Scholl, A. E. and A. H. Taylor. 2010. Fire regimes, forest change, and self-organization in an old-growth mixed-conifer forest, Yosemite National Park, USA. *Ecological Applications* 20:362–380.

Schoolcraft, H. R. 1821. *Journal of a Tour Into the Interior of Missouri and Arkansas from Potosi, or Mine a Burton, in Missouri Territory, in a Southwest Direction, toward the Rocky Mountains: Performed in the Years 1818 and 1819*. London: Richard Phillips and Company.

Schooley, R. L. and L. C. Branch. 2006. Space use by round-tailed muskrats in isolated wetlands. *Journal of Mammalogy* 87:495–500.

Schooley, R. L. and L. C. Branch. 2009. Enhancing the area-isolation paradigm: Habitat heterogeneity and metapopulation dynamics of a rare wetland mammal. *Ecological Applications* 19:1708–1722.

Schreiber, S. and V. H. Rudolf. 2008. Crossing habitat boundaries: Coupling dynamics of ecosystems through complex life cycles. *Ecology Letters* 11:576–587.

Schumacher, F. X. and B. B. Day. 1939. The influence of precipitation upon the width of annual rings of certain timber trees. *Ecological Monographs* 9:387–429.

Schwartz, M. W. 1994. Natural distribution and abundance of forest species and communities in northern Florida. *Ecology* 75:687–705.

Schwarz, G. F. 1907. *The Longleaf Pine in Virgin Forest: A Silvical Study*. New York, NY: John Wiley & Sons.

Schwanz, L. E. and F. J. Janzen. 2008. Climate change and temperature-dependent sex determination: Can individual plasticity in nesting phenology prevent extreme sex ratios? *Physiological and Biochemical Zoology* 81:826–834.

Seamon, P. A., R.L.Myers, L.E. Robbins, and G.S. Seamon.1989. Wiregrass reproduction and community restoration. *Natural Areas Journal* 9:264–265.

Seastedt, T. R., W. J. Parton, and D. S. Ojima. 1992. Mass loss and nitrogen dynamics of decaying litter of grasslands: The apparent low nitrogen immobilization potential of root detritus. *Canadian Journal of Botany* 70:384–391.

Sedjo, R. A. 2001. The role of forest plantations in the world's future timber supply. *Forestry Chronicle* 77:221–225.

Semlitsch, R. D. 1998. Biological delineation of terrestrial buffer zones for pond-breeding salamanders. *Conservation Biology* 12:1113–1119.

Semlitsch, R. D. 2002. Critical elements for biologically based recovery plans of aquatic-breeding amphibians. *Conservation Biology* 16:619–629.

Semlitsch, R. D. and J. R. Bodie. 1998. Are small, isolated wetlands expendable? *Conservation Biology* 12:1129–1133.

Semlitsch, R. D. and J. R. Bodie. 2003. Biological criteria for buffer zones around wetlands and riparian habitats for amphibians and reptiles. *Conservation Biology* 17:1219–1228.

Semlitsch, R. D., C. A. Conner, D. J. Hocking, T. A. Rittenhouse, and E. B. Harper. 2008. Effects of timber harvesting on pond-breeding amphibian persistence: Testing the evacuation hypothesis. *Ecological Applications* 18:283–289.

Semlitsch, R. D., D. E. Scott, J. H. K. Pechmann, and J. W. Gibbons. 1996. Structure and dynamics of an amphibian community: Evidence from a 16-year study of a natural pond. Pages 217–248 *in* M. L. Cody, and J. A. Smallwood, editors. *Long-Term Studies of Vertebrate Communities*. San Diego, CA: Academic Press.

Semlitsch, R. D., B. D. Todd, S. M. Blomquist, A. J. Calhoun, J. W. Gibbons, J. P. Gibbs, G. J. Graeter, E. B. Harper, D. J. Hocking, M. L. Hunter, Jr., D. A. Patrick, T. A. G. Rittenhouse, and B. B. Rothermel. 2009. Effects of timber harvest on amphibian populations: Understanding mechanisms from forest experiments. *Bioscience* 59:853–862.

Seni, G. and J. F. Elder. 2010. Ensemble methods in data mining: Improving accuracy through combining predictions. *Synthesis Lectures on Data Mining and Knowledge Discovery* 2:1–126.

Sgrò, C. M., A. J. Lowe, and A. A. Hoffmann. 2011. Building evolutionary resilience for conserving biodiversity under climate change. *Evolutionary Applications* 4:326–337.

Shands, W. E. and R. G. Healy. 1977. *The Lands Nobody Wanted*. Washington, DC: The Conservation Foundation.

Shankman, D. and K. M. Wills. 1995. Pre-European settlement forest communities of the Talladega Mountains, Alabama. *Southeastern Geographer* 35:118–131.

Sharitz, R. R. 2003. Carolina bay wetlands: Unique habitats of the southeastern U.S. *Wetlands* 23:550–562.

Sharitz, R. R. and J. W. Gibbons. 1982. The ecology of southeastern shrub bogs (pocosins) and Carolina bays: A community profile. FWS/OBS-82/04. U.S. Fish and Wildlife Service, Division of Biological Services, Washington, DC.

Sharitz, R. R. and C. A. Gresham. 1998. Pocosins and Carolina bays. Pages 343–377 *in* M. G. Messina, and W. H. Conner, editors. *Southern Forested Wetlands Ecology and Management*. Boca Raton, FL: Lewis Publishers.

Sharma, A. 2012. Role of uneven-aged silviculture and the soil seed bank in restoration of longleaf pine-slash pine (*Pinus palustris-Pinus elliottii*) ecosystems, PhD dissertation. University of Florida, Gainesville, FL.

Sharma, A., S. Jose, K. K. Bohn, and M. G. Andreu. 2012. Effects of reproduction methods and overstory species composition on understory light availability in longleaf pine–slash pine ecosystems. *Forest Ecology and Management* 284:23–33.

Sheffield, M. C. P., J. L. Gagnon, S. B. Jack, and D. J. McConville. 2003. Phenological patterns of mature longleaf pine (*Pinus palustris* Miller) under two different soil moisture regimes. *Forest Ecology and Management* 179:157–167.

Shepherd, B. J., D. L. Miller, and M. Thetford. 2012. Fire season effects on flowering characteristics and germination of longleaf pine (*Pinus palustris*) savanna grasses. *Restoration Ecology* 20:268–276.

Sherriff, R. L., R. V. Platt, T. T. Veblen, T. L. Schoennagel, and M. H. Gartner. 2014. Historical, observed, and modeled wildfire severity in montane forests of the Colorado Front Range. *PLOS ONE* 9:e106971. doi:10.1371/journal.pone.0106971.

Sherriff, R. L. and T. T. Veblen. 2006. Ecological effects of changes in fire regimes in *Pinus ponderosa* ecosystems in the Colorado Front Range. *Journal of Vegetation Science* 17:705–718.

Sherriff, R. L. and T. T. Veblen. 2007. A spatially explicit reconstruction of historical fire occurrence in the ponderosa pine zone of the Colorado Front Range. *Ecosystems* 10:311–323.

Shoo, L. P., D. H. Olson, S. K. McMenamin, K. A. Murray, M. Van Sluys, M. A. Donnelly, D. Stratford, J. Terhivuo, A. Merino-Viteri, S. M. Herbert, and P. J. Bishop. 2011. Engineering a future for amphibians under climate change. *Journal of Applied Ecology* 48:487–492.

Siddig, A. A. H., A. M. Ellison, A. Ochs, C. Villar-Leeman, and M. K. Lau. 2016. How do ecologists select and use indicator species to monitor ecological change? Insights from 14 years of publication in *Ecological Indicators*. *Ecological Indicators* 60:223–230.

Siemann, E., J. A. Carrillo, C. A. Gabler, R. Zipp, and W. E. Rogers. 2009. Experimental test of the impacts of feral hogs on forest dynamics and processes in the southeastern U.S. *Forest Ecology and Management* 258:546–553.

Simberloff, D. 1998. Flagships, umbrellas, and keystones: Is single-species management passé in the landscape era? *Biological Conservation* 83:247–257.

Simkin S. M. and W. K. Michener. 2005. Faunal soil disturbance regime of a longleaf pine ecosystem. *Southeastern Naturalist* 4:133–152.

Simkin, S. M., W. K. Michener, and R. Wyatt. 2004. Mound microclimate, nutrients and seedling survival. *American Midland Naturalist* 152:12–24.

Simpson, K. J., B. S. Ripley, P. Christin, C. M. Belcher, C. E. R. Lehmann, G. H. Thomas, and C. P. Osborne. 2016. Determinants of flammability in savanna grass species. *Journal of Ecology* 104:138–148.

Sims, S. E., J. J. Hendricks, R. J. Mitchell, K. A. Kuehn, and S. D. Pecot. 2007. Nitrogen decreases and precipitation increases ectomycorrhizal extramatrical mycelia production in a longleaf pine forest. *Mycorrhiza* 17:299–309.

Siry, J. P. 2002. Intensive timber management practices. Pages 327–340 *in* D. N. Wear, and J. G. Greis, editors. *Southern Forest Resource Assessment*. General Technical Report SRS-53. USDA Forest Service, Southern Research Station, Asheville, NC.

Sitters, J., C. L. Atkinson, N. Guelzow, P. Kelly, and L. L. Sullivan. 2015. Spatial stoichiometry: Cross-ecosystem material flows and their impact on recipient ecosystems and organisms. *Oikos* 124:920–930.

Sitton, T. 1995. *Backwoodsmen: Stockmen and Hunters along a Big Thicket River Valley*. Norman, OK: University of Oklahoma Press.

Skelley, P. E. and P. W. Kovarik. 2001. Insect surveys in the southeast: Investigating a relictual entomofauna. *Florida Entomologist* 84:552–555.

Skelly, D. K., E. E. Werner, and S. A. Cortwright. 1999. Long-term distributional dynamics of a Michigan amphibian assemblage. *Ecology* 80:2326–2337.

Slama, C. C., C. Theurer, and S. W. Henriksen. 1980. *Manual of Photogrammetry*. Falls Church, VA: American Society of Photogrammetry.

Slater, G. L., J. D. Lloyd, and R. W. Snow. 2013. Re-introduction of brown-headed nuthatch & eastern bluebird to South Florida pine rocklands, USA. Page 63 *in* P. S. Soorae, editor. *Global Re-introduction Perspectives: 2013*. Gland: IUCN, SSC Re-introduction Specialist Group.

Slatton, K. C., M. Coleman, W. Carter, R. Shrestha, and M. Sartori. 2004. Control methods for merging ALSM and ground-based laser point clouds acquired under forest canopies. *Proceedings of SPIE, Fourth International Asia-Pacific Environmental Remote Sensing Symposium* 5661:96–103.

Smith, D. M. 1986. *The Practice of Silviculture*, 8th edition. New York, NY: John Wiley & Sons.

Smith, B. D. 1989. Origins of agriculture in eastern North America. *Science* 246:1566–1571.

Smith, M. A. and D. M. Green. 2005. Dispersal and the metapopulation paradigm in amphibian ecology and conservation: Are all amphibian populations metapopulations? *Ecography* 28:110–128.

Smith, W. K. and Z. C. Berry. 2013. Sunflecks? *Tree Physiology* 33:233–237. doi:10.1093/treephys/tpt005.

Smith, L. L., J. M. Linehan, J. M. Stober, M. J. Elliott, and J. B. Jensen. 2009. An evaluation of distance sampling for large-scale gopher tortoise surveys in Georgia, USA. *Applied Herpetology* 6:355–368.

Smith, C. R., P. V. McCormick, S. W. Golladay, and A. P. Covich. 2015. Invertebrate assemblage changes indicative of reduced flow in an agricultural watershed. *Proceedings of the 2015 Georgia Water Resources Conference*, 28–29 April 2015, Athens, GA. http://gwri.gatech.edu/sites/default/files/files/docs/2015/6.4.1smith.pdf (accessed 4 February 2017).

Smith, L. L., D. A. Steen, L. M. Conner, and J. C. Rutledge. 2013. Effects of predator exclusion on nest and hatchling survival in the gopher tortoise. *Journal of Wildlife Management* 77:352–358.

Smith, L. L., D. A. Steen, J. M. Stober, M. C. Freeman, S. W. Golladay, L. M. Conner, and J. Cochrane. 2006. The vertebrate fauna of Ichauway, Baker County, GA. *Southeastern Naturalist* 5:599–620.

Smith, S. L., A. A. Sher, and T. A. Grant III. 2007. Genetic diversity in restoration materials and the impacts of seed collection in Colorado's restoration plant production industry. *Restoration Ecology* 15:369–374.

Smith, L. L., J. M. Stober, H. E. Balbach, and W. D. Meyer. 2009. *Gopher Tortoise Survey Handbook*. U.S. Army Corps of Engineers, Construction Engineering Research Laboratory ERDC/CERL TR-09-7.

Smith, A., W. T. Tinkham, D. P. Roy, L. Boschetti, R. L. Kremens, S. S. Kumar, A. M. Sparks, and M. J. Falkowski. 2013. Quantification of fuel moisture effects on biomass consumed derived from fire radiative energy retrievals. *Geophysical Research Letters* 40:6298–6302.

Snider, A. T. and J. K. Bowler. 1992. *Longevity of Reptiles and Amphibians in North American Collections*, 2nd edition. Herpetological Circular No. 21. Oxford, OH:, Society for the Study of Amphibians and Reptiles.

Snodgrass, J. W., J. W. Ackerman, A. L. Bryan, Jr., and J. Burger. 1999. Influence of hydroperiod, isolation, and heterospecifics on the distribution of aquatic salamanders (Siren and Amphiuma) among depression wetlands. *Copeia* 1999:107–113.

Society for Ecological Restoration (SER). 2004. SER primer on ecological restoration, version 2. http://www.ser.org/resource/resmgr/custompages/publications/SER_Primer/ser_primer.pdf (accessed 4 February 2017).

Soller, D. R. and H. H. Mills. 1991. Surficial geology and geomorphology. Pages 290–308 *in* W. Horton,Jr., and V. A. Zullo, editors. *The Geology of the Carolinas*. Knoxville, TN: The University of Tennessee Press.

Solomon, K. R., D. B. Baker, R. P. Richards, K. R. Dixon, S. J. Klaine, T. W. La Point, R. J. Kendall, C. P. Weisskopf, J. M. Giddings, J. P. Giesy, and L. W. Hall, Jr. 1996. Ecological risk assessment of atrazine in North American surface waters. *Environmental Toxicology and Chemistry* 15:31–76.

Song, C. and L. E. Band. 2004. MVP: A model to simulate the spatial patterns of photosynthetically active radiation under discrete forest canopies. *Canadian Journal of Forest Research* 34:1192–1203.

Song C., G. Katul, R. Oren, L. E. Band, C. L. Tague, P. C. Stoy, and H. R. McCarthy. 2009. Energy, water and carbon fluxes in a loblolly pine stand: Results from uniform and gappy canopy models with comparisons to eddy flux data. *Journal of Geophysical Research Biogeosciences* 114:G04021.

Sørensen, L. H. 1974. Rate of decomposition of organic matter in soil as influenced by repeated air drying-rewetting and repeated additions of organic material. *Soil Biology and Biochemistry* 6:287–292.

Sorice, M. G., W. Haider, J. R. Conner, and R. B. Ditton. 2011. Incentive structure of and private landowner participation in an endangered species conservation program. *Conservation Biology* 25:587–596.

Sorrie, B. A. and A. S. Weakley. 2001. Coastal plain plant endemics: Phytogeographic patterns. *Castanea* 66:50–82.

Sorrie, B. A. and A. S. Weakley. 2006. Conservation of the endangered *Pinus palustris* ecosystem based on coastal plain centers of plant endemism. *Applied Vegetation Science* 9:59–66.

South Carolina Natural Resources Conservation Service (SC NRCS). 2015. Cultivar and ecotype recommendations for partridge pea and switchgrass (guidance for CRP, CP-36). https://efotg.sc.egov.usda.gov/references/public/SC/Cultivar_and_ecotype_recommendations_for_Partridge_Pea_and_Switchgrass_Nov_2015.pdf (accessed 4 February 2017).

Sovada, M. A., A. B. Sargeant, and J. W. Grier. 1995. Differential effects of coyotes and red foxes on duck nest success. *Journal of Wildlife Management* 59:1–9.

Spurr, S. H. 1960. *Photogrammetry and Photo-interpretation. With a Section on Applications to Forestry.* Whitefish, MT: Kessinger Publishing, LLC.

Staller, E. L., W. E. Palmer, J. P. Carroll, R. P. Thornton, and D. C. Sisson. 2005. Identifying predators at northern bobwhite nests. *Journal of Wildlife Management* 69:124–132.

Stambaugh, M. C. and R. P. Guyette. 2004. Long-term growth and climate response of shortleaf pine at the Missouri Ozark Forest Ecosystem Project. Pages 448–458 *in* D. A. Yaussy, D. M. Hiz, R. P. Long, and P. C. Goebel, editors. *Proceedings of the 14th Central Hardwood Forest Conference*, 16–19 March 2004, Wooster, Ohio, USA. General Technical Report GTR-NE-316. USDA Forest Service, Northeastern Research Station, Newtown Square, PA.

Stambaugh, M. C., R. P. Guyette, and J. M. Marschall. 2011. Longleaf pine fire scars reveal new detail of a frequent fire regime. *Journal of Vegetation Science* 22:1094–1104.

Stambaugh, M. C., R. P. Guyette, and J. Marschall. 2013. Fire history in the Cherokee Nation of Oklahoma. *Human Ecology* 41:749–758.

Stambaugh, M. C. and R. M. Muzika. 2007. Successional trends of six mature shortleaf pine forests in Missouri. Pages 59–67 *in* J. M. Kabrick, D. C. Dey, and D. Gwayze, editors. *Shortleaf Pine Restoration and Ecology in the Ozarks. Proceedings of a Symposium*, 7–9 November 2006, Springfield, Missouri, USA. General Technical Report Proceedings NRS-P-15. USDA Forest Service, Northern Research Station, Newtown Square, PA.

Stambaugh, M. C., R. M. Muzika, and R. P. Guyette. 2002. Disturbance characteristics and overstory composition of an old-growth shortleaf pine (*Pinus echinata* Mill.) forest in the Ozark Highlands, Missouri, USA. *Natural Areas Journal* 22:108–119.

Stangel, P. W., M. R. Lennartz, and M. H. Smith. 1992. Genetic variation and population structure of red-cockaded woodpeckers. *Conservation Biology* 6:283–292.

Stanturf, J. A. and S. L. Goodrick. 2013. Fire. Pages 509–542 *in* D. N. Wear, and J. G. Greis, editors. The southern forest futures research project. General Technical Report SRS-GTR-178. USDA Forest Service, Southern Research Station, Asheville, NC.

Stanturf, J. A., S. L. Goodrick, and K. W. Outcalt. 2007. Disturbance and coastal forests: A strategic approach to forest management in hurricane impact zones. *Forest Ecology and Management* 250:119–135.

Stanturf, J. A., B. J. Palik, and R. K. Dumroese. 2014. Contemporary forest restoration: A review emphasizing function. *Forest Ecology and Management* 331:292–323.

Stapanian, M. A., T. E. Lewis, C. J. Palmer, and M. M. Amos. 2016. Assessing accuracy and precision for field and laboratory data: A perspective in ecosystem restoration. *Restoration Ecology* 24:18–26.

Starr, G., C. L. Staudhammer, and H. W. Loescher. 2014. Time series analysis of forest carbon dynamics: Recovery of *Pinus palustris* physiology following a prescribed fire. *New Forests* 46: 63–90.

Starr, G., C. L. Staudhammer, H. W. Loescher, R. J. Mitchell, A. Whelan, J. K. Hiers, and J. J. O'Brien. 2015. Time series analysis of forest carbon dynamics: Recovery of *Pinus palustris* physiology following a prescribed fire. *New Forests* 46:63–90.

Starr, G., C. L. Staudhammer, S. Wiesner, S. Kunwor, H. W. Loescher, A. F. Baron, A. Whelan, R. J. Mitchell, and L. R. Boring. 2016. Carbon dynamics of *Pinus palustris* ecosystems following drought. *Forests* 7:98. doi:10.3390/ f705009.

Staver, A. C., W. J. Bond, W. D. Stock, S. J. Van Rensburg, and M. S. Waldram. 2009. Browsing and fire interact to suppress tree density in an African savanna. *Ecological Applications* 19:1909–1919.

Stebbins, G. L. 1981. Coevolution of grasses and herbivores. *Annals of the Missouri Botanical Garden* 68:75–86.

Steel, Z. L., H. D. Safford, and J. H. Viers. 2015. The fire frequency-severity relationship and the legacy of fire suppression in California forests. *Ecosphere* 6:Article 8.

Steen, D. A., L. M. Conner, L. L, Smith, L. Provencher, J. K. Hiers, S. Pokswinski, B. S. Helms, and C. Guyer. 2013. Bird assemblage response to restoration of fire-suppressed longleaf pine sandhills. *Ecological Applications* 23:134–147.

Steen D. A., J. P. Gibbs, K. A. Buhlmann, J. L. Carr, B. W. Compton, J. D. Congdon, J. S. Doody, J. C. Godwin, K. L. Holcomb, D. R. Jackson, F. J. Janzen, G. Johnson, M. T. Jones, J. T. Lamer, T. A. Langen, M. V. Plummer, J. W. Rowe, R. A. Saumure, J. K. Tucker, and D. S. Wilson. 2012. Terrestrial habitat requirements of nesting freshwater turtles. *Biological Conservation* 150:121–128.

Steen, D. A., J. M. Linehan, and L. L. Smith. 2010. Multiscale habitat selection and refuge use of common kingsnakes, *Lampropeltis getula*, in southwestern Georgia. *Copeia* 2010:227–231.

Steen, D. A., L. L. Smith, L. M. Conner, A. R. Litt, L. Provencher, J. K. Hiers, S. Pokswinski, and C. Guyer. 2013. Reptile assemblage response to restoration of fire-suppressed longleaf pine sandhills. *Ecological Applications* 23:148–158.

Steen, D. A., A. D. Steen, S. Pokswinski, S. P. Graham, and L. L. Smith. 2010. Snakes using stumpholes and windfall tree-associated subterranean structures in longleaf pine forests. *Reptiles & Amphibians* 17:49–51.

Steen, D. A., S. C. Sterrett, S. A. Miller, and L. L. Smith. 2007. Terrestrial movements and microhabitat selection of overwintering subadult eastern mud turtles (*Kinosternon subrubrum*) in Southwest Georgia. *Journal of Herpetology* 41:532–535.

Steen, D. A., D. J. Stevenson, J. C. Beane, J. D. Willson, M. J. Aresco, J. C. Godwin, S. P. Graham, L. L. Smith, J. M. Howze, D. C. Rudolph, J. B. Pierce, J. R. Lee, B. B. Gregory, J. Jensen, S. H. Stiles, J. A. Stiles, N. H. Nazdrowicz, and C. Guyer. 2013. Terrestrial movements of the red-bellied mudsnake (*Farancia abacura*) and rainbow snake (*F. erytrogramma*). *Herpetological Review* 44:208–213.

Stephens, S. L., S. W. Bigelow, R. D. Burnett, B. M. Collins, C. V. Gallagher, J. Keane, D. A. Kelt, M. P. North, L. J. Roberts, P. A. Stine, and D. H. Van Vuren. 2014. California spotted owl, songbird, and small mammal responses to landscape fuel treatments. *BioScience* 64:893–906.

Stephens, S. L. and B. M. Collins. 2004. Fire regimes of mixed conifer forest in the north-central Sierra Nevada at multiple spatial scales. *Northwest Science* 78:12–23.

Stephens, S. L. and M. A. Finney. 2002. Prescribed fire mortality of Sierra Nevada mixed conifer tree species: Effects of crown damage and forest floor combustion. *Forest Ecology and Management* 162:261–271.

Stephens, S. L., M. A. Finney, and H. Schantz. 2004. Bulk density and fuel loads of ponderosa pine and white fir forest floors: Impacts of leaf morphology. *Northwest Science* 78:93–110.

Stephens, S. L., J. M. Lydersen, B. M. Collins, D. L. Fry, and M. D. Meyer. 2015. Historical and current landscape-scale ponderosa pine and mixed conifer forest structure in the southern Sierra Nevada. *Ecosphere* 6:Article 79.

Stephens, S. L. and L. W. Ruth. 2005. Federal forest-fire policy in the United States. *Ecological Applications* 15:532–542.

Stephenson, N. L. 1990. Climatic control of vegetation distribution: The role of the water balance. *American Naturalist* 135:649–670.

Stephenson, N. L. 1998. Actual evapotranspiration and deficit: Biologically meaningful correlates of vegetation distribution across spatial scales. *Journal of Biogeography* 25:855–870.

Stevens, G. C. and B. J. Enquist. 1999. Macroecological limits to the abundance and distribution of *Pinus*. Pages 183–192 *in* D. M. Richardson, editor. *Ecology and Biogeography of Pinus*. Cape Town: Cambridge University Press.

Stewart, O. C. 2002. *Forgotten Fires: Native Americans and the Transient Wilderness*. Norman, OK: University of Oklahoma Press.

Stiles, S. H. 2013. Seasonal habitat and shelter selection by reintroduced eastern indigo snakes in Conecuh National Forest, Alabama, MS thesis. Auburn University, Auburn, AL.

Stoddard, H. L. 1931. *The Bobwhite Quail: Its Habits, Preservation and Increase*. New York, NY: Charles Scribner's Sons.

Stoddard, H. L. 1935. Use of controlled fire in southeastern upland game management. *Journal of Forestry* 33:346–351.

Stoddard, H. L. 1969. *Memoirs of a Naturalist*. Norman, OK: University of Oklahoma Press.

Stout, I. J. and W. R. Marion. 1993. Pine flatwoods and xeric pine forests of the southern (lower) coastal plain. Pages 373–446 *in* H. Martin, S. G. Boyce, and A. C. Echternacht, editors. *Biodiversity of the southern United States: Lowland Terrestrial Communities*. New York, NY: John Wiley & Sons.

Streng, D. R., J. S. Glitzenstein, and W. J. Platt. 1993. Evaluating effects of season of burn in longleaf pine forests: A critical literature review and some results from an ongoing long-term study. Pages 227–263 *in* S. M. Hermann, editor. *The Longleaf Pine Ecosystem: Ecology, Restoration and Management. Proceedings 18th Tall Timbers Fire Ecology Conference*, 30 May-2 June 1991, Tallahassee, Florida, USA. Tall Timbers Research Station, Tallahassee, FL.

Strom, B. L., R. A. Goyer, L. L. Ingram, G. D. L. Boyd, and L. H. Lott. 2002. Oleoresin characteristics of progeny of loblolly pines that escaped attack by the southern pine beetle. *Forest Ecology and Management* 158:169–178.

Stuber, O. S. 2013. The relationship between land use and the ecological integrity of isolated wetlands in the Dougherty Plain, Georgia, USA, MS thesis. University of Georgia, Athens, GA.

Stuber, O. S., L. K. Kirkman, J. Hepinstall-Cymerman, and G. I. Martin. 2016. The ecological condition of geographically isolated wetlands in the southeastern United States: The relationship between landscape level assessments and macrophyte assemblages. *Ecological Indicators* 62:191–200.

Stuble, K. L., L. K. Kirkman, and C. R. Carroll. 2010. Are red imported fire ants facilitators of native seed dispersal? *Biological Invasions* 12:1661–1669.

Subalusky, A. L., L. A. Fitzgerald, and L. L. Smith. 2009. Ontogenetic niche shifts in the American alligator establish functional connectivity between aquatic systems. *Biological Conservation* 142:1507–1514.

Suding, K. N. 2011. Toward an era of restoration in ecology: Successes, failures, and opportunities ahead. *Annual Review of Ecology, Evolution, and Systematics* 42:465–487.

Suding, K. N., K. L. Gross, and G. R. Houseman. 2004. Alternative states and positive feedbacks in restoration ecology. *Trends in Ecology and Evolution* 19:46–53.

Sullivan, B. K., E. M. Nowak, and M. A. Kwiatkowski. 2015. Problems with mitigation translocation of herpetofauna. *Conservation Biology* 29:12–18.

Sun, G., P. V. Caldwell, A. P. Georgakakos, S. Arumugam, J. Cruise, R. T. McNider, A. Terando, P. A. Conrads, J. Feldt, V. Misra, L. Romolo, T. C. Rasmussen, S. G. McNulty, and D. A. Marion. 2013. Impacts of climate change and variability on water resources in the southeastern U.S. Pages 210–236 *in* K. T. Ingram, K. Dow, and L. Carter, editors. *Southeastern Regional Technical Report to the National Climate Change Assessment*, Chap. 10. Washington, DC: Island Press.

Sun, G., P. V. Caldwell, and S. G. McNulty. 2015. Modeling the potential role of forest thinning in maintaining water supplies under a changing climate across the conterminous United States. *Hydrological Processes* 29:5016–5030.

Sun, G., P. V. Caldwell, A. Noormets, E. Cohen, S. G. McNulty, E. Treasure, J. C. Domec, Q. Mu, J. Xiao, R. John, and J. Chen. 2011. Upscaling key ecosystem functions across the conterminous United States by a water-centric ecosystem model. *Journal of Geophysical Research* 116:1–16.

Sun, G., S. G. McNulty, J. Lu, D. M. Amatya, Y. Liang, and R. K. Kolka. 2005. Regional annual yield from forest lands and its response to potential deforestation across the southeastern United States. *Journal of Hydrology* 308:258–268.

Sun, G., S. G. McNulty, J. A. Moore Myers, and E. C. Cohen. 2008. Impacts of multiple stresses on water demand and supply across the southeastern United States. *Journal of American Water Resources Association* 44:1441–1457.

Sun, G. and J. M. Vose. 2016. Forest management challenges for sustaining water resources in the Anthropocene. *Forests* 7:1–13.

Sutherland, W. J. 1996. *Ecological Census Techniques: A Handbook*. Cambridge: Cambridge University Press.

Sutter, R. D., J. J. Bachant, D. R. Gordon, and A. R. Litt. 2001. An assessment of the desired future conditions for focal conservation targets on Eglin Air Force Base. Report to Natural Resources Division, Eglin Air Force Base, Niceville, Florida. The Nature Conservancy, Gainesville, FL.

Sutter R. D. and R. Kral. 1994. The ecology, status, and conservation of two non-alluvial wetland communities in the south Atlantic and eastern Gulf coastal plain, USA. *Biological Conservation* 68:235–243.

Sutter, R. D., S. Wainscott, J. R. Boetsch, C. J. Palmer, and D. J. Rugg. 2015. Practical guidance for integrating data management into long-term ecological monitoring projects. *The Wildlife Society Bulletin* 39:451–463.

Swank, W. T. and J. E. Douglass. 1974. Streamflow greatly reduced by converting deciduous hardwood stands to pine. *Science* 185:857–859.

Swank, W. T. and N. H. Miner. 1968. Conversion of hardwood-covered watersheds to white pine reduces water yield. *Water Resources Research* 4:947–954.

Swanton, J. R. 1979. *The Indians of the Southeastern United States*. Washington, DC: Smithsonian Institution Press.

Swenson, N. G., P. Anglada-Cordero, and J. A. Barone. 2011. Deterministic tropical tree community turnover: Evidence from patterns of functional beta diversity along an elevational gradient. *Proceedings of the Royal Society B-Biological Sciences* 278:877–884.

Sword, M. A. and J. D. Haywood. 1999. Effects of crown scorch on longleaf pine fine roots. Pages 223–227 *in* J. D. Haywood, editor. *Proceedings of the 10th Biennial Southern Silvicultural Research Conference*. General Technical Report SRS-30. USDA Forest Service, Southern Research Station, Asheville, NC.

Tague, C. L. and L. E. Band. 2004. RHESSys: Regional hydro-ecologic simulation system-an object-oriented approach to spatially distributed modeling of carbon, water, and nutrient cycling. *Earth Interactions* 8:1–42.

Tague, C. L., N. G. McDowell, and C. D. Allen. 2013. An integrated model of environmental effects on growth, carbohydrate balance, and mortality of *Pinus ponderosa* forests in the southern Rocky Mountains. *PLOS ONE* 8:e80286.

Tai, A. P. K., M. Val Martin, and C. L. Heald. 2014. Threat to future global food security from climate change and ozone air pollution. *Nature Climate Change* 4:817–821.

Taillie, P. J., M. N. Peterson, and C. E. Moorman. 2015. The relative importance of multiscale factors in the distribution of Bachman's sparrow and the implications for ecosystem conservation. *The Condor* 117:137–146.

Talmage, R. V. and G. D. Buchanan. 1954. The geographical distribution of the armadillo in the United States. *Texas Journal of Science* 6:142–150.

Taulman, J. F. and L. W. Robbins. 1996. Recent range expansion and distributional limits of the nine-banded armadillo in the United States. *Journal of Biogeography* 23:635–648.

Taulman, J. F. and L. W. Robbins. 2014. Range expansion and distributional limits of the nine-banded armadillo in the United States: An update of Taulman & Robbins (1996). *Journal of Biogeography* 41:1626–1630.

Taylor, A. H. 2000. Fire regimes and forest changes in mid and upper montane forests of the southern Cascades. *Journal of Biogeography* 27:87–104.

Taylor, A. H. 2004. Identifying forest reference conditions on early cut-over lands, Lake Tahoe Basin, USA. *Ecological Applications* 14:1903–1920.

Teeter, L. and G. Somers. 2005. Longleaf ecosystem restoration: Long rotation economics. Pages 129–139 *in* J. S. Kush, editor. *Proceedings of the Fifth Longleaf Alliance Regional Conference*, 12–15 October 2004, Hattiesburg, Mississippi, USA. The Longleaf Alliance, Andalusia, AL.

Terando, A. J., J. K. Costanza, C. Belyea, R. R. Dunn, A. J. McKerrow, and J. Collazo. 2014. The southern megalopolis: Using the past to predict the future of urban sprawl in the Southeast U.S. *PLOS ONE* 9:e102261.

Terhune, T. M., D. C. Sisson, and H. L. Stribling. 2006. The efficacy of relocating wild northern bobwhites prior to breeding season. *Journal of Wildlife Management* 70:914–921.

Thaxton, J. M. and W. J. Platt. 2006. Small-scale fuel variation alters fire intensity and shrub abundance in a pine savanna. *Ecology* 87:1331–1337.

The Conservation Fund. 2016. Coastal headwaters forest—longleaf conservation and restoration. http://www.conservationfund.org/projects/coastal-headwaters-forest-longleaf-conservation-and-restoration (accessed 4 February 2017).

The Longleaf Alliance. 2016. The longleaf alliance, Andalusia, AL. http://www.longleafalliance.org/restoring-and-managing/restoration/determine-the-starting-point (accessed 4 February 2017).

The Nature Conservancy (TNC). 2009. Global ecoregions, major habitat types, biogeographical realms, and The Nature Conservancy terrestrial assessment units as of 14 December 2009. http://maps.tnc.org/files/metadata/TerrEcos.xml (accessed 4 February 2017).

Theobald, D. M. and W. H. Romme. 2007. Expansion of the U.S. wildland-urban interface. *Landscape and Urban Planning* 83:340–354.

Thornton, D. H., M. E. Sunquist, and M. B. Main. 2004. Ecological separation within newly sympatric populations of coyotes and bobcats in south-central Florida. *Journal of Mammalogy* 85:973–982.

Thurber, J. M. and R. O. Peterson. 1993. Effects of population density and pack size on the foraging ecology of gray wolves. *Journal of Mammalogy* 74:879–889.

Thurgate, N. Y. and J. H. Pechmann, 2007. Canopy closure, competition, and the endangered dusky gopher frog. *Journal of Wildlife Management* 71:1845–1852.

Tiegs, S. D., K. A. Berven, D. J. Carmack, and K. A. Capps. 2015. Stoichiometric implications of a biphasic life cycle. *Oecologia* 180:853–863. doi:10.1007/s00442-015-3504-2.

Tilman, D. 1994. Competition and biodiversity in spatially structured habitats. *Ecology* 75:2–16.

Tilman, D. and S. Pacala. 1993. The maintenance of species richness in plant communities. Pages 13–25 *in* R. E. Ricklefs, and D. Schluter, editors. *Species Diversity in Ecological Communities*. Chicago, IL: The University of Chicago Press.

Timmerman, W. W. and W. H. Martin. 2003. Conservation Guide to the Eastern Diamondback Rattlesnake, *Crotalus adamanteus*. SSAR, Herpetological Circular No. 32.

Tiner, R. W. 2003. Geographically isolated wetlands of the United States. *Wetlands* 23:494–516.

Todd, D., J. Platt, and J. Hodges. 1995. The impact and value of tree improvement in the South. Pages 7–15 *in* R. J. Weir, and A. V. Hatcher, editors. Sponsored Publication No. 45. *Proceedings of the 23rd Southern Forest Tree Improvement Conference*, 20–22 June 1995, Asheville, NC.

Toumey, J. W. and C. F. Korstian. 1937. *Foundations of Silviculture upon an Ecological Basis*, 2nd edition. New York, NY: John Wiley & Sons.

Trainor, A. M., J. R. Walters, W. F. Morris, J. Sexton, and A. Moody. 2013. Empirical estimation of dispersal resistance surfaces: A case study with red-cockaded woodpeckers. *Landscape Ecology* 28:755–767.

Trainor, A. M., J. R. Walters, D. Urban, and A. Moody. 2013. Evaluating the effectiveness of a Safe Harbor Program for connecting wildlife populations. *Animal Conservation* 16:610–620.

Trani, M. K. 2002. Terrestrial ecosystems. Pages 1–45 *in* D. N. Wear, and J. G. Greis, editors. *Southern Forest Resource Assessment*. General Technical Report SRS-53. USDA Forest Service, Southern Research Station, Asheville, NC.

Treseder, K. K. 2004. A meta-analysis of mycorrhizal responses to nitrogen, phosphorus and atmospheric CO_2 in field studies. *New Phytologist* 164:347–355.

Tucker, J., G. E. Hill, and N. Holler. 2003. Longleaf pine restoration: Implications for landscape-level effects on bird communities in the lower Gulf coastal plain. *Southern Journal of Applied Forestry* 27:107–121.

Turesson, G. 1922. The genotypical response of the plant species to the habitat. *Hereditas* 3:211–350.

Turner, C. L. and W. E. Grant. 1987. Effects of removal of *Sigmodon hispidus* on microhabitat utilization by *Baiomys taylori* and *Reithrodontomys fulvescens*. *Journal of Mammalogy* 68:80–85.

Turner, M. G., W. H. Romme, and R. H. Gardner. 1999. Prefire heterogeneity, fire severity, and early postfire plant reestablishment in subalpine forests of Yellowstone National Park, Wyoming. *International Journal of Wildland Fire* 9:21–36.

Ukkola, A. M., I. C. Prentice, T. F. Keenan, A. I. J. van Dijk, N. R. Viney, R. B. Myneni, and J. Bi. 2016. Reduced streamflow in water-stressed climates consistent with CO_2 effects on vegetation. *Nature Climate Change* 6:75–78.

Underhill, J. L., Y. Dickinson, A. Rudney, and J. Thinnes. 2014. Silviculture of the Colorado Front Range landscape restoration initiative. *Journal of Forestry* 112:484–493.

Underwood, E. C., J. H. Viers, J. F. Quinn, and M. North. 2010. Using topography to meet wildlife and fuels treatment objectives in fire-suppressed landscapes. *Environmental Management* 46:809–819.

Urbanski, S. P. 2014. Wildland fire emissions, carbon, and climate: Emission factors. *Forest Ecology and Management* 317:51–60.

Urbanski, S. P., W. M. Hao, and S. Baker. 2008. Chemical composition of wildland fire emissions. *Developments in Environmental Science* 8:79–107.

U.S. Department of Agriculture (USDA). 2015. Summary Report: 2012. National Resources Inventory, Natural Resources Conservation Service, Washington, D.C., and Center for Survey Statistics and Methodology, Iowa State University, Ames, IA. https://www.nrcs.usda.gov/Internet/FSE_DOCUMENTS/nrcseprd396218.pdf (accessed 4 February 2017).

U.S. Department of Agriculture (USDA). 2016. Agricultural Conservation Easement Program. Washington, DC. https://www.nrcs.usda.gov/wps/portal/nrcs/main/national/programs/easements/acep/ (accessed 4 February 2017).

U.S. Department of Defense (U.S. DOD) and Environmental Protection Agency (EPA). 2015. Clean water rule: Definition of "Waters of the United States". *Federal Register* 80(124):37054–37127.

U.S. Environmental Protection Agency. Air Pollution Training Institute. (U.S. EPA APTI). 2016. Health effects of ozone in patients with asthma and other chronic respiratory disease. https://www.epa.gov/ozone-pollution-and-your-patients-health/health-effects-ozone-patients-asthma-and-other-chronic (accessed 4 February 2017).

U.S. Environmental Protection Agency (U.S. EPA). 1998. Interim air quality policy on wildland and prescribed fires. U.S. Environmental Protection Agency, Office of Air Quality Planning and Standards, Research Triangle Park, Durham, NC. https://www.frames.gov/files/4613/6018/1437/Interim_air_quality_policy_1998.pdf (accessed 4 February 2017).

U.S. Environmental Protection Agency (U.S. EPA). 2007. Treatment of data influenced by exceptional events: Final rule. Federal Register 72, no. 55 (March 22, 2007): 13560 . https://www.gpo.gov/fdsys/pkg/FR-2007-03-22/pdf/E7-5156.pdf. Accessed 11 May 2017.

U.S. Environmental Protection Agency (U.S. EPA). 2015a. 2015 National Ambient Air Quality Standards (NAAQS) for Ozone. U.S. Environmental Protection Agency, Washington, DC. https://www.epa.gov/ozone-pollution/2015-national-ambient-air-quality-standards-naaqs-ozone (accessed 4 February 2017).

U.S. Environmental Protection Agency (U.S. EPA). 2015b. Proposed Exceptional Events Rule Revisions. U.S. Environmental Protection Agency, Washington, DC. https://www.epa.gov/sites/production/files/2015-11/documents/ee_nprm_11-20-15_80_fr_72840.pdf (accessed 4 February 2017).

U.S. Environmental Protection Agency (U.S. EPA). 2015c. Connectivity of streams and wetlands to downstream waters: A review and synthesis of the scientific evidence (Final Report). EPA/600/R-14/475F. U.S. Environmental Protection Agency, Washington, DC.

U.S. Environmental Protection Agency (U.S. EPA). 2016a. NAAQS Table. https://www.epa.gov/criteria-air-pollutants/naaqs-table (accessed December 2016).

U.S. Environmental Protection Agency. (U.S. EPA). 2016b. Particulate matter (PM) basics. https://www.epa.gov/pm-pollution/particulate-matter-pm-basics (accessed 4 February 2017).

U.S. Environmental Protection Agency (U.S. EPA). 2016c. Treatment of data influenced by exceptional events: Final rule. Federal Register 81, no. 191 (October 3, 2016): 68216. https://www.epa.gov/sites/production/files/2016-09/documents/exceptional_events_rule_revisions_2060-as02_final.pdf. Accessed 11 May 2017.

U.S. Fish and Wildlife Service (U.S. FWS). 2003. *Recovery Plan for the Red-Cockaded Woodpecker (Picoides borealis), Second Revision.* Atlanta, GA: U.S. Fish and Wildlife Service, Southeast Region. https://www .fws.gov/rcwrecovery/files/RecoveryPlan/finalrecoveryplan.pdf. (accessed 4 February 2017).

U.S. Fish and Wildlife Service (U.S. FWS). 2011. Endangered and threatened wildlife and plants; 12-month finding on a petition to list the striped newt as threatened. *Federal Register* 76(109):32911–32929. U.S. Fish and Wildlife Service, Southeast Region, Atlanta, GA.

U.S. Fish and Wildlife Service (U.S. FWS). 2013. Alabama Department of Conservation and Natural Resources, Georgia Department of Natural Resources, Florida Fish and Wildlife Conservation Commission, Louisiana Department of Wildlife and Fisheries, Mississippi Department of Wildlife Fisheries & Parks, South Carolina Department of Natural Resources. 2013. Range-Wide Conservation Strategy for the Gopher Tortoise.

U.S. Fish and Wildlife Service (U.S. FWS). 2014. Technical/agency draft recovery plan for the dusky Gopher frog (*Rana sevosa*). U.S. Fish and Wildlife Service, Southeast Region, Atlanta, GA.

U.S. Geological Survey (USGS). 2016. National Gap Analysis Program Protected Areas Data Portal. https://gapanalysis.usgs.gov/padus/ (accessed 13 April 2017).

U.S. Department of Agriculture, Forest Service (USDA FS). 2015. The rising cost of fire operations: Effects on the Forest Service's non-fire work. USDA Forest Service, Washington, DC. http://www.fs.fed.us/sites/default/files/2015-Rising-Cost-Wildfire-Operations.pdf (accessed 4 February 2017).

U.S. Department of Agriculture, Forest Service (USDA FS). 2016. Francis Marion National Forest Revised Land Management Plan. August 2016. R8-MB 149A. Southern Region, Atlanta, GA.

U.S. Department of Agriculture, Natural Resources Conservation Service (USDA NRCS). 2001. Summary Report: 1997 Natural Resources Inventory (revised Dec 2001). USDA Natural Resources Conservation Service, Washington, DC.

U.S. Department of Agriculture, Natural Resources Conservation Service (USDA NRCS). 2008. Using Farm Bill Programs for Pollinator Conservation. Technical Note No. 78. USDA Natural Resources Conservation Service, Washington, DC. https://plants.usda.gov/pollinators/Using_Farm_Bill_Programs_for_Pollinator_Conservation.pdf (accessed 4 February 2017).

U.S. Department of Agriculture, Natural Resources Conservation Service (USDA NRCS). 2016. The PLANTS Database 2016. USDA Natural Resources Conservation, Service National Plant Data Team, Greensboro, NC. http://plants.usda.gov (accessed 4 February 2017).

U.S. Geological Survey. 1974. Bethany quadrangle, Georgia and Elmodel quadrangle, Georgia [maps]. 1:24,000. 7.5 Minute Series. Reston, VA: United States Department of the Interior, USGS.

Valladares, F. and Ü. Niinemets. 2008. Shade tolerance, a key plant feature of complex nature and consequences. *Annual Review of Ecology, Evolution, and Systematics* 39:237–257.

van Aardt, J. 2000. Spectral separability among six southern tree species, MS thesis. Virginia Polytechnic Institute and State University, Blacksburg, VA.

Van Auken, O. W. 2000. Shrub invasions of North American semiarid grasslands. *Annual Review of Ecology and Systematics* 31:197–215.

Vance, R. B. 1932. *Human Geography of the South: A Study in Regional Resources and Human Adequacy.* Chapel Hill, NC: University of North Carolina Press.

Van De Gevel, S. L., J. L. Hart, H. D. Grissino-Mayer, and K. W. Robinson. 2007. Tree-ring dating of old-growth longleaf pine (*Pinus palustris* Mill.) logs from an exposed timber crib dam, Hope Mills, North Carolina, USA. *Tree-Ring Research* 65:69–80.

van den Avyle, M. J. and J. W. Evans. 1990. Temperature selection by striped bass in a Gulf of Mexico coastal river system. *North American Journal of Fisheries Management* 10:58–66.

van der Maarel, E. and M. T. Sykes. 1993. Small-scale plant species turnover in a limestone grassland: The carousel model and some comments on the niche concept. *Journal of Vegetation Science* 4:179–188.

van Eerden, B. P. 1998. Studies on the reproductive biology of wiregrass (*Aristida stricta* Michaux) in the Carolina Sandhills, MS thesis. University of Georgia, Athens, GA.

van Kooten, G. C., C. S. Binkley, and G. Delcourt. 1995. Effect of carbon taxes and subsidies on optimal forest rotation age and supply of carbon services. *American Journal of Agricultural Economics* 77:365–374.

Van Langevelde, F., C. A. D. M Van De Vijver, L. Kumar, J. Van De Koppel, N. De Ridder, J. J. Van Ande, A. K. Skidmore, J. W. Hearne, L. Stroosnijder, W. J. Bond, H. H. T. Prins, and M. Rietkerk. 2003. Effects of fire and herbivory on the stability of savanna ecosystems. *Ecology* 84:337–350.

Van Lear, D. H., W. D. Carroll, P. R. Kapeluck, and R. Johnson. 2005. History and restoration of the longleaf pine-grassland ecosystem: Implications for species at risk. *Forest Ecology and Management* 211:150–165.

van Mantgem, P. and M. Schwartz. 2003. Bark heat resistance of small trees in Californian mixed conifer forests: Testing some model assumptions. *Forest Ecology and Management* 178:341–352.

van Mantgem, P. J., A. C. Caprio, N. L. Stephenson, and A. J. Das. 2016. Does prescribed fire promote resistance to drought in low elevation forests of the Sierra Nevada, California, USA? *Fire Ecology* 12:13–25.

Van Wagner, C. E. 1968. The line intersect method in forest fuel sampling. *Forest Science* 14:20–26.

Van Wagner, C. E. 1971. Two solitudes in forest fire research. Information Report PS-X-29. Canadian Forest Service, Ottawa, Canada.

Vargas, R., S. E. Trumbore, and M. F. Allen. 2009. Evidence of old carbon used to grow new fine roots in a tropical forest. *New Phytologist* 182:710–718.

Varner, J. M., M. Arthur, S. Clark, D. Dey, J. Hart, and C. Schweitzer. 2016. Fire in eastern North American oak ecosystems: Filling the gaps. Fire Ecology 12:1–6. doi:10.4996/fireecology.1202001.

Varner, J. M., D. R. Gordon, F. E. Putz, and J. K. Hiers. 2005. Restoring fire to long-unburned Pinus palustris ecosystems: Novel fire effects and consequences for long-unburned ecosystems. *Restoration Ecology* 13:536–544.

Varner, J. M., III, J. K. Hiers, R. D. Ottmar, D. R. Gordon, F. E. Putz, and D. D. Wade. 2007. Overstory tree mortality resulting from reintroducing fire to long-unburned longleaf pine forests: The importance of duff moisture. *Canadian Journal of Forest Research* 37:1349–1358.

Varner, J. M., J. M. Kane, J. K. Hiers, J. K. Kreye, and J. W. Veldman. 2016. Suites of fire-adapted traits in the southeastern USA oaks: Multiple strategies for persistence. *Fire Ecology* 12:48–64. doi: 10.4996/fireecology.1202048.

Varner, J. M., J. M. Kane, J. K. Kreye, and E. Engber. 2015. The flammability of forest and woodland litter: A synthesis. *Current Forestry Reports* 1:91–99. doi:10.1007/s40725-015-0012-x.

Varner, J. M., III, and J. S. Kush. 2004. Remnant old-growth longleaf pine (*Pinus palustris* Mill.) savannas and forests of the southeastern USA: Status and threats. *Natural Areas Journal* 24:141–149.

Varner, J. M. and N. Pederson. 2004. The dendrochronological potential of Suwannee Basin tree species. Unpublished report to Suwanee River Water Management District, Live Oak, FL.

Varner, J. M., F. E. Putz, J. J. O'Brien, J. K. Hiers, R. J. Mitchell, and D. R. Gordon. 2009. Post-fire tree stress and growth following smoldering duff fires. *Forest Ecology and Management* 258:2467–2474.

Veblen, T. T., T. Kitzberger, and J. Donnegan. 2000. Climatic and human influences on fire regimes in ponderosa pine forests in the Colorado Front Range. *Ecological Applications* 10:1178–1195.

Veblen, T. T. and D. C. Lorenz. 1991. *The Colorado Front Range: A Century of Ecological Change*. Salt Lake City, UT: University of Utah Press.

Vekemans, X. and O. J. Hardy. 2004. New insights from fine-scale spatial genetic structure analyses in plant populations. *Molecular Ecology* 13:921–935.

Veldman, J. W., L. A. Brudvig, E. I. Damschen, J. L. Orrock, W. B. Mattingly, and J. L. Walker. 2014. Fire frequency, agricultural history and the multivariate control of pine savanna understory plant diversity. *Journal of Vegetation Science* 25:1438–1449.

Veldman, J. W., E. Buisson, G. Durigan, E. G. Fernandes, S. Le Stradic, G. Mahy, D. Negreiros, G. E. Overbeck, R. G. Veldman, N. P. Zaloumis, F. E. Putz, and W. J. Bond. 2015. Toward an old-growth concept for grasslands, savannas, and woodlands. *Frontiers in Ecology and the Environment* 13:154–162.

Veldman, J. W., W. B. Mattingly, and L. A. Brudvig. 2013. Understory plant communities and the functional distinction between savanna trees, forest trees, and pines. *Ecology* 94:424–434.

Vellend, M., D. S. Srivastava, K. M. Anderson, C. D. Brown, J. E. Jankowski, E. J. Kleynhans, N. J. B. Kraft, A. D. Letaw, A. A. M. Macdonald, J. E. Maclean, I. H. Myers-Smith, A. R. Norris, and X. Xue. 2014. Assessing the relative importance of neutral stochasticity in ecological communities. *Oikos* 123:1420–1430.

Veno, P. A. 1976. Successional relationships of five plant communities. *Ecology* 57:498–508.

Verhoeven, J. T., B. Arheimer, C. Yin, and M. M. Hefting. 2006. Regional and global concerns over wetlands and water quality. *Trends in Ecology and Evolution* 21:96–103.

Vines, T. H., A. Y. Albert, R. L. Andrew, F. Débarre, D. G. Bock, M. T. Franklin, and D. J. Rennison. 2014. The availability of research data declines rapidly with article age. *Current Biology* 24:94–97.

Viney, N. R. 1991. A review of fine fuel moisture modelling. *International Journal of Wildland Fire* 1:215–234.

Vitková, L., Á. N. Dhubháin, and A. Pommerening. 2016. Agreement in tree marking: What is the uncertainty of human tree selection in selective forest management? *Forest Science* 62:288–296.

Vogel, K. P., M. R. Schmer, and R. B. Mitchell. 2005. Plant adaptation regions: Ecological and climatic classification of plant materials. *Rangeland Ecology and Management* 58:315–319.

Vogl, R. J. 1973. Fire in the southeastern grasslands. Pages 175–198. *Proceeedings of the Tall Timbers Fire Ecology Conference*, Tall Timbers Research Station, Tallahassee, FL.

Vogl, R. J. 1979. Some basic principles of grassland fire management. *Environmental Management* 3:51–57.

Vose J. M. and K. D. Klepzig. 2014. *Climate Change Adaptation and Mitigation Management Options*. Boca Raton, FL: CRC Press.

Vose, J. M., C. F. Miniat, C. H. Luce, H. Asbjornsen, P. V. Caldwell, J. L. Campbell, G. E. Grant, D. J. Isaak, S. P. Loheide, and G. Sun. 2016. Ecohydrological implications of drought for forests in the United States. *Forest Ecology and Management* 380:335–345. doi:10.1016/j.foreco.2016.03.025.

Vose, J. M., C. F. Miniat, G. Sun, and P. V. Caldwell. 2015. Potential implications for expansion of freeze-tolerant eucalyptus plantations on water resources in the southern United States. *Forest Science* 61:509–521.

Vose, J. M., D. L. Peterson, and T. Patel-Weynand. 2012. Effects of climatic variability and change on forest ecosystems: A comprehensive science synthesis for the U.S. forest sector. General Technical Report PNW-GTR-870. USDA Forest Service, Pacific Northwest Research Station, Portland, OR.

Vose, J. M., G. Sun, C. R. Ford, M. Bredemeier, K. Ostsuki, A. Wei, Z. Zhang, and L. Zang. 2011. Forest ecohydrological research in the 21st century: What are the critical needs? *Ecohydrology* 4:146–158.

Vreeland, J. K., D. R. Diefenbach, and B. D. Wallingford. 2004. Survival rates, mortality causes, and habitats of Pennsylvania white-tailed deer fawns. *Wildlife Society Bulletin* 32:542–553.

Wade, D., S. Miller, J. Stowe, and J. Brenner. 2006. Rx fire laws: Tools to protect fire: The "ecological imperative." Pages 233–262 *in* M. B. Dickinson, editor. *Conference on Fire in Eastern Oak Forests: Delivering Science to Land Managers*, 15–17 November 2005, Columbus, Ohio, USA. General Technical Report NRS-P-1. USDA Forest Service, Northern Research Station, Newtown Square, PA.

Wade, D. and H. Mobley. 2007. Managing smoke at the wildland-urban interface. General Technical Report SRS-103. USDA Forest Service, Southern Research Station, Asheville, NC.

Wade, D. D. and J. D. Lunsford. 1989. A guide for prescribed fire in southern forests. Technical Publication R8-TP-11. USDA Forest Service, Southern Region, Atlanta, GA.

Wahlenberg, W. G. 1946. *Longleaf Pine: Its Use, Ecology, Regeneration, Protection, Growth, and Management*, 1st edition. Washington, DC: Charles Lathrop Pack Forestry Foundation.

Wahlenberg, W. G. 1960. *Loblolly Pine: Its Use, Ecology, Regeneration, Protection, Growth and Management*. Durham, NC: Duke University Publishing.

Wahlenberg, W. G., S. W. Greene, and H. R. Reed. 1939. *Effects of Fire and Cattle Grazing on Longleaf Pine Lands as Studies at McNeill*, Miss. Technical Bulletin No. 683. Washington, DC: U.S. Department of Agriculture.

Wakely, P. C. 1954. *Planting the Southern Pines*. Agricultural Monograph 18. Washington, DC: USDA Forest Service.

Waldrop, T. A. and S. L. Goodrick. 2012. *Introduction to Prescribed Fires in Southern Ecosystems. Science Update SR-054*. Asheville, NC: USDA Forest Service, Southern Research Station.

Waldrop, T. A., D. H. Van Lear, F. T. Lloyd, and W. R. Harms. 1987. Long-term studies of prescribed burning in loblolly pine forests of the southeastern coastal plain. U.S. Forest Service General Technical Report SE-45. USDA Forest Service, Southeastern Forest Experiment Station, Asheville, NC.

Waldrop, T. A., D. L. White, and S. M. Jones. 1992. Fire regimes for pine-grassland communities in the southeastern United States. *Forest Ecology and Management* 47:195–210.

Walker, L. C. 1954. Early scrub-oak control helps longleaf pine seedlings. *Journal of Forestry* 52:939–940.

Walker, J. 1993. Rare vascular plant taxa associated with the longleaf pine ecosystems: Patterns in taxonomy and ecology. Pages 105–125 *in* S. M. Hermann, editor. *The Longleaf Pine Ecosystem, Ecology, Restoration and Management. Proceedings of the 18th Tall Timbers Fire Ecology Conference*, 30 May–2 June 1991, Tallahassee, Florida, USA. Tall Timbers Research Station, Tallahassee, FL.

Walker, J. L. and S. Cohen. 2009. Regenerating longleaf pine on hydric soils: Short- and long-term effects on native ground-layer vegetation. Final Report SI-1303. Strategic Environmental Research and Development Program, Research Triangle Park, NC.

Walker, J. and G. Hernandez. 2010. Ecoregions and endemism to define gene conservation guidelines for longleaf pine ground-layer restoration. *Poster Presented at the Longleaf Alliance 8th Biennial Regional Conference*, 12–15 October 2010, Columbia, South Carolina, USA. Longleaf Alliance Report No. 16. The Longleaf Alliance, Andalusia, AL.

Walker, J. and R. K. Peet. 1983. Composition and species diversity of pine-wiregrass savannas of the Green Swamp, North Carolina. *Vegetatio* 55:163–179.

Walker, J. L. and A. M. Silletti. 2006. Restoring the ground layer of longleaf pine ecosystems. Pages 297–333 *in* S. Jose, E. J. Jokela, and D. L. Miller, editors. *The Longleaf Pine Ecosystem: Ecology, Silviculture, and Restoration*. New York, NY: Springer.

Wallace, L. L. 1987. Mycorrhizae in grasslands: Interactions of ungulates, fungi, and drought. *New Phytologist* 105:619–632.

Wallach, A. D., I. Izhaki, J. D. Toms, W. J. Ripple, and U. Shanas. 2015. What is an apex predator? *Oikos* 124:1453–1461.

Wallander, H., L. O. Nilsson, D. Hagerberg, and E. Baath. 2001. Estimation of the biomass and seasonal growth of external mycelium of ectomycorrhizal fungi in the field. *New Phytologist* 151:753–760.

Walls, S. C., W. J. Barichivich, M. E. Brown, D. E. Scott, and B. R. Hossack. 2013. Influence of drought on salamander occupancy of isolated wetlands on the southeastern coastal plain of the United States. *Wetlands* 33:345–354.

Walmsley, J. D. and D. L. Godbold. 2010. Stump harvesting for bioenergy—a review of the environmental impacts. *Forestry* 83:17–38.

Walter, H. 1939. Grasland, savanne und Busch der arideren Teile Afrikas in iherer ökologischen Bedingtheit. *Jahrbücher für wissenschaftliche Botanik* 87:750–860.

Walter, H. 1971. *Ecology of Tropical and Subtropical Vegetation*. Edinburgh: Oliver and Boyd.

Walters, J. R., C. K. Copeyon, and J. H. Carter. 1992. Test of the ecological basis of cooperative breeding in red-cockaded woodpeckers. *The Auk* 109:90–97.

Wang, C. and N. F. Glenn. 2009. Estimation of fire severity using pre-and post-fire LiDAR data in sagebrush steppe rangelands. *International Journal of Wildland Fire* 18:848–856.

Wang, G. G., L. S. Pile, B. O. Knapp and H. Hu. 2016. Longleaf pine adaptation to fire: Is early height growth pattern critical to fire survival? Pages 214–218 *in* C. J. Schweitzer, W. K. Clatterbuck, and C. M. Oswalt, editors. *Proceedings of the 18th Biennial Southern Silvicultural Research Conference*. e-General Technical Report SRS-212. USDA Forest Service, Southern Research Station, Asheville, NC.

Ward, D. E. and C. C. Hardy. 1991. Smoke emissions from wildland fires. *Environment International* 17:117–134.

Ward, D., K. Wiegand, and S. Getzin. 2013. Walter's two-layer hypothesis revisited: Back to the roots! *Oecologia* 172:617–630.

Ware, C. 2014. Longleaf pine maintenance class conditions: A guide to assess optimal forest habitat conditions for associated plant and wildlife species. America's Longleaf Restoration Initiative Longleaf Partnership Council. http://www.americaslongleaf.org/media/14299/final-lpc-maintenance-condition-class-metrics-oct-2014-high-res.pdf (accessed 4 February 2017).

Ware, S., C. Frost, and P. D. Doerr. 1993. Southern mixed hardwood forest: The former longleaf pine forest. Pages 447–493 *in* W. H. Martin, editor. *Biodiversity of the Southeastern United States: Lowland Terrestrial Communities*. New York, NY: Wiley.

Warner, R. E., J. W. Walk, and J. R. Herkert. 2012. Managing farmlands for wildlife. Pages 157–168 *in* N. J. Silvy, editor. *The Wildlife Techniques Manual: Management*. Bethesda, MD: The Wildlife Society.

Washington, W. M., J. W. Weatherly, G. A. Meehl, A. J. Semtner, Jr., T. W. Bettge, A. P. Craig, W. G. Strand, Jr., J. Arblaster, V. B. Wayland, R. James, and Y. Zhang. 2000. Parallel climate model (PCM) control and transient simulations. *Climate Dynamics* 16:755–774.

Watt, K. M. and S. W. Golladay. 1999. Organic matter dynamics in seasonally inundated, forested wetlands of the Gulf coastal plain. *Wetlands* 19:139–148.

Watts, W. A. 1980. Late-Quaternary vegetation history at White Pond on the inner coastal plain of South Carolina. *Quaternary Research* 13:187–199.

Watts, W. A., E. C. Grimm, and T. C. Hussey. 1996. Mid-Holocene forest history of Florida and the coastal plain of Georgia and South Carolina. Pages 28–38 *in* K. E. Sassaman, and D. G. Anderson, editors. *Archeology of the Mid-Holocene Southeast*. Gainesville, FL: University Press of Florida.

Watts, W. A. and B. C. S. Hansen. 1994. Pre-Holocene and Holocene pollen records of vegetation history from the Florida peninsula and their climatic implications. *Palaeogeography, Palaeoclimatology, Palaeoecology* 109:163–176.

Watts, W. A., B. C. S. Hansen, and E. C. Grimm. 1992. Camel Lake: A 40,000-yr record of vegetational and forest history from northwest Florida. *Ecology* 1992:1056–1066.

Watts, W. A. and M. Stuiver. 1980. Late Wisconsin climate of northern Florida and the origin of species-rich deciduous forest. *Science* 210:325–327.

Way, A. G. 2011. *Conserving Southern Longleaf: Herbert Stoddard and the Rise of Ecological Land Management*. Athens, GA: The University of Georgia Press.

Wayman, R. B. and M. North. 2007. Initial response of a mixed-conifer understory plant community to burning and thinning restoration treatments. *Forest Ecology and Management* 239:32–44.

Wear, D. N. 2011. Forecasts of county-level land uses under three future scenarios: A technical document supporting the Forest Service 2010 RPA Assessment. General Technical Report SRS-141. USDA Forest Service, Southern Research Station, Asheville, NC.

Wear, D. N. 2013. Forecasts of land uses. Pages 45–71 *in* D. N. Wear, and J. G. Greis, editors. *The Southern Forest Futures Project: Technical Report*. General Technical Report SRS-GTR-178. USDA Forest Service, Southern Research Station, Asheville, NC.

Wear, D. N. 2014. Forest ownership, policy and economics. Pages 67–80 *in* S. Kant, and J. Alavalapati, editors. *Handbook of Forest Resource Economics*. New York, NY: Routledge.

Wear, D. N. and J. Coulston. 2015. From sink to source: Regional variation in U.S. forest carbon futures. *Scientific Reports* 5:16518. doi:10.1038/srep16518.

Wear, D. N. and J. G. Greis. 2002a. Southern forest resource assessment: Summary of findings. *Journal of Forestry* 100:6–14.

Wear, D. N. and J. G. Greis. 2002b. Southern forest resource assessment. General Technical Report SRS-53. USDA Forest Service, Southern Research Station, Asheville, NC.

Wear, D. N. and J. G. Greis. 2012. The southern forest futures project: Summary report. General Technical Report SRS-168. USDA Forest Service, Southern Research Station, Asheville, NC.

Wear, D. N. and J. G. Greis. 2013. The southern forest futures project: Technical Report. General Technical Report SRS-GTR-178. USDA Forest Service, Southern Research Station, Asheville, NC.

Wear, D. N. and D. H. Newman. 2004. The speculative shadow over timberland values in the U.S. South. *Journal of Forestry* 102:25–31.

Wear, D. N., J. Prestemon, and M. Foster. 2016. U.S. forest products in the global economy. *Journal of Forestry* 114:483–493. doi:10.5849/jof.15-091.

Wear, D. N., J. Prestemon, R. Huggett, and D. Carter. 2013. Markets. Pages 183–212 *in* D. N. Wear, and J. G. Greis, editors. *The Southern Forest Futures Project: Technical Report*. General Technical Report SRS-GTR-178. USDA Forest Service, Southern Research Station, Asheville, NC.

Weaver, H. 1943. Fire as an ecological factor in the ponderosa pine region of the Pacific slope. *Journal of Forestry* 41:7–15.

Weaver, H. 1959. Ecological changes in the ponderosa pine forest of the Warm Springs Indian Reservation in Oregon. *Journal of Forestry* 57:15–20.

Webb, D. S. 1974. *Pleistocene Mammals of Florida*. Gainesville, FL: University of Florida Press.

Weddell, D. J. and L. M. Ware. 1935. The effect of fires of different frequencies on the survival of different species of pines. Alabama Agricultural Experiment Station Report:28–29.

Weiher, E., D. Freund, T. Bunton, A. Stefanski, T. Lee, and S. Bentivenga. 2011. Advances, challenges and a developing synthesis of ecological community assembly theory. *Royal Society Philosophical Transactions B-Biological Sciences* 366:2403–2413.

Weixelman, D. A., D. C. Zamudio, K. A. Zamudio, and R. J. Tausch. 1997. Classifying ecological types and evaluating site degradation. *Journal of Range Management* 50:315–321.

Wellendorf, S. D., W. E. Palmer, and P. T. Bromley. 2004. Estimating calling rates of northern bobwhite coveys and measuring abundance. *Journal of Wildlife Management* 68:672–682.

Wells, B. W. 1928. Plant communities of the coastal plain of North Carolina and their successional relations. *Ecology* 9:230–242.

Wells, B. W. and I. V. Shunk. 1928. *A Southern Upland Grass-Sedge Bog*. Technical Bulletin No. 32. Raleigh, NC: North Carolina Agricultural Experiment Station.

Wells, B. W. and I. V. Shunk. 1931. The vegetation and habitat factors of the coarser sands of the North Carolina coastal plain: An ecological study. *Ecological Monographs* 1:465–520.

Wenk, E. S., G. G. Wang, and J. L. Walker. 2011. Within-stand variation in understorey vegetation affects fire behavior in longleaf pine xeric sandhills. *International Journal of Wildland Fire* 20:866–875.

Wenk, E. S., G. G. Wang, and J. L. Walker. 2013. Understory fuel variation at the Carolina Sandhills National Wildlife Refuge. Pages 351–356 *in* J. M. Guldin, editor. *Proceedings of the 15th Biennial Southern Silvicultural Research Conference*. eGeneral Technical Report SRS-GTR-175. USDA Forest Service, Southern Research Station, Asheville, NC.

Werner, E. E., J. F. Gilliam, D. J. Hall, and G. G. Mittelbach. 1983. An experimental test of the effects of predation risk on habitat use in fish. *Ecology* 64:1540–1548.

Westerling, A. L., H. G. Hidalgo, D. R. Cayan, and T. W. Swetnam. 2006. Warming and earlier spring increase western U.S. forest fire activity. *Science* 313:940–942.

Westgate M. J., G. E. Likens, and D. B. Lindenmayer. 2013. Adaptive management of biological systems: A review. *Biological Conservation* 158:128–139.

Wheeler, P. R. 1969. The South's third forest. Southern Resource Analysis Committee, John Squires, Chairman. Forest Farmer's Association, Atlanta, GA.

Whelan, A., R. Mitchell, C. Staudhammer, and G. Starr. 2013. Cyclic occurrence of fire and its role in carbon dynamics along an edaphic moisture gradient in longleaf pine ecosystems. *PLOS ONE* 8:e54045.

Whelan, A., G. Starr, C. L. Staudhammer, H. W. Loescher, and R. J. Mitchell. 2015. Effects of drought and prescribed fire on energy exchange in longleaf pine ecosystems. *Ecosphere* 6:Article 128.

Whigham, D. F. and T. E. Jordan. 2003. Isolated wetlands and water quality. *Wetlands* 23:541–549.

White, P. S., M. D. MacKenzie, and R. T. Busing. 1985. Natural disturbance and gap phase dynamics in southern Appalachian spruce-fir forests. *Canadian Journal of Forest Research* 15:233–240.

White, P. S. and J. L. Walker. 1997. Approximating nature's variation: Selecting and using reference information in restoration ecology. *Restoration Ecology* 5:338–349.

Whitehead, D. R. 1981. Late-Pleistocene vegetational changes in northeastern North Carolina. *Ecological Monographs* 51:451–471.

Whitehead, D. R. and M. C. Sheehan. 1985. Holocene vegetational changes in the Tombigbee River valley, eastern Mississippi. *American Midland Naturalist* 113:122–137.

Wiedinmyer, C. and M. D. Hurteau. 2010. Prescribed fire as a means of reducing forest carbon emissions in the western United States. *Environmental Science Technology* 44:1926–1932.

Wiegert, R. G. 1972. Avian versus mammalian predation on a population of cotton rats. *Journal of Wildlife Management* 36:1322–1327.

Wiens, J., R. Sutter, M. Anderson, J. Blanchard, A. Barnett, N. Aguilar-Amuchastegui, C. Avery, and S. Laine. 2009. Selecting and conserving lands for biodiversity: The role of remote sensing. *Remote Sensing of Environment* 113:1370–1381.

Wiggers, M. S., J. K. Hiers, A. Barnett, R. S. Boyd, and L. K. Kirkman. 2017. Seed heat tolerance and germination of six legume species native to a fire-prone longleaf pine forest. *Plant Ecology* 218:151–171.

Wiggers, M. S., L. K. Kirkman, R. S. Boyd, and J. K. Hiers. 2013. Fine-scale variation in surface fire environment and legume germination in the longleaf pine ecosystem. *Forest Ecology and Management* 310:54–63.

Wigmosta, M. S., L. W. Vail, and D. P. Lettenmaier. 1994. A distributed hydrology-vegetation model for complex terrain. *Water Resources Research* 30:1665–1679.

Wikstrom, J. H. and J. R. Alley. 1967. Cost control in timber growing on the National Forests of the Northern Region. Research Paper INT-42. USDA Forest Service, Intermountain Forest Range Experiment Station, Ogden, UT.

Wilkins, R. N., G. T. Tanner, R. Mulholland, and D. G. Neary. 1993. Use of hexazinone for understory restoration of a successionally advanced xeric sandhill in Florida. *Ecological Engineering* 2:31–48.

Wilkinson, D. M. 2001. Is local provenance important in habitat creation? *Journal of Applied Ecology* 38:1371–1373.

Willard, D. A. 1994. Palynological record from the North Atlantic region at 3 Ma: Vegetational distribution during a period of global warmth. *Review of Palaeobotany and Palynology* 83:275–297.

Willard, D. A., T. M. Cronin, S. E. Ishman, and R. J. Litwin. 1993. Terrestrial and marine records of climatic and environmental changes during the Pliocene in subtropical Florida. *Geology* 21:679–6982.

Willcox, E. V. and W. M. Giuliano. 2010. Seasonal effects of prescribed burning and roller chopping on saw palmetto in flatwoods. *Forest Ecology and Management* 259:1580–1585.

Williams, R. E. 1952. Better management on longleaf pine forest ranges. *Journal of Range Management* 5:135–140.

Williams, M. 1989. *Americans and Their Forests: A Historical Geography*. New York, NY: Cambridge University Press.

Williams, G. W. 2003. The beginnings of the national forests in the south: Protection of watersheds. *Paper presented at the 14th Annual Environment Virginia Conference*, 29 April–1 May 2003, Lexington, VA. http://www.foresthistory.org/ASPNET/Policy/WeeksAct/ProtectionofWatersheds_Williams.pdf (accessed 4 February 2017).

Williams, B. K. 2011. Adaptive management of natural resources framework and issues. *Journal of Environmental Management* 92:1346–1353.

Williams, B. K. and G. S. Boomer. 2012. A typology of adaptive management. Chapter 4 in Adaptive management: Structured decision-making for recurrent decisions. U.S. Fish and Wildlife Service, National Conservation Training Center, Shepherdstown, West Virginia, USA. http://nctc.fws.gov/courses/ALC/ALC3176/resources/index.html (accessed 4 February 2017).

Williams, B. K. and E. D. Brown. 2012. *Adaptive Management: The U.S. Department of Interior Applications Guide*. Washington, DC: Adaptive Management Working Group, U.S. Department of the Interior.

Williams, C., B. P. Flower, D. W. Hastings, T. P. Guilderson, K. A. Quinn, and E. A. Goddard. 2010. Deglacial abrupt climate change in the Atlantic Warm Pool: A Gulf of Mexico perspective. *Paleoceanography* 25:PA4221. doi:10.1029/2010PA001928.

Williams, B. K., R. C. Szaro, and C. D. Shapiro. 2009. *Adaptive Management: The U.S. Department of the Interior Technical Guide*. Washington, DC: Adaptive Management Working Group, U.S. Department of the Interior.

Williamson, G. B. and E. M. Black. 1981. High temperature of forest fires under pines as a selective advantage over oaks. *Nature* 293:643–644.

Williston, H. L., J. G. Guthrie, and C. A. Hood. 1989. Managing and harvesting longleaf pine for specialty products. Pages 209–214 *in* R. M. Farrar, editor. *Proceedings of the Symposium on the Management of Longleaf Pine*, 4–6 April 1989, Long Beach, Mississippi, USA. General Technical Report SO-75. USDA Forest Service, Southern Forest Experiment Station, New Orleans, LA.

Willson, J. D., C. T. Winne, M. E. Dorcas, and J. W. Gibbons. 2006. Post-drought responses of semi-aquatic snakes inhabiting an isolated wetland: Insights on different strategies for persistence in a dynamic habitat. *Wetlands* 26:1071–1078.

Willyard, A., J. Syring, D. S. Gernandt, A. Liston, and R. Cronn. 2007. Fossil calibration of molecular divergence infers a moderate mutation rate and recent radiations for *Pinus*. *Molecular Biology and Evolution* 24:90–101.

Wilson, M. F. 1993. Mammals as seed-dispersal mutualists in North America. *Oikos* 67:159–176.

Wilson, C. A., R. J. Mitchell, L. R. Boring, and J. J. Hendricks. 2002. Soil nitrogen dynamics in a fire-maintained forest ecosystem: Results over 3-year burn interval. *Soil Biology and Biochemistry* 34:679–689.

Wilson, C. A., R. J. Mitchell, J. J. Hendricks, and L. R. Boring. 1999. Patterns and controls of ecosystem function in longleaf pine-wiregrass savannas: II. Nitrogen dynamics. *Canadian Journal of Forest Research* 29:752–760.

Wimberly, M. C., Y. Zhang, and J. A. Stanturf. 2006. Digital forestry in the wildland-urban interface. Pages 201–222 *in* G. Shao, and K. M. Reynolds, editors. *Computer Applications in Sustainable Forest Management*. Dordrecht: Springer.

Wing, B. M., M. W. Ritchie, K. Boston, W. B. Cohen, A. Gitelman, and M. J. Olsen. 2012. Prediction of understory vegetation cover with airborne lidar in an interior ponderosa pine forest. *Remote Sensing of Environment* 124:730–741.

Winter, G., C. A. Vogt, and S. McCaffrey. 2004. Examining social trust in fuels management strategies. *Journal of Forestry* 102:8–15.

Wood, D. R., L. W. Burger, Jr., J. L. Bowman, and C. L. Hardy. 2004. Avian community response to pine-grassland restoration. *Wildlife Society Bulletin* 32:819–828.

Wright, A. H. 1932. *Life Histories of the Frogs of Okefinokee Swamp, Georgia. North American* Salientia (Anura), Vol. 2. Ithaca, NY: McMillan Company.

Wright, C. S. 2013. Fuel consumption models for pine flatwoods fuel types in the southeastern United States. *Southern Journal of Applied Forestry* 37:148–159.

Wright, C. S. and J. K. Agee. 2004. Fire and vegetation history in the eastern Cascade Mountains, Washington. *Ecological Applications* 14:443–459.

Wright, H. A. and A.W. Bailey. 1982. *Fire Ecology, United States and Canada*. New York, NY: John Wiley & Sons.

Wright, A., S. A. Schnitzer, and P. B. Reich. 2014. Living close to your neighbors: The importance of both competition and facilitation in plant communities. *Ecology* 95:2213–2223.

Wright, A., S. A. Schnitzer, and P. B. Reich. 2015. Daily environmental conditions determine the competition–facilitation balance for plant water status. *Journal of Ecology* 103:648–656.

Wullschleger, S. D., F. C. Meinzer, and R. A. Vertessy. 1998. A review of whole-plant water use studies in trees. *Tree Physiology* 18:499–512.

Wyant, J. G., P. N. Omi, and R. D. Laven. 1986. Fire induced tree mortality in a Colorado ponderosa pine Douglas-fir stand. *Forest Science* 32:49–59.

Yamamoto, S. I. 2000. Forest gap dynamics and tree regeneration. *Journal of Forest Research* 5:223–229.

Yanquoi, F. B. 1992. Economic guides for managing longleaf pine (*Pinus palustris* Mill) in the South, MS thesis. Auburn University, Auburn, AL.

Yarrow, G. K. and D. T. Yarrow. 1999. *Managing Wildlife on Private Lands in Alabama and the Southeast.* Birmingham, AL: Sweetwater Press.

Yates, R. F. K. and F. P. Day, Jr. 1983. Decay rates and nutrient dynamics in confined and unconfined leaf litter in the Great Dismal Swamp. *American Midland Naturalist* 110:37–45.

Yellick, A. H., D. L. Jacob, E. S. DeKeyser, C. L. Hargiss, L. M. Meyers, M. Ell, L. T. Kissoon-Charles, and M. L. Otte. 2016. Multi-element composition of soils of seasonal wetlands across North Dakota, USA. *Environmental Monitoring and Assessment* 188:17.

Yilmaz, N., W. Gill, A. B. Donaldson, and R. E. Lucero. 2008. Problems encountered in fluctuating flame temperature measurements by thermocouple. *Sensors* 8:7882–7893.

Youngblood, A. 2010. Thinning and burning in dry coniferous forests of the western United States: Effectiveness in altering diameter distributions. *Forest Science* 56:46–59.

Zagalikis, G., A. D. Cameron, and D. R. Miller. 2005. The application of digital photogrammetry and image analysis techniques to derive tree and stand characteristics. *Canadian Journal of Forest Research* 35:1224–1237.

Zahner, R. 1989. Tree-ring series related to stand and environmental factors in south Alabama longleaf pine stands. Pages 193–197 *in* J. H. Miller, editor. *Proceedings of the Fifth Biennial Southern Silvicultural Research Conference*, 1–3 November 1988, Memphis, Tennessee, USA. General Technical Report SO-74. USDA Forest Service, Southern Forest Experiment Station, New Orleans, LA.

Zald, H. S. J., A. N. Gray, M. North, and R. A. Kern. 2008. Initial tree regeneration responses to fire and thinning treatments in a Sierra Nevada mixed-conifer forest, USA. *Forest Ecology and Management* 256:168–179.

Zarnowitz, J. E. and D. A. Manuwal. 1985. The effects of forest management on cavity-nesting birds in northwestern Washington. *Journal of Wildlife Management* 49:255–263.

Zedler, J. B. 2003. Wetlands at your service: Reducing impacts of agriculture at the watershed scale. *Frontiers in Ecology and the Environment* 1:65–72.

Zedler, J. B. and S. Kercher. 2004. Causes and consequences of invasive plants in wetlands: Opportunities, opportunists, and outcomes. *Critical Reviews in Plant Sciences* 23:431–452.

Zedler, J. B. and S. Kercher. 2005. Wetland resources: Status, trends, ecosystem services, and restorability. *Annual Review of Environment and Resources* 30:39–74.

Zhang, D., B. J. Butler, and R. V. Nagubadi. 2012. Institutional timberland ownership in the U.S. south: Magnitude, location, dynamics, and management. *Journal of Forestry* 110:355–361.

Zhang, D. and S. R. Mehmood. 2002. Safe harbor for the red-cockaded woodpecker: Private forest landowners share their views. *Journal of Forestry* 100:24–29.

Zhu, A. X., L. E. Band, B. Dutton, and T. Nimlos. 1996. Automated soil inference under fuzzy logic. *Ecological Modelling* 90:123–145.

Zhu, A. X., F. Qi, A. Moore, and J. E. Burt. 2010. Prediction of soil properties using fuzzy membership. *Geoderma* 158:199–206.

Zinkhan, F. C. 1988. Forestry projects, modern portfolio theory, and discount rate selection. *Southern Journal of Applied Forestry* 12:132–135.

Zinkhan, F. C., W. R. Sizemore, G. H. Mason, and T. J. Ebner. 1992. *Timberland Investments: A Portfolio Perspective.* Portland, OR: Timber Press.

Zobel, M., M. Otsus, J. Liira, M. Moora, and T. Mols. 2000. Is small-scale species richness limited by seed availability or microsite availability? *Ecology* 81:3274–3282.

Index

D

E

F

G

H

I

K

L

M

N

O

P

Q

R

For Product Safety Concerns and Information please contact our EU representative GPSR@taylorandfrancis.com
Taylor & Francis Verlag GmbH, Kaufingerstraße 24, 80331 München, Germany

www.ingramcontent.com/pod-product-compliance
Ingram Content Group UK Ltd.
Pitfield, Milton Keynes, MK11 3LW, UK
UKHW050800180926
475701UK00008B/1155
9780367657734